CHAOS
AN INTRODUCTION TO DYNAMICAL SYSTEMS

KATHLEEN T. ALLIGOOD

TIM D. SAUER

JAMES A. YORKE

C H A O S   An Introduction to Dynamical Systems

Springer

New York
Berlin
Heidelberg
Barcelona
Budapest
Hong Kong
London
Milan
Paris
Santa Clara
Singapore
Tokyo

CHAOS

An Introduction to Dynamical Systems

KATHLEEN T. ALLIGOOD
George Mason University

TIM D. SAUER
George Mason University

JAMES A. YORKE
University of Maryland

 Springer

Textbooks in Mathematical Sciences

Series Editors:

Thomas F. Banchoff
Brown University

Keith Devlin
St. Mary's College

Gaston Gonnet
ETH Zentrum, Zürich

Jerrold Marsden
California Institute of Technology

Stan Wagon
Macalester College

COVER: Rene Magritte, *Golconde 1953.* © 1996 C. Herscovici, Brussels/Artists Rights Society (ARS), New York. Used by permission of ARS.

Library of Congress Cataloging-in-Publication Data
Alligood, Kathleen T.
    Chaos - an introduction to dynamical systems / Kathleen Alligood,
    Tim Sauer, James A. Yorke.
       p.   cm. — (Textbooks in mathematical sciences)
    Includes bibliographical references and index.
    1. Differentiable dynamical systems.   2. Chaotic behavior in
systems.   I. Sauer, Tim.   II. Yorke, James A.   III. Title.   IV. Series.
    QA614.8.A44   1996
    003'.85—dc20   95-51304
                                                                         CIP

Printed on acid-free paper.

Production managed by Frank Ganz; manufacturing supervised by Jeffrey Taub.
Photocomposed by Integre Technical Publishing Co., Inc., Albuquerque, NM.
Printed and bound by R.R. Donnelley & Sons, Harrisonburg, VA.
Printed in the United States of America.

9 8 7 6 5 4 3 2 1

ISBN 0-387-94677-2 Springer-Verlag New York Berlin Heidelberg   SPIN 10527282

Introduction

BACKGROUND

Sir Isaac Newton brought to the world the idea of modeling the motion of physical systems with equations. It was necessary to invent calculus along the way, since fundamental equations of motion involve velocities and accelerations, which are derivatives of position. His greatest single success was his discovery that the motion of the planets and moons of the solar system resulted from a single fundamental source: the gravitational attraction of the bodies. He demonstrated that the observed motion of the planets could be explained by assuming that there is a gravitational attraction between any two objects, a force that is proportional to the product of masses and inversely proportional to the square of the distance between them. The circular, elliptical, and parabolic orbits of astronomy were no longer fundamental determinants of motion, but were approximations of laws specified with differential equations. His methods are now used in modeling motion and change in all areas of science.

Subsequent generations of scientists extended the method of using differential equations to describe how physical systems evolve. But the method had a limitation. While the differential equations were sufficient to determine the behavior—in the sense that solutions of the equations did exist—it was frequently difficult to figure out what that behavior would be. It was often impossible to write down solutions in relatively simple algebraic expressions using a finite number of terms. Series solutions involving infinite sums often would not converge beyond some finite time.

When solutions could be found, they described very regular motion. Generations of young scientists learned the sciences from textbooks filled with examples of differential equations with regular solutions. If the solutions remained in a bounded region of space, they settled down to either (A) a steady state, often due to energy loss by friction, or (B) an oscillation that was either periodic or quasiperiodic, akin to the clocklike motion of the moon and planets. (In the solar system, there were obviously many different periods. The moon traveled around the earth in a month, the earth around the sun in about a year, and Jupiter around the sun in about 11.867 years. Such systems with multiple incommensurable periods came to be called quasiperiodic.)

Scientists knew of systems which had more complicated behavior, such as a pot of boiling water, or the molecules of air colliding in a room. However, since these systems were composed of an immense number of interacting particles, the complexity of their motions was not held to be surprising.

Around 1975, after three centuries of study, scientists in large numbers around the world suddenly became aware that there is a third kind of motion, a type (C) motion, that we now call "chaos". The new motion is erratic, but not simply quasiperiodic with a large number of periods, and not necessarily due to a large number of interacting particles. It is a type of behavior that is possible in very simple systems.

A small number of mathematicians and physicists were familiar with the existence of a third type of motion prior to this time. James Clerk Maxwell, who studied the motion of gas molecules in about 1860, was probably aware that even a system composed of two colliding gas particles in a box would have neither motion type A nor B, and that the long term behavior of the motions would for all practical purposes be unpredictable. He was aware that very small changes in the initial motion of the particles would result in immense changes in the trajectories of the molecules, even if they were thought of as hard spheres.

Maxwell began his famous study of gas laws by investigating individual collisions. Consider two atoms of equal mass, modeled as hard spheres. Give the atoms equal but opposite velocities, and assume that their positions are selected at random in a large three-dimensional region of space. Maxwell showed that if they collide, all directions of travel will be equally likely after the collision. He recognized that small changes in initial positions can result in large changes in outcomes. In a discussion of free will, he suggested that it would be impossible to test whether a leopard has free will, because one could never compute from a study of its atoms what the leopard would do. But the chaos of its atoms is limited, for, as he observed, "No leopard can change its spots!"

Henri Poincaré in 1890 studied highly simplified solar systems of three bodies and concluded that the motions were sometimes incredibly complicated. (See Chapter 2). His techniques were applicable to a wide variety of physical systems. Important further contributions were made by Birkhoff, Cartwright and Littlewood, Levinson, Kolmogorov and his students, among others. By the 1960s, there were groups of mathematicians, particularly in Berkeley and in Moscow, striving to understanding this third kind of motion that we now call chaos. But only with the advent of personal computers, with screens capable of displaying graphics, have scientists and engineers been able to see that important equations in their own specialties had such solutions, at least for some ranges of parameters that appear in the equations.

In the present day, scientists realize that chaotic behavior can be observed in experiments and in computer models of behavior from all fields of science. The key requirement is that the system involve a nonlinearity. It is now common for experiments whose previous anomalous behavior was attributed to experiment error or noise to be reevaluated for an explanation in these new terms. Taken together, these new terms form a set of unifying principles, often called dynamical systems theory, that cross many disciplinary boundaries.

The theory of dynamical systems describes phenomena that are common to physical and biological systems throughout science. It has benefited greatly from the collision of ideas from mathematics and these sciences. The goal of scientists and applied mathematicians is to find nature's unifying ideas or laws and to fashion a language to describe these ideas. It is critical to the advancement of science that exacting standards are applied to what is meant by knowledge. Beautiful theories can be appreciated for their own sake, but science is a severe taskmaster. Intriguing ideas are often rejected or ignored because they do not meet the standards of what is knowledge.

The standards of mathematicians and scientists are rather different. Mathematicians prove theorems. Scientists look at realistic models. Their approaches are somewhat incompatible. The first papers showing chaotic behavior in computer studies of very simple models were distasteful to both groups. The mathematicians feared that nothing was proved so nothing was learned. Scientists said that models without physical quantities like charge, mass, energy, or acceleration could not be relevant to physical studies. But further reflection led to a change in viewpoints. Mathematicians found that these computer studies could lead to new ideas that slowly yielded new theorems. Scientists found that computer studies of much more complicated models yielded behaviors similar to those of the simplistic models, and that perhaps the simpler models captured the key phenomena.

Finally, laboratory experiments began to be carried out that showed unequivocal evidence of unusual nonlinear effects and chaotic behavior in very familiar settings. The new dynamical systems concepts showed up in macroscopic systems such as fluids, common electronic circuits and low-energy lasers that were previously thought to be fairly well understood using the classical paradigms. In this sense, the chaotic revolution is quite different than that of relativity, which shows its effects at high energies and velocities, and quantum theory, whose effects are submicroscopic. Many demonstrations of chaotic behavior in experiments are not far from the reader's experience.

In this book we study this field that is the uncomfortable interface between mathematics and science. We will look at many pictures produced by computers and we try to make mathematical sense of them. For example, a computer study of the driven pendulum in Chapter 2 reveals irregular, persistent, complex behavior for ten million oscillations. Does this behavior persist for one billion oscillations? The only way we can find out is to continue the computer study longer. However, even if it continues its complex behavior throughout our computer study, we cannot guarantee it would persist forever. Perhaps it stops abruptly after one trillion oscillations; we do not know for certain. We can prove that there exist initial positions and velocities of the pendulum that yield complex behavior forever, but these choices are conceivably quite atypical. There are even simpler models where we know that such chaotic behavior does persist forever. In this world, pictures with uncertain messages remain the medium of inspiration.

There is a philosophy of modeling in which we study idealized systems that have properties that can be closely approximated by physical systems. The experimentalist takes the view that only quantities that can be measured have meaning. Yet we can prove that there are beautiful structures that are so infinitely intricate that they can never be seen experimentally. For example, we will see immediately in Chapters 1 and 2 the way chaos develops as a physical parameter like friction is varied. We see infinitely many periodic attractors appearing with infinitely many periods. This topic is revisited in Chapter 12, where we show

how this rich bifurcation structure, called a cascade, exists with mathematical certainty in many systems. This is a mathematical reality that underlies what the experimentalist can see. We know that as the scientist finds ways to make the study of a physical system increasingly tractable, more of this mathematical structure will be revealed. It is there, but often hidden from view by the noise of the universe. All science is of course dependent on simplistic models. If we study a vibrating beam, we will generally not model the atoms of which it is made. If we model the atoms, we will probably not reflect in our model the fact that the universe has a finite age and that the beam did not exist for all time. And we do not include in our model (usually) the tidal effects of the stars and the planets on our vibrating beam. We ignore all these effects so that we can isolate the implications of a very limited list of concepts.

It is our goal to give an introduction to some of the most intriguing ideas in dynamics, the ideas we love most. Just as chemistry has its elements and physics has its elementary particles, dynamics has its fundamental elements: with names like attractors, basins, saddles, homoclinic points, cascades, and horseshoes. The ideas in this field are not transparent. As a reader, your ability to work with these ideas will come from your own effort. We will consider our job to be accomplished if we can help you learn what to look for in your own studies of dynamical systems of the world and universe.

ABOUT THE BOOK

As we developed the drafts of this book, we taught six one semester classes at George Mason University and the University of Maryland. The level is aimed at undergraduates and beginning graduate students. Typically, we have used parts of Chapters 1–9 as the core of such a course, spending roughly equal amounts of time on iterated maps (Chapters 1–6) and differential equations (Chapters 7–9). Some of the maps we use as examples in the early chapters come from differential equations, so that their importance in the subject is stressed. The topics of stable manifolds, bifurcations, and cascades are introduced in the first two chapters and then developed more fully in the Chapters 10, 11, and 12, respectively. Chapter 13 on time series may be profitably read immediately after Chapter 4 on fractals, although the concepts of periodic orbit (of a differential equation) and chaotic attractor will not yet have been formally defined.

The impetus for advances in dynamical systems has come from many sources: mathematics, theoretical science, computer simulation, and experimental science. We have tried to put this book together in a way that would reflect its wide range of influences.

We present elaborate dissections of the proofs of three deep and important theorems: The Poincaré-Bendixson Theorem, the Stable Manifold Theorem, and the Cascade Theorem. Our hope is that including them in this form tempts you to work through the nitty-gritty details, toward mastery of the building blocks as well as an appreciation of the completed edifice.

Additionally, each chapter contains a special feature called a Challenge, in which other famous ideas from dynamics have been divided into a number of steps with helpful hints. The Challenges tackle subjects from period-three implies chaos, the cat map, and Sharkovskii's ordering through synchronization and renormalization. We apologize in advance for the hints we have given, when they are of no help or even mislead you; for one person's hint can be another's distraction.

The Computer Experiments are designed to present you with opportunities to explore dynamics through computer simulation, the venue through which many of these concepts were first discovered. In each, you are asked to design and carry out a calculation relevant to an aspect of the dynamics. Virtually all can be successfully approached with a minimal knowledge of some scientific programming language. Appendix B provides an introduction to the solution of differential equations by approximate means, which is necessary for some of the later Computer Experiments.

If you prefer not to work the Computer Experiments from scratch, your task can be greatly simplified by using existing software. Several packages are available. *Dynamics: Numerical Explorations* by H.E. Nusse and J.A. Yorke (Springer-Verlag 1994) is the result of programs developed at the University of Maryland. *Dynamics*, which includes software for Unix and PC environments, was used to make many of the pictures in this book. We can also recommend *Differential and Difference Equations through Computer Experiments* by H. Kocak (Springer-Verlag, 1989) for personal computers. A sophisticated package designed for Unix platforms is *dstool*, developed by J. Guckenheimer and his group at Cornell University, currently free by anonymous ftp. In the absence of special purpose software, general purpose scientific computing environments such as Matlab, Maple, and Mathematica will do nicely.

The Lab Visits are short reports on carefully selected laboratory experiments that show how the mathematical concepts of dynamical systems manifest themselves in real phenomena. We try to impart some flavor of the setting of the experiment and the considerable expertise and care necessary to tease a new secret from nature. In virtually every case, the experimenters' findings far surpassed what we survey in the Lab Visit. We urge you to pursue more accurate and detailed discussions of these experiments by going straight to the original sources.

ACKNOWLEDGEMENTS

In the course of writing this book, we received valuable feedback from colleagues and students too numerous to mention. Suggestions that led to major improvements in the text were made by Clark Robinson, Ittai Kan, Karen Brucks, Miguel San Juan, and Brian Hunt, and from students Leon Poon, Joe Miller, Rolando Castro, Guocheng Yuan, Reena Freedman, Peter Calabrese, Michael Roberts, Shawn Hatch, Joshua Tempkin, Tamara Gibson, Barry Peratt, and Ed Fine.

Kathleen T. Alligood
Tim D. Sauer
Fairfax, VA

James A. Yorke
College Park, MD

Contents

One-Dimensional Maps

T HE FUNCTION $f(x) = 2x$ is a rule that assigns to each number x a number twice as large. This is a simple mathematical model. We might imagine that x denotes the population of bacteria in a laboratory culture and that $f(x)$ denotes the population one hour later. Then the rule expresses the fact that the population doubles every hour. If the culture has an initial population of 10,000 bacteria, then after one hour there will be $f(10,000) = 20,000$ bacteria, after two hours there will be $f(f(10,000)) = 40,000$ bacteria, and so on.

A **dynamical system** consists of a set of possible states, together with a rule that determines the present state in terms of past states. In the previous paragraph, we discussed a simple dynamical system whose states are population levels, that change with time under the rule $x_n = f(x_{n-1}) = 2x_{n-1}$. Here the variable n stands for time, and x_n designates the population at time n. We will require that the rule be **deterministic**, which means that we can determine the present state (population, for example) *uniquely* from the past states.

No randomness is allowed in our definition of a deterministic dynamical system. A possible mathematical model for the price of gold as a function of time would be to predict today's price to be yesterday's price plus or minus one dollar, with the two possibilities equally likely. Instead of a dynamical system, this model would be called a **random**, or **stochastic**, **process**. A typical realization of such a model could be achieved by flipping a fair coin each day to determine the new price. This type of model is not deterministic, and is ruled out by our definition of dynamical system.

We will emphasize two types of dynamical systems. If the rule is applied at discrete times, it is called a **discrete-time** dynamical system. A discrete-time system takes the current state as input and updates the situation by producing a new state as output. By the state of the system, we mean whatever information is needed so that the rule may be applied. In the first example above, the state is the population size. The rule replaces the current population with the population one hour later. We will spend most of Chapter 1 examining discrete-time systems, also called maps.

The other important type of dynamical system is essentially the limit of discrete systems with smaller and smaller updating times. The governing rule in that case becomes a set of differential equations, and the term **continuous-time** dynamical system is sometimes used. Many of the phenomena we want to explain are easier to describe and understand in the context of maps; however, since the time of Newton the scientific view has been that nature has arranged itself to be most easily modeled by differential equations. After studying discrete systems thoroughly, we will turn to continuous systems in Chapter 7.

1.1 ONE-DIMENSIONAL MAPS

One of the goals of science is to predict how a system will evolve as time progresses. In our first example, the population evolves by a single rule. The output of the rule is used as the input value for the next hour, and the same rule of doubling is applied again. The evolution of this dynamical process is reflected by composition of the function f. Define $f^2(x) = f(f(x))$ and in general, define $f^k(x)$ to be the result of applying the function f to the initial state k times. Given an initial value of x, we want to know about $f^k(x)$ for large k. For the above example, it is clear that if the initial value of x is greater than zero, the population will grow without bound. This type of expansion, in which the population is multiplied by a constant factor per unit of time, is called **exponential growth**. The factor in this example is 2.

WHY STUDY MODELS?

We study models because they suggest how real-world processes behave. In this chapter we study extremely simple models.

Every model of a physical process is at best an idealization. The goal of a model is to capture some feature of the physical process. The feature we want to capture now is the patterns of points on an orbit. In particular, we will find that the patterns are sometimes simple, and sometimes quite complicated, or "chaotic", even for simple maps.

The question to ask about a model is whether the behavior it exhibits is *because* of its simplifications or if it captures the behavior *despite* the simplifications. Modeling reality too closely may result in an intractable model about which little can be learned. Model building is an art. Here we try to get a handle on possible behaviors of maps by considering the simplest ones.

The fact that real habitats have finite resources lies in opposition to the concept of exponential population increase. From the time of Malthus (Malthus, 1798), the fact that there are limits to growth has been well appreciated. Population growth corresponding to multiplication by a constant factor cannot continue forever. At some point the resources of the environment will become compromised by the increased population, and the growth will slow to something less than exponential.

In other words, although the rule $f(x) = 2x$ may be correct for a certain range of populations, it may lose its applicability in other ranges. An improved model, to be used for a resource-limited population, might be given by $g(x) = 2x(1 - x)$, where x is measured in millions. In this model, the initial population of 10,000 corresponds to $x = .01$ million. When the population x is small, the factor $(1 - x)$ is close to one, and $g(x)$ closely resembles the doubling function $f(x)$. On the other hand, if the population x is far from zero, then $g(x)$ is no longer proportional to the population x but to the product of x and the "remaining space" $(1 - x)$. This is a nonlinear effect, and the model given by $g(x)$ is an example of a **logistic** growth model.

Using a calculator, investigate the difference in outcomes imposed by the models $f(x)$ and $g(x)$. Start with a small value, say $x = 0.01$, and compute $f^k(x)$ and $g^k(x)$ for successive values of k. The results for the models are shown in Table 1.1. One can see that for $g(x)$, there is computational evidence that the population approaches an eventual limiting size, which we would call a steady-state population for the model $g(x)$. Later in this section, using some elementary calculus, we'll see how to verify this conjecture (Theorem 1.5).

There are obvious differences between the behavior of the population size under the two models, $f(x)$ and $g(x)$. Under the dynamical system $f(x)$, the starting population size $x = 0.01$ results in arbitrarily large populations as time progresses. Under the system $g(x)$, the same starting size $x = 0.01$ progresses in a strikingly similar way at first, approximately doubling each hour. Eventually, however, a limiting size is reached. In this case, the population saturates at $x = 0.50$ (one-half million), and then never changes again.

So one great improvement of the logistic model $g(x)$ is that populations can have a finite limit. But there is a second improvement contained in $g(x)$. If

n	$f^n(x)$	$g^n(x)$
0	0.0100000000	0.0100000000
1	0.0200000000	0.0198000000
2	0.0400000000	0.0388159200
3	0.0800000000	0.0746184887
4	0.1600000000	0.1381011397
5	0.3200000000	0.2380584298
6	0.6400000000	0.3627732276
7	1.2800000000	0.4623376259
8	2.5600000000	0.4971630912
9	5.1200000000	0.4999839039
10	10.2400000000	0.4999999995
11	20.4800000000	0.5000000000
12	40.9600000000	0.5000000000

Table 1.1 Comparison of exponential growth model $f(x) = 2x$ to logistic growth model $g(x) = 2x(1 - x)$.
The exponential model explodes, while the logistic model approaches a steady state.

we use starting populations other than $x = 0.01$, the same limiting population $x = 0.50$ will be achieved.

↪ **COMPUTER EXPERIMENT 1.1**

Confirm the fact that populations evolving under the rule $g(x) = 2x(1 - x)$ prefer to reach the population 0.5. Use a calculator or computer program, and try starting populations x_0 between 0.0 and 1.0. Calculate $x_1 = g(x_0)$, $x_2 = g(x_1)$, etc. and allow the population to reach a limiting size. You will find that the size $x = 0.50$ eventually "attracts" any of these starting populations.

Our numerical experiment suggests that this population model has a natural built-in carrying capacity. This property corresponds to one of the many ways that scientists believe populations should behave—that they reach a steady-state which is somehow compatible with the available environmental resources. The limiting population $x = 0.50$ for the logistic model is an example of a fixed point of a discrete-time dynamical system.

Definition 1.1 A function whose domain (input) space and range (output) space are the same will be called a **map**. Let x be a point and let f be a map. The **orbit** of x under f is the set of points $\{x, f(x), f^2(x), \ldots\}$. The starting point x for the orbit is called the **initial value** of the orbit. A point p is a **fixed point** of the map f if $f(p) = p$.

For example, the function $g(x) = 2x(1 - x)$ from the real line to itself is a map. The orbit of $x = 0.01$ under g is $\{0.01, 0.0198, 0.0388, \ldots\}$, and the fixed points of g are $x = 0$ and $x = 1/2$.

1.2 COBWEB PLOT: GRAPHICAL REPRESENTATION OF AN ORBIT

For a map of the real line, a rough plot of an orbit—called a cobweb plot—can be made using the following graphical technique. Sketch the graph of the function f together with the diagonal line $y = x$. In Figure 1.1, the example $f(x) = 2x$ and the diagonal are sketched. The first thing that is clear from such a picture is the location of fixed points of f. At any intersection of $y = f(x)$ with the line $y = x$,

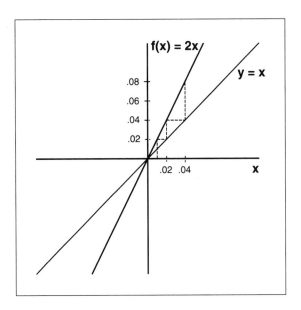

Figure 1.1   An orbit of f(x) = 2x.
The dotted line is a cobweb plot, a path that illustrates the production of a trajectory.

the input value x and the output $f(x)$ are identical, so such an x is a fixed point. Figure 1.1 shows that the only fixed point of $f(x) = 2x$ is $x = 0$.

Sketching the orbit of a given initial condition is done as follows. Starting with the input value $x = .01$, the output $f(.01)$ is found by plotting the value of the function above .01. In Figure 1.1, the output value is .02. Next, to find $f(.02)$, it is necessary to consider .02 as the new input value. In order to turn an output value into an input value, draw a horizontal line from the input–output pair $(.01, .02)$ to the diagonal line $y = x$. In Figure 1.1, there is a vertical dotted line segment starting at $x = .01$, representing the function evaluation, and then a horizontal dotted segment which effectively turns the output into an input so that the process can be repeated.

Then start over with the new value $x = .02$, and draw a new pair of vertical and horizontal dotted segments. We find $f(f(.01)) = f(.02) = .04$ on the graph of f, and move horizontally to move output to the input position. Continuing in this way, a graphical history of the orbit $\{.01, .02, .04, \ldots\}$ is constructed by the path of dotted line segments.

EXAMPLE 1.2

A more interesting example is the map $g(x) = 2x(1 - x)$. First we find fixed points by solving the equation $x = 2x(1 - x)$. There are two solutions, $x = 0$

COBWEB PLOT

A cobweb plot illustrates convergence to an attracting fixed point of $g(x) = 2x(1 - x)$. Let $x_0 = 0.1$ be the initial condition. Then the first iterate is $x_1 = g(x_0) = 0.18$. Note that the point (x_0, x_1) lies on the function graph, and (x_1, x_1) lies on the diagonal line. Connect these points with a horizontal dotted line to make a path. Then find $x_2 = g(x_1) = 0.2952$, and continue the path with a vertical dotted line to (x_1, x_2) and with a horizontal dotted line to (x_2, x_2). An entire orbit can be mapped out this way.

In this case it is clear from the geometry that the orbit we are following will converge to the intersection of the curve and the diagonal, $x = 1/2$. What happens if instead we start with $x_0 = 0.8$? These are examples of simple cobweb plots. They can be much more complicated, as we shall see later.

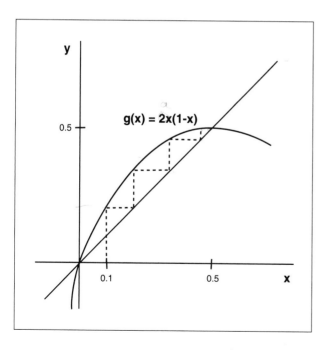

Figure 1.2 **A cobweb plot for an orbit of $g(x) = 2x(1 - x)$.**
The orbit with initial value .1 converges to the sink at .5.

and $x = 1/2$, which are the two fixed points of g. Contrast this with a linear map which, except for the case of the identity $f(x) = x$, has only one fixed point $x = 0$. What is the behavior of orbits of g? The graphical representation of the orbit with initial value $x = 0.1$ is drawn in Figure 1.2. It is clear from the figure that the orbit, instead of diverging to infinity as in Figure 1.1, is converging to the fixed point $x = 1/2$. Thus the orbit with initial condition $x = 0.1$ gets stuck, and cannot move beyond the fixed point $x = 0.5$. A simple rule of thumb for following the graphical representation of an orbit: If the graph is above the diagonal line $y = x$, the orbit will move to the right; if the graph is below the line, the orbit moves to the left.

EXAMPLE 1.3

Let f be the map of $\mathbb{R}$ given by $f(x) = (3x - x^3)/2$. Figure 1.3 shows a graphical representations of two orbits, with initial values $x = 1.6$ and 1.8, respectively. The former orbit appears to converge to the fixed point $x = 1$ as the map is iterated; the latter converges to the fixed point $x = -1$.

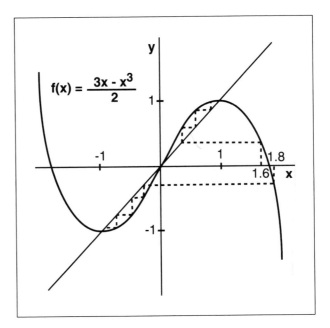

Figure 1.3 A cobweb plot for two orbits of $f(x) = (3x - x^3)/2$.
The orbit with initial value 1.6 converges to the sink at 1; the orbit with initial value 1.8 converges to the sink at -1.

Fixed points are found by solving the equation $f(x) = x$. The map has three fixed points, namely $-1, 0$, and 1. However, orbits beginning near, but not precisely on, each of the fixed points act differently. You may be able to convince yourself, using the graphical representation technique, that initial values near -1 stay near -1 upon iteration by the map, and that initial values near 1 stay near 1. On the other hand, initial values near 0 depart from the area near 0. For example, to four significant digits, $f(.1) = 0.1495, f^2(.1) = 0.2226, f^5(.1) = 0.6587$, and so on. The problem with points near 0 is that f magnifies them by a factor larger than one. For example, the point $x = .1$ is moved by f to approximately $.1495$, a magnification factor of 1.495. This magnification factor turns out to be approximately the derivative $f'(0) = 1.5$.

1.3 STABILITY OF FIXED POINTS

With the geometric intuition gained from Figures 1.1, 1.2, and 1.3, we can describe the idea of stability of fixed points. Assuming that the discrete-time system exists to model real phenomena, not all fixed points are alike. A stable fixed point has the property that points near it are moved even closer to the fixed point under the dynamical system. For an unstable fixed point, nearby points move away as time progresses. A good analogy is that a ball at the bottom of a valley is stable, while a ball balanced at the tip of a mountain is unstable.

The question of stability is significant because a real-world system is constantly subject to small perturbations. Therefore a steady state observed in a realistic system must correspond to a stable fixed point. If the fixed point is unstable, small errors or perturbations in the state would cause the orbit to move away from the fixed point, which would then not be observed.

Example 1.3 gave some insight into the question of stability. The derivative of the map at a fixed point p is a measure of how the distance between p and a nearby point is magnified or shrunk by f. That is, the points 0 and $.1$ begin exactly $.1$ units apart. After applying the rule f to both points, the distance separating the points is changed by a factor of approximately $f'(0)$. We want to call the fixed point 0 "unstable" when points very near 0 tend to move away from 0.

The concept of "near" is made precise by referring to all real numbers within a distance ϵ of p as the **epsilon neighborhood** $N_\epsilon(p)$. Denote the real line by $\mathbb{R}$. Then $N_\epsilon(p)$ is the interval of numbers $\{x \in \mathbb{R} : |x - p| < \epsilon\}$. We usually think of ϵ as a small, positive number.

Definition 1.4 Let f be a map on $\mathbb{R}$ and let p be a real number such that $f(p) = p$. If all points sufficiently close to p are attracted to p, then p is called a

sink or an **attracting fixed point**. More precisely, if there is an $\epsilon > 0$ such that for all x in the epsilon neighborhood $N_\epsilon(p)$, $\lim_{k \to \infty} f^k(x) = p$, then p is a sink. If all points sufficiently close to p are repelled from p, then p is called a **source** or a **repelling fixed point**. More precisely, if there is an epsilon neighborhood $N_\epsilon(p)$ such that each x in $N_\epsilon(p)$ except for p itself eventually maps outside of $N_\epsilon(p)$, then p is a source.

In this text, unless otherwise stated, we will deal with functions for which derivatives of all orders exist and are continuous functions. We will call this type of function a **smooth** function.

Theorem 1.5 *Let f be a (smooth) map on $\mathbb{R}$, and assume that p is a fixed point of f.*

1. *If $|f'(p)| < 1$, then p is a sink.*
2. *If $|f'(p)| > 1$, then p is a source.*

Proof: PART 1. Let a be any number between $|f'(p)|$ and 1; for example, a could be chosen to be $(1 + |f'(p)|)/2$. Since

$$\lim_{x \to p} \frac{|f(x) - f(p)|}{|x - p|} = |f'(p)|,$$

there is a neighborhood $N_\varepsilon(p)$ for some $\varepsilon > 0$ so that

$$\frac{|f(x) - f(p)|}{|x - p|} < a$$

for x in $N_\varepsilon(p)$.

In other words, $f(x)$ is closer to p than x is, by at least a factor of a (which is less than 1). This implies two things: First, if $x \in N_\epsilon(p)$, then $f(x) \in N_\varepsilon(p)$; that means that if x is within ε of p, then so is $f(x)$, and by repeating the argument, so are $f^2(x), f^3(x)$, and so forth. Second, it follows that

$$|f^k(x) - p| \le a^k |x - p| \tag{1.1}$$

for all $k \ge 1$. Thus p is a sink.

✎ **EXERCISE T1.1**

Show that inequality (1.1) holds for $k = 2$. Then carry out the mathematical induction argument to show that it holds for all $k = 1, 2, 3, \ldots$

✎ **EXERCISE T1.2**

Use the ideas of the proof of Part 1 of Theorem 1.5 to prove Part 2.

Note that the proof of part 1 of Theorem 1.5 says something about the rate of convergence of $f^k(x)$ to p. The fact that $|f^k(x) - p| \leq a^k|x - p|$ for $a < 1$ is described by saying that $f^k(x)$ converges exponentially to p as $k \to \infty$.

Our definition of a fixed-point sink requires only that the sink attract some epsilon neighborhood $(p - \epsilon, p + \epsilon)$ of nearby points. As far as the definition is concerned, the radius ϵ of the neighborhood, although nonzero, could be extremely small. On the other hand, sinks often attract orbits from a large set of nearby initial conditions. We will refer to the set of initial conditions whose orbits converge to the sink as the basin of the sink.

With Theorem 1.5 and our new terminology, we can return to Example 1.2, an example of a logistic model. Setting $x = g(x) = 2x(1 - x)$ shows that the fixed points are 0 and $1/2$. Taking derivatives, we get $g'(0) = 2$ and $g'(1/2) = 0$. Theorem 1.5 shows that $x = 1/2$ is a sink, which confirms our suspicions from Table 1.1. On the other hand, $x = 0$ is a source. Points near 0 are repelled from 0 upon application of g. In fact, points near 0 are repelled at an exponential magnification factor of approximately 2 (check this number with a calculator). These points are attracted to the sink $x = 1/2$.

What is the basin of the sink $x = 1/2$ in Example 1.2? The point 0 does not belong, since it is a fixed point. Also, 1 does not belong, since $g(1) = 0$ and further iterations cannot budge it. However, all initial conditions from the interval $(0, 1)$ will produce orbits that converge to the sink. You should sketch a graph of $g(x)$ as in Figure 1.1 and use the idea of the cobweb plot to convince yourself of this fact.

There is a second way to show that the basin of $x = 1/2$ is $(0, 1)$, which is quicker and trickier but far less general. That is to use algebra (not geometry) to compare $|g(x) - 1/2|$ to $|x - 1/2|$. If the former is smaller than the latter, it means the orbit is getting closer to $1/2$. The algebra says:

$$|g(x) - 1/2| = |2x(1 - x) - 1/2|$$

$$= 2|x - 1/2||x - 1/2| \tag{1.2}$$

Now we can see that if $x \in (0, 1)$, the multiplier $2|x - 1/2|$ is smaller than one. Any point x in $(0, 1)$ will have its distance from $x = 1/2$ decreased on each iteration by g. Notice that the algebra also tells us what happens for initial conditions outside of $(0, 1)$: they will never converge to the sink $x = 1/2$. Therefore the basin of the sink is exactly the open interval $(0, 1)$. Informally, we could also say that the basin of infinity is $(-\infty, 0) \cup (1, \infty)$, since the orbit of each initial condition in this set tends toward $-\infty$.

Theorem 1.5 also clarifies Example 1.3, which is the map $f(x) = (3x - x^3)/2$. The fixed points are $-1, 0$, and 1, and the derivatives are $f'(-1) = f'(1) = 0$, and $f'(0) = 1.5$. By the theorem, the fixed points -1 and 1 are attracting fixed points, and 0 is a repelling fixed point.

Let's try to determine the basins of the two sinks. Example 1.3 is already significantly more complicated than Example 1.2, and we will have to be satisfied with an incomplete answer. We will consider the sink $x = 1$; the other sink has very similar properties by the symmetry of the situation.

First, cobweb plots (see Figure 1.3) convince us that the interval $I_1 = (0, \sqrt{3})$ of initial conditions belongs to the basin of $x = 1$. (Note that $f(\sqrt{3}) = f(-\sqrt{3}) = 0$.) So far it is similar to the previous example. Have we found the entire basin? Not quite. Initial conditions from the interval $I_2 = [-2, -\sqrt{3})$ map to $(0, 1]$, which we already know are basin points. (Note that $f(-2) = 1$.) Since points that map to basin points are basin points as well, we know that the set $[-2, -\sqrt{3}) \cup (0, \sqrt{3})$ is included in the basin of $x = 1$. Now you may be willing to admit that the basin can be quite a complicated creature, because the graph shows that there is a small interval I_3 of points to the right of $x = 2$ that map into the interval $I_2 = [-2, -\sqrt{3})$, and are therefore in the basin, then a small interval I_4 to the left of $x = -2$ that maps into I_3, and so forth ad infinitum. These intervals are all separate (they don't overlap), and the gaps between them consist of similar intervals belonging to the basin of the other sink $x = -1$. The intervals I_n get smaller with increasing n, and all of them lie between $-\sqrt{5}$ and $\sqrt{5}$. Since $f(\sqrt{5}) = -\sqrt{5}$ and $f(-\sqrt{5}) = \sqrt{5}$, neither of these numbers is in the basin of either sink.

✎ **EXERCISE T1.3**

Solve the inequality $|f(x) - 0| > |x - 0|$, where $f(x) = (3x - x^3)/2$. This identifies points whose distance from 0 increases on each iteration. Use the result to find a large set of initial conditions that do not converge to any sink of f.

There is one case that is not covered by Theorem 1.5. The stability of a fixed point p cannot be determined solely by the derivative when $|f'(p)| = 1$ (see Exercise 1.2).

So far we have seen the important role of fixed points in determining the behavior of orbits of maps. If the fixed point is a sink, it provides the final state for the orbits of many nearby initial conditions. For the linear map $f(x) = ax$ with $|a| < 1$, the sink $x = 0$ attracts all initial conditions. In Examples 1.2 and 1.3, the sinks attract large sets of initial conditions.

✎ **EXERCISE T1.4**

Let p be a fixed point of a map f. Given some $\epsilon > 0$, find a geometric condition under which all points x in $N_\epsilon(p)$ are in the basin of p. Use cobweb plot analysis to explain your reasoning.

1.4 PERIODIC POINTS

Changing a, the constant of proportionality in the logistic map $g_a(x) = ax(1 - x)$, can result in a picture quite different from Example 1.2. When $a = 3.3$, the fixed points are $x = 0$ and $x = 23/33 = .\overline{69} = .696969\ldots$, both of which are repellers. Now that there are no fixed points around that can attract orbits, where do they go? Use a calculator to convince yourself that for almost every choice of initial condition, the orbit settles into a pattern of alternating values $p_1 = .4794$ and $p_2 = .8236$ (to four decimal place accuracy). Some typical orbits are shown in Table 1.2. The orbit with initial condition 0.2 is graphed in Figure 1.4. This figure shows typical behavior of an orbit converging to a period-2 sink $\{p_1, p_2\}$. It is attracted to p_1 every two iterates, and to p_2 on alternate iterates.

There are actually two important parts of this fact. First, there is the apparent coincidence that $g(p_1) = p_2$ and $g(p_2) = p_1$. Another way to look at this is that $g^2(p_1) = p_1$; thus p_1 is a fixed point of $h = g^2$. (The same could be said for p_2.) Second, this periodic oscillation between p_1 and p_2 is stable, and attracts orbits. This fact means that periodic behavior will show up in a physical system modeled by g. The pair $\{p_1, p_2\}$ is an example of a periodic orbit.

Definition 1.6 Let f be a map on $\mathbb{R}$. We call p a **periodic point of period k** if $f^k(p) = p$, and if k is the smallest such positive integer. The orbit with initial point p (which consists of k points) is called a **periodic orbit of period k**. We will often use the abbreviated terms **period-k point** and **period-k orbit**.

Notice that we have defined the period of an orbit to be the *minimum* number of iterates required for the orbit to repeat a point. If p is a periodic point of period 2 for the map f, then p is a fixed point of the map $h = f^2$. However, the converse is not true. A fixed point of $h = f^2$ may also be a fixed point of a lower

n	$g^n(x)$	$g^n(x)$	$g^n(x)$
0	0.2000	0.5000	0.9500
1	0.5280	0.8250	0.1568
2	0.8224	0.4764	0.4362
3	0.4820	0.8232	0.8116
4	0.8239	0.4804	0.5047
5	0.4787	0.8237	0.8249
6	0.8235	0.4792	0.4766
7	0.4796	0.8236	0.8232
8	0.8236	0.4795	0.4803
9	0.4794	0.8236	0.8237
10	0.8236	0.4794	0.4792
11	0.4794	0.8236	0.8236
12	0.8236	0.4794	0.4795
13	0.4794	0.8236	0.8236
14	0.8236	0.4794	0.4794

Table 1.2 Three different orbits of the logistic model $g(x) = 3.3x(1 - x)$. Each approaches a period-2 orbit.

iterate of f, specifically f, and so may not be a periodic point of period two. For example, if p is a fixed point of f, it will be a fixed point of f^2 but not, according to our definition, a period-two point of f.

EXAMPLE 1.7

Consider the map defined by $f(x) = -x$ on $\mathbb{R}$. This map has one fixed point, at $x = 0$. Every other real number is a period-two point, because f^2 is the identity map.

EXERCISE T1.5

The map $f(x) = 2x^2 - 5x$ on $\mathbb{R}$ has fixed points at $x = 0$ and $x = 3$. Find a period-two orbit for f by solving $f^2(x) = x$ for x.

What about the stability of periodic orbits? As in the fixed point case, points near the periodic orbit can be trapped or repelled by the orbit. The key fact is that a periodic point for f is a fixed point for f^k. We can use Theorem 1.5 to investigate the stability of a periodic orbit. For a period-k orbit, we apply Theorem 1.5 to the map f^k instead of f.

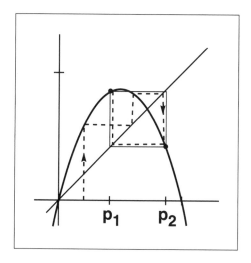

Figure 1.4 Orbit converging to a period-two sink.
The dashed lines form a cobweb plot showing an orbit which moves toward the sink
orbit $\{p_1, p_2\}$.

Definition 1.8   Let f be a map and assume that p is a period-k point. The
period-k orbit of p is a **periodic sink** if p is a sink for the map f^k. The orbit of p is
a **periodic source** if p is a source for the map f^k.

It is helpful to review the chain rule of calculus, which shows how to expand
the derivative of a composition of functions:

$$(f \circ g)'(x) = f'(g(x))g'(x) \tag{1.3}$$

Our current interest in the chain rule is for $f = g$, in which case we have $(f^2)'(x) =$
$f'(f(x))f'(x)$. If x happens to be a period-two point for f, the chain rule is saying
something quite simple: the derivative of f^2 at a point of a period-two orbit
is simply the product of the derivatives of f at the two points in the orbit. In
particular, the derivative of f^2 is the same, when evaluated at either point of the
orbit. This agreement means that it makes sense to talk about the stability of a
period-two orbit.

Now the period-two behavior of $g(x) = 3.3x(1 - x)$ we found in Table 1.2
can be completely explained. The periodic orbit $\{.4794, .8236\}$ will be a sink
as long as the derivative $(g^2)'(p_1) = g'(p_1)g'(p_2) = (g^2)'(p_2)$ is smaller than 1 in
absolute value. An easy calculation shows this number to be $g'(.4794)g'(.8236) =$
-0.2904.

If instead we consider yet another version of the logistic map, $g(x) = 3.5x(1 - x)$, the situation is again changed. The fixed points are $x = 0$ and $x = 5/7$. Checking derivatives, $g'(0) = 3.5$ and $g'(5/7) = -1.5$, so they are sources. The orbit $\{3/7, 6/7\}$ is a period-two orbit for g. Check that $(g^2)'$ at each of the orbit points is $-5/4$, so that this period-two orbit repels nearby points. Now where do points end up?

⇨ **C O M P U T E R   E X P E R I M E N T   1 . 2**

Write a computer program with the goal of redoing Table 1.2 for the logistic map $g_a(x) = ax(1 - x)$, using $a = 3.5$. What periodic behavior wins out in the long run? Try several different initial conditions to explore the basin of the attracting periodic behavior. Then try different values of $a < 3.57$ and report your results.

Now that we have some intuition from period-two orbits, we note that the situation is essentially the same for higher periods. Let $\{p_1, \ldots, p_k\}$ denote a period-k orbit of f. The chain rule says that

$$(f^k)'(p_1) = (f(f^{k-1}))'(p_1)$$
$$= f'(f^{k-1}(p_1))(f^{k-1})'(p_1)$$
$$= f'(f^{k-1}(p_1))f'(f^{k-2}(p_1)) \cdots f'(p_1)$$
$$= f'(p_k)f'(p_{k-1}) \cdots f'(p_1). \tag{1.4}$$

STABILITY TEST FOR PERIODIC ORBITS

The periodic orbit $\{p_1, \ldots, p_k\}$ is a sink if

$$|f'(p_k) \cdots f'(p_1)| < 1$$

and a source if

$$|f'(p_k) \cdots f'(p_1)| > 1.$$

This formula tells us that the derivative of the kth iterate f^k of f at a point of a period-k orbit is the product of the derivatives of f at the k points of the orbit. In particular, stability is a collective property of the periodic orbit, in that $(f^k)'(p_i) = (f^k)'(p_j)$ for all i and j.

1.5 THE FAMILY OF LOGISTIC MAPS

We are beginning to get an overall view of the family $g_a(x) = ax(1 - x)$ associated with the logistic model. When $0 \leq a < 1$, the map has a sink at $x = 0$, and we will see later that every initial condition between 0 and 1 is attracted to this sink. (In other words, with small reproduction rates, small populations tend to die out.) The graph of the map is shown in Figure 1.5(a).

If $1 < a < 3$, the map, shown in Figure 1.5(b), has a sink at $x = (a - 1)/a$, since the magnitude of the derivative is less than 1. (Small populations grow to a steady state of $x = (a - 1)/a$.) For a greater than 3, as in Figure 1.5(c), the fixed point $x = (a - 1)/a$ is unstable since $|g_a'(x)| > 1$, and a period-two sink takes its place, which we saw in Table 1.2 for $a = 3.3$. When a grows above $1 + \sqrt{6} \approx 3.45$, the period-two sink also becomes unstable.

✎ **EXERCISE T1.6**

Verify the statements in the previous paragraph by solving for the fixed points and period-two points of $g_a(x)$ and evaluating their stability.

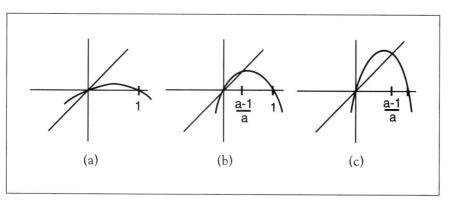

(a) (b) (c)

Figure 1.5 The logistic family.
(a) The origin attracts all initial conditions in $[0, 1]$. (b) The fixed point at $(a - 1)/a$ attracts all initial conditions in $(0, 1)$. (c) The fixed point at $(a - 1)/a$ is unstable.

⊃   COMPUTER EXPERIMENT 1.3

Use your logistic map program to investigate the long-run behavior of g_a for a near $a_* = 1 + \sqrt{6}$. Repeat Table 1.2 for values of a slightly smaller than a_*. What qualitative or quantitative conclusions can be made about the speed of convergence to the period-two orbit as a gets closer to a_*? What happens to iterations beginning at a period-two point for a slightly larger than a_*?

For slightly larger values of a, the story of the periodic points of $g_a(x)$ becomes significantly more complicated. Many new periodic orbits come into existence as a is increased from 3.45 to 4. Figure 1.6 shows the limiting behavior of orbits for values of a in the range $1 < a < 4$. This computer-generated picture was made by repeating the following procedure: (1) Choose a value of a, starting with $a = 1$, (2) Choose x at random in $[0,1]$, (3) Calculate the orbit of x under $g_a(x)$, (4) Ignore the first 100 iterates and plot the orbit beginning with iterate 101. Then increment a and begin the procedure again. The points that are plotted will (within the resolution of the picture) approximate either fixed or periodic sinks or other attracting sets. This figure is called a **bifurcation diagram** and shows the birth, evolution, and death of attracting sets. The term "bifurcation" refers to significant changes in the set of fixed or periodic points or other sets of dynamic interest. We will study bifurcations in detail in Chapter 11.

We see, for example, that the vertical slice $a = 3.4$ of Figure 1.6 intersects the diagram in the two points of a period-two sink. For a slightly larger than 3.45, there appears to be a period-four sink. In fact, there is an entire sequence of periodic sinks, one for each period 2^n, $n = 1, 2, 3, \ldots$. Such a sequence is called a "period-doubling cascade". The phenomenon of cascades is the subject of Chapter 12. Figure 1.7 shows portions of the bifurcation diagram in detail. Magnification near a period-three sink, in Figure 1.7(b) hints at further period-doublings that are invisible in Figure 1.6.

For other values of the parameter a, the orbit appears to randomly fill out the entire interval $[0, 1]$, or a subinterval. A typical cobweb plot formed for $a = 3.86$ is shown in Figure 1.8. These attracting sets, called "chaotic attractors", are harder to describe than periodic sinks. We will try to unlock some of their secrets in later chapters. As we shall see, it is a characteristic of chaotic attractors that they can abruptly appear or disappear, or change size discontinuously. This phenomenon, called a "crisis", is apparent at various a values. In particular, at $a = 4$, there is a

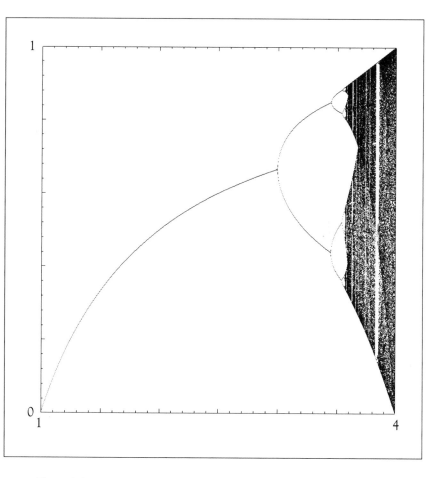

1

0

1   4

Figure 1.6 Bifurcation diagram of $g_a(x) = ax(1 - x)$.
The fixed point that exists for small values of a gives way to a period-two orbit at the
"bifurcation point" $a = 3$, which in turn leads to more and more complicated orbits
for larger values of a. Notice that the fixed point is only plotted while it is a sink.
When the period-two orbit appears, the fixed point is no longer plotted because it
does not attract orbits.

crisis at which the chaotic attractor disappears. For $a > 4$, there is no attracting
set.

The successive blow-ups of the bifurcation diagrams reveal another interesting feature, that of "periodic windows". The period-three window, for example,
is apparent in Figure 1.7(a) and is shown in magnified form in Figure 1.7(b). This
refers to a set of parameter values for which there is a periodic sink, in this case
a period-three sink. Since a period-three point of g_a is a fixed point of the third
iterate g_a^3, the creation of the period-three sink can be seen by viewing the de-

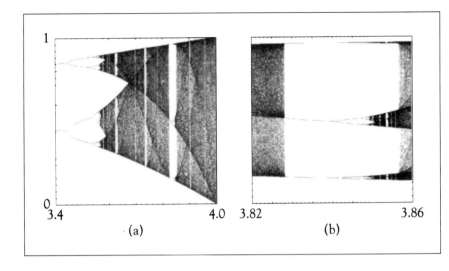

1

0
3.4   4.0 3.82   3.86

(a)   (b)

Figure 1.7 Magnifications of the logistic bifurcation diagram.
(a) Horizontal axis is $3.4 \leq a \leq 4.0$ (b) Horizontal axis is $3.82 \leq a \leq 3.86$.

velopment of the graph of g_a^3 as a moves from 3.82 to 3.86. This development is shown in Figure 1.9.

In Figure 1.9(a), the period-three orbit does not exist. This parameter value $a = 3.82$ corresponds to the left end of Figure 1.7(b). In Figure 1.9(b), the period-three orbit has been formed. Of course, since each point of a period-three orbit of g

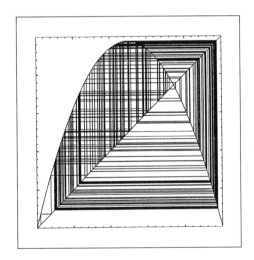

Figure 1.8 Cobweb plot for the logistic map.
A single orbit of the map $g(x) = 3.86x(1 - x)$ shows complicated behavior.

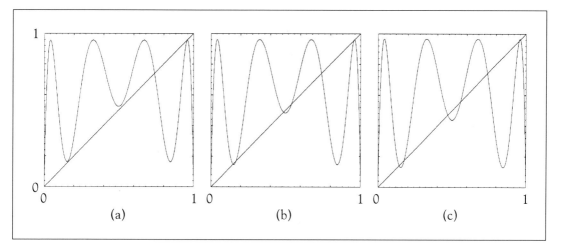

Figure 1.9 Graphs of the third iteration $g^3(x)$ of the logistic map $g_a(x) = ax(1 − x)$.

Three different parameter values are shown: (a) $a = 3.82$ (b) $a = 3.84$ (c) $a = 3.86$.

is a fixed point of g^3, the period-three orbit will appear as three intersections with the diagonal $y = x$. As you can see from the figure, the shape of the graph forces two period-three orbits to be created simultaneously. This is called a saddle-node bifurcation, or alternatively, a tangency bifurcation. The "node" is the sink, which is the set of three points at which the graph intersects the diagonal in negative slope. (Can you explain why the three negative slopes are exactly equal? Use the chain rule.) The fact that it is a sink corresponds to the fact that the negative slopes are between $−1$ and 0. The "saddle" is a period-three source consisting of the three upward sloping points. A vertical slice through the middle of Figure 1.7(b) shows that all initial conditions are attracted to the period-three sink. In Figure 1.9(c), the period-three sink has turned into a source. This parameter value $a = 3.86$ corresponds to the right side of Figure 1.7(b).

There are many more features of Figure 1.7 that we have to leave unexplained for now. The demise of the period-three sink as an attractor coincides with a so-called period-doubling bifurcation, which creates a period-six sink, which then meets a similar fate. There are periodic windows of arbitrarily high period. We will try to unlock some of the deeper mysteries of bifurcations in Chapter 11.

What happens to the bifurcation diagram if different x values are selected? (Recall that for each a, the orbit of one randomly chosen initial x is computed.) Surprisingly, nothing changes. The diagram looks the same no matter what initial condition is picked at random between 0 and 1, since there is at most one attracting fixed or periodic orbit at each parameter value. As we shall see, however, there are many unstable, hence unseen, periodic orbits for larger a.

1.6 THE LOGISTIC MAP $G(x) = 4x(1 - x)$

In the previous sections we studied maps from the logistic family $g(x) = ax(1 - x)$. For $a = 2.0, 3.3$, and 3.5, we found the existence of sinks of period 1, 2, and 4, respectively. Next, we will focus on one more case, $a = 4.0$, which is so interesting that it gets its own section. The reason that it is so interesting is that it has no sinks, which leads one to ask where orbits end up.

The graph of $G(x) = g_4(x) = 4x(1 - x)$ is shown in Figure 1.10(a). Although the graph is a parabola of the type often studied in elementary precalculus courses, the map defined by G has very rich dynamical behavior. To begin with, the diagonal line $y = x$ intersects $y = G(x) = 4x(1 - x)$ in the points $x = 0$ and $x = 3/4$, so there are two fixed points, both unstable. Does G have any other periodic orbits?

One way to look for periodic orbits is to sketch the graph of $y = G^n(x)$. Any period-two point, for example, will be a fixed point of $G^2(x)$. Therefore we can find periodic points graphically.

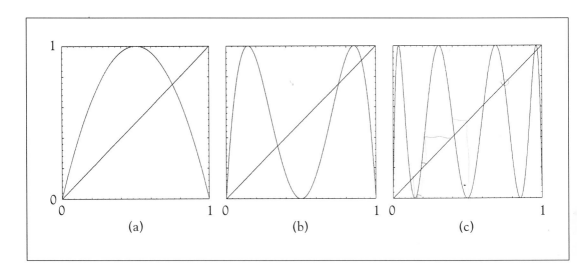

$$(a) \qquad\qquad (b) \qquad\qquad (c)$$

Figure 1.10 Graphs of compositions of the logistic map.
(a) the logistic map $G(x) = 4x(1 - x)$. (b) The map $G^2(x)$. (c) The map $G^3(x)$.

The graph of $y = G^2(x)$ is shown in Figure 1.10(b). It is not hard to verify by hand the general shape of the graph. First, note that the image of $[0,1]$ under G is $[0,1]$, so the graph stays entirely within the unit square. Second, note that $G(1/2) = 1$ and $G(1) = 0$ implies that $G^2(1/2) = 0$. Further, since $G(a_1) = 1/2$ for some a_1 between 0 and $1/2$, it follows that $G^2(a_1) = 1$. Similarly, there is another number a_2 such that $G^2(a_2) = 1$.

It is clear from Figure 1.10(b) that G^2 has four fixed points, and therefore G has four points that have period either one or two. Two of these points are already known to us—they are fixed points for G. The new pair of points, p_1 and p_2, make up a period-two orbit: that is, $G(p_1) = p_2$ and $G(p_2) = p_1$. This reasoning should have you convinced that the period-two orbit exists. The next exercise asks you to explicitly find p_1 and p_2.

✎ **EXERCISE T1.7**

Find the period-two orbit of $G(x) = 4x(1 - x)$.

Does G have any period-three points? There is a point b_1 between 0 and a_1 for which $G(b_1) = a_1$. This implies that $G^3(b_1) = 1$. The same holds for three other points in $[0,1]$, so $y = G^3(x)$ has four relative maxima of height 1 in $[0,1]$. Since $G(1) = 0$, G^3 has roots at $x = 0, a_1, 1/2, a_2$, and 1, which separate the maxima. The graph of G^3 is shown in Figure 1.10(c).

The map G^3 has eight fixed points, two of which were known to be the fixed points 0 and $3/4$ of G. The period-two points of G are not fixed points of G^3. (Why not?) There remain six more points to account for, which must form two period-three orbits. You should be able to prove to yourself in a similar way that G^4 has $16 = 2^4$ fixed points, all in $[0, 1]$. With each successive iteration of G, the number of fixed points of the iterate is doubled. In general, we see that G^k has 2^k fixed points, all in $[0, 1]$. Of course, for $k > 1$, G has fewer than 2^k points of period-k. (Remember that the definition of period-k for the point p is that k is the smallest positive integer for which $f^k(p) = p$.) For example, $x = 0$ is a period one point and therefore not a period-k point for $k > 1$, although it is one of the 2^k fixed points of G^k.

✎ **EXERCISE T1.8**

Let $G(x) = 4x(1 - x)$. Prove that for each positive integer k, there is an orbit of period-k.

Period k	Number of fixed points of G^k	Number of fixed points of G^k due to lower period orbits	Orbits of period k
1	2	0	2
2	4	2	1
3	8	2	2
4	16	4	3
$\vdots$	$\vdots$	$\vdots$	$\vdots$

Table 1.3 The periodic table for the logistic map.
The nth iterate of the map $G(x) = 4x(1 - x)$ has 2^n fixed points, which are periodic orbits for G.

The number of orbits of the map for each period can be tabulated in the map's **periodic table** . For the logistic map it begins as shown in Table 1.3. The first column is the period k, and the second column is the number of fixed points of G^k, which is 2^k, as seen in Figure 1.10. The third column keeps track of fixed points of G^k which correspond to orbits of lower period than k. When these are subtracted away from the entry in the second column, the result is the number of period-k points, which is divided by k to get the number of period-k orbits.

✎ **EXERCISE T1.9**

Let $G(x) = 4x(1 - x)$.

(a) Decide whether the fixed points and period-two points of G are sinks.

(b) Continue the periodic table for G begun in Table 1.3. In particular, how many periodic orbits of (minimum) period k does G have, for each $k \le 10$?

Is this what we mean by "chaos"? Not exactly. The existence of infinitely many periodic orbits does not in itself imply the kind of unpredictability usually associated with chaotic maps, although it does hint at the rich structure present. Chaos is identified with nonperiodicity and sensitive dependence on initial conditions, which we explore in the next section.

1.7 SENSITIVE DEPENDENCE ON INITIAL CONDITIONS

EXAMPLE 1.9

Consider the map $f(x) = 3x \pmod 1$ on the unit interval. The notation y (mod 1) stands for the number $y + n$, where n is the unique integer that makes $0 \leq y + n < 1$. For example, $14.92 \pmod 1 = .92$ and $-14.92 \pmod 1 = .08$. For a positive number y, this is the fractional part of y. See Figure 1.11(a) for a graph of the map. Because of the breaks at $x = 1/3, 2/3$, this function is not continuous.

This map is not continuous, however the important property that we are interested in is not caused by the discontinuity . It may be more natural to view f as a map on the circle of circumference one. Glue together the ends of the unit interval to form a circle, as in Figure 1.11(b). If we consider $f(x)$ as a map from this circle to itself, it is a continuous map. In Figure 1.11(b), we show the image of the subinterval $[0, 1/2]$ on the circle. Whether we think of f as a discontinuous map on the unit interval or as a continuous map on the circle makes no difference for the questions we will try to answer below.

We call a point x **eventually periodic** with period p for the map f if for some positive integer N, $f^{n+p}(x) = f^n(x)$ for all $n \geq N$, and if p is the smallest

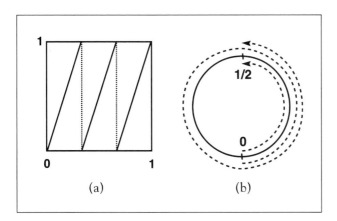

Figure 1.11 The 3x mod 1 map.
(a) The map $f(x) = 3x \pmod 1$ is discontinuous on the unit interval. (b) When the points 0 and 1 are identified, turning the unit interval into a circle, the map is continuous. The inner dashed semicircle is the subinterval $[0, 1/2]$, and the outer dashed curve is its image under the map. If x and y are two points that are close together on the circle, then $f(x)$ and $f(y)$ will be 3 times further apart than x and y.

such positive integer. This says exactly that the orbit of x eventually maps directly onto a periodic orbit. For example, $x = 1/3$ is an eventually periodic point, since it maps under f to the period one orbit 0.

✎ **EXERCISE T1.10**

Show that a point x is eventually periodic for Example 1.9 if and only if x is a rational number.

✎ **EXERCISE T1.11**

Construct the periodic table for f in Example 1.9 (follow the form given by Table 1.3).

The 3x mod 1 map demonstrates the main characteristic of chaos: sensitive dependence on initial conditions. This refers to the property that pairs of points, which begin as close together as desired, will eventually move apart. Table 1.4 shows the beginning of two separate orbits whose initial conditions differ by .0001. In fact, no matter how close they begin, the difference between two nearby orbits is magnified by a factor of 3 on each iteration. This idea is important enough to be assigned a formal definition.

n	$f^n(x_0)$	$f^n(y_0)$
0	0.25	0.2501
1	0.75	0.7503
2	0.25	0.2509
3	0.75	0.7527
4	0.25	0.2581
5	0.75	0.7743
6	0.25	0.3229
7	0.75	0.9687
8	0.25	0.9061
9	0.75	0.7183
10	0.25	0.1549

Table 1.4 Comparison of the orbits of two nearly equal initial conditions under the 3x mod 1 map.

The orbits become completely uncorrelated in fewer than 10 iterates.

Definition 1.10 Let f be a map on $\mathbb{R}$. A point x_0 has **sensitive dependence on initial conditions** if there is a nonzero distance d such that some points arbitrarily near x_0 are eventually mapped at least d units from the corresponding image of x_0. More precisely, there exists $d > 0$ such that any neighborhood N of x_0 contains a point x such that $|f^k(x) - f^k(x_0)| \geq d$ for some nonnegative integer k. Sometimes we will call such a point x_0 a **sensitive point**.

Ordinarily, the closer x is to x_0, the larger k will need to be. The point x will be sensitive if it has neighbors as close as desired that eventually move away the prescribed distance d for some sufficiently large k.

✎ **EXERCISE T1.12**

Consider the $3x \bmod 1$ map of the unit interval $[0, 1]$. Define the distance between a pair of points x, y to be either $|x - y|$ or $1 - |x - y|$, whichever is smaller. (We are measuring with the "circle metric", in the sense of Figure 1.11, corresponding to the distance between two points on the circle.)

(a) Show that the distance between any pair of points that lie within $1/6$ of one another is tripled by the map. (b) Find a pair of points whose distance is not tripled by the map. (c) Show that to prove sensitive dependence for any point, d can be taken to be any positive number less than $1/2$ in Definition 1.10, and that k can be chosen to be the smallest integer greater than $\ln(d/|x - x_0|)/\ln 3$.

✎ **EXERCISE T1.13**

Prove that for any map f, a source has sensitive dependence on initial conditions.

1.8 ITINERARIES

The fact that the logistic map $G(x) = 4x(1 - x)$ has periodic orbits of every period is one indication of its complicated dynamics. An even more important reflection of this complexity is sensitive dependence on initial conditions, which is the hallmark of chaos.

In this section we will show for the logistic map $G = g_4$ that for any initial point in the unit interval and any preset distance $\delta > 0$, no matter how small, there is a second point within δ units of the first so that their two orbits will map at least $d = 1/4$ units apart after a sufficient number of iterations. Since $1/4$

unit is 25% of the length of the unit interval, it is fair to say that the two initial conditions which began very close to one another are eventually moved by the map so they are no longer close, by any reasonable definition of "close".

In order to investigate sensitive dependence, we introduce the concept of the **itinerary** of an orbit. This is a bookkeeping device that allows much of the information of an orbit to be coded in terms of discrete symbols.

For the logistic map, assign the symbol **L** to the left subinterval $[0, 1/2]$, and **R** to the right subinterval $[1/2, 1]$. Given an initial value x_0, we construct its itinerary by listing the subintervals, **L** or **R**, that contain x_0 and all future iterates. For example, the initial condition $x_0 = 1/3$ begets the orbit $\{\frac{1}{3}, \frac{8}{9}, \frac{32}{81}, \ldots\}$, whose itinerary begins **LRL**.... For the initial condition $x_0 = \frac{1}{4}$, the orbit is $\{\frac{1}{4}, \frac{3}{4}, \frac{3}{4}, \ldots\}$,

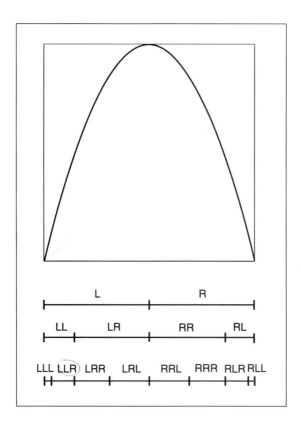

Figure 1.12 Schematic itineraries for $G(x) = 4x(1-x)$.
The rules: (1) an interval ending in **L** splits into two subintervals ending in **LL** and **LR** if there are an even number of **R**'s; the order is switched if there are an odd number of **R**'s, (2) an interval ending in **R** splits into two subintervals ending in **RL** and **RR** if there are an even number of **R**'s; the order is switched if there are an odd number of **R**'s

which terminates in the fixed point $x = \frac{3}{4}$. The itinerary for this orbit is **LRR**...,
which we abbreviate by **L$\overline{\text{R}}$**; the overbar indicates that the **R** repeats indefinitely.

Notice that there is a special orbit, or group of orbits, for which the itinerary
is not uniquely defined. That is because the intervals **L** and **R** overlap at $x = 1/2$.
In particular, consider the initial condition $x_0 = 1/2$. The corresponding orbit
is $\{1/2, 1, 0, 0, \ldots\}$, which can be assigned itinerary **RR$\overline{\text{L}}$** or **LR$\overline{\text{L}}$**. This particular
orbit (and some others like it) are assigned two different names under this naming
system. Except for the case of orbits which land precisely on $x = 1/2$ at some
point of the orbit (and therefore end up mapping onto the fixed point 0), the
itinerary is uniquely defined.

Once we are given this way of assigning an itinerary to each orbit, we can
map out, on the unit interval, the locations of points that have certain itineraries.
Of course, an itinerary is in general an infinite sequence, but we could ask: what
is the set of points whose itinerary begins with, say, **LR**? These points share the
property of beginning in the **L** subinterval and being mapped to the **R** subinterval
by one iterate of the map. This set, which we could call the **LR** set, is shown in
Figure 1.12, along with a few other similar sets.

We would like to identify the sets of all initial points whose itineraries begin
with a specified sequence of symbols. For example, the set of initial conditions
whose itinerary begins with **LR** forms a subinterval of the unit interval. The
subintervals in Figure 1.12 give information about the future behavior of the
initial conditions lying in them. Another example is the subinterval marked **LRL**,
which consists of orbits that start out in the interval **L** $= [0, 1/2]$, whose first
iterate lands in **R** $= [1/2, 1]$, and whose second iterate lands in $[0, 1/2]$. For
example, $x = 1/3$ lies in **LRL**. Likewise, $x = 1/4$ lies in **LRR** because its first and
second iterate are in **R**.

✎ **EXERCISE T1.14**

(a) Find a point that lies in the subinterval **LLR**. (You are asked for a specific
number.) (b) For each subinterval corresponding to a sequence of length 3,
find a point in the subinterval.

You may see some patterns in the arrangement of the subintervals of Figure
1.12. It turns out that the rule for dividing an interval, say **LR**, into its two
subintervals is the following: Count the number of **R**'s in the sequence (one in
this case). If odd, the interval is divided into **LRR, LRL** in that order. If even, the
L subinterval precedes the **R** subinterval. With this information, the reader can
continue Figure 1.12 schematically to finer and finer levels.

✎ **EXERCISE T1.15**

Continue the schematic diagram in Figure 1.12 to subintervals corresponding to length 4 sequences.

✎ **EXERCISE T1.16**

Let x_0 be a point in the subinterval RLLRRRLRLR. (a) Is x_0 less than, equal to, or greater than $1/2$? (b) Same question for $f^6(x_0)$.

A graphical way of specifying the possible itineraries for the logistic map is shown in Figure 1.13. We call this the **transition graph** for the subintervals L and R. An arrow from L to R, for example, means that the image $f(L)$ contains the interval R. For every path through this graph with directed edges (arrows), there exists an orbit with an itinerary satisfying the sequence of symbols determined by the path. It is clear from Figure 1.12 that the image of each of the intervals L and R contains both L and R, so the transition graph for the logistic map is fully connected (every symbol is connected to every other symbol by an arrow). Since the graph is fully connected, all possible sequences of L and R are possible.

The concept of itineraries make it easy to explain what we mean by sensitive dependence on initial conditions. In specifying the first k symbols of the itinerary, we have 2^k choices. If k is large, then most of the 2^k subintervals are forced to be rather small, since the sum of their lengths is 1. It is a fact (that we will prove in Chapter 3) that each of the 2^k subintervals is shorter than $\pi/2^{k+1}$ in length.

Consider any one of these small subintervals for a large value of k, corresponding to some sequence of symbols $S_1 \cdots S_k$, where each S_i is either R or L. This subinterval in turn contains subintervals corresponding to the sequences $S_1 \cdots S_k LL$, $S_1 \cdots S_k LR$, $S_1 \cdots S_k RR$ and $S_1 \cdots S_k RL$. If we choose one point from each, we have four initial conditions that lie within $\pi/2^{k+1}$ (since they all lie

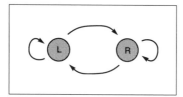

Figure 1.13 Transition graph for the logistic map $G(x) = 4x(1 - x)$.
The leftmost arrow tells us that f maps the interval L over itself, i.e., that $f(L)$ contains L. The top arrow says that $f(L)$ contains R, and so forth.

in $S_1 \cdots S_k$), but which map k iterates later to subintervals LL, LR, RR, and RL, respectively. (If this step isn't clear, it may help to recall Exercise T1.16.) In Figure 1.12, the width of the LR and RR subintervals are greater than $1/4$, so that the LR and RL subintervals, for example, lie over $1/4$ unit apart.

It is now possible to see why *every* point in $[0, 1]$ has sensitive dependence on initial conditions under the logistic map G. To find a neighbor close to x_0 that eventually separates by a distance of at least $d = 1/4$, identify which subinterval of level $k + 2$ that x_0 belongs to, say $S_1 \cdots S_k$LR. Then it is always possible to identify a subinterval within $\pi/2^{k+1}$ which maps $1/4$ unit away after k iterates, such as $S_1 \cdots S_k$RL. Therefore every point exhibits sensitive dependence with neighbors that are arbitrarily close.

We illustrate for $k = 1000$: There is a pair of initial conditions within $2^{-1001} \approx 10^{-300}$ that eventually are mapped at least $1/4$ unit apart. This is an expansion of 1000 factors of 2 in 1000 iterates, for an average multiplicative expansion rate of approximately 2. In Chapter 3 we will introduce the term "Lyapunov number", which will quantify the average multiplicative separation rate of a map, which is in this case $2^{1000/1000} \approx 2$ per iterate. The fact that this number is greater than 1 will mean that repeated expansion is occurring.

The impact of sensitive dependence is that changes in initial measurements or errors in calculation along the orbit can result in quite different outcomes. The consequences of this behavior were not fully appreciated until the advent of computers and the computer simulation of dynamical models.

⇨   COMPUTER EXPERIMENT 1.4

Use a computer program to illustrate sensitive dependence for the logistic map $G(x) = 4x(1 - x)$. Start with two different initial conditions that are very close together, and iterate G on each. The two orbits should stay near one another for a while, and then separate for good. By collecting statistics on your experiments, try to quantify how many iterations are required for the points to move apart, say $1/2$ unit, when the initial separation is .01, .001, etc. Does the location of the initial pair matter?

☞ CHALLENGE 1

Period Three Implies Chaos

In Chapter 1 we have studied periodic orbits for continuous maps and the idea of sensitive dependence on initial conditions. In Challenge 1 you will prove the fact that the existence of a period-three orbit alone implies the existence of a large set of sensitive points. The set is infinite (in fact, uncountably infinite, a concept we will study in more detail later in the book). This surprising fact was discovered by T.Y. Li and J.A. Yorke (Li and Yorke, 1975).

A chaotic orbit is a bounded, non-periodic orbit that displays sensitive dependence. When we give a precise definition of chaos, we will find that the discussion is simplified if we require a stronger definition of sensitivity, namely that chaotic orbits separate exponentially fast from their neighbors as the map is iterated.

A much simpler fact about continuous maps is that the existence of a period-three orbit implies that the map has periodic orbits of all periods (all integers). See Exercise T3.9 of Chapter 3. This fact doesn't say anything directly about sensitive dependence, although it guarantees that the map has rather complicated dynamical behavior.

We show a particular map f in Figure 1.14 that has a period-three orbit, denoted $\{A, B, C\}$. That is, $f(A) = B$, $f(B) = C$, and $f(C) = A$. We will discover that there are infinitely many points between A and C that exhibit sensitive dependence on initial conditions. To simplify the argument, we will use an assumption that is explicitly drawn into Figure 1.14: the map $f(x)$ is **unimodal**, which means that it has only one critical point. (A **critical point** for a function $f(x)$ is a point for which $f'(x) = 0$ or where the derivative does not exist.) This assumption, that $f(x)$ has a single maximum, is not necessary to prove sensitive dependence—in fact sensitive dependence holds for any continuous map with a period-three orbit. The final step of Challenge 1 asks you to extend the reasoning to this general case.

The existence of the period-three orbit in Figure 1.14 and the continuous nature of f together guarantee that the image of the interval $[A, B]$ covers $[B, C]$; that is, that $f[A, B] \supseteq [B, C]$. Furthermore, $f[B, C] = [A, C]$. We will try to repeat our analysis of itineraries, which was successful for the logistic map, for this new map. Let the symbol L represent the interval $[A, B]$, and R represent $[B, C]$.

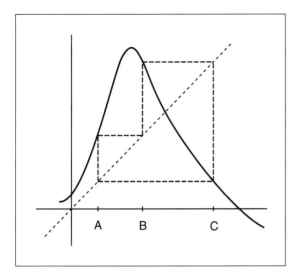

Figure 1.14 A map with a period-three orbit.
The dashed lines follow a cobweb orbit, from A to B to C to A.

Unlike the logistic map example, notice that in the itinerary of an orbit, **L** must be followed by **R**, although **R** can be followed by either **L** or **R**. Some of the itinerary subintervals are shown schematically in Figure 1.15.

A second difference from the logistic map example is that there may be gaps in the interval, as shown in Figure 1.15. For example, points just to the left of B are mapped to the right of C, and therefore out of the interval [A, C]; we do not include these points in our analysis. (A more sophisticated analysis might include these points, but we can demonstrate sensitive dependence without considering these orbits.) To simplify our analysis, we will not assign an itinerary to points that map outside [A, C].

The corresponding transition graph is shown in Figure 1.16. The transition graph tells us that every finite sequence of the symbols **L** and **R** corresponds to a subinterval of initial conditions x, as long as there are never two consecutive **L**'s in the sequence. (This follows from the fact that the left-hand interval [A, B] does not map over itself.)

The proof that period three implies chaos is given below in outline form. In each part, you are expected to fill in a reason or argument.

Step 1 Let d denote the length of the subinterval **RR**. Denote by $J = S_1 \cdots S_k R$ any subinterval that ends in **R**. (Each S_i denotes either **R** or **L**.) Show

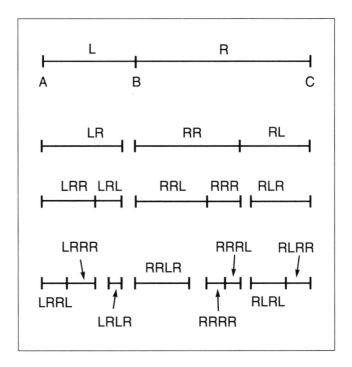

Figure 1.15 Schematic itineraries for period-three map.
The rules: (1) an interval ending in **R** splits into two subintervals ending in **RR** and
RL; the order is switched if there are an even number of **R**'s, (2) an interval ending
in **L** contains a shorter subinterval ending in **LR**, and a gap on the left (for an odd
number of **R**'s) or the right (for an even number).

that J contains a pair of points, one in each of the subintervals $S_1 \cdots S_k RRL$ and
$S_1 \cdots S_k RLR$, that eventually map at least d units apart.

Step 2   Show that inside J there are 3 subintervals of type $S_1 \cdots S_k RS_{k+2}$
$S_{k+3}R$. Explain why at least 2 of them must have length that is less than half the
length of J.

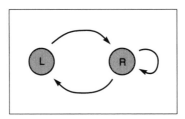

Figure 1.16 Transition graph for map with period-three orbit.
The three arrows imply that $f(L) \supseteq R$, $f(R) \supseteq R$, and $f(R) \supseteq L$.

Step 3 Combine the two previous steps. Show that a subinterval J of form $S_1 \cdots S_k R$ must contain a subinterval J_1 of form $S_1 \cdots S_k R S_{k+2} S_{k+3} R L R$ with the following property. Each point x of J_1 has a neighbor y within length$(J)/2$ whose pairwise distance upon further iteration eventually exceeds d.

Step 4 Let $h = C - A$ be the length of the original interval. Show that for each positive integer k there are 2^k disjoint subintervals (denoted by sequences of $5k + 1$ symbols) of length less than $2^{-k}h$, each of which contain a point that has sensitive dependence on initial conditions. Therefore, there are infinitely many sensitive points.

Step 5 Quantify the number of sensitive points you have located in the following way. Show that there is a one-to-one correspondence between the sensitive points found in Step 4 and binary numbers between 0 and 1 (infinite sequences of form $.a_1 a_2 a_3 \cdots$, where each a_i is a 0 or 1). This means that the set of sensitive points is uncountable, a concept we will meet in Chapter 4.

Step 6 Our argument is based on Figure 1.14, where $f(A) = B, f(B) = C, f(C) = A$, and where $A < B < C$. How many other distinct "cases" need to be considered? Does a similar argument work for these cases? What changes are necessary?

Step 7 Explain how to modify the arguments above to work for the case where f is any continuous map with a period-three orbit. (Begin by identifying one-piece subintervals of $[A, B]$ and $[B, C]$ that are mapped onto $[A, B]$ and $[B, C]$.)

Postscript. The subintervals described in the previous argument, although many in number, may comprise a small proportion of all points in the interval $[A, C]$. For example, the logistic map $g(x) = 3.83x(1 - x)$ has a period-three sink. Since there is a period-three orbit (its stability does not matter), we now know that there are many points that exhibit sensitive dependence with respect to their neighbors. On the other hand, the orbits of most points in the unit interval converge to one or another point in this periodic attractor under iteration by g^3. These points do *not* exhibit sensitive dependence. For example, points that lie a small distance from one of the points p of the period-three sink will be attracted toward p, as we found in Theorem 1.5. The distances between points that start out near p decrease by a factor of approximately $|(g^3)'(p)|$ with each three iterates. These nearby points do not separate under iteration. There are, however, infinitely many points whose orbits do not converge to the period-three sink. It is these points that exhibit sensitive dependence.

EXERCISES

1.1. Let $l(x) = ax + b$, where a and b are constants. For which values of a and b does l have an attracting fixed point? A repelling fixed point?

1.2. (a) Let $f(x) = x - x^2$. Show that $x = 0$ is a fixed point of f, and describe the dynamical behavior of points near 0.

(b) Let $g(x) = \tan x$, $-\pi/2 < x < \pi/2$. Show that $x = 0$ is a fixed point of g, and describe the dynamical behavior of points near 0.

(c) Give an example of a function h for which $h'(0) = 1$ and $x = 0$ is an attracting fixed point.

(d) Give an example of a function h for which $h'(0) = 1$ and $x = 0$ is a repelling fixed point.

1.3. Let $f(x) = x^3 + x$. Find all fixed points of f and decide whether they are sinks or sources. You will have to work without Theorem 1.5, which does not apply.

1.4. Let $x_1 < \cdots < x_8$ be the eight fixed points of $G^3(x)$, where $G(x) = 4x(1 - x)$, as in 1.10(c). Clearly, $x_1 = 0$.

(a) For which i is $x_i = 3/4$?

(b) Group the remaining six points into two orbits of three points each. It may help to consult Figure 1.10(c). The most elegant solution (that we know of) uses the chain rule.

1.5. Is the period-two orbit of the map $f(x) = 2x^2 - 5x$ on $\mathbb{R}$ a sink, a source, or neither? See Exercise T1.5.

1.6. Define the map $f(x) = 2x$ (mod 1) on the unit interval $[0, 1]$. Let L denote the subinterval $[0, 1/2]$ and R the subinterval $[1/2, 1]$.

(a) Draw a chart of the itineraries of f as in Figure 1.12.

(b) Draw the transition graph for f.

(c) Establish sensitive dependence for orbits under this map. Show that each point has neighbors arbitrarily near that eventually map at least $1/2$ unit apart.

1.7. Define the tent map on the unit inverval $[0, 1]$ by

$$T(x) = \begin{cases} 2x & \text{if } 0 \leq x \leq 1/2 \\ 2(1 - x) & \text{if } 1/2 \leq x \leq 1 \end{cases}.$$

(a) Divide the unit interval into two appropriate subintervals and repeat parts (a)–(c) of Exercise 1.6 for this map.

(b) Complete a periodic table for f, similar to the one in Table 1.3, for periods less than or equal to 10. In what ways, if any, does it differ from the periodic table for the logistic map G?

1.8. Let $f(x) = 4x(1 - x)$. Prove that there are points in $I = [0, 1]$ that are not fixed points, periodic points, or eventually periodic points of f.

1.9. Define $x_{n+1} = (x_n + 2)/(x_n + 1)$.

(a) Find $L = \lim_{n \to \infty} x_n$ for $x_0 \geq 0$.

(b) Describe the set of all negative x_0 for which the limit $\lim_{n \to \infty} x_n$ either exists and is not equal to the L in part (a) or does not exist (for example, $x_0 = -1$).

1.10. For the map $g(x) = 3.05x(1 - x)$, find the stability of all fixed points and period-two points.

1.11. Let f be a one-to-one smooth map of the real line to itself. One-to-one means that if $f(x_1) = f(x_2)$, then $x_1 = x_2$. A function f is called **increasing** if $x_1 < x_2$ implies $f(x_1) < f(x_2)$, and **decreasing** if $x_1 < x_2$ implies $f(x_1) > f(x_2)$.

(a) Show that f is increasing for all x or f is decreasing for all x.

(b) Show that every orbit $\{x_0, x_1, x_2, \ldots\}$ of f^2 satisfies either $x_0 \geq x_1 \geq x_2 \geq \ldots$ or $x_0 \leq x_1 \leq x_2 \leq \ldots$.

(c) Show that every orbit of f^2 either diverges to $+\infty$ or $-\infty$ or converges to a fixed point of f^2.

(d) What does this imply about convergence of the orbits of f?

1.12. The map $g(x) = 2x(1 - x)$ has negative values for large x. Population biologists sometimes prefer maps that are positive for positive x.

(a) Find out for what value of a the map $h(x) = axe^{-x}$ has a **superstable fixed point** x_0, which means that $h(x_0) = x_0$ and $h'(x_0) = 0$.

(b) Investigate the orbit starting at $x_0 = 0.1$ for this value of a using a calculator. How does the behavior of this orbit differ if a is increased by 50%?

(c) What is the range of $a \geq 1$ for which $h(x)$ has a positive sink?

1.13. Let $f : [0, \infty) \to [0, \infty)$ be a smooth function, $f(0) = 0$, and let $p > 0$ be a fixed point such that $f'(p) \geq 0$. Assume further that $f'(x)$ is decreasing. Show that all positive x_0 converge to p under f.

1.14. Let $f(x) = x^2 + x$. Find all fixed points of f. To where are orbits under f attracted? *too easy?*

1.15. Prove the following explicit formula for any orbit $\{x_0, x_1, x_2, \ldots\}$ of the logistic map $G(x) = 4x(1 - x)$:

$$x_n = \frac{1}{2} - \frac{1}{2} \cos(2^n \arccos(1 - 2x_0)).$$

Caution: Not all explicit formulas are useful.

1.16. Let f be a map defined on the real line, and assume p is a fixed point. Let $\epsilon > 0$ be a given number. Find a condition that guarantees that every initial point x in the interval $(p - \epsilon, p + \epsilon)$ satisfies $f^n(x) \to p$ as $n \to \infty$.

1.17. Let $f(x) = 4x(1 - x)$. Prove that LLL . . . L, the interval of initial values x in $[0, 1]$ such that $0 < f^i(x) < 1/2$ for $0 \le i < k$, has length $[1 - \cos(\pi/2^k)]/2$.

Boom, Bust, and Chaos in the Beetle Census

DAMAGE DUE TO flour beetles is a significant cost to the food processing industry. One of the major goals of entomologists is to gain insight into the population dynamics of beetles and other insects, as a way of learning about insect physiology. A commercial application of population studies is the development of strategies for population control.

A group of researchers recently designed a study of population fluctuation in the flour beetle *Tribolium*. The newly hatched larva spends two weeks feeding before entering a pupa stage of about the same length. The beetle exits the pupa stage as an adult. The researchers proposed a discrete map that models the three separate populations. Let the numbers of larvae, pupae, and adults at any given time t be denoted L_t, P_t, and A_t, respectively. The output of the map is three numbers: the three populations L_{t+1}, P_{t+1}, and A_{t+1} one time unit later. It is most convenient to take the time unit to be two weeks. A typical model for the three beetle populations is

$$L_{t+1} = bA_t$$
$$P_{t+1} = L_t(1 - \mu_l)$$
$$A_{t+1} = P_t(1 - \mu_p) + A_t(1 - \mu_a), \qquad (1.5)$$

where b is the birth rate of the species (the number of new larvae per adult each time unit), and where μ_l, μ_p, and μ_a are the death rates of the larva, pupa, and adult, respectively.

We call a discrete map with three variables a three-dimensional map, since the state of the population at any given time is specified by three numbers L_t, P_t, and A_t. In Chapter 1, we studied one-dimensional maps, and in Chapter 2 we move on to higher dimensional maps, of which the beetle population model is an example.

Tribolium adds an interesting twist to the above model: cannibalism caused by overpopulation stress. Under conditions of overcrowding, adults will eat pupae

Costantino, R.F., Cushing, J.M., Dennis, B., Desharnais, R.A., Experimentally induced transitions in the dynamic behavior of insect populations. Nature **375**, 227–230 (1995).

and unhatched eggs (future larvae); larvae will also eat eggs. Incorporating these refinements into the model yields

$$L_{t+1} = bA_t \exp(-c_{ea}A_t - c_{el}L_t)$$

$$P_{t+1} = L_t(1 - \mu_l)$$

$$A_{t+1} = P_t(1 - \mu_p) \exp(-c_{pa}A_t) + A_t(1 - \mu_a). \tag{1.6}$$

The parameters $c_{el} = 0.012$, $c_{ea} = 0.009$, $c_{pa} = 0.004$, $\mu_l = 0.267$, $\mu_p = 0$, and $b = 7.48$ were determined from population experiments. The mortality rate of the adult was determined from experiment to be $\mu_a = 0.0036$.

The effect of calling the exterminator can be modeled by artificially changing the adult mortality rate. Figure 1.17 shows a bifurcation diagram from Equations (1.6). The horizontal axis represents the mortality rate μ_a. The asymptotic value of L_t—found by running the model for a long time at a fixed μ_a and recording the resulting larval population—is graphed vertically.

Figure 1.17 suggests that for relatively low mortality rates, the larval population reaches a steady state (a fixed point). For $\mu_a > .1$ (representing a death rate of 10% of the adults over each 2 week period), the model shows oscillation between two widely-different states. This is a "boom-and-bust" cycle, well-known to population biologists. A low population (bust) leads to uncrowded living con-

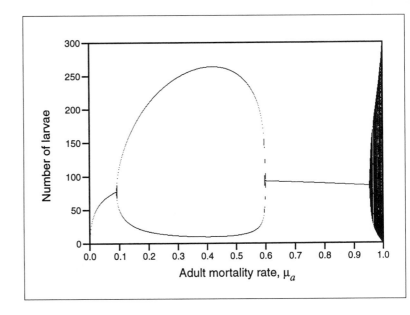

Figure 1.17 Bifurcation diagram for the model equations (1.6).
The bifurcation parameter is μ_a, the adult mortality rate.

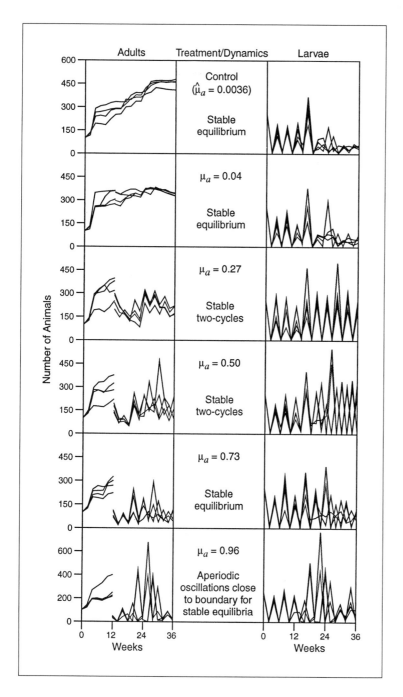

Figure 1.18 Population as a function of time.

Four replicates of the experiment for each of six different adult mortality rates are plotted together.

ditions and runaway growth (boom) at the next generation. At this point the limits to growth (cannibalism, in this system) take over, leading to a catastrophic decline and repeat of the cycle.

The period-doubling bifurcation near $\mu_a = 0.1$ is followed by a period-halving bifurcation at $\mu_a \approx 0.6$. For very high adult mortality rates (near 100%), we see the complicated, nonperiodic behavior characteristic of the logistic map.

It is one thing to find "chaos" in a mathematical model. A much more significant finding would show that the model is true enough to a real-world system that its chaotic behavior can be reproduced in the laboratory. The experimenters put several hundred beetles and 20 grams of food in each of several half-pint milk bottles. They recorded the populations for 18 consecutive two-week periods. Five different adult mortality rates, $\mu_a = 0.0036$ (the natural rate), 0.04, 0.27, 0.50, 0.73, and 0.96 were enforced in different bottles, by periodically removing the requisite number of adult beetles to artificially reach that rate. Each of the five experiments was replicated in four separate bottles.

Figure 1.18 shows the population counts taken from the experiment. Populations of adults from the four separate bottles are graphed together in the boxes on the left. The four curves in the box are the adult population counts for the four bottles as a function of time. The boxes on the right are similar but show the population counts for the larvae. During the first 12 weeks, the populations were undisturbed, so that the natural adult mortality rate applied; after that, the artificial mortality rates were imposed by removing or adding adult bettles as needed.

The population counts from the experiment agree remarkably well with the computer simulations from Figure 1.18. The top two sets of boxes represent $\mu_a = 0.0036$ and 0.04, which appear experimentally to be sinks, or stable equilibria, as predicted by Figure 1.18. The period-two sink predicted also can be seen in the populations for $\mu_a = 0.27$ and 0.50. For $\mu_a = 0.96$, the populations seem to be governed by aperiodic oscillations.

Two-Dimensional Maps

IN CHAPTER 1 we developed the fundamental properties of one-dimensional dynamics. The concepts of periodic orbits, stability, and sensitive dependence of orbits are most easily understood in that context.

In this chapter we will begin the transition from one-dimensional to many dimensional dynamics. The discussion centers around two-dimensional maps, since much of the new phenomena present in higher dimensions appears there in its simplest form. For example, we will expand our classification of one-dimensional fixed points as sinks and sources to include saddles, which are contracting in some directions and expanding in others.

2.1 MATHEMATICAL MODELS

Maps are important because they encode the behavior of deterministic systems. We have discussed population models, which have a single input (the present population) and a single output (next year's population). The assumption of determinism is that the output can be uniquely determined from the input. Scientists use the term "state" for the information being modeled. The state of the population model is given by a single number—the population—and the **state space** is one-dimensional. We found in Chapter 1 that even a model where the state runs along a one-dimensional axis, as $G(x) = 4x(1 - x)$, can have extremely complicated dynamics.

The changes in temperature of a warm object in a cool room can be modeled as a one-dimensional map. If the initial temperature difference between the object and the room is $D(0)$, the temperature difference $D(t)$ after t minutes is

$$D(t) = D(0)e^{-Kt}, \tag{2.1}$$

where $K > 0$ depends on the specific heat of the object. We can call the model

$$f(x) = e^{-K}x \tag{2.2}$$

the "time-1 map", because application of this map advances the system one minute. Since e^{-K} is a constant, f is a linear one-dimensional map, the easiest type we studied in Chapter 1. It is clear that the fixed point $x = 0$ is a sink because $|e^{-K}| < 1$. More generally, we could consider the map that advances the system T time units, the **time-T map**. For any fixed time T, the time-T map for this example is the linear one-dimensional map $f(x) = e^{-KT}x$, also written as

$$x \longmapsto e^{-KT}x. \tag{2.3}$$

Formula (2.1) is derived from Newton's law of cooling, which is the differential equation

$$\frac{dD}{dt} = -kD, \tag{2.4}$$

with initial value D_0. There is a basic assumption that the body is a good conductor so that at any instant, the temperature throughout the body is uniform. Equation (2.2) is an example of the derivation of a map as the time-T map of a differential equation.

In the examples above, the state space is one-dimensional, meaning that the information needed to advance the model in time is a single number. This is in effect the definition of state in mathematical modeling: the amount of information needed to advance the model. Nothing else is relevant for the purposes of the model, so the state is completely sufficient to describe the conditions of the system. In the case of a one-dimensional map or a single differential equation, the state is a single number. In the case of a two-dimensional map or a system of two coupled ordinary differential equations, it is a pair of numbers. The state is essentially the initial information needed for the dynamical system model to operate and respond unambiguously with an orbit. In partial differential equations models, the state space may be infinite-dimensional. In such a case, infinitely many numbers are needed to specify the current state. For a vibrating string, modeled by a partial differential equation called the wave equation, the state is the entire real-valued function describing the shape of the string.

You are already familiar with systems that need more than one number to specify the current condition of the system. For the system consisting of a projectile falling under Newton's laws of motion, the state of the system at a given time can be specified fully by six numbers. If we know the position $\vec{p} = (x, y, z)$ and velocity $\vec{v} = (v_x, v_y, v_z)$ of the projectile at time $t = 0$, then the state at any future time t is uniquely determined as

$$x(t) = x(0) + v_x(0)t$$

$$y(t) = y(0) + v_y(0)t$$

$$z(t) = z(0) + v_z(0)t - gt^2/2$$

$$v_x(t) = v_x(0)$$

$$v_y(t) = v_y(0)$$

$$v_z(t) = v_z(0) - gt, \tag{2.5}$$

where the constant g represents the acceleration toward earth due to gravity. (In meters-kilograms-seconds units, $g \approx 9.8$ m/sec^2.) Formula (2.5) is valid as long as the projectile is aloft. The assumptions built into this model are that gravity alone is the only force acting on the projectile, and that the force of gravity is constant. The formula is again derived from a differential equation due to Newton, this time the law of motion

$$F = ma. \tag{2.6}$$

Here the gravitational force is

$$F = \frac{GMm}{r^2}, \tag{2.7}$$

where M is the mass of the earth, m is the mass of the projectile, G is the gravitational constant, and r is the distance from the center of the earth to the projectile. Since r is constant to good approximation as long as the projectile is near the surface of the earth, g can be calculated as GM/r^2.

The position of the falling projectile evolves according to a set of six equations (2.5). The identification of a projectile's motion as a system with a six-dimensional state is one of the great achievements of Newtonian physics. However, it is rare to find a dynamical system that has an explicit formula like (2.5) that describes the evolution of the state's future. More often we know a formula that describes the new state in terms of the previous and/or current state.

A map is a formula that describes the new state in terms of the previous state. We studied many such examples in Chapter 1. A differential equation is a formula for the instantaneous *rate of change* of the current state, in terms of the current state. An example of a differential equation model is the motion of a projectile far from the earth's surface. For example, a satellite falling to earth must follow the gravitational equation (2.7) but with a *nonconstant* distance r. That makes the acceleration a function of the position of the satellite, rendering equations (2.5) invalid.

A system consisting of two orbiting masses interacting through gravitational acceleration can be expressed as a differential equation. Using Newton's law of motion (2.6) and the gravitation formula (2.7), the motions of the two masses can be derived as a function of time as in (2.5). We say that such a system is "analytically solvable". The resulting formulas show that the masses follow elliptical orbits around the combined center of mass of the two bodies.

On the other hand, a system of three or more masses interacting exclusively through gravitational acceleration is not analytically solvable. The so-called **three-body problem**, for example, has an 18-dimensional state space. To solve the equations needed to advance the dynamics, one must know the three positions and three velocities of the point masses, a total of six three-dimensional vectors, or 18 numbers. Although there are no exact formulas of type (2.5) in this case, one can use computational methods to approximate the solution of the differential equations resulting from Newton's laws of motion to get an idea of the complicated behavior that results.

At one time it was not known that there are no such exact formulas. In 1889, to commemorate the 60th birthday of King Oscar II of Sweden and Norway, a contest was held to produce the best research in celestial mechanics pertaining to the stability of the solar system, a particularly relevant n-body problem. The winner was declared to be Henri Poincaré, a professor at the University of Paris.

Poincaré submitted an entry full of seminal ideas. In order to make progress on the problem, he made two simplifying assumptions. First, he assumed that there are three bodies all moving in a plane. Second, he assumed that two of the bodies were massive and that the third had negligible mass in comparison, so small that it did not affect the motion of the other two. We can imagine two stars and a small asteroid. In general, the two large stars would travel in ellipses, but Poincaré made another assumption, that the initial conditions were chosen such that the two moved in circles, at constant speed, circling about their combined center of mass.

It is simplest to observe the trajectory of the asteroid in the rotating co-ordinate system in which the two stars are stationary. Imagine looking down on the plane in which they are moving, rotating yourself with them so that the two appear fixed in position. Figure 2.1 shows a typical path of the asteroid. The horizontal line segment in the center represents the (stationary) distance between the two larger bodies: The large one is at the left end of the segment and the medium one is at the right end. The tiny body moves back and forth between the two larger bodies in a seemingly unpredictable manner for a long time. The asteroid gains speed as it is ejected toward the right with sufficient momentum so

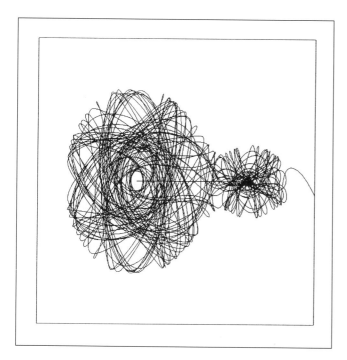

Figure 2.1   A trajectory of a tiny mass in the three-body problem.
Two larger bodies are in circular motion around one another. This view is of a rotating coordinate system in which the two larger bodies lie at the left and right ends of the horizontal line segment. The tiny mass is eventually ejected toward the right. Other trajectories starting close to one of the bodies can be forever trapped.

that it never returns. Other pictures, with different initial conditions, could be shown in which the asteroid remains forever close to one of the stars.

This three-body system is called the "planar restricted three-body problem", but we will refer to it as the three-body problem. Poincaré's method of analysis was based on the fact that the motion of the small mass could be studied, in a rather nonobvious manner, by studying the orbit of a plane map (a function from $\mathbb{R}^2$ to $\mathbb{R}^2$). He discovered the crucial ideas of "stable and unstable manifolds", which are special curves in the plane (see Section 2.6 for definitions).

On the basis of his entry Poincaré was declared the winner. However, he did not fully understand the nature of the stable and unstable manifolds at that time. These manifolds can cross each other, in so-called homoclinic points. Poincaré was confused by these points. Either he thought they didn't exist or he didn't understand what happened when they did cross. This error made his general conclusions about the nature of the trajectories totally wrong. His error was detected after he had been declared the winner but before his entry was published. He eventually realized that the existence of homoclinic points implied that there was incredibly complicated motion near those points, behavior we now call "chaotic".

Poincaré worked feverishly to revise his winning entry. The article "Sur les équations de la dynamique et le problème des trois corps" (on the equations of dynamics and the three-body problem) was published in 1890. In this 270-page work, Poincaré established convincingly that due to the possibility of homoclinic crossings, no general exact formula exists, beyond Newton's differential equations arising from (2.6) and (2.7), for making predictions of the positions of the three bodies in the future.

Poincaré succeeded by introducing qualitative methods into an area of study that had long been dominated by highly refined quantitative methods. The quantitative methods essentially involved developing infinite series expansions for the positions and velocities of the gravitational bodies. These series expansions were known to be problematic for representing near-collisions of the bodies. Poincaré was able to show through his geometric reasoning that these infinite series expansions could not converge in general.

One of Poincaré's most important innovations was a simplified way to look at complicated continuous trajectories, such as those resulting from differential equations. Instead of studying the entire trajectory, he found that much of the important information was encoded in the points in which the trajectory passed through a two-dimensional plane. The order of these intersection points defines a plane map. Figure 2.2 shows a schematic view of a trajectory C. The plane S is defined by $x_3 = $ constant. Each time the trajectory C pierces S in a downward

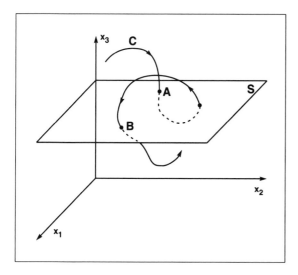

Figure 2.2   A Poincaré map derived from a differential equation.
The map G is defined to encode the downward piercings of the plane S by the
solution curve C of the differential equation so that G(A) = B, and so on.

direction, as at points **A** and **B** in Figure 2.2, we record the point of piercing
on the plane **S**. We can label these points by the coordinates (x_1, x_2). Let **A**
represent the kth downward piercing of the plane and **B** the $(k + 1)$th downward
piercing. The **Poincaré map** is the two-dimensional map G such that G(**A**) = **B**.
The Poincaré map is similar in principle to the time-T map we considered above,
though different in detail. While the time-T map is stroboscopic (it logs the value
of a variable at equal time intervals), the Poincaré map records plane piercings,
which need not be equally spaced in time. Although Figure 2.2 shows a plane,
more general surfaces can be used. The plane or surface is called a **surface of
section**.

Given **A**, the differential equations can be solved with **A** as an initial
value, and the solution followed until the next downward piercing at **B**. Thus **A**
uniquely determines **B**. This ensures that the map G is well-defined. This map
can be iterated to find all subsequent piercings of **S**. In general, the Poincaré
map technique reduces a k-dimensional, continuous-time dynamical system to a
$(k - 1)$-dimensional map. Much of the dynamical behavior of the trajectory **C** is
present in the two-dimensional map G. For example, the trajectory **C** is periodic
(forms a closed curve, which repeats the same dynamics forever) if and only if the
plane map G has a periodic orbit.

Now we can explain how the differential equations of the three-body problem shown in Figure 2.1 led Poincaré to a plane map. In our rotating coordinate

system (where the two stars are fixed at either end of the line segment) the differential equations involve the position (x, y) and velocity $(\dot{x}, \dot{y})$ of the asteroid. These four numbers determine the current state of the asteroid.

There is a constant of the motion, called the Hamiltonian. For this problem it is a function in the four variables that is constant with respect to time. It is like total energy of the asteroid, which never changes. The following fact can be shown regarding this Hamiltonian: for $y = 0$ and for any particular x and $\dot{x}$, the Hamiltonian reduces to $\dot{y}^2 - C$, where $C \geq 0$. Thus when the vertical component of the position of the asteroid satisfies $y = 0$, the vertical velocity component of the asteroid is restricted to the two possible values $\pm\sqrt{C}$.

Because of this fact, it makes sense to consider the Poincaré map, using upward piercings of the surface of section $y = 0$. The variable y is the vertical component of the position, so this corresponds to an upward crossing of the horizontal line segment in Figure 2.1. There are two "branches" of the two-dimensional surface $y = 0$, corresponding to the two possible values of $\dot{y}$ mentioned above. (The two dimensions correspond to independent choices of the numbers x and $\dot{x}$.) We choose one branch, say the one that corresponds to the positive value of $\dot{y}$, for our surface of section. What we actually do is follow the solution of the differential equation, computing $x, y, \dot{x}, \dot{y}$ as we go, and at the instant when y goes from negative to positive, we check the current $\dot{y}$ from the differential equation. If $\dot{y} > 0$, then an upward crossing of the surface has occurred, and $x, \dot{x}$ are recorded. This defines a Poincaré map.

Starting the system at a particular value of $(x, \dot{x})$, where y is zero and is moving from negative to positive, signalled by $\dot{y} > 0$, we get the image $F(x, \dot{x})$ by recording the new $(x, \dot{x})$ the next time this occurs. The Poincaré map F is a two-dimensional map. What Poincaré realized should now be clear to us. Even this restricted version of the full three-body problem contains much of the complicated behavior possible in two-dimensional maps, including chaotic dynamics caused by homoclinic crossings, shown in Figure 2.24. Understanding these complications will lead us to the study of stable and unstable manifolds, in Section 2.6 at first, and then in more detail in Chapter 10.

Work in the twentieth century has continued to reflect the philosophy that much of the chaotic phenomena present in differential equations can be approached, through reduction by time-T maps and Poincaré maps, by studying discrete-time dynamics. As you can gather from the three-body example, Poincaré maps are seldom simple to evaluate, even by computer. The French astronomer M. Hénon showed in 1975 that much of the interesting phenomena present in Poincaré maps of differential equations can be found as well in a two-dimensional quadratic map, which is much easier to simulate on a computer. The version of Hénon's map that we will study is

$$\mathbf{f}(x, y) = (a - x^2 + by, x). \tag{2.8}$$

Note that the map has two inputs, x, y, and two outputs, the new x, y. The new y is just the old x, but the new x is a nonlinear function of the old x and y. The letters a and b represent parameters that are held fixed as the map is iterated.

The nonlinear term x^2 in (2.8) is just about as unobtrusive as could be achieved. Hénon's remarkable discovery is that this "barely nonlinear" map displays an impressive breadth of complex phenomena. In its way, the Hénon map is to two-dimensional dynamics what the logistic map $G(x) = 4x(1 - x)$ is to one-dimensional dynamics, and it continues to be a catalyst for deeper understanding of nonlinear systems.

For now, set $a = 1.28$ and $b = -0.3$. (Here we diverge from Hénon, whose most famous example has $b = 0.3$ instead.) If we start with the initial condition $(x, y) = (0, 0)$ and iterate this map, we find that the orbit moves toward a period-two sink. Figure 2.3(a) shows an analysis of the results of iteration with general initial values. The picture was made by starting with a 700×700 grid of points

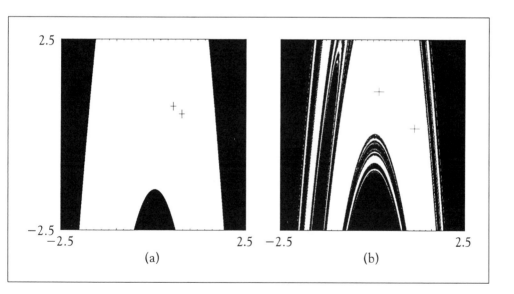

(a)   (b)

Figure 2.3   A square of initial conditions for the Hénon map with $b = -0.3$.
Initial values whose trajectories diverge to infinity upon repeated iteration are colored black. The crosses show the location of a period-two sink, which attracts the white initial values. (a) For $a = 1.28$, the boundary of the basin is a smooth curve. (b) For $a = 1.4$, the boundary is "fractal" .

in $[-2.5, 2.5] \times [-2.5, 2.5]$ as initial values. The map (2.8) was iterated by computer for each initial value until the orbit either converges to the period-two sink, or diverges toward infinity. Points in black represent initial conditions whose orbits diverge to infinity, and points in white represent initial values whose orbits converge to the period-two orbit. The black points are said to belong to the basin of infinity, and the white points to the basin of the period-two sink. The boundary of the basin consists of a smooth curve, which moves in and out of the rectangle of initial values shown here.

✎ **EXERCISE T2.1**

Check that the period-two orbit of the Hénon map (2.8) with $a = 1.28$ and $b = -0.3$ is approximately $\{(0.7618, 0.5382), (0.5382, 0.7618)\}$. We will see how to find these points in Section 2.5.

⇨ **COMPUTER EXPERIMENT 2.1**

Write a program to iterate the Hénon map (2.8). Set $a = 1.28$ and $b = -0.3$ as in Figure 2.3(a). Using the initial condition $(x, y) = (0, 0)$, create the period-two orbit, and view it either by printing a list of numbers or by plotting (x, y) points. Change a to 1.2 and repeat. How does the second orbit differ from the first? Find as accurately as possible the value of a between 1.2 and 1.28 at which the orbit behavior changes from the first type to the second.

If we instead set $a = 1.4$, with $b = -0.3$, and repeat the process, we see quite a different picture. First, the points of the period-two sink have distanced themselves a bit from one another. More interesting is that the boundary of the basin is no longer a smooth curve, but is in a sense infinitely complicated. This is fractal structure, which we shall discuss in detail in Chapter 4.

Next we want to show how a two-dimensional map can be derived from a differential equations model of a pendulum. Figure 2.4 shows a pendulum swinging under the influence of gravity. We will assume that the pendulum is free to swing through 360 degrees. Denote by θ the angle of the pendulum with respect to the vertical, so that $\theta = 0$ corresponds to straight down. Therefore θ and $\theta + 2\pi$ should be considered the same position of the pendulum.

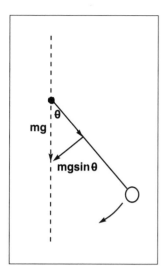

Figure 2.4 The pendulum under gravitational acceleration.
The force of gravity causes the pendulum bob to accelerate in a direction perpendicular to the rod. Here θ denotes the angle of displacement from the downward position.

We will use Newton's law of motion $F = ma$ to find the pendulum equation. The motion of the pendulum bob is contrained to be along a circle of radius l, where l is the length of the pendulum rod. If θ is measured in radians, then the component of acceleration tangent to the circle is $l\ddot{\theta}$, because the component of position tangent to the circle is $l\theta$. The component of force along the direction of motion is $mg \sin \theta$. It is a restoring force, meaning that it is directed in the opposite direction from the displacement of the variable θ. We will denote the first and second time derivatives of θ by $\dot{\theta}$ (the angular velocity) and $\ddot{\theta}$ (the angular acceleration) in what follows. The differential equation governing the frictionless pendulum is therefore

$$ml\ddot{\theta} = F = -mg \sin \theta, \qquad (2.9)$$

according to Newton's law of motion.

From this equation we see that the pendulum requires a two-dimensional state space. Since the differential equation is second order, the two initial values $\theta(0)$ and $\dot{\theta}(0)$ at time $t = 0$ are needed to specify the solution of the equation after time $t = 0$. It is not enough to specify $\theta(0)$ alone. Knowing $\theta(0) = 0$ means that the pendulum is in the straight down configuration at time $t = 0$, but we can't predict what will happen next without knowing the angular velocity at that time. For example, if the angle θ is 0, so that the pendulum is hanging straight down, we need to know whether the angular velocity $\dot\theta$ is positive or negative to tell whether the bob is moving right or left. The same can be said for knowing the angular velocity alone. Specifying θ and $\dot\theta$ together at time $t = 0$ will uniquely specify $\theta(t)$. Thus the state space is two-dimensional, and the state consists of the two numbers $(\theta, \dot\theta)$.

To simplify matters we will use a pendulum of length $l = g$, and to the resulting equation $\ddot\theta = -\sin\theta$ we add the damping term $-c\dot\theta$, corresponding to friction at the pivot, and a periodic term $\rho\sin t$ which is an external force constantly providing energy to the pendulum. The resulting equation, which we

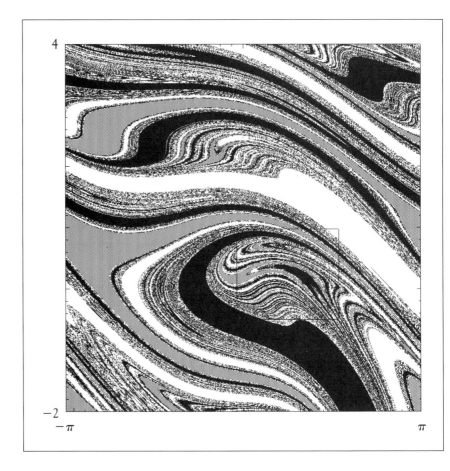

Figure 2.5 Basins of three coexisting attracting fixed points.
Parameters for the forced damped pendulum are $c = 0.2$, $\rho = 1.66$. The basins are shown in black, gray, and white. Each initial value $(\theta, \dot\theta)$ is plotted according to the sink to which it is attracted. Since the horizontal axis denotes angle in radians, the right and left edge of the picture could be glued together, creating a cylinder. The rectangle shown is magnified in figure 2.6.

call the **forced damped pendulum** model, is

$$\ddot{\theta} = -c\dot{\theta} - \sin\theta + \rho\sin t. \tag{2.10}$$

This differential equation includes friction (the friction constant is c) and the force of gravity, which pulls the pendulum bob down, as well as the sinusoidal force $\rho\sin t$, which accelerates the bob first clockwise and then counterclockwise, continuing in a periodic way. This periodic forcing guarantees that the pendulum will keep swinging, provided ρ is nonzero. The term $\rho\sin t$ has period 2π.

Because of the periodic forcing, if $\theta(t)$ is a solution of (2.10), then so is $\theta(t + 2\pi)$, and in fact so is $\theta(t + 2\pi N)$ for each integer N. Assume that $\theta(t)$ is a solution of the differential equation (2.10), and define $u(t) = \theta(t + 2\pi)$. Why is $u(t)$ also a solution of (2.10)? Note first that $\dot{u}(t) = \dot{\theta}(t)$ and $\ddot{u}(t) = \ddot{\theta}(t)$. Second, since $\theta(t)$ is assumed to be a solution for all t, (2.10) implies

$$\ddot{\theta}(t + 2\pi) = -c\dot{\theta}(t + 2\pi) - \sin\theta(t + 2\pi) + \rho\sin(t + 2\pi). \tag{2.11}$$

Since the nonhomogenous term of the differential equation is periodic with period 2π, $\sin t = \sin(t + 2\pi)$, it follows that

$$\ddot{u}(t) = -c\dot{u}(t) - \sin u(t) + \rho\sin t. \tag{2.12}$$

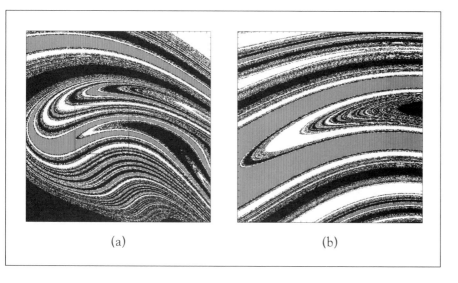

(a)   (b)

Figure 2.6 Detail views of the pendulum basin.
(a) Magnification of the rectangle shown in Figure 2.5. (b) Magnification of the rectangle shown in (a).

Note that this argument depends on the fact that $u(t)$ is a translation of $\theta(t)$ by exactly a multiple of 2π. The argument fails if we choose, for example, $u(t) = \theta(t + \pi/2)$ (assuming $\rho \neq 0$).

We conclude from this fact that **the time-2π map of the forced damped pendulum is well-defined**. If $(\theta_1, \dot\theta_1)$ is the result of starting with initial conditions $(\theta_0, \dot\theta_0)$ at time $t = 0$ and following the differential equation for 2π time units, then $(\theta_1, \dot\theta_1)$ will also be the result of starting with initial conditions $(\theta_0, \dot\theta_0)$ at time $t = 2\pi$ (or $4\pi, 6\pi, \ldots$) and following the differential equation for 2π time units. This fact allows us to study many of the important features by studying the time-2π map of the pendulum. Because the forcing is periodic with period 2π, the action of the differential equation is the same between $2N\pi$ and $2(N + 1)\pi$ for each integer N. Although the state equations for this system are differential equations, we can learn a lot of information about it by viewing snapshots taken each 2π time units.

When the pendulum is started at time $t = 0$, its behavior will be determined by the initial values of θ and $\dot\theta$. The differential equation uniquely determines the values of θ and $\dot\theta$ at later times, such as $t = 2\pi$. If we write $(\theta_0, \dot\theta_0)$ for the initial values and $(\theta_1, \dot\theta_1)$ as the values at time 2π, we can define the time-2π map F by

$$F(\theta_0, \dot\theta_0) = (\theta_1, \dot\theta_1). \tag{2.13}$$

Just because we give the time-2π map a name does not mean that there is a simple formula for computing it. Analyzing the time-2π map is different from analyzing the Hénon map, in the sense that there is no simple expression for the former map. The differential equation must be solved from time 0 to time 2π in order to iterate the map. For this example, investigation must be carried out largely by computer.

Figure 2.5 shows the basins of three coexisting attractors for the time-2π map of the forced damped pendulum. Here we have set the forcing parameter $\rho = 1.66$ and the damping parameter $c = 0.2$. The picture was made by solving the differential equation for an initial condition representing each pixel, and coloring the pixel white, gray, or black depending on which sink orbit attracts the orbit.

The three attractors are one fixed point and two period-two orbits. There are five other fixed points (not shown) that are not attractors. This system displays both great simplicity, in that the stable behaviors (sinks) are periodic orbits of low period, and great complexity, in that the boundaries between the three basins are infinitely-layered, or fractal.

Figure 2.6 shows further detail of the pendulum basins. Part (a) zooms in on the rectangle in Figure 2.5; part (b) is a further magnification. Note that the level of complication does not decrease upon magnification. The fractal structure continues on finer and finer scales. Color Plates 3–10 show an alternate setting of parameters for which four basins exist.

For other sets of parameter values, there are apparently no sinks for the forced damped pendulum, fixed or periodic. Figure 2.7 shows a long orbit of the pendulum with parameter settings $c = 0.05, \rho = 2.5$. One-half million points are shown. If these points were erased, and the next one-half million points were plotted, the picture would look the same. There are many fixed points and periodic orbits that coexist with this orbit. Some of them are shown in Figure 2.23(a).

4.5

−3.5

−π   π

Figure 2.7   A single orbit of the forced damped pendulum with $c = 0.05, \rho = 2.5$.

Different initial values yield essentially the same pattern, unless the initial value is an unstable periodic orbit, of which there are several (see Figure 2.23).

2.2 SINKS, SOURCES, AND SADDLES

We introduced the term "sink" in our discussion of one-dimensional maps to refer to a fixed point or periodic orbit that attracts an ϵ-neighborhood of initial values. A source is a fixed point that repels a neighborhood. These definitions make sense in higher-dimensional state spaces without alteration. In the plane, for example, the neighborhoods in question are disks (interiors of circles).

Definition 2.1 The **Euclidean length** of a vector $\mathbf{v} = (x_1, \ldots, x_m)$ in $\mathbb{R}^m$ is $|\mathbf{v}| = \sqrt{x_1^2 + \cdots + x_m^2}$. Let $\mathbf{p} = (p_1, p_2, \ldots, p_m) \in \mathbb{R}^m$, and let ϵ be a positive number. The **ϵ-neighborhood** $N_\epsilon(\mathbf{p})$ is the set $\{\mathbf{v} \in \mathbb{R}^m : |\mathbf{v} - \mathbf{p}| < \epsilon\}$, the set of points within Euclidean distance ϵ of $\mathbf{p}$. We sometimes call $N_\epsilon(\mathbf{p})$ an **ϵ-disk** centered at $\mathbf{p}$.

Definition 2.2 Let $\mathbf{f}$ be a map on $\mathbb{R}^m$ and let $\mathbf{p}$ in $\mathbb{R}^m$ be a fixed point, that is, $\mathbf{f}(\mathbf{p}) = \mathbf{p}$. If there is an $\epsilon > 0$ such that for all $\mathbf{v}$ in the ϵ-neighborhood $N_\epsilon(\mathbf{p})$, $\lim_{k \to \infty} \mathbf{f}^k(\mathbf{v}) = \mathbf{p}$, then $\mathbf{p}$ is a **sink** or **attractor**. If there is an ϵ-neighborhood $N_\epsilon(\mathbf{p})$ such that each $\mathbf{v}$ in $N_\epsilon(\mathbf{p})$ except for $\mathbf{p}$ itself eventually maps outside of $N_\epsilon(\mathbf{p})$, then $\mathbf{p}$ is a **source** or **repeller**.

Figure 2.8 shows schematic views of a sink and a source for a two-dimensional map, along with a typical disk neighborhood and its image under the map. Along with the sink and source, a new type of fixed point is shown in Figure 2.8(c), which cannot occur in a one-dimensional state space. This type of fixed point, which we will call a saddle, has at least one attracting direction and at least one repelling direction. A saddle exhibits sensitive dependence on initial conditions, because of the neighboring initial conditions that escape along the repelling direction.

EXAMPLE 2.3

Consider the two-dimensional map

$$\mathbf{f}(x, y) = (-x^2 + 0.4y, x). \tag{2.14}$$

This is a version of the Hénon map considered earlier in this chapter, with the parameters set at $a = 0$ and $b = 0.4$.

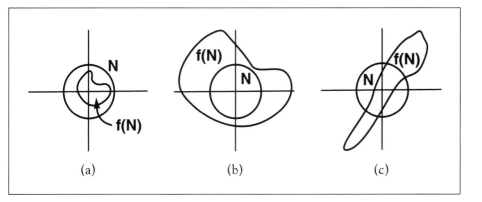

Figure 2.8 Local dynamics near a fixed point.
The origin is (a) a sink, (b) a source, and (c) a saddle. Shown is a disk N and its
iterate under the map $\mathbf{f}$.

✎ EXERCISE T2.2

Show that the map in (2.14) has exactly two fixed points, $(0, 0)$ and
$(-0.6, -0.6)$.

Figure 2.9 shows the two fixed points. Around each is drawn a small disk N
of radius 0.3. Also shown are the images $\mathbf{f}(N)$ and $\mathbf{f}^2(N)$ of each disk. The fixed
point $(0, 0)$ is a sink, and the fixed point $(-0.6, -0.6)$ is a saddle. Each time the
map is iterated, the disks shrink to 40% of their previous size. Therefore $\mathbf{f}(N)$ is
40% the size of N, and $\mathbf{f}^2(N)$ is 16% the size of N. We will explain the origin of
these numbers in Remark 2.15.

Although saddles, as well as sources, are unstable fixed points (they are
sensitive to initial conditions), they play surprising roles in the dynamics. In
Figure 2.10(a), the basin of the sink $(0, 0)$ is shown in white. The entire square
is the box $[-2, 2] \times [-2, 2]$, and the sink is the cross at the center. Not all of
the white basin is shown: it has infinite area. The points in black diverge under
iteration by $\mathbf{f}$; they are in the basin of infinity. You may wonder about the final
disposition of the points along the boundary between the two basins. Do they
go in or out? The answer is: neither. In Figure 2.10(b), the set of points that
converge to the saddle $(-0.6, -0.6)$ is plotted, along with the saddle denoted by
the cross. Although not an attractor, the saddle evidently plays a decisive role in
determining which points go to which basin.

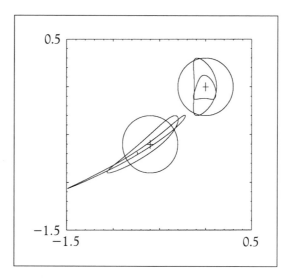

Figure 2.9 Local dynamics near fixed points of the Hénon map.
The crosses mark two fixed points of the Hénon map $\mathbf{f}$ with $a = 0$, $b = 0.4$, in the square $[-1.5, 0.5] \times [-1.5, 0.5]$. Around each fixed point a circle is drawn along with its two forward images under $\mathbf{f}$. On the left is a saddle: the images of the disk are becoming increasingly long and thin. On the right the images are shrinking, signifying a sink.

Attractors, as well as basins, can be more complicated than those shown in Figure 2.10. Consider the Hénon map (2.8) with $a = 2$ and $b = -0.3$. In Figure 2.11 the dark area is again the basin of infinity, while the white set is the basin for the two-piece attractor that looks like two hairpin curves.

Our goal in the next few sections is to find ways of identifying sinks, sources and saddles from the defining equations of the map. In Chapter 1 we found that the key to deciding the stability of a fixed point is the derivative at the point. Since the derivative determines the tangent line, or best linear approximation near the point, it determines the amount of shrinking/stretching in the vicinity of the point. The same mechanism is operating in higher dimensions. The action of the dynamics in the vicinity of a fixed point is governed by the best linear approximation to the map at the point. This best linear approximation is given by the Jacobian matrix, a matrix of partial derivatives calculated from the map. We will define the Jacobian matrix in Section 2.5. To find out what it can tell us, we need to fully understand linear maps first. For linear maps, the Jacobian matrix is equal to the map itself.

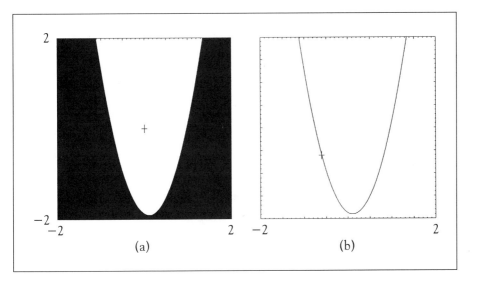

Figure 2.10 Basins of attraction for the Hénon map with $a = 0, b = 0.4$.
(a) The cross marks the fixed point $(0, 0)$. The basin of the fixed point $(0, 0)$ is shown in white; the points in black diverge to infinity. (b) The initial conditions that are on the boundary between the white and black don't converge to $(0, 0)$ or infinity; instead they converge to the saddle $(-0.6, -0.6)$, marked with a cross. This set of boundary points is the stable manifold of the saddle (to be discussed in Section 2.6).

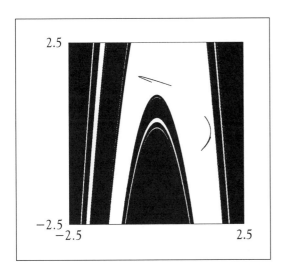

Figure 2.11 Attractors for the Hénon map with $a = 2, b = -0.3$.
Initial values in the white region are attracted to the hairpin attractor inside the white region. On each iteration, the points on one piece of the attractor map to the other piece. Orbits from initial values in the black region diverge to infinity.

2.3 LINEAR MAPS

The linear maps on $\mathbb{R}^2$ are those of the particularly simple form $\mathbf{v} \mapsto A\mathbf{v}$, where A is a 2×2 matrix:

$$A \begin{pmatrix} x \\ y \end{pmatrix} = \begin{pmatrix} a_{11} & a_{12} \\ a_{21} & a_{22} \end{pmatrix} \begin{pmatrix} x \\ y \end{pmatrix} = \begin{pmatrix} a_{11}x + a_{12}y \\ a_{21}x + a_{22}y \end{pmatrix}. \tag{2.15}$$

Definition 2.4 A map $A(\mathbf{v})$ from $\mathbb{R}^m$ to $\mathbb{R}^m$ is **linear** if for each $a, b \in \mathbb{R}$, and $\mathbf{v}, \mathbf{w} \in \mathbb{R}^m$, $A(a\mathbf{v} + b\mathbf{w}) = aA(\mathbf{v}) + bA(\mathbf{w})$. Equivalently, a linear map $A(\mathbf{v})$ can be represented as multiplication by an $m \times m$ matrix.

Every linear map has a fixed point at the origin. This is analogous to the one-dimensional linear map $f(x) = ax$. The stability of the fixed point will be investigated the same way as in Chapter 1. If all of the points in a neighborhood of the fixed point $(0, 0)$ approach the fixed point when iterated by the map, we consider the fixed point to be an attractor.

In some cases the dynamics for a two-dimensional map resemble one-dimensional dynamics. Recall that λ is an **eigenvalue** of the matrix A if there is a nonzero vector $\mathbf{v}$ such that

$$A\mathbf{v} = \lambda\mathbf{v}.$$

The vector $\mathbf{v}$ is called an **eigenvector**. Notice that if $\mathbf{v}_0$ is an eigenvector with eigenvalue λ, we can write down a special trajectory

$$\mathbf{v}_{n+1} = A\mathbf{v}_n$$

that satisfies

$$\mathbf{v}_1 = A\mathbf{v}_0 = \lambda\mathbf{v}_0$$

$$\mathbf{v}_2 = A\lambda\mathbf{v}_0 = \lambda A\mathbf{v}_0 = \lambda^2\mathbf{v}_0,$$

and in general $\mathbf{v}_n = \lambda^n\mathbf{v}_0$. Hence the map behaves like the one-dimensional map $x_{n+1} = \lambda x_n$.

We will begin by looking at three important examples of linear maps on $\mathbb{R}^2$. In fact, the three different types of 2×2 matrices we will encounter will be more than just good examples, they will be all possible examples, up to change of coordinates.

EXAMPLE 2.5

[Distinct real eigenvalues.] Let $\mathbf{v} = (x, y)$ denote a two-dimensional vector, and let $A(\mathbf{v})$ be the map on $\mathbb{R}^2$ defined by

$$A(x, y) = (ax, by).$$

Each input is a two-dimensional vector; so is each output. Any linear map can be represented by multiplication by a matrix, and following tradition we use A also to represent the matrix. Thus

$$A(\mathbf{v}) = A\mathbf{v} = \begin{pmatrix} a & 0 \\ 0 & b \end{pmatrix} \begin{pmatrix} x \\ y \end{pmatrix}. \tag{2.16}$$

The eigenvalues of the matrix A are a and b, with associated eigenvectors $(1, 0)$ and $(0, 1)$, respectively. For the purposes of this example, we will assume that they are not equal, although most of what we say now will not depend on that fact. Part of the importance of this example comes from the fact that a large class of linear maps can be expressed in the form (2.16), if the right coordinate system is used. For example, it is shown in Appendix A that any 2×2 matrix with distinct real eigenvalues takes this form when its eigenvectors are used to form the basis vectors of the coordinate system.

For the map in Example 2.5, the result of iterating the map n times is represented by the matrix

$$A^n = \begin{pmatrix} a^n & 0 \\ 0 & b^n \end{pmatrix}. \tag{2.17}$$

The unit disk is mapped into an ellipse with semi-major axes of length $|a|^n$ along the x-axis and $|b|^n$ along the y-axis. An epsilon disk $N_\epsilon(0, 0)$ would become an ellipse with axes of length $\epsilon|a|^n$ and $\epsilon|b|^n$. For example, suppose that a and b are smaller than 1 in absolute value. Then this ellipse shrinks toward the origin as $n \to \infty$, so $(0, 0)$ is a sink for A. If $|a|, |b| > 1$, then the origin is a source.

On the other hand, if $|a| > 1 > |b|$, we see dynamical behavior that is not seen in one-dimensional maps. As n is increased, the ellipse grows in the x-direction and shrinks in the y-direction, essentially growing to look like the x-axis as $n \to \infty$. In Figure 2.12, we plot the unit disk and its two iterates under A where we set $a = 2$ and $b = 1/2$. In this case, the origin is neither a sink nor a source. If the ellipses formed by successive iterates of the map grow without bound along one direction and shrink to zero along another, we will call the origin a saddle.

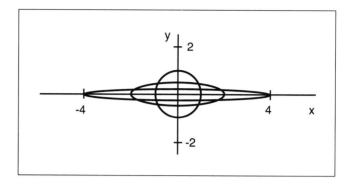

Figure 2.12 The unit disk and two images of the unit disk under a linear map.
The origin is a saddle fixed point.

Points act as if they were moving along the surface of a saddle under the influence of gravity. A cowboy who spills coffee on his saddle will see it run toward the center along the front-to-back axis of the saddle (the y-axis in Figure 2.13) and run away from the center along the side-to-side axis (the x-axis in Figure 2.13). Presumably, a drop situated at the exact center of the saddle would stay there (our assumption is that the horse is not moving).

We see the same behavior for the iteration of points in Figure 2.14, which illustrates the linear map represented by the matrix

$$A = \begin{pmatrix} 2 & 0 \\ 0 & 0.5 \end{pmatrix}. \tag{2.18}$$

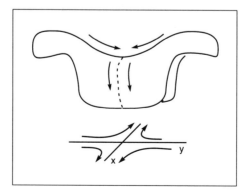

Figure 2.13 Dynamics near a saddle point.
Points in the vicinity of a saddle fixed point (here the origin in the xy-plane) move as if responding to the influence of gravity on a saddle.

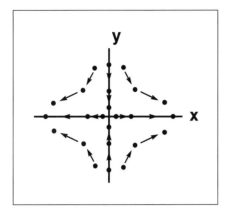

Figure 2.14 Saddle dynamics.
Successive images of points near a saddle fixed point are shown.

A typical point (x_0, y_0) maps to $(2x_0, \frac{1}{2}y_0)$, and then to $(4x_0, \frac{1}{4}y_0,)$, and so on. Notice that the product of the x- and y- coordinates is the constant quantity $x_0 y_0$, so that orbits shown in Figure 2.14 traverse the hyperbola $xy = $ constant $= x_0 y_0$. More generally, for a linear map A on $\mathbb{R}^2$, the origin is a saddle if and only if iteration of the unit disk results in ellipses whose two axis lengths converge to zero and infinity, respectively.

A simplification can be made when analyzing small neighborhoods under linear maps. Because linearity implies $A(\mathbf{v}) = |\mathbf{v}|A(\frac{\mathbf{v}}{|\mathbf{v}|})$, the image of a vector $\mathbf{v}$ can be found by mapping the unit vector in the direction of $\mathbf{v}$, followed by scalar multiplication by the magnitude $|\mathbf{v}|$. The effect of the map on a small disk neighborhood of the origin is just a scaled-down version of the effect on the unit disk $N = N_1(0,0) = \{\mathbf{v} : |\mathbf{v}| < 1\}$. As a result we will often restrict our attention to the effect of the matrix on the unit disk. For example, the image of the unit disk centered at the origin under multiplication by any matrix is a filled ellipse centered at the origin. If the radius of the disk is r instead of 1, the resulting ellipse will also be changed precisely by a factor of r. (The semi-major axes will be changed by a factor of r.)

EXAMPLE 2.6

[Repeated eigenvalue.] For an example where the eigenvalues are not distinct, let

$$A = \begin{pmatrix} a & 1 \\ 0 & a \end{pmatrix}. \tag{2.19}$$

The eigenvalues of an upper triangular matrix are its diagonal entries, so the matrix has a repeated eigenvalue of a. Check that

$$A^n = a^{n-1} \begin{pmatrix} a & n \\ 0 & a \end{pmatrix}. \tag{2.20}$$

Therefore the effect of A^n on vectors is

$$A^n \begin{pmatrix} x \\ y \end{pmatrix} = a^{n-1} \begin{pmatrix} ax + ny \\ ay \end{pmatrix}. \tag{2.21}$$

✎ **EXERCISE T2.3**

(a) Verify equation (2.20). (b) Use equation (2.21) to show that the fixed point $(0, 0)$ is a sink if $|a| < 1$ and a source if $|a| > 1$.

EXAMPLE 2.7

[Complex eigenvalues.] Let

$$A = \begin{pmatrix} a & -b \\ b & a \end{pmatrix}. \tag{2.22}$$

This matrix has no real eigenvalues. The eigenvalues of this matrix are $a - bi$ and $a + bi$, where $i = \sqrt{-1}$. The corresponding eigenvectors are $(1, -i)$ and $(1, i)$, respectively. Fortunately, this information can be interpreted in terms of real vectors. A more intuitive way to look at this matrix follows from multiplying and dividing by $r = \sqrt{a^2 + b^2}$. Then

$$A = r \begin{pmatrix} a/r & -b/r \\ b/r & a/r \end{pmatrix} = \sqrt{a^2 + b^2} \begin{pmatrix} \cos \theta & -\sin \theta \\ \sin \theta & \cos \theta \end{pmatrix}. \tag{2.23}$$

Here we used the fact that any pair of numbers c, s such that $c^2 + s^2 = 1$ can be written as $c = \cos \theta$ and $s = \sin \theta$ for some angle θ. The angle θ can be identified as $\theta = \arctan(b/a)$. It is now clear that multiplication by this matrix rotates points about the origin by an angle θ, and multiplies the distances by $\sqrt{a^2 + b^2}$. Therefore it is a combination of a rotation and a dilation.

✎ **EXERCISE T2.4**

Verify that multiplication by A rotates a vector by $\arctan(b/a)$ and stretches by a factor of $\sqrt{a^2 + b^2}$.

In summary, the effect of multiplication by A on the length of a vector is contraction/expansion by a factor of $\sqrt{a^2 + b^2}$. It follows that the stability result is the same as in the previous two cases. If the magnitude of the eigenvalues is less than 1, the origin is a sink; if greater than 1, a source.

2.4 COORDINATE CHANGES

Now that we have some experience with iterating linear maps, we return to the fundamental issue of how a matrix represents a linear map. Changes of coordinates can simplify stability calculations for higher-dimensional maps.

A vector in $\mathbb{R}^m$ can be represented in many different ways, depending on the coordinate system chosen. Choosing a coordinate system is equivalent to choosing a basis of $\mathbb{R}^m$; the coordinates of a vector are simply the coefficients which express the vector in that basis. Changing the basis of $\mathbb{R}^m$ requires changing the matrix representing the linear map $A(\mathbf{v})$. In particular, let S be a square matrix whose columns are the new basis vectors. Then the matrix $S^{-1}AS$ represents the linear map in the new basis. A matrix of form $S^{-1}AS$, where S is a nonsingular matrix, is **similar** to A.

Similar matrices have the same set of eigenvalues and the same determinant. The **determinant** $\det(A) = a_{11}a_{22} - a_{12}a_{21}$ is a measure of area transformation by the matrix A. If R represents a two-dimensional region of area c, then the set $A(R)$ has area $\det(A) \cdot c$. It stands to reason that area transformation should be independent of the choice of coordinates. See Appendix A for justification of these statements and for a thorough discussion of changes of coordinates.

Matrices that are similar have the same dynamical properties when viewed as maps, since they only differ by the coordinate system used to view them. For example, the property that a small neighborhood of the fixed point origin is attracted to the origin is independent of the choice of coordinates. If $(0,0)$ is a sink under A, it remains so under $S^{-1}AS$. This puts us in position to analyze the dynamics of all linear maps on $\mathbb{R}^2$, because of the following fact: All 2×2 matrices are similar to one of Examples 2.5, 2.6, 2.7. See Appendix A for a proof of this fact.

Since similar matrices have identical eigenvalues, deciding the stability of the origin for a linear map $A(\mathbf{v})$ is as simple as computing the eigenvalues of a matrix representation A. For example, if the eigenvalues a and b of A are real and distinct, then A is similar to the matrix

$$A_2 = \begin{pmatrix} a & 0 \\ 0 & b \end{pmatrix}. \tag{2.24}$$

Therefore the map has this matrix representation in some coordinate system. Referring to Example 2.5, we see that the origin is a sink if $|a|, |b| < 1$ and a source if $|a|, |b| > 1$. The same analysis works for matrices with repeated eigenvalues, or a pair of complex eigenvalues. Summing up, we have proved the $m = 2$ version of the following theorem.

> **Theorem 2.8** Let $A(\mathbf{v})$ be a linear map on $\mathbb{R}^m$, which is represented by the matrix A (in some coordinate system). Then
>
> 1. The origin is a sink if all eigenvalues of A are smaller than one in absolute value;
> 2. The origin is a source if all eigenvalues of A are larger than one in absolute value.

In dimensions two and greater, we must also consider linear maps of mixed stability, i.e., those for which the origin is a saddle.

> **Definition 2.9** Let A be a linear map on $\mathbb{R}^m$. We say A is **hyperbolic** if A has no eigenvalues of absolute value one. If a hyperbolic map A has at least one eigenvalue of absolute value greater than one and at least one eigenvalue of absolute value smaller than one, then the origin is called a **saddle**.

Thus there are three types of hyperbolic maps: ones for which the origin is a sink, ones for which the origin is a source, and ones for which the origin is a saddle. Hyperbolic linear maps are important objects of study because they have well-defined expanding and contracting directions.

2.5 NONLINEAR MAPS AND THE JACOBIAN MATRIX

So far we have discussed linear maps, which always have a fixed point at the origin. We now want to discuss nonlinear maps, and in particular how to determine the stability of fixed points.

Our treatment of stability in Chapter 1 is relevant to this case. Theorem 1.5 showed that whether a fixed point of a one-dimensional nonlinear map is a sink or source depends on its "linearization", or linear part, near the fixed point. In the one-dimensional case the linearization is given by the derivative at the fixed point. If p is a fixed point and h is a small number, then the change in the output of the map at $p + h$, compared to the output at p, is well approximated by the linear map $L(h) = Kh$, where K is the constant number $f'(p)$. In other words,

$$f(p + h) \approx f(p) + hf'(p). \tag{2.25}$$

Our proof of Theorem 1.5 was based on the fact that the error in this approximation was of size proportional to h^2. This can be made as small as desired by restricting attention to sufficiently small h. If $|f'(p)| < 1$, the fixed point p is a sink, and if $|f'(p)| > 1$, it is a source. The situation is very similar for nonlinear maps in higher dimensions. The place of the derivative in the above discussion is taken by a matrix.

Definition 2.10   Let $\mathbf{f} = (f_1, f_2, \ldots, f_m)$ be a map on $\mathbb{R}^m$, and let $\mathbf{p} \in \mathbb{R}^m$. The **Jacobian matrix** of $\mathbf{f}$ at $\mathbf{p}$, denoted $D\mathbf{f}(\mathbf{p})$, is the matrix

$$D\mathbf{f}(\mathbf{p}) = \begin{pmatrix} \dfrac{\partial f_1}{\partial x_1}(\mathbf{p}) & \cdots & \dfrac{\partial f_1}{\partial x_m}(\mathbf{p}) \\ \vdots & & \vdots \\ \dfrac{\partial f_m}{\partial x_1}(\mathbf{p}) & \cdots & \dfrac{\partial f_m}{\partial x_m}(\mathbf{p}) \end{pmatrix}$$

of partial derivatives evaluated at $\mathbf{p}$.

Given a vector $\mathbf{p}$ and a small vector $\mathbf{h}$, the increment in $\mathbf{f}$ due to $\mathbf{h}$ is approximated by the Jacobian matrix times the vector $\mathbf{h}$:

$$\mathbf{f}(\mathbf{p} + \mathbf{h}) - \mathbf{f}(\mathbf{p}) \approx D\mathbf{f}(\mathbf{p}) \cdot \mathbf{h}, \tag{2.26}$$

where again the error in the approximation is proportional to $|\mathbf{h}|^2$ for small $\mathbf{h}$. If we assume that $\mathbf{f}(\mathbf{p}) = \mathbf{p}$, then for a small change $\mathbf{h}$, the map moves $\mathbf{p} + \mathbf{h}$ approximately $D\mathbf{f}(\mathbf{p}) \cdot \mathbf{h}$ away from $\mathbf{p}$. That is, $\mathbf{f}$ magnifies a small change $\mathbf{h}$ in input to a change $D\mathbf{f}(\mathbf{p}) \cdot \mathbf{h}$ in output.

As long as this deviation remains small (so that $|\mathbf{h}|^2$ is negligible and our approximation is valid), the action of the map near $\mathbf{p}$ is essentially the same as the linear map $\mathbf{h} \mapsto A\mathbf{h}$, where $A = D\mathbf{f}(\mathbf{p})$, with fixed point $\mathbf{h} = 0$. Small disk neighborhoods centered at $\mathbf{h} = 0$ (corresponding to disks around $\mathbf{p}$) map to regions approximated by ellipses whose axes are determined by A. In that case, we can appeal to Theorem 2.8 for information about linear stability for higher-dimensional maps in order to understand the nonlinear case.

The following theorem is an extension of Theorems 1.5 and 2.8 to higher dimensional nonlinear maps. It determines the stability of a map at a fixed point based on the Jacobian matrix at that point. The proof is omitted.

Theorem 2.11   *Let f be a map on $\mathbb{R}^m$, and assume $\mathbf{f}(\mathbf{p}) = \mathbf{p}$.*

1. *If the magnitude of each eigenvalue of $\mathbf{Df}(\mathbf{p})$ is less than 1, then $\mathbf{p}$ is a sink.*
2. *If the magnitude of each eigenvalue of $\mathbf{Df}(\mathbf{p})$ is greater than 1, then $\mathbf{p}$ is a source.*

Just as linear maps of $\mathbb{R}^m$ for $m > 1$ can have some directions in which orbits diverge from 0 and some in which orbits converge to 0, so fixed points of nonlinear maps can attract points in some directions and repel points in others.

Definition 2.12   Let f be a map on $\mathbb{R}^m$, $m \geq 1$. Assume that $\mathbf{f}(\mathbf{p}) = \mathbf{p}$. Then the fixed point $\mathbf{p}$ is called **hyperbolic** if none of the eigenvalues of $\mathbf{Df}(\mathbf{p})$ has magnitude 1. If $\mathbf{p}$ is hyperbolic and if at least one eigenvalue of $\mathbf{Df}(\mathbf{p})$ has magnitude greater than 1 and at least one eigenvalue has magnitude less than 1, then $\mathbf{p}$ is called a **saddle**. (For a periodic point of period k, replace $\mathbf{f}$ by $\mathbf{f}^k$.)

Saddles are unstable. If even one eigenvalue of $\mathbf{Df}(\mathbf{p})$ has magnitude greater than 1, then $\mathbf{p}$ is unstable in the sense previously described: Almost any perturbation of the orbit away from the fixed point will be magnified under iteration. In a small epsilon neighborhood of $\mathbf{p}$, $\mathbf{f}$ behaves very much like a linear map with an eigenvalue that has magnitude greater than 1; that is, the orbits of most points near $\mathbf{p}$ diverge from $\mathbf{p}$.

EXAMPLE 2.13

The Hénon map

$$\mathbf{f}_{a,b}(x, y) = (a - x^2 + by, x), \tag{2.27}$$

where a and b are constants, has at most two fixed points. Setting $a = 0$ and $b = 0.4$, $\mathbf{f}$ has the two fixed points $(0, 0)$ and $(-0.6, -0.6)$. The Jacobian matrix $\mathbf{Df}$ is

$$\mathbf{Df}(x, y) = \begin{pmatrix} -2x & b \\ 1 & 0 \end{pmatrix}. \tag{2.28}$$

Evaluated at $(0, 0)$, the Jacobian matrix is

$$\mathbf{Df}(0, 0) = \begin{pmatrix} 0 & 0.4 \\ 1 & 0 \end{pmatrix},$$

with eigenvalues $\pm\sqrt{0.4}$, approximately equal to 0.632 and -0.632. Evaluated at $(-0.6, -0.6)$, the Jacobian is

$$\mathbf{Df}(-0.6, -0.6) = \begin{pmatrix} 1.2 & 0.4 \\ 1 & 0 \end{pmatrix},$$

with eigenvalues approximately equal to 1.472 and -0.272. Thus $(0, 0)$ is a sink and $(-0.6, -0.6)$ is a saddle.

For the parameter values $a = 0.43, b = 0.4$, there is a period-two orbit for the map. Check that $\{(0.7, -0.1), (-0.1, 0.7)\}$ is such an orbit. In order to check the stability of this orbit, we need to compute the Jacobian matrix of $\mathbf{f}^2$ evaluated at $(0.7, -0.1)$. Because of the chain rule, we can do this without explicitly forming $\mathbf{f}^2$, since $\mathbf{Df}^2(x) = \mathbf{Df}(\mathbf{f}(x)) \cdot \mathbf{Df}(x)$. We compute

$$\mathbf{Df}^2((0.7, -0.1)) = \mathbf{Df}((-0.1, 0.7)) \cdot \mathbf{Df}((0.7, -0.1))$$

$$= \begin{pmatrix} -2(-0.1) & 0.4 \\ 1 & 0 \end{pmatrix} \begin{pmatrix} -2(0.7) & 0.4 \\ 1 & 0 \end{pmatrix}$$

$$= \begin{pmatrix} 0.12 & 0.08 \\ -1.4 & 0.4 \end{pmatrix}.$$

The eigenvalues of this Jacobian matrix are approximately $0.26 \pm 0.30i$, which are complex numbers of magnitude ≈ 0.4, so the period-two orbit is a sink.

Note that the same eigenvalues are obtained by evaluating

$$\mathbf{Df}^2((-0.1, 0.7)) = \mathbf{Df}((0.7, -0.1)) \cdot \mathbf{Df}((-0.1, 0.7)),$$

which means that stability is a property of the periodic orbit as a whole, not of the individual points of the orbit. This is true because the eigenvalues of a product AB of two matrices are identical to the eigenvalues of BA, as shown in the Appendix A. This result compares with (1.4) of Chapter 1.

Remark 2.14   For a map on $\mathbb{R}^m$, there is a more general statement of this fact. Assume there is a periodic orbit $\{\mathbf{p}_1, \ldots, \mathbf{p}_k\}$ of period k. By Lemma A.2 of Appendix A, the set of eigenvalues of a product of several matrices is unchanged under a cyclic permutation of the order of the product. Using the chain rule,

$$\mathbf{Df}^k(\mathbf{p}_1) = \mathbf{Df}(\mathbf{p}_k) \cdot \mathbf{Df}(\mathbf{p}_{k-1}) \cdots \mathbf{Df}(\mathbf{p}_1). \qquad (2.29)$$

The eigenvalues of the $m \times m$ Jacobian matrix evaluated at $\mathbf{p}_1$, $\mathbf{Df}^k(\mathbf{p}_1)$, will determine the stability of the period-k orbit. But one should also be able to determine the stability by examining the eigenvalues of $\mathbf{Df}^k(\mathbf{p}_r)$, where $\mathbf{p}_r$ is one of the other points in the periodic orbit. Applying the chain rule as above, we find that

$$\mathbf{Df}^k(\mathbf{p}_r) = \mathbf{Df}(\mathbf{p}_{r-1}) \cdot \mathbf{Df}(\mathbf{p}_{r-2}) \cdots \mathbf{Df}(\mathbf{p}_1) \cdot \mathbf{Df}(\mathbf{p}_k) \cdots \mathbf{Df}(\mathbf{p}_r). \qquad (2.30)$$

According to Lemma A.2, the eigenvalues of (2.29) and (2.30) are identical. This guarantees that the eigenvalues are shared by the periodic orbit, and can be measured by multiplying together the k Jacobian matrices starting at any of the k points.

A more systematic study can be made of the fixed points and period-two points of the Hénon map. Let the parameters a and b be arbitrary. Then all fixed points satisfy

$$x = a - x^2 + by$$

$$y = x, \tag{2.31}$$

which is equivalent to the equation $x = a - x^2 + bx$, or

$$x^2 + (1 - b)x - a = 0. \tag{2.32}$$

Using the quadratic formula, we see that fixed points exist as long as

$$4a > -(1 - b)^2 \tag{2.33}$$

If (2.33) is satisfied, there are exactly two fixed points, whose x-coordinates are found from the quadratic formula and whose y-coordinate is the same as the x-coordinate.

To look for period-two points, set $(x, y) = \mathbf{f}^2(x, y)$:

$$x = a - (a - x^2 + by)^2 + bx$$

$$y = a - x^2 + by. \tag{2.34}$$

Solving the second equation for y and substituting into the first, we get an equation for the x-coordinate of a period-two point:

$$0 = (x^2 - a)^2 + (1 - b)^3 x - (1 - b)^2 a$$

$$= (x^2 - (1 - b)x - a + (1 - b)^2)(x^2 + (1 - b)x - a). \tag{2.35}$$

We recognize the factor on the right from Equation (2.32): Zeros of it correspond to fixed points of $\mathbf{f}$, which are also fixed points of $\mathbf{f}^2$. In fact, it was the knowledge that (2.32) must be a factor which was the trick that allowed us to write (2.35) in factored form. The period-two orbit is given by the zeros of the left factor, if they exist.

✎   **EXERCISE T2.5**

Prove that the Hénon map has a period-two orbit if and only if $4a > 3(1 - b)^2$.

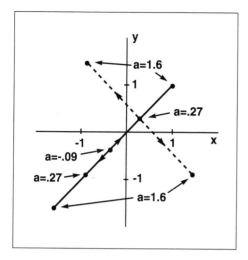

Figure 2.15 Fixed points and period-two points for the Hénon map with *b* fixed at 0.4.
The solid line denotes the trails of the two fixed points as *a* moves from -0.09, where the two fixed points are created together, to 1.6 where they have moved quite far apart. The fixed point that moves diagonally upward is attracting for $-0.09 < a < 0.27$; the other is a saddle. The dashed line follows the period-two orbit from its creation when $a = 0.27$, at the site of the (previously) attracting fixed point, to $a = 1.6$.

Figure 2.15 shows the fixed points and period-two points of the Hénon map for $b = .4$ and for various values of *a*. We understand why the fixed points lie along the diagonal line $y = x$, but why do the period-two orbits lie along a line, as shown in Figure 2.15?

EXERCISE T2.6

(a) If (x_1, y_1) and (x_2, y_2) are the two fixed points of the Hénon map (2.27) with some fixed parameters *a* and *b*, show that $x_1 - y_1 = x_2 - y_2 = 0$ and $x_1 + x_2 = y_1 + y_2 = b - 1$.

(b) If $\{(x_1, y_1), (x_2, y_2)\}$ is the period-two orbit, show that $x_1 + y_1 = x_2 + y_2 = x_1 + x_2 = y_1 + y_2 = 1 - b$. In particular the period-two orbit lies along the line $x + y = 1 - b$, as seen in Figure 2.15.

Figure 2.16 shows a bifurcation diagram for the Hénon map for the case $b = 0.4$. For each fixed value $0 \le a \le 1.25$ along the horizontal axis, the *x*-coordinates of the attracting set are plotted vertically. The information in Figure

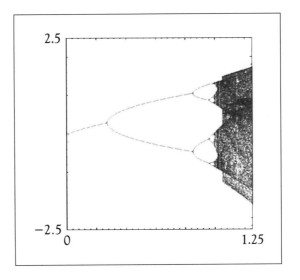

Figure 2.16   Bifurcation diagram for the Hénon map, $b = 0.4$.
Each vertical slice shows the projection onto the x-axis of an attractor for the map for a fixed value of the parameter a.

2.15 is recapitulated here. At $a = 0.27$, a period-doubling bifurcation occurs, when the fixed point loses stability and a period-two orbit is born. The period-two orbit is a sink until $a = 0.85$, when it too doubles its period. In the next exercise, you will be asked to use the equations we developed here to verify some of these facts.

✎   **EXERCISE T2.7**

Set $b = 0.4$.

(a) Prove that for $-0.09 < a < 0.27$, the Hénon map **f** has one sink fixed point and one saddle fixed point.

(b) Find the largest magnitude eigenvalue of the Jacobian matrix at the first fixed point when $a = 0.27$. Explain the loss of stability of the sink.

(c) Prove that for $0.27 < a < 0.85$, **f** has a period-two sink.

(d) Find the largest magnitude eigenvalue of $\mathbf{Df}^2$, the Jacobian of $\mathbf{f}^2$ at the period-two orbit, when $a = 0.85$.

For $b = 0.4$ and $a > 0.85$, the attractors of the Hénon map become more complex. When the period-two orbit becomes unstable, it is immediately replaced with an attracting period-four orbit, then a period-eight orbit, etc. Figure 2.17

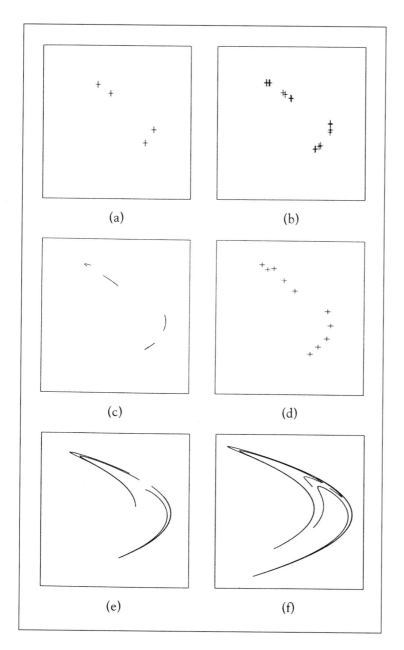

Figure 2.17 Attractors for the Hénon map with $b = 0.4$.
Each panel displays a single attracting orbit for a particular value of the parameter
a. (a) $a = 0.9$, period 4 sink. (b) $a = 0.988$, period 16 sink. (c) $a = 1.0$, four-piece
attractor. (d) $a = 1.0293$, period-ten sink. (e) $a = 1.045$, two-piece attractor. The
points of an orbit alternate between the pieces. (f) $a = 1.2$, two pieces have merged
to form one-piece attractor.

shows a number of these attractors. An example is the "period-ten window" at $a = 1.0293$, barely detectable as a vertical white gap in Figure 2.16.

⇨   **COMPUTER EXPERIMENT 2.2**

Make a bifurcation diagram like Figure 2.16, but for $b = -0.3$, and for $0 \le a \le 2.2$. For each a, choose the initial point $(0, 0)$ and calculate its orbit. Plot the x-coordinates of the orbit, starting with iterate 101 (to allow time for the orbit to approximately settle down to the attracting orbit). Questions to answer: Does the resulting bifurcation diagram depend on the choice of initial point? How is the picture different if the y-coordinates are plotted instead?

Periodic points are the key to many of the properties of a map. For example, trajectories often converge to a periodic sink. Periodic saddles and sources, on the other hand, do not attract open neighborhoods of initial values as sinks do, but are important in their own ways, as will be seen in later chapters.

Remark 2.15   The theme of this section has been the use of the Jacobian matrix for determining stability of periodic orbits of nonlinear maps, in the way that the map matrix itself is used for linear maps. There are other important uses for the Jacobian matrix. The magnitude of its determinant measures the transformation of areas for nonlinear maps, at least locally.

For example, consider the Hénon map (2.27). The determinant of the Jacobian matrix (2.28) is fixed at $-b$ for all $\mathbf{v}$. For the case $a = 0, b = 0.4$, the map $\mathbf{f}$ transforms area near each point $\mathbf{v}$ at the rate $|\det(\mathbf{Df}(\mathbf{v}))| = |-b| = 0.4$. Each plane region is transformed by $\mathbf{f}$ into a region that is 40% of its original size. The circle around each fixed point in Figure 2.9, for example, has forward images which are $.4 = 40\%$ and $(.4)^2 = .16 = 16\%$, respectively.

Most of the plane maps we will deal with are invertible, meaning that their inverses exist.

Definition 2.16   A map $\mathbf{f}$ on $\mathbb{R}^m$ is **one-to-one** if $\mathbf{f}(\mathbf{v}_1) = \mathbf{f}(\mathbf{v}_2)$ implies $\mathbf{v}_1 = \mathbf{v}_2$.

Recall that functions are well-defined by definition, i.e. $\mathbf{v}_1 = \mathbf{v}_2$ implies $\mathbf{f}(\mathbf{v}_1) = \mathbf{f}(\mathbf{v}_2)$. Two points do not get mapped together under a one-to-one map. It follows that if $\mathbf{f}$ is a one-to-one map, then its **inverse map** $\mathbf{f}^{-1}$ is a function. The

INVERSE MAPS

A function is a uniquely-defined assignment of a range point for each domain point. (If the domain and range are the same set, we call the function a map.) Several domain points may map to the same range point. For $f_1(x, y) = (x^2, y^2)$, the points $(2, 2)$, $(2, -2)$, $(-2, 2)$ and $(-2, -2)$ all map to $(4, 4)$. On the other hand, for $f_2(x, y) = (x^3, y^3)$, this never happens. A point (a, b) is the image of $(a^{1/3}, b^{1/3})$ only. Thus f_2 is a one-to-one map, and f_1 is not.

An inverse map f^{-1} automatically exists for any one-to-one map f. The domain of f^{-1} is the image of f. For the example $f_2(x, y) = (x^3, y^3)$, the inverse is $f_2^{-1}(x, y) = (x^{1/3}, y^{1/3})$.

To compute an inverse map, set $v_1 = f(v)$ and solve for v in terms of v_1. We demonstrate using $f(x, y) = (x + 2y, x^3)$. Set

$$x_1 = x + 2y$$

$$y_1 = x^3$$

and solve for x and y. The result is

$$x = y_1^{1/3}$$

$$y = (x_1 - y_1^{1/3})/2,$$

so that the inverse map is $f(x, y) = (y_1^{1/3}, (x_1 - y_1^{1/3})/2)$.

nverse map is characterized by the fact that $f(v) = w$ if and only if $v = f^{-1}(w)$. Because one-to-one implies the existence of an inverse, a one-to-one map is also called an **invertible** map.

EXERCISE T2.8

Show that the Hénon map (2.27) with $b \neq 0$ is invertible by finding a formula for the inverse. Is the map one-to-one if $b = 0$?

2.6 Stable and Unstable Manifolds

A saddle fixed point is unstable, meaning that most initial values near it will move away under iteration of the map. However, unlike the case of a source, not all nearby initial values will move away. The set of initial values that converge to the saddle will be called the stable manifold of the saddle. We start by looking at a simple linear example.

EXAMPLE 2.17

For the linear map $f(x, y) = (2x, y/2)$, the origin is a saddle fixed point. The dynamics of this map were shown in Figure 2.14. It is clear from that figure that points along the y-axis converge to the saddle 0; all other points diverge to infinity. Unless the initial value has x-coordinate 0, the x-coordinate will grow (by a factor of 2 per iterate) and get arbitrarily large.

A convenient way to view the direction of the stable manifold in this case is in terms of eigenvectors. The linear map

$$f(v) = Av = \begin{pmatrix} 2 & 0 \\ 0 & 0.5 \end{pmatrix} \begin{pmatrix} x \\ y \end{pmatrix}$$

has eigenvector $\begin{pmatrix} 1 \\ 0 \end{pmatrix}$, corresponding to the (stretching) eigenvalue 2, and $\begin{pmatrix} 0 \\ 1 \end{pmatrix}$, corresponding to the (shrinking) eigenvalue $1/2$. The latter direction, the y-axis, is the "incoming" direction, and is the stable manifold of 0. We will call the x-axis the "outgoing" direction, the unstable manifold of 0. Another way to describe the unstable manifold in this example is as the stable manifold under the inverse of the map $f^{-1}(x, y) = ((1/2)x, 2y)$.

Definition 2.18   Let f be a smooth one-to-one map on $\mathbb{R}^2$, and let p be a saddle fixed point or periodic saddle point for f. The **stable manifold** of p, denoted $S(p)$, is the set of points v such that $|f^n(v) - f^n(p)| \to 0$ as $n \to \infty$. The **unstable manifold** of p, denoted $U(p)$, is the set of points v such that $|f^{-n}(v) - f^{-n}(p)| \to 0$ as $n \to \infty$.

EXAMPLE 2.19

The linear map $f(x, y) = (-2x + \frac{5}{2}y, -5x + \frac{11}{2}y)$ has a saddle fixed point at 0 with eigenvalues 0.5 and 3. The corresponding eigenvectors are $(1, 1)$ and

WHAT IS A MANIFOLD?

An n-dimensional manifold is a set that locally resembles Euclidean space $\mathbb{R}^n$. By "resembles" we could mean a variety of things, and in fact, various definitions of manifold have been proposed. For our present purposes, we will mean resemblance in a topological sense. A small piece of a manifold should look like a small piece of $\mathbb{R}^n$.

A 1-dimensional manifold is locally a curve. Every short piece of a curve can be formed by stretching and bending a piece of a line. The letters D and O are 1-manifolds. The letters A and X are not, since each contains a point for which no small neighborhood looks like a line segment. These bad points occur at the meeting points of separate segments, like the center of the letter X.

Notable 2-manifolds are the surface of oranges and doughnuts. The space-time continuum of the universe is often described as a 4-manifold, whose curvature due to relativity is an active topic among cosmologists.

In the strict definition of manifold, each point of a manifold must have a neighborhood around itself that looks like $\mathbb{R}^n$. Thus the letters L and U fail to be 1-manifolds because of their endpoints—small neighborhoods of them look like a piece of half-line (say the set of nonnegative real numbers), not a line, since there is nothing on one side. This type of set is called a **manifold with boundary**, although technically it is not a manifold. A Möbius band is a 2-manifold with boundary because the edge looks locally like a piece of half-plane, not a plane. Whole oranges and doughnuts are 3-manifolds with boundary.

One of the goals of Chapter 10 is to explain why a stable or unstable manifold is a topological manifold. Stable and unstable manifolds emanate from two opposite sides of a fixed point or periodic orbit. At a saddle point in the plane, they together make an "X" through the fixed point, although individually they are manifolds.

(1, 2), respectively. [According to Appendix A, there is a linear change of coordinates giving the map $h(u_1, u_2) = (0.5u_1, 3u_2)$.] Points lying on the line $y = x$ undergo the dynamics $v \rightarrow 0.5v$ on each iteration of the map. This line is the stable manifold of 0 for f. Points lying on the line $y = 2x$ (the line in the direction of eigenvector $(1, 2)$) undergo $v \rightarrow 3v$ under f: this is the unstable manifold. These sets are illustrated in Figure 2.18.

EXAMPLE 2.20

Let $f(x, y) = (2x + 5y, -0.5y)$. The eigenvalues of f are 2 and -0.5, with corresponding eigenvectors $(1, 0)$ and $(2, -1)$. Points on the line in the direction of the vector $(2, -1)$ undergo $v \rightarrow -0.5v$ on each iteration of f. As a result, successive images flip from one side of the origin to the other along the line. This flipping behavior of orbits about the fixed point is shown in Figure 2.19. It is characteristic of all fixed points for which the Jacobian has negative eigenvalues, even when the map is nonlinear. A saddle with at least one negative eigenvalue is sometimes called a **flip saddle**.

EXAMPLE 2.21

The invertible nonlinear map $f(x, y) = (x/2, 2y - 7x^2)$ has a fixed point at $0 = (0, 0)$. To analyze the stability of this fixed point we evaluate

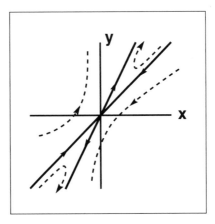

Figure 2.18   Stable and unstable manifolds for regular saddle.
The stable manifold is the solid inward-directed line; the unstable manifold is the solid outward-directed line. Every initial condition leads to an orbit diverging to infinity except for the stable manifold of the origin.

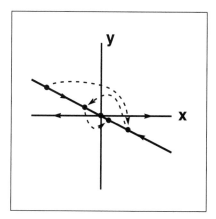

Figure 2.19 Stable and unstable manifolds for flip saddle.
Flipping occurs along the stable manifold (inward-directed line). The unstable manifold is the x-axis.

$$\mathbf{Df}(0, 0) = \begin{pmatrix} 0.5 & 0 \\ 0 & 2 \end{pmatrix}.$$

The origin is a saddle fixed point. The eigenvectors lie on the two coordinate axes. What is the relation of these eigenvector directions to the stable and unstable manifolds?

In the case of linear maps, the stable and unstable manifolds coincide with the eigenvector directions. For a saddle of a general nonlinear map, the stable manifold is tangent to the shrinking eigenvector direction, and the unstable manifold is tangent to the stretching eigenvector direction. Since $\mathbf{f}(0, y) = (0, 2y)$, the y-axis can be seen to be part of the unstable manifold. For any point not on the y-axis, the absolute value of the x-coordinate is nonzero and increases under iteration by $\mathbf{f}^{-1}$; in particular, it doesn't converge to the origin. Thus the y-axis is the entire unstable manifold, as shown in Figure 2.20. The stable manifold of $\mathbf{0}$, however, is described by the parabola $y = 4x^2$; i.e., $S(0) = \{(x, 4x^2) : x \in \mathbb{R}\}$.

EXERCISE T2.9

Consider the saddle fixed point $\mathbf{0}$ of the map $f(x, y) = (x/2, 2y - 7x^2)$ from Example 2.21.

    (a) Find the inverse map $\mathbf{f}^{-1}$.

    (b) Show that the set $S = \{(x, 4x^2) : x \in \mathbb{R}\}$ is invariant under f, that is, if $\mathbf{v}$ is in S, then $f(\mathbf{v})$ and $\mathbf{f}^{-1}(\mathbf{v})$ are in S.

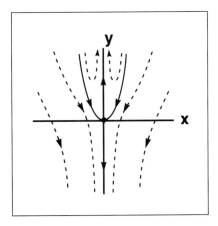

Figure 2.20   Stable and unstable manifolds for the nonlinear map of Example 2.21.
The stable manifold is a parabola tangent to the x-axis at 0; the unstable manifold is the y-axis.

(c) Show that each point in S converges to **0** under **f**.

(d) Show that no points outside of S converge to **0** under **f**.

When a map is linear, the stable and unstable manifolds of a saddle are always linear subspaces. In the case of linear maps on $\mathbb{R}^2$, for example, they are lines. For nonlinear maps, as we saw in Example 2.21, they can be curves. The nonlinear examples we have looked at so far are not typical; usually, formulas for the stable and unstable manifolds cannot be found directly. Then we must rely on computational techniques to approximate their locations. (We describe one such technique in Chapter 10.) One thing you may have noticed about the stable manifolds in these examples is that they are always one-dimensional: lines or curves. Just as in the linear case, the stable and unstable manifolds of saddles in the plane are always one-dimensional sets. This fact is not immediately obvious—it is proved as part of the Stable Manifold Theorem in Chapter 10. We will also see that stable and unstable manifolds of saddles have a tremendous influence on the underlying dynamics of a system. In particular, their relative positions can determine whether or not chaos occurs.

We will leave the investigation of the mysteries of stable and unstable manifolds to Chapter 10. Here we give a small demonstration of the subtlety and importance of these manifolds. Drawing stable and unstable manifolds of the Hénon map can illuminate Figure 2.3, which showed the basin of the period-two

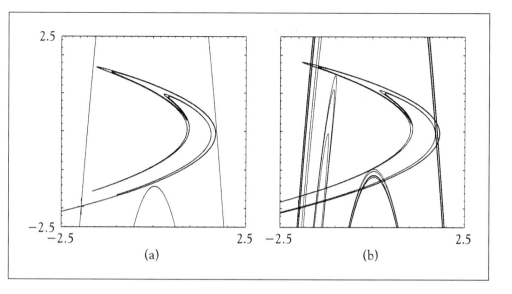

2.5

−2.5

−2.5   2.5 −2.5   2.5

(a)   (b)

Figure 2.21   Stable and unstable manifolds for a saddle point.
The stable manifolds (mainly vertical) and unstable manifolds (more horizontal)
are shown for the saddle fixed point (marked with a cross in the lower left corner) of
the Hénon map with $b = -0.3$. Note the similarity of the unstable manifold with
earlier figures showing the Hénon attractor. (a) For $a = 1.28$, the leftward piece of
the unstable manifold moves off to infinity, and the rightward piece initially curves
toward the sink, but oscillates around it in an erratic way. The rightward piece is
contained in the region bounded by the two components of the stable manifold.
(b) For $a = 1.4$, the manifolds have crossed one another.

sink under two different parameter settings. In Figure 2.21, portions of the stable
and unstable manifolds of a saddle point near $(-2, -2)$ are drawn. In each case,
the upward and downward piece of the stable manifold, which is predominantly
vertical, forms the boundary of the basin of the period-two sink. (Compare with
Figure 2.3.) For a larger value of the parameter a, as in Figure 2.21(b), the stable
and unstable manifolds intersect, and the basin boundary changes from simple to
complicated.

EXAMPLE 2.22

Figure 2.22 shows the relation of the stable and unstable manifolds to the
basin of the two-piece attractor for the Hénon map with $a = 2, b = -0.3$. This
basin was shown earlier in Figure 2.11. The stable manifold of the saddle fixed
point in the lower left corner forms the boundary of the attractor basin; the
attractor lies along the unstable manifold of the saddle.

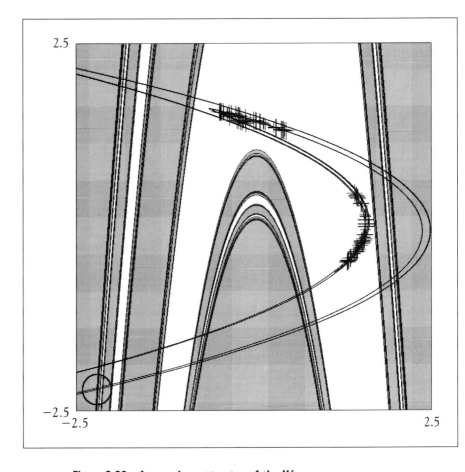

2.5

−2.5
−2.5

2.5

Figure 2.22 A two-piece attractor of the Hénon map.
The crosses mark 100 points of a trajectory lying on a two-piece attractor. The basin of attraction of this attractor is white; the shaded points are initial conditions whose orbits diverge to ∞. The saddle fixed point circled at the lower left is closely related to the dynamics of the attractor. The stable manifold of the saddle, shown in black, forms the boundary of the basin of the attractor. The attractor lies along the unstable manifold of the saddle, which is also in black.

EXAMPLE 2.23

Figure 2.23(a) shows 18 fixed points (large crosses) and 38 period-two orbits (small crosses) for the time-2π map of the forced damped pendulum (2.10) with $c = 0.05, \rho = 2.5$. The orbits were found by computer approximation methods; there may be more. None of these orbits are sinks; they coexist with an complicated attracting orbit shown in Figure 2.7. Exactly half of the 56 orbits shown

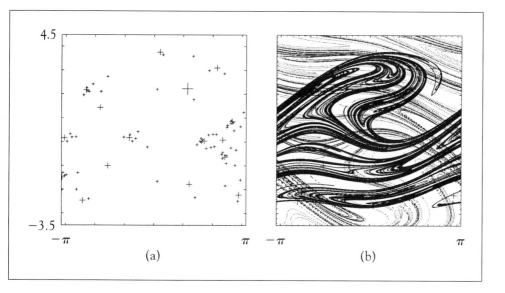

Figure 2.23   The forced damped pendulum.
(a) Periodic orbits for the pendulum with parameters $c = 0.05, \rho = 2.5$. The large crosses denote 18 fixed points, and the small crosses, 38 period-two orbits. (b) The stable and unstable manifolds of the largest cross in (a). The unstable manifold is drawn in black; compare to Figure 2.7. The stable manifold is drawn in gray dashed curves. The manifolds overlay the periodic orbits from (a)—note that without exception these orbits lie close to the unstable manifold.

are flip saddles; the rest are regular saddles. The largest cross in Figure 2.23(a) is singled out, and its stable and unstable manifolds are drawn in Figure 2.23(b).

Exercise 10.6 of Chapter 10 states that a stable manifold cannot cross itself, nor can it cross the stable manifold of another fixed point. However, there is no such restriction for a stable manifold crossing an unstable manifold.

The discovery that stable and unstable manifolds of a fixed point can intersect was made by Poincaré. He made this observation in the process of fixing his entry to King Oscar's contest. (In his original entry he made the assumption that they could not cross.) Realizing this possibility was a watershed in the knowledge of dynamical systems, whose implications are still being worked out today.

Poincaré was surprised to see the extreme complexity that such an intersection causes in the dynamics of the map. If $\mathbf{p}$ is a fixed or periodic point, and if $\mathbf{h}_0 \neq \mathbf{p}$ is a point of intersection of the stable and unstable manifold of $\mathbf{p}$, then $\mathbf{h}_0$ is called a **homoclinic point**. For starters, an intersection of the stable and unstable manifolds of a single fixed point (called a **homoclinic** intersection) immediately

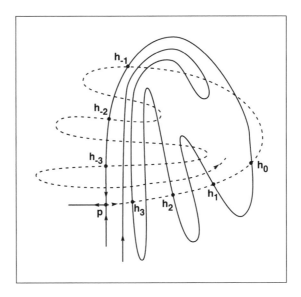

Figure 2.24 A schematic view of a homoclinic point h_0.
The stable manifold (solid curve) and unstable manifold (dashed curve) of the saddle fixed point **p** intersect at h_0, and therefore also at infinitely many other points. This figure only hints at the complexity. Poincaré showed that if a circle was drawn around any homoclinic point, there would be infinitely many homoclinic points inside the circle, no matter how small its radius.

forces infinitely many such intersections. Poincaré drew diagrams similar to Figure 2.24, which displays the infinitely many intersections that logically follow from the intersection h_0.

To understand the source of this complexity, first notice that a stable manifold, by definition, is an invariant set under the map **f**. This means that if h_0 is a point on the stable manifold of a fixed point **p**, then so are $h_1 = f(h_0)$ and $h_{-1} = f^{-1}(h_0)$. This is easy to understand: if the orbit of h_0 eventually converges to **p** under **f**, then so must the orbits of h_1 and h_{-1}, being one step ahead and behind of h_0, respectively. In fact, if h_0 is a point on a stable manifold of **p**, then so is the entire (forward and backward) orbit of h_0. By the same reasoning, the unstable manifold of a fixed point is also invariant.

Once a point like h_0 in Figure 2.24 lies on both the stable and unstable manifolds of a fixed point, then the entire orbit of h_0 must lie on both manifolds, because both manifolds are invariant. Remember that the stable manifold is directed toward the fixed point, and the unstable manifold leads away from it. The result is a configuration drawn schematically in Figure 2.24, and generated by a computer for the Hénon map in Figure 2.21(b).

The key fact about a homoclinic intersection point is that it essentially spreads the sensitive dependence on initial conditions—ordinarily situated at a single saddle fixed point—throughout a widespread portion of state space. Figures 2.21 and 2.24 give some insight into this process. In Chapter 10, we will return to study this mechanism for manufacturing chaos.

2.7 MATRIX TIMES CIRCLE EQUALS ELLIPSE

Near a fixed point v_0, we have seen that the dynamics essentially reduce to a single linear map $A = \mathbf{Df}(v_0)$. If a map is linear then its action on a small disk neighborhood of the origin is just a scaled–down version of the effect on the unit disk. We found that the magnitudes of the eigenvalues of A were decisive for classifying the fixed point. The same is true for a period-k orbit; in that case the appropriate matrix A is a product of k matrices.

In case the orbit is not periodic (which is one of our motivating situations), there is no magic matrix A. The local dynamics in the vicinity of the orbit is ruled, even in its linear approximation, by an infinite product of usually nonrepeating $\mathbf{Df}(v_0)$. The role of the eigenvalues of A is taken over by Lyapunov numbers, which measure contraction and expansion. When we develop Lyapunov numbers for many-dimensional maps (Chapter 5), it is this infinite product that we will have to measure or approximate in some way.

To visualize what is going on in cases like this, it helps to have a way to calculate the image of a disk from the matrix representing a linear map. For simplicity, we will choose the disk of radius one centered at the origin, and a square matrix. The image will be an ellipse, and matrix algebra explains how to find that ellipse.

The technique (again) involves eigenvalues. The image of the unit disk N under the linear map A will be determined by the eigenvectors and eigenvalues of AA^T, where A^T denotes the transpose matrix of A (formed by exchanging the rows and columns of A). The eigenvalues of AA^T are nonnegative for any A. This fact can be found in Appendix A, along with the next theorem, which shows how to find the explicit ellipse AN.

Theorem 2.24 *Let N be the unit disk in $\mathbb{R}^m$, and let A be an $m \times m$ matrix. Let $s_1^2, \ldots, s_m^2$ and $v_1, \ldots, v_m$ be the eigenvalues and unit eigenvectors, respectively, of the $m \times m$ matrix AA^T. Then*

1. *$v_1, \ldots, v_m$ are mutually orthogonal unit vectors; and*
2. *the axes of the ellipse AN are $s_i v_i$ for $1 \leq i \leq m$.*

Check that in Example 2.5, the map A gives $s_1 = a$, $s_2 = b$, while v_1 and v_2 are the x and y unit vectors, repectively. Therefore a and b are the lengths of the axes of the ellipse AN. For the nth iterate of A, represented by the matrix A^n, we find ellipse axes of length a^n and b^n for A^nN, the nth image of the unit disk.

In Example 2.5, the axes of the ellipse AN are easy to find. Each axis is an eigenvector not only of AA^T but also of A, whose length is the corresponding eigenvalue of A. In general (for nonsymmetric matrices), the eigenvectors of A do not give the directions along which the ellipse lies, and it is necessary to use Theorem 2.24. To see how Theorem 2.24 applies in general, we'll return for a look at our three important examples.

EXAMPLE 2.25

[Distinct real eigenvalues.] Let

$$A(x) = Ax = \begin{pmatrix} .8 & .5 \\ 0 & 1.3 \end{pmatrix} \begin{pmatrix} x \\ y \end{pmatrix}. \tag{2.36}$$

The eigenvalues of the matrix A are 0.8 and 1.3, with corresponding eigenvectors $\begin{pmatrix} 1 \\ 0 \end{pmatrix}$ and $\begin{pmatrix} 1 \\ 1 \end{pmatrix}$, respectively. From this it is clear that the fixed point at the origin is a saddle—the two eigenvectors give directions along which the fixed point attracts and repels, respectively. The attracting direction is illustrated by

$$A^n \begin{pmatrix} 1 \\ 0 \end{pmatrix} = (0.8)^n \begin{pmatrix} 1 \\ 0 \end{pmatrix} = \begin{pmatrix} (0.8)^n \\ 0 \end{pmatrix}, \tag{2.37}$$

and the repelling direction by

$$A^n \begin{pmatrix} 1 \\ 1 \end{pmatrix} = (1.3)^n \begin{pmatrix} 1 \\ 1 \end{pmatrix} = \begin{pmatrix} (1.3)^n \\ (1.3)^n \end{pmatrix}. \tag{2.38}$$

The stable manifold of the origin saddle point is $y = 0$, and the unstable manifold is $y = x$. Points along the x-axis move directly toward the origin under iteration by A, and points along the line $y = x$ move toward infinity. Since we know the nth iterate of the unit circle is an ellipse with one growing direction and one shrinking direction, we know that in the limit the ellipses become long and thin. The ellipses A^nN representing higher iterates of the unit disk gradually line up along the dominant eigenvector $\begin{pmatrix} 1 \\ 1 \end{pmatrix}$ of A.

The first few images of the unit disk under the map A can be found using Theorem 2.24, and are graphed in Figure 2.25. For an application of Theorem

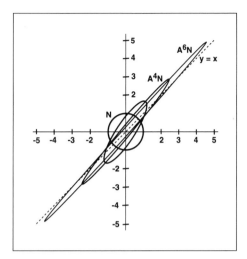

Figure 2.25   Successive images of the unit circle N for a saddle fixed point.
The image of a circle under a linear map is an ellipse. Successive images are therefore
also ellipses, which in this example line up along the expanding eigenspace.

2.24, we calculate the first iterate of the unit disk under A. Since

$$AA^T = \begin{pmatrix} .8 & .5 \\ 0 & 1.3 \end{pmatrix} \begin{pmatrix} .8 & 0 \\ .5 & 1.3 \end{pmatrix} = \begin{pmatrix} .89 & .65 \\ .65 & 1.69 \end{pmatrix}, \tag{2.39}$$

the unit eigenvectors of AA^T are (approximately) $\begin{pmatrix} .873 \\ -.488 \end{pmatrix}$ and $\begin{pmatrix} .488 \\ .873 \end{pmatrix}$, with
eigenvalues .527 and 2.053, respectively. Taking square roots, we see that the
ellipse AN has principal axes of lengths $\approx \sqrt{.527} \approx .726$ and $\approx \sqrt{2.053} \approx$
1.433. The ellipse AN, along with A^4N and A^6N, is illustrated in Figure 2.25.

EXAMPLE 2.26

[Repeated eigenvalue.] Even in the sink case, the ellipse A^nN can grow a
little in some direction before shrinking for large n. Consider the example

$$A = \begin{pmatrix} \frac{2}{3} & 1 \\ 0 & \frac{2}{3} \end{pmatrix}. \tag{2.40}$$

✎ EXERCISE T2.10

Use Theorem 2.24 to calculate the axes of the ellipse AN from (2.40). Then
verify that the ellipses A^nN shrink to the origin as $n \to \infty$.

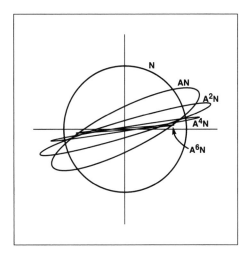

Figure 2.26 Successive images of the unit circle N under a linear map A in the case of repeated eigenvalues.
The nth iterate A^nN of the circle lies wholly inside the circle for $n \geq 6$. In this case, the origin is a sink.

The first few iterates of the unit circle N are graphed in Figure 2.26. The ellipse AN sticks out of the unit disk N. Further iteration by A continues to roll the ellipse to lie parallel to the x-axis and to eventually shrink it to the origin, as the calculation of Exercise T2.10 requires.

EXAMPLE 2.27

[Complex eigenvalues.] Let

$$A = \begin{pmatrix} a & -b \\ b & a \end{pmatrix}. \tag{2.41}$$

The eigenvalues of this matrix are $a \pm bi$. Calculating AA^T yields

$$AA^T = \begin{pmatrix} a^2 + b^2 & 0 \\ 0 & a^2 + b^2 \end{pmatrix}, \tag{2.42}$$

so it follows that the image of the unit disk N by A is again a disk of radius $\sqrt{a^2 + b^2}$. The matrix A rotates the disk by $\arctan b/a$ and stretches by a factor of $\sqrt{a^2 + b^2}$ on each iteration. The stability result is the same as in the previous two cases: if the absolute value of the eigenvalues is less than 1, the origin is a sink; if greater than 1, a source.

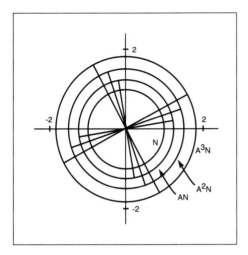

Figure 2.27 Successive images of the unit circle N.
The origin is a source with complex eigenvalues.

In Figure 2.27, the first few iterates A^nN of the unit disk are graphed for $a = 1.2, b = 0.2$. Since $a^2 + b^2 > 1$, the origin is a source. The radii of the images of the disk grow at the rate of $\sqrt{1.2^2 + .2^2} \approx 1.22$ per iteration, and the disks turn counterclockwise at the rate of $\arctan(.2/1.2) \approx 9.5$ degrees per iteration.

☞ CHALLENGE 2

Counting the Periodic Orbits of Linear Maps on a Torus

IN CHAPTER 1, we investigated the properties of the linear map $f(x) = 3x$ (mod 1). This map is discontinuous on the interval $[0, 1]$, but continuous when viewed as a map on a circle. We found that the map had infinitely many periodic points, and we discussed ways to count these orbits.

We will study a two-dimensional map with some of the same properties in Challenge 2. Consider the map $\mathbf{S}$ defined by a 2×2 matrix

$$A = \begin{pmatrix} a & b \\ c & d \end{pmatrix}$$

with integer entries a, b, c, d, where we define $\mathbf{S}(\mathbf{v}) = A\mathbf{v}$ (mod 1). The domain for the map will be the unit square $[0, 1] \times [0, 1]$. Even if $A\mathbf{v}$ lies outside the unit square, $\mathbf{S}(\mathbf{v})$ lies inside if we count modulo one. In general, $\mathbf{S}$ will fail to be continuous, in the same way as the map in Example 1.9 of Chapter 1.

For example, assume

$$A = \begin{pmatrix} 2 & 1 \\ 1 & 1 \end{pmatrix}. \tag{2.43}$$

Consider the image of the point $\mathbf{v} = (x, 1/2)$ under $\mathbf{S}$. For x slightly less than $1/2$, the image $\mathbf{S}(\mathbf{v})$ lies just below $(1/2, 1)$. For x slightly larger than $1/2$, the image $\mathbf{S}(\mathbf{v})$ lies just above $(1/2, 0)$, quite far away. Therefore $\mathbf{S}$ is discontinuous at $(1/2, 1/2)$.

We solved this problem for the $3x$ mod 1 map in Chapter 1 by sewing together the ends of the unit interval to make a circle. Is there a geometric object for which S can be made continuous? The problem is that when the image value 1 is reached (for either coordinate x or y), the map wants to restart at the image value 0.

The **torus** $\mathbb{T}^2$ is constructed by identifying the two pairs of opposite sides of the unit square in $\mathbb{R}^2$. This results in a two-dimensional object resembling a doughnut, shown in Figure 2.28. We have simultaneously glued together the x-axis at 0 and 1, and the y-axis at 0 and 1. The torus is the natural domain for maps that are formed by integer matrices modulo one.

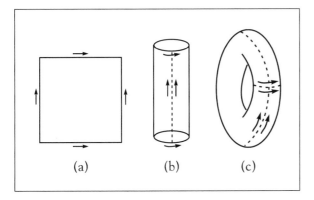

Figure 2.28   Construction of a torus in two easy steps.
(a) Begin with unit square. (b) Identify (glue together) vertical sides. (c) Identify horizontal sides.

Given a 2×2 matrix A, we can define a map from the torus to itself by multiplying the matrix A times a vector (x, y), followed by taking the output vector modulo 1. If the matrix A has integer entries, then the map so defined is continuous on the torus. In the following steps we derive some fundamental properties of torus maps, and then specialize to the particular map (2.43), called the cat map.

Assume in Steps 1–6 that A has integer entries, and that the determinant of A, $\det(A) = ad - bc$, is nonzero.

Step 1   Show that the fact that A is a 2×2 matrix with integer entries implies that the torus map $\mathbf{S}(\mathbf{v}) = A\mathbf{v} \pmod 1$ is a continuous map on the torus $\mathbb{T}^2$. (You will need to explain why the points $(0, y)$ and $(1, y)$, which are identified together on the torus, map to the same point on the torus. Similarly for $(x, 0)$ and $(x, 1)$.)

Step 2   (a) Show that $A \begin{pmatrix} x + n_1 \\ y + n_2 \end{pmatrix} = A \begin{pmatrix} x \\ y \end{pmatrix} \pmod 1$ for any integers n_1, n_2.

(b) Show that $\mathbf{S}^2(\mathbf{v}) = A^2\mathbf{v} \pmod 1$.

(c) Show that $\mathbf{S}^n(\mathbf{v}) = A^n\mathbf{v} \pmod 1$ for any positive integer n.

Step 2 says that in computing the nth iterate of $\mathbf{S}$, you can wait until the end to apply the modulo 1 operation.

A real number r is **rational** if $r = p/q$, where p and q are integers. A number that is not rational is called **irrational**. Note that the sum or product of two rational numbers is rational, the sum of a rational and an irrational is irrational, and the product of a rational and an irrational is irrational unless the rational is zero.

Step 3 Assume that A has no eigenvalue equal to 1. Show that $S(v) = v$ implies that both components of v are rational numbers.

Step 4 Assume that A has no eigenvalues of magnitude one. Since the eigenvalues of A^n are the nth powers of the eigenvalues of A (see Appendix A), this assumption guarantees that for all n, the matrix A^n does not have an eigenvalue equal to 1. Show that a point $v = (x, y)$ in the unit square is eventually periodic if and only if its coordinates are rational. (The $3x \pmod 1$ map of Chapter 1 had a similar property.) [*Hint:* Use Step 3 to show that if any component is irrational, then v cannot be eventually periodic. If both components of v are rational, show that there are only a finite number of possibilities for iterates of v.]

Step 5 Show that the image of the map S covers the square precisely $|\det(A)|$ times. More precisely, if v_0 is a point in the square, show that the number of solutions of $S(v) = v_0$ is $|\det A|$. [*Hint:* Draw the image under A of the unit square in the plane. It is a parallelogram with one vertex at the origin and three other vertices with integer coordinates. The two sides with the origin as vertex are the vectors (a, c) and (b, d). The area of the parallelogram is therefore $|\det(A)|$. Show that the number of solutions of $Av = v_0 \pmod 1$ is the same for all x_0 in the square. Therefore the parallelogram can be cut up by mod 1 slices and placed onto the square, covering it $|\det(A)|$ times. See Figure 2.29.]

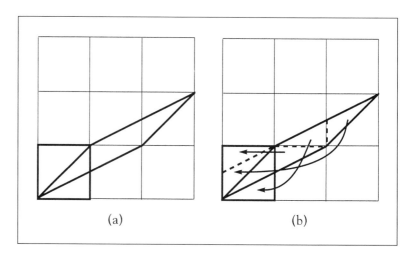

(a)   (b)

Figure 2.29 Dynamics of the cat map.
(a) The unit square and its image under the cat map (2.43). (b) Since $\det(A) = 1$, the image of the unit square, modulo 1, covers the unit square exactly once.

Step 6 Define the **trace** of A by $\mathrm{tr}(A) = a + d$, the sum of the diagonal entries. Prove that the number of fixed points of S on the torus is

$$|\det A - \mathrm{tr}\, A + 1|,$$

as long as that number is not zero. [*Hint:* Apply Step 5 to the matrix $A - I$.]

Now we study a particular choice of A. Define the matrix

$$A = \begin{pmatrix} 2 & 1 \\ 1 & 1 \end{pmatrix}$$

as in Equation (2.43). The resulting map S, called the "cat map", is shown in Figure 2.30.

Step 7 Let F_n denote the nth Fibonacci number, where $F_0 = F_1 = 1$, and where $F_n = F_{n-1} + F_{n-2}$ for $n \geq 2$. Prove that

$$A^n = \begin{pmatrix} F_{2n} & F_{2n-1} \\ F_{2n-1} & F_{2n-2} \end{pmatrix}.$$

Step 8 Find all fixed points and period-two orbits of the cat map. [*Hint:* For the period-2 points, find all solutions of

$$5\frac{a}{b} + 3\frac{c}{d} = m + \frac{a}{b}$$
$$3\frac{a}{b} + 2\frac{c}{d} = n + \frac{c}{d} \qquad\qquad (2.44)$$

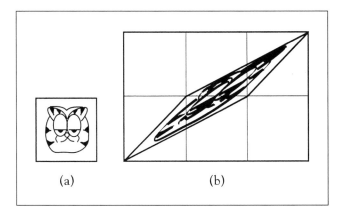

(a) (b)

Figure 2.30 An illustration of multiplication by the matrix A in (2.43).
(a) Cat in unit square. (b) Image of the unit square under the matrix A.

where a, b, c, d, m, n are integers.] Answers: $(0,0)$ is the only fixed point; the period-2 orbits are:

$$\left\{ \begin{pmatrix} .2 \\ .4 \end{pmatrix}, \begin{pmatrix} .8 \\ .6 \end{pmatrix} \right\} \text{ and } \left\{ \begin{pmatrix} .4 \\ .8 \end{pmatrix}, \begin{pmatrix} .6 \\ .2 \end{pmatrix} \right\}.$$

Step 9   Use Step 6 to find a formula for the number of fixed points of S^n. (Answer: $|(F_{2n} - 1)(F_{2n-2} - 1) - F_{2n-1}^2|$.)

The formula in Step 9 can be checked against Figure 2.31 for low values of n. For example, Figure 2.31(a) shows a plot of the fixed points of S^4 in the unit square. There are 45 points, each with coordinates expressible as $(i/15, j/15)$ for some integers i, j. One of them is a period-one point of S (the origin), and four of them are the period-two points (two different orbits) of S found in Step 8. That leaves a total of 10 different period-four orbits. In counting, remember that the points are defined modulo 1, so that a point on the boundary of the square will also appear as the same point on the opposite boundary. The period-five points in Figure 2.31(b) are somewhat easier to count; they are the 120 points of form $(i/11, j/11)$ where $0 \le i, j \le 10$, omitting the origin, which is a period-one point. There are 24 period-five orbits. The period-three points show up as the medium-sized crosses in Figure 2.31(c). They are the 15 points of form $(i/4, j/4)$ where $0 \le i, j \le 3$, again omitting the origin. Can you guess the denominators of the period-six points? See Step 12 for the answer.

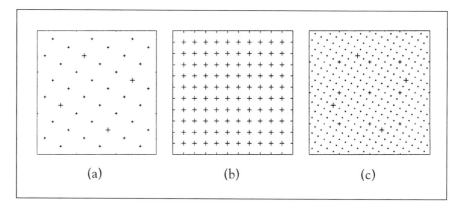

(a)   (b)   (c)

Figure 2.31   Periodic points of the cat map on the torus.
(a) Period-four points (small crosses), period-two points (large crosses). (b) Period-five points. (c) Period-six points (small crosses), period-three points (medium crosses), period-two points (large crosses)

Step 10 Prove the identities $F_n F_{n-2} - F_{n-1}^2 = (-1)^n$ and $F_{2n} = F_n^2 + F_{n-1}^2$ for Fibonacci numbers, and use them to simplify the formula for the number of fixed points of S^n to $(F_n + F_{n-2})^2 - 2 - 2(-1)^n$.

Step 11 Write out the periodic table for S (list number of periodic points for each period) for periods up to 10. See Table 1.3 of Chapter 1 for the form of a periodic table. Compare this with Figure 2.31 where applicable. Show that S has periodic points of all periods.

Step 12 Here is the formula for the denominator of the period n orbits. The orbits consist of points a/b where

$$b = \begin{cases} F_n + F_{n-2} & \text{if } n \text{ is odd and } n > 1 \\ 5F_{n-1} & \text{if } n \text{ is even} \end{cases}$$

For example, the denominator of the period-two points is $5F_1 = 5$, and the denominator of the period-three points is $F_3 + F_1 = 4$. Confirm this with the answers from Step 8 and Figure 2.31(a). Prove this formula for general n. [Thanks to Joe Miller for simplifying this formula.]

Postscript. How robust are the answers you derived in Challenge 2? If you're still reading, you have done an exhaustive accounting of the periodic orbits for the cat map (2.43). Does this accounting apply only to the cat map, or are the same formulas valid for cat-like maps?

Using a few advanced ideas, you can show that the results apply as well to maps in a neighborhood of the cat map in "function space", including nearby nonlinear maps. Exactly the same number of orbits of each period exist for these maps. The reasoning is as follows. As we will see in Chapter 11, as a map is perturbed (say, by moving a parameter a), a fixed point $\mathbf{p}$ of $\mathbf{f}_a^k$ cannot appear or disappear (as a moves) without an eigenvalue of the matrix $D\mathbf{f}_a^k(\mathbf{p})$ crossing through 1. (At such a crossing, the Jacobian of $\mathbf{f}_a^k - I$ is singular and the implicit function theorem is violated, allowing a solution of $\mathbf{f}^k - I = 0$ to appear or disappear as the map is perturbed.)

The eigenvalues of the cat map A are $e_1 = (3 + \sqrt{5})/2 \approx 2.618$ and $e_2 = (3 - \sqrt{5})/2 \approx 0.382$, and the eigenvalues of A^k are e_1^k and e_2^k. Since eigenvalues move continuously as a parameter in the matrix is varied, any map whose Jacobian entries are close to those of the cat map will have eigenvalues close to $|e_1|$ and $|e_2|$, bounded away from 1, and so the eigenvalues of $D\mathbf{f}^k(\mathbf{p})$ will not equal 1. This rules out the appearance and disappearance of periodic points for linear and nonlinear maps sufficiently similar to the cat map, showing that the same formulas derived in Challenge 2 hold for them as well. Nonlinear maps of the torus close to the cat map such as

$$f(x, y) = (2x + y + a\cos 2\pi x, x + y + b\sin 2\pi y)$$

where $|a|, |b|$ are sufficiently small have this property, and so have the same number of periodic points of each period as the cat map.

Exercises

2.1. For each of the following linear maps, decide whether the origin is a sink, source, or saddle.

(a) $\begin{pmatrix} 4 & 30 \\ 1 & 3 \end{pmatrix}$ (b) $\begin{pmatrix} 1 & 1/2 \\ 1/4 & 3/4 \end{pmatrix}$ (c) $\begin{pmatrix} -0.4 & 2.4 \\ -0.4 & 1.6 \end{pmatrix}$

2.2. Find $\lim\limits_{n \to \infty} \begin{pmatrix} 4.5 & 8 \\ -2 & -3.5 \end{pmatrix}^n \begin{pmatrix} 6 \\ 9 \end{pmatrix}$.

2.3. Let $g(x, y) = (x^2 - 5x + y, x^2)$. Find and classify the fixed points of g as sinks, sources, or saddles.

2.4. Find and classify all fixed points and period-two orbits of the Hénon map (2.27) with

(a) $a = -0.56$ and $b = -0.5$

(b) $a = 0.21$ and $b = 0.6$

2.5. Let $f(x, y, z) = (x^2 y, y^2, xz + y)$ be a map on $\mathbb{R}^3$. Find and classify the fixed points of f.

2.6. Let $f(x, y) = (\sin \frac{\pi}{3} x, \frac{y}{2})$. Find all fixed points and their stability. Where does the orbit of each initial value go?

2.7. Set $b = 0.3$ in the Hénon map (2.27). (a) Find the range of parameters a for which the map has one fixed sink and one saddle fixed point. (b) Find the range of parameters a for which the map has a period-two sink.

2.8. Calculate the image ellipse of the unit disk under each of the following maps.

(a) $\begin{pmatrix} 2 & 0.5 \\ 2 & -0.5 \end{pmatrix}$ (b) $\begin{pmatrix} 2 & 1 \\ -2 & 2 \end{pmatrix}$

What are the areas of these ellipses?

2.9. Find the inverse map for the cat map defined in (2.43). Check your answer by composing with the cat map.

2.10. (a) Find a 2×2 matrix $A \neq I$ with integer entries, a rational number x, and an irrational number y such that $S(x, y) = (x, y)$. Here S is the mod 1 map associated to A. (b) Same question but require x and y to be irrational. According to Step 4 of Challenge 2, each of your answers must have an eigenvalue equal to 1.

Is the Solar System Stable?

KING OSCAR'S CONTEST in 1889 was designed to answer the question of whether the solar system is stable, once and for all. The actual result clarified just how difficult the question is. If the contest were repeated today, there would be no greater hope of producing a definitive answer, despite (or one might say, because of) all that has been learned about the problem in the intervening century.

Poincaré's entry showed that in the presence of homoclinic intersections, there is sensitive dependence on initial conditions. If this exists in our solar system, then long-term predictability is severely compromised. The positions, velocities, and masses of the planets of the solar system are known with considerably more precision than was known in King Oscar's time. However, even these current measurements fall far short of the accuracy needed to make long-term predictions in the presence of sensitive dependence on initial conditions. Two key questions are: (1) whether chaos exists in planetary trajectories, and (2) if there are chaotic trajectories, whether the chaos is sufficiently pronounced to cause ejection of a planet from the system, or a planetary collision.

The question of whether chaos exists in the solar system has led to innovations in theory, algorithms, computer software, and hardware in an attempt to perform accurate long-term solar system simulations. In 1988, Sussman and Wisdom reported on an 845 Myr (Myr denotes one million years) integration of the gravitational equations for the solar system. This integration was performed in a special-purpose computer that they designed for this problem, called the Digital Orrery, which has since been retired to the Smithsonian Institution in

Sussman, G. J., Wisdom, J., "Numerical evidence that the motion of Pluto is chaotic." Science **241**, 433-7 (1988).

Laskar, J., "A numerical experiment on the chaotic behaviour of the solar system." Nature **338**, 237-8 (1989).

Sussman, G.J., Wisdom, J., "Chaotic evolution of the solar system." Science **257**, 56-62 (1992).

Laskar, J., Robutel, P., "The chaotic obliquity of the planets." Nature **361**, 608-612 (1993).

Touma, J., Wisdom, J., "The chaotic obliquity of Mars." Science **259**, 1294-1297 (1993).

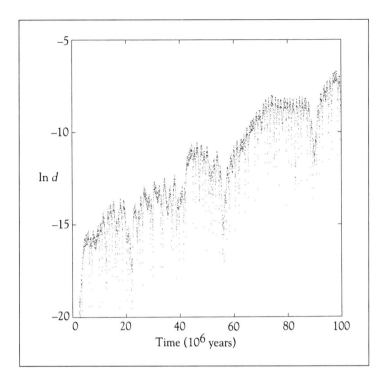

Figure 2.32 Comparison of two computer simulations.
The difference in the position of Pluto in two simulations with slightly different initial conditions is plotted as a function of time. The vertical scale is ln of the difference, in units of AU (astronomical units).

Washington, D.C. A surprising result of the integration was an indication of sensitive dependence in the orbit of Pluto.

Figure 2.32 shows exponential divergence of nearby trajectories over a 100 Myr simulation, done by Sussman and Wisdom in 1992 with a successor of the Digital Orrery, built as a collaboration between MIT and Hewlett-Packard. They made two separate computer runs of the simulated solar system. The runs were identical except for a slight difference in the initial condition for Pluto. The curve shows the distance between the positions of Pluto for the two solar system simulations.

The plot in Figure 2.32 is semilog, meaning that the vertical axis is logarithmic. Assume that the distance d between the Plutos in the two almost parallel universes grows exponentially with time, say as $d = ae^{kt}$. Then $\log d = \log a + kt$, which is a linear relation between $\log d$ and t. If we plot $\log d$ versus t, we will get a line with slope k. This is what Figure 2.32 shows in rough terms. In Chapter 3 the slope k will be called the Lyapunov exponent. A careful look at the figure yields the slope to be approximately $1/12$, meaning that the distance between nearby initial conditions is multiplied by a factor of $e^{1/12}$ each million years, or by a factor of $e \approx 2.718$ each 12 million years. We can say that the exponential separation time for Pluto is about 12 million years on the basis of this simulation.

Laskar, working at the time at the Bureau des Longitudes in Paris, used quite different techniques to simulate the solar system *without* Pluto, and concluded that the exponential separation time for this system is on the order of 5 million years. He attributed the chaotic behavior to resonances among the inner planets Mercury, Venus, Earth, and Mars. Recent simulations by Sussman and Wisdom have also arrived at the approximate value of 5 million years for some of the inner planets.

To put these times in perspective, note that $e^{19} \approx 1.5 \times 10^8$, which is the number of kilometers from the sun to the earth, or one astronomical unit. Therefore a difference or uncertainty of 1 km in a measured position could grow to an uncertainty of 1 astronomical unit in about 19 time units, or $19 \times 5 \approx 95$ Myrs. The solar system has existed for several billions of years, long enough for this to happen many times over.

The finding of chaos in solar system trajectories does not in itself mean that the solar system is on the verge of disintegration, or that Earth will soon cross the path of Venus. It does, however, establish a limit on the ability of celestial mechanics to predict such an event in the far future.

A more recent conclusion may have an impact on life on Earth within a scant few millions of years. The two research groups mentioned above published articles within one week of one another in 1993 regarding the erratic obliquity of planets in the solar system. The obliquity of a planet is the tilt of the spin axis with respect to the "plane" of the solar system. The obliquity of Earth is presently about 23.3°, with estimated variations over time of 1° either way. Large variations in the obliquity, for example those that would turn a polar icecap toward the sun for extended periods, would have a significant effect on climate.

The existence of a large moon has a complicating and apparently stabilizing effect on the obliquity of Earth. A more straightforward calculation can be made for Mars. Laskar and Robutel found that in a 45 Myr simulation, the obliquity of Mars can oscillate erratically by dozens of degrees, for some initial configurations. For other initial conditions, the oscillations are regular.

Figure 2.33(a) shows the variation of the obliquity of Mars for 80 Myrs into the past, from a computer simulation due to Touma and Wisdom. This simulation takes the present conditions as initial conditions and moves backwards in time. Figure 2.33(b) is a magnification of part (a) for the last 10 Myrs only. It shows an abrupt transition about 4 Myrs ago, from oscillations about 35° obliquity to oscillations about 25° obliquity.

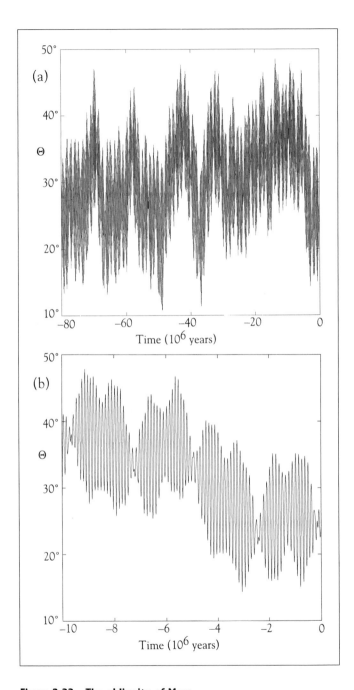

Figure 2.33 The obliquity of Mars.

(a) The result of a computer simulation of the solar system shows that the obliquity of Mars undergoes erratic variation as a function of time. (b) Detail from (a), showing only the last 10 million years. There is an abrupt transition about 4 million years ago.

No one knows whether the graphs shown here are correct over a several-million-year time range. Perhaps core samples from the poles of Mars (not yet obtained) could shed light on this question. If the initial position, velocity, mass, and other physical parameters of Mars used in the simulations were changed from the values used in the simulations, the results would be different, because of the sensitive dependence of the problem. Any calculation of the obliquity or the position of Mars for many millions of years can only be considered as representative of the wide range of possibilities afforded by chaotic dynamics.

While sensitive dependence on initial conditions causes unpredictability at large time scales, it can provide opportunity at shorter, predictable time scales. One of the first widely available English translations of Poincaré's writings on celestial mechanics was commissioned by the U.S. National Aeronautics and Space Administration (NASA). Several years ago, their scientists exploited the sensitive dependence of the three-body problem to achieve a practical goal.

In 1982, NASA found itself unable to afford to send a satellite to the Giacobini-Zinner comet, which was scheduled to visit the vicinity of Earth's orbit in 1985. It would pass quite far from the current Earth position (50 million miles), but near Earth's orbit. This led to the possibility that a satellite already near Earth's orbit could be sent to the correct position with a relatively low expenditure of energy.

A 1000-lb. satellite called ISEE-3 had been launched in 1978 to measure the solar wind and to count cosmic rays. ISEE-3 was parked in a "halo orbit", centered on the Lagrange point L_1. The halo orbit is shown in Color Plate 13. A Lagrange point is a point of balance between the gravitational pull of the Earth and Sun. In the rotating coordinate system in which the Earth and Sun are fixed, the L_1 point is an unstable equilibrium. Lagrange points are useful because little energy is required to orbit around them.

The ISEE-3 satellite was nearing the end of its planned mission, and had a limited amount of maneuvering fuel remaining. NASA scientists renamed the satellite the International Comet Explorer (ICE) and plotted a three-year trajectory for the Giacobini-Zinner comet. The new trajectory took advantage of near collisions with the Moon, or "lunar swingbys", to make large changes in the satellite's trajectory with small fuel expenditures. The first thruster burn, on June 10, 1982, changed the satellite's velocity by less than 10 miles per hour. In all, 37 burns were needed to make small changes in the trajectory, resulting in 5 lunar swingbys.

Dynamical motion of gravitational bodies is especially sensitive at a swingby. Since the distance between the two bodies is small, the forces between them is relatively large. This is where small changes can have a large effect.

The first lunar swingby occurred on March 30, 1983. A schematic picture of the satellite trajectory is shown in Color Plate 14. The 4 other near collisions with the moon are shown in the following Color Plates 15–16. During the final swingby on Dec. 22, 1983, denoted by 5 in the figure, the satellite passed within 80 miles of the surface of the moon on its way toward the comet. ICE passed through the tail of the Giacobini-Zinner comet on Sept. 11, 1985, exactly as planned.

CHAPTER THREE

Chaos

THE CONCEPT of an unstable steady state is familiar in science. It is not possible in practice to balance a ball on the peak of a mountain, even though the configuration of the ball perfectly balanced on the peak is a steady state. The problem is that the trajectory of any initial position of the ball near, but not exactly at, the steady state, will evolve away from the steady state. We investigated sources and saddles, which are unstable fixed points of maps, in Chapters 1 and 2.

What eventually happens to the ball placed near the peak? It moves away from the peak and settles in a valley at a lower altitude. The valley represents a stable steady state. One type of behavior for an initial condition that begins near an unstable steady state is to move away and be attracted by a stable steady state, or perhaps a stable periodic state.

We have seen this behavior in maps of the real line. Consider an initial condition that is near a source p of a map f. At the beginning of such an orbit, unstable behavior is displayed. Exponential separation means that the distance between the orbit point and the source increases at an exponential rate. Each iteration multiplies the distance between them by $|f'(p)| > 1$. We say that the exponential rate of separation is $|f'(p)|$ per iterate. That is, at least at first, small separations grow. After some wandering, the orbit may be attracted to a sink q. As it nears the sink, the orbit will display convergent behavior—the distance between the orbit point and the sink will change by the factor $|f'(q)| < 1$. As the orbit nears the attractor, small distances shrink.

It is common to see behavior like this, in which unstable behavior is transient and gives way eventually to stable behavior in the long run. But there is no reason that an initial condition starting near a source is forced to end up attracted to a sink or periodic sink. Perhaps no stable states exist, as in the example of the logistic map $G(x) = 4x(1 - x)$ we discussed in Chapter 1.

A chaotic orbit is one that forever continues to experience the unstable behavior that an orbit exhibits near a source, but that is not itself fixed or periodic. It never manages to find a sink to be attracted to. At any point of such an orbit, there are points arbitrarily near that will move away from the point during further iteration. This sustained irregularity is quantified by Lyapunov numbers and Lyapunov exponents. We will define the Lyapunov number to be the average per-step divergence rate of nearby points along the orbit, and the Lyapunov exponent to be the natural logarithm of the Lyapunov number. Chaos is defined by a Lyapunov exponent greater than zero.

In this chapter, we will study elementary properties of Lyapunov exponents and exhibit some maps for which they can be explicitly calculated. For example, we will see that for the logistic map, the Lyapunov exponent of most orbits (but not all) is $\ln 2$. We'll also develop a fixed point theorem for detecting fixed and periodic points, which will be used in Challenge 3 to establish a remarkable fact called Sharkovskii's Theorem.

3.1 LYAPUNOV EXPONENTS

We learned in Chapter 1 that for fixed points of discrete dynamical systems, stability is heavily influenced by the derivative of the map. For example, if x_1 is a fixed point of a one-dimensional map f and $f'(x_1) = a > 1$, then the orbit of each point x near x_1 will separate from x_1 at a multiplicative rate of approximately a per iteration, until the orbit of x moves significantly far away from x_1. That is, the distance between $f^n(x)$ and $f^n(x_1) = x_1$ will be magnified by approximately $a > 1$ for each iteration of f.

For a periodic point of period k, we have to look at the derivative of the kth iterate of the map, which, by the chain rule, is the product of the derivatives at the k points of the orbit. Suppose this product of derivatives is $A > 1$. Then the orbit of each neighbor x of the periodic point x_1 separates from x_1 at a rate of approximately A after each k iterates. This is a cumulative amount of separation—it takes k iterations of the map to separate by a distance A. It makes sense to describe the average multiplicative rate of separation as $A^{1/k}$ per iterate.

The term Lyapunov number is introduced to quantify this average multiplicative rate of separation of points x very close to x_1. (The Lyapunov exponent will be simply the natural logarithm of the Lyapunov number.) A Lyapunov number of 2 (or equivalently, a Lyapunov exponent of $\ln 2$) for the orbit of x_1 will mean that the distance between the orbit of x_1 and the orbit of a nearby point x doubles each iteration, on the average. For a periodic point x_1 of period k, this is the same as saying that

$$|(f^k)'(x_1)| = |f'(x_1)||f'(x_2)| \cdots |f'(x_k)| = 2^k.$$

But we want to consider this concept even when x_1 is not a fixed point or periodic point. A Lyapunov number of $\frac{1}{2}$ would mean this distance would be halved on each iteration, and the orbits of x and x_1 would move rapidly closer.

The significance of the concept of Lyapunov number is that it can be applied to nonperiodic orbits. A characteristic of chaotic orbits is sensitive dependence on initial conditions—the eventual separation of the orbits of nearby initial conditions as the system moves forward in time. In fact, our definition of a chaotic orbit is one that does not tend toward periodicity and whose Lyapunov number is greater than 1.

In order to formally define Lyapunov number and Lyapunov exponent for a general orbit, we follow the analogy of the periodic case, and consider the product of the derivatives at points along the orbit. We begin by restricting our attention to one-dimensional maps.

Definition 3.1 Let f be a smooth map of the real line $\mathbb{R}$. The **Lyapunov number** $L(x_1)$ of the orbit $\{x_1, x_2, x_3, \ldots\}$ is defined as

$$L(x_1) = \lim_{n \to \infty} (|f'(x_1)| \ldots |f'(x_n)|)^{1/n},$$

if this limit exists. The **Lyapunov exponent** $h(x_1)$ is defined as

$$h(x_1) = \lim_{n \to \infty} (1/n)[\ln |f'(x_1)| + \cdots + \ln |f'(x_n)|],$$

if this limit exists. Notice that h exists if and only if L exists, and $\ln L = h$.

Remark 3.2 Lyapunov numbers and exponents are undefined for some orbits. In particular, an orbit containing a point x_i with $f'(x_i) = 0$ causes the Lyapunov exponent to be undefined.

✎ **EXERCISE T3.1**

Show that if the Lyapunov number of the orbit of x_1 under the map f is L, then the Lyapunov number of the orbit of x_1 under the map f^k is L^k, whether or not x_1 is periodic.

It follows from the definition that the Lyapunov number of a fixed point x_1 for a one-dimensional map f is $|f'(x_1)|$, or equivalently, the Lyapunov exponent of the orbit is $h = \ln |f'(x_1)|$. If x_1 is a periodic point of period k, then it follows that the Lyapunov exponent is

$$h(x_1) = \frac{\ln |f'(x_1)| + \cdots + \ln |f'(x_k)|}{k}.$$

The point is that for a periodic orbit, the Lyapunov number $e^{h(x_1)}$ describes the average local stretching, on a per-iterate basis, near a point on the orbit.

Definition 3.3 Let f be a smooth map. An orbit $\{x_1, x_2, \ldots x_n, \ldots\}$ is called **asymptotically periodic** if it converges to a periodic orbit as $n \to \infty$; this means that there exists a periodic orbit $\{y_1, y_2, \ldots, y_k, y_1, y_2, \ldots\}$ such that

$$\lim_{n \to \infty} |x_n - y_n| = 0.$$

Any orbit that is attracted to a sink is asymptotically periodic. The orbit with initial condition $x = 1/2$ of $G(x) = 4x(1 - x)$ is also asymptotically periodic, since after two iterates it coincides with the fixed point $x = 0$. The term **eventually periodic** is used to describe this extreme case, where the orbit lands precisely on a periodic orbit.

Theorem 3.4 Let f be a map of the real line $\mathbb{R}$. If the orbit $\{x_1, x_2, \ldots\}$ of f satisfies $f'(x_i) \neq 0$ for all i and is asymptotically periodic to the periodic orbit $\{y_1, y_2, \ldots\}$, then the two orbits have identical Lyapunov exponents, assuming both exist.

Proof: We use the fact that a sequence average converges to the sequence limit; that is, if s_n is an infinite sequence of numbers with $\lim_{n \to \infty} s_n = s$, then

$$\lim_{n \to \infty} \frac{1}{n} \sum_{i=1}^{n} s_i = s.$$

Assume $k = 1$ to begin with, so that y_1 is a fixed point. Since $\lim_{n\to\infty} x_n = y_1$, the fact that the derivative f' is a continuous function implies that

$$\lim_{n\to\infty} f'(x_n) = f'(\lim_{n\to\infty} x_n) = f'(y_1).$$

Moreover, since $\ln |x|$ is a continuous function for positive x,

$$\lim_{n\to\infty} \ln |f'(x_n)| = \ln |\lim_{n\to\infty} f'(x_n)| = \ln |f'(y_1)|.$$

This equation gives us the limit of an infinite sequence. Using the fact that the sequence average converges to the sequence limit, we see that

$$h(x_1) = \lim_{n\to\infty} \frac{1}{n} \sum_{i=1}^{n} \ln |f'(x_i)| = \ln |f'(y_1)| = h(y_1).$$

Now assume that $k > 1$, so that y_1 is not necessarily a fixed point. Then y_1 is a fixed point for f^k, and the orbit of x_1 is asymptotically periodic under f^k to the orbit of y_1. From what we proved above, the Lyapunov exponent of the orbit of x_1 under f^k is $\ln |f^{k'}(y_1)|$. By Exercise T3.1, the Lyapunov exponent of x_1 under f is $\frac{1}{k} \ln |(f^k)'(y_1)| = h(y_1)$. $\qquad\qquad\square$

✎ **EXERCISE T3.2**

Find the Lyapunov exponent shared by most bounded orbits of $g(x) = 2.5x(1 - x)$. Begin by sketching $g(x)$ and considering the graphical representation of orbits. What are the possible bounded asymptotic behaviors? Do all bounded orbits have the same Lyapunov exponents?

⇨ **COMPUTER EXPERIMENT 3.1**

Write a program to calculate the Lyapunov exponent of $g_a(x) = ax(1 - x)$ for values of the parameter a between 2 and 4. Graph the results as a function of a.

3.2 CHAOTIC ORBITS

In Section 3.1 we defined the Lyapunov exponent h of an orbit to be the natural log of the average per-step stretching of the orbit. We were able to calculate h

in certain special cases: for a fixed point or periodic orbit we could express h in terms of derivatives, and orbits converging to a periodic orbit share the same Lyapunov exponent. More interesting cases involve bounded orbits that are not asymptotically periodic. When such an orbit has a positive Lyapunov exponent, it is a chaotic orbit.

Definition 3.5   Let f be a map of the real line $\mathbb{R}$, and let $\{x_1, x_2, \ldots\}$ be a bounded orbit of f. The orbit is **chaotic** if

1. $\{x_1, x_2, \ldots\}$ is not asymptotically periodic.
2. the Lyapunov exponent $h(x_1)$ is greater than zero.

EXAMPLE 3.6

The map $f(x) = 2x \pmod 1$ on the real line $\mathbb{R}$ exhibits positive Lyapunov exponents and chaotic orbits. See Figure 3.1(a). The map is not continuous, and therefore not differentiable, at $x = \frac{1}{2}$. We restrict our attention to orbits that never map to the point $\frac{1}{2}$. For these orbits, it is easy to compute the Lyapunov exponent as

$$\lim_{n\to\infty} \frac{1}{n} \sum_{i=1}^{n} \ln |f'(x_i)| = \lim_{n\to\infty} \frac{1}{n} \sum_{i=1}^{n} \ln 2 = \ln 2.$$

Therefore each orbit that forever avoids $\frac{1}{2}$ and is not asymptotically periodic is a chaotic orbit, with Lyapunov exponent $\ln 2$.

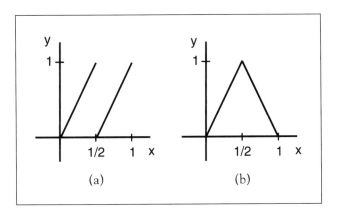

Figure 3.1   Two simple maps with chaotic orbits.
(a) $2x \pmod 1$ map. (b) The tent map.

The action of this map on a real number in $[0, 1]$ can be expressed easily if we consider the binary expansion of the number (see accompanying box). The map f applied to a number x expressed in a binary expansion chops off the leftmost bit:

$$1/5 = .0011\overline{0011}$$

$$f(1/5) = .011\overline{0011}$$

$$f^2(1/5) = .11\overline{0011}$$

$$f^3(1/5) = .1\overline{0011}$$

$$f^4(1/5) = .\overline{0011}$$

Notice that $x = 1/5$ is a period-four orbit of f.

Since the map is so simple in the binary system, we can see immediately which points in $[0, 1]$ are periodic—they are the points with a repeating binary

---

BINARY NUMBERS

The binary expansion of a real number x has form

$$x = .b_1 b_2 b_3 \ldots,$$

where each b_i represents the 2^{-i}-contribution to x. For example,

$$\frac{1}{4} = 0 \cdot 2^{-1} + 1 \cdot 2^{-2} + 0 \cdot 2^{-3} + 0 \cdot 2^{-4} + \cdots = .01\overline{0}$$

and

$$\frac{1}{5} = 0 \cdot 2^{-1} + 0 \cdot 2^{-2} + 1 \cdot 2^{-3} + 1 \cdot 2^{-4} + \cdots = \overline{.0011}$$

where the overbar means infinite repetition.

To compute the binary expansion of a number between 0 and 1, multiply the number by 2 (using the decimal system if that's easiest for you) and take the integer part (if any) as the first binary digit (bit). Repeat the process, multiply the remainder by 2, and take the integer part as the second bit, and so on. Actually, we are applying the $2x$ (mod 1) map, recording a 1 bit when the mod truncation is necessary, and a 0 bit if not.

expansion, like $x = 1/5$. The eventually periodic points are those with an eventually repeating binary expansion, like $x = 1/4 = .01\bar{0}$ or $x = 1/10 = .0\overline{0011}$. The only way to have an asymptotically repeating expansion is for it to be eventually repeating. We conclude that any number in $[0, 1]$ whose binary expansion is not eventually repeating represents a chaotic orbit. These are exactly the initial points that are not rational.

EXAMPLE 3.7

Let $f(x) = (x + q)$ (mod 1), where q is an irrational number. Although f is not continuous as a map of the unit interval, when viewed as a map of the circle (by gluing together the unit interval at the ends 0 and 1), it rotates each point through a fixed angle and so is continuous.

There are no periodic orbits, and therefore no asymptotically periodic orbits. Each orbit wanders densely throughout the circle, and yet no orbit is chaotic. The Lyapunov exponent of any orbit is 0. A bounded orbit that is not asymptotically periodic and that does not exhibit sensitive dependence on initial conditions is called **quasiperiodic**.

✎ EXERCISE T3.3

Let $f(x) = (x + q)$ (mod 1), where q is irrational. Verify that f has no periodic orbits and that the Lyapunov exponent of each orbit is 0.

In the remainder of this section we establish the fact that the tent map has infinitely many chaotic orbits. Since the Lyapunov exponent of each orbit for which it is defined is $\ln 2$, proving chaos reduces to checking for the absence of asymptotic periodicity, which we do through itineraries.

EXAMPLE 3.8

The tent map

$$T(x) = \begin{cases} 2x & \text{if } x \leq 1/2 \\ 2(1 - x) & \text{if } 1/2 \leq x \end{cases}$$

on the unit interval $[0, 1]$ also exhibits positive Lyapunov exponents. The tent map is sketched in Figure 3.1(b). Notice the similarity of its shape to that of the logistic map. It is the logistic map "with a corner".

In analogy with the treatment of logistic map, we set $L = [0, 1/2]$ and $R = [1/2, 1]$. The transition graph of T is shown in Figure 3.2(a), and the itineraries

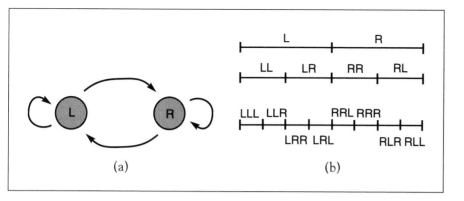

(a)

(b)

Figure 3.2   Tent map symbolic dynamics.
(a) Transition graph and (b) schematic iteneraries for the tent map T.

of T are shown in Figure 3.2(b). Recall that the subinterval **RRL**, for example, denotes the set of points x that satisfy x is in **R**, $T(x)$ is in **R**, and $T^2(x)$ is in **L**.

Because of the uniform shape of $T(x)$, there is a uniformity in the lengths of subintervals having a given finite itinerary. The set of points with itinerary $S_1 \ldots S_k$ has length 2^{-k}, independent of the choice of symbols. Figures 3.2(a) and (b) are identical to the corresponding figures in Chapter 1 for the logistic map $G(x) = 4x(1 - x)$, except that for the tent map the level-k itinerary subintervals are all of equal length.

✎ **EXERCISE T3.4**

Explain why each infinite itinerary of the tent map T represents the orbit of exactly one point in $[0, 1]$.

We conclude that if the orbits that contain $x = 1/2$ are ignored, then the orbits of T are in one-to-one correspondence with infinite sequences of two symbols.

✎ **EXERCISE T3.5**

Explain why an eventually periodic orbit must have an eventually repeating itinerary.

Using the correspondence between orbits of T and their itineraries, chaos can be shown to exist in the tent map.

Theorem 3.9 *The tent map T has infinitely many chaotic orbits.*

Proof: Since the absolute value of the slope of T is 2 whenever it exists (for $x \neq 1/2$), the Lyapunov exponent of an orbit of T is $\ln 2$ whenever it is defined. Any orbit that avoids $1/2$ and is not asymptotically periodic is therefore a chaotic orbit.

Any asymptotically periodic orbit of the tent map must be eventually periodic. The reason is that the derivative of T^k at a period-k orbit is 2^k, so all periodic orbits are sources and attract no orbits. According to Exercise T3.5, an eventually periodic orbit must have an eventually repeating itinerary. There are infinitely many nonrepeating itineraries that correspond to distinct chaotic orbits. □

3.3 CONJUGACY AND THE LOGISTIC MAP

In the previous section we established the fact that the tent map T has chaotic orbits. In this section we see that the logistic map $G(x) = 4x(1 - x)$ also has chaotic orbits, and in particular has a chaotic orbit that fills up (is dense in) the unit interval.

Calculations for the Lyapunov exponent of the tent map were extremely easy: since the slope is exactly 2 at every point (except the point of nondifferentiability), the exponential separation factor is $\ln 2$ at every iteration. The logistic map is more challenging; clearly the slope varies from iteration to iteration along typical orbits. The logistic map is a smooth map (no points of nondifferentiability) with the same general shape as the tent map. Our strategy will be to show that the similarity extends far enough that the logistic map, as well as the tent map, has chaotic orbits. The concept of conjugacy is a way of making the similarity explicit.

Figure 3.3 compares the two maps. They each have a critical point at $x = 1/2$, which maps to 1 and then 0. Each has a fixed point at zero, and one other fixed point: the tent map at $x = 2/3$ and the logistic map at $x = 3/4$. Each has a single period-two orbit; the tent map has $\{0.4, 0.8\}$, and the logistic map has $\{(5 - \sqrt{5})/8, (5 + \sqrt{5})/8\}$. In each case, the period-two orbit lies in the same relation to the other points; the left-hand point lies between the origin and the critical point, and the right-hand point lies between the other fixed point and 1.

More coincidences arise when we examine the stability of these orbits. The derivative of T at its fixed point $x = 2/3$ and the derivative of G at its fixed point $x = 3/4$ are both -2, so both points are sources. The derivative of T^2 at its period-two point $x = 0.4$ is $T'(0.4)T'(0.8) = (2)(-2) = -4$; likewise, the

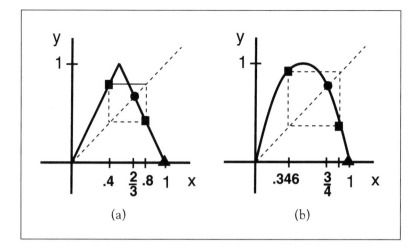

(a) (b)

Figure 3.3 Similarities in tent map and logistic map.
Both (a) the tent map and (b) the logistic map have a fixed point (small circle) to
the right of the critical point $x = 0.5$. For each, the midpoint maps to 1 and then
to 0. Each has a single period-two orbit (small squares).

derivative of G^2 at its period-two point is

$$G'\left(\frac{5+\sqrt{5}}{8}\right)G'\left(\frac{5-\sqrt{5}}{8}\right) = \left[4 - 8\left(\frac{5+\sqrt{5}}{8}\right)\right]\left[4 - 8\left(\frac{5-\sqrt{5}}{8}\right)\right]$$

$$= (-1 - \sqrt{5})(-1 + \sqrt{5})$$

$$= 1 - 5 = -4.$$

Is the same sort of thing true for period-three orbits? (There should be two
such orbits, according to the periodic table, Table 1.3 of Chapter 1.) For period-
100 orbits? How far can we expect to push the similarity? At its extreme, it could
be expressed as follows: For each point x in the tent map domain $[0, 1]$, there is
a specified companion point $C(x)$ in the logistic map domain $[0, 1]$ that imitates
its dynamics exactly. (Think of C as the "companion map".) By that we mean
that the two image points of x and $C(x)$ by their respective maps, the two points
$T(x)$ and $G(C(x))$, *are also* companions. That would mean $CT(x) = GC(x)$. If
there exists such a function C, then dynamical phenomena seen for the tent map
will be mimicked by its companions, or "conjugates", in the logistic map. If C is
one-to-one, then the reverse is true as well.

Definition 3.10 The maps f and g are **conjugate** if they are related by
a continuous one-to-one change of coordinates, that is, if $C \circ f = g \circ C$ for a
continuous one-to-one map C.

The conjugacy map C should be viewed as a correspondence between two systems for assigning coordinates, similar to a translation from one language to another. For example, we will show that the logistic map G and the tent map T are conjugate by the one-to-one continuous map $C(x) = (1 - \cos \pi x)/2$, which is shown in Figure 3.4. Notice that C is one-to-one from $[0, 1]$ to $[0, 1]$, which are the domains of G and T.

To verify that G and T are conjugate by the conjugacy C, we need to check that $C(T(x)) = G(C(x))$ for each x in $[0, 1]$. We will show how to do this for $0 \le x \le 1/2$, and leave the other half to an exercise. The right hand side is

$$G(C(x)) = 4C(x)(1 - C(x))$$
$$= 4 \left(\frac{1 - \cos \pi x}{2} \right) \left(\frac{1 + \cos \pi x}{2} \right)$$
$$= 1 - \cos^2 \pi x = \sin^2 \pi x. \tag{3.1}$$

For the left side, we make use of the fact that $T(x) = 2x$ for $0 \le x \le 1/2$, so that

$$C(T(x)) = \frac{1 - \cos \pi T(x)}{2}$$
$$= \frac{1 - \cos 2\pi x}{2} = \sin^2 \pi x, \tag{3.2}$$

where the last equality follows from the double angle formula for cos. Hence $G \circ C(x) = C \circ T(x)$ for $0 \le x \le 1/2$.

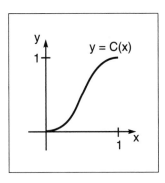

Figure 3.4   The conjugacy map.
The graph of $C(x) = (1 - \cos \pi x)/2$ is continuous and one-to-one. Since $CT(x) = GC(x)$ for all $0 \le x \le 1$, it is a conjugacy between the tent map and logistic map.

✎ **EXERCISE T3.6**

Check that $C(T(x)) = G(C(x))$ for all x in $[1/2, 1]$, completing the verification of the conjugacy.

The meaning of the conjugacy between G and T is that there are two ways to look at the map G—two ways to get from the domain of G to the range of G. One way is to evaluate $G(x)$; the second way is to find the inverse image of x under the one-to-one correspondence (which is $C^{-1}(x)$), apply the map T, and then return to the original system by applying C. This amounts, of course, to saying $G(x) = CTC^{-1}(x)$, which you proved in Exercise T3.6. In the second method, the work of evaluating the map is done by T, and C is used to translate x in and out of the language used by T.

The purpose of such a roundabout way of evaluating G is that T might be easier to handle than G. Notice also that we can do multiple iterations of G this way, with only one translation to and from T-language:

$$G^n = CTC^{-1}CTC^{-1} \cdots CTC^{-1} = CT^nC^{-1}.$$

So if we have a lot of information about high iterates of T (which we do), we may be able to make conclusions about high iterates of G.

Figure 3.5(a) illustrates the directions of the maps. The conjugacy says that if you begin with a number x in the upper left corner of the diagram, either choice of direction to the lower right corner arrives at the same result. Either go across first and then down (which represents $C(T(x))$), or go down first and then across (which represents $G(C(x))$). Ending at the same number either way means that $C(T(x)) = G(C(x))$. Figure 3.5(b) shows the correspondence between coordinate systems in a more concrete way, for a typical x.

The fact that C is one-to-one means that there is a one-to-one correspondence between a point x being mapped by T and the point $C(x)$ being mapped by G. Moreover, much of the behavior of one map corresponds to similar behavior in the other. Suppose for instance that x is a fixed point for T, so that $T(x) = x$. Then $C(x)$ is a fixed point for G, since $GC(x) = CT(x) = C(x)$.

✎ **EXERCISE T3.7**

Show that if x is a period-k point for T, then $C(x)$ is a period-k point for G.

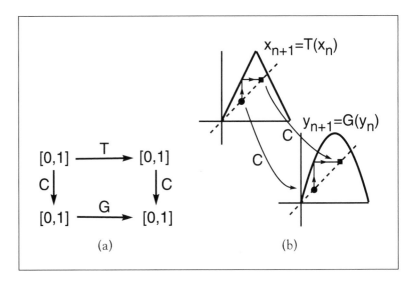

Figure 3.5 Two illustrations of the fact that the maps *T* and *G* are conjugate.
(a) Both ways of composing maps from upper left to lower right are equal. (b) If
$y_n = C(x_n)$ implies $y_{n+1} = C(x_{n+1})$, then C is a conjugacy between the x-map and
the y-map.

The similar behavior of conjugate maps extends to stretching and shrinking
information. The chain rule says that

$$C'(T(x))T'(x) = G'(C(x))C'(x). \tag{3.3}$$

Now suppose that x is a fixed point for T and $C'(x) \neq 0$. Then $T(x) = x$ implies
that $T'(x) = G'(C(x))$, meaning that the corresponding fixed points x for the
map T and $C(x)$ for the map G have the same derivative, and thus the same
stability. To get the same conclusion for period-two orbits, apply the conjugacy
to $T^2(x)$ to get $CT(T(x)) = GC(T(x)) = GGC(x)$. Then apply the chain rule
to the result $CT^2(x) = G^2C(x)$ to get

$$C'(T^2(x))(T^2)'(x) = (G^2)'(C(x))C'(x).$$

If x is a period-two point for T, and assuming $C'(x) \neq 0$, we have

$$(T^2)'(x) = (G^2)'(C(x)),$$

meaning that the stability characteristics of the period-two orbit of T and the
corresponding period-two orbit of G are identical. This fact holds in general for
periodic orbits.

Theorem 3.11 Let f and g be conjugate maps, that is $gC(x) = Cf(x)$ for all x. If x is a period-k point for f, then $C(x)$ is a period-k point for g. If also C' is never zero on the periodic orbit of f, then

$$(g^k)'(C(x)) = (f^k)'(x).$$

✎ **EXERCISE T3.8**

Use the chain rule (1.3) of Chapter 1 to provide a proof of Theorem 3.11.

The content of Theorem 3.11 is that the derivative at the fixed point $x = 2/3$ of T and $C(x) = 3/4$ of G are identical, as we stated above. These points are both sources, since $G'(3/4) = T'(2/3) = -2$. Note that Theorem 3.11 requires the assumption that the derivative C' is not zero on the orbit in question. For example, the theorem does not apply to the other fixed point $x = 0$ of T and its companion $C(0) = 0$ of G. The reason is that $C'(0) = 0$. The derivatives $T'(0) = 2$ and $G'(0) = 4$ disagree. A further application of Theorem 3.11 is the following fact:

All periodic points of the logistic map G are sources.

For any period-k point other than zero, the theorem applies and states that the magnitude of the derivative of G^k is 2^k, the same as the corresponding period-k point of T. Therefore all period-k points are sources with sensitivity to initial conditions.

Conjugacies contain a great deal of information. If $\{x_1, \ldots, x_k\}$ is a periodic orbit of the logistic map G, its stability is determined by the derivative of G^k evaluated at x_1, which is the product $G'(x_1) \cdots G'(x_k)$, according to the chain rule. The derivatives of G on $[0, 1]$ range between 0 and 4 (in magnitude)—there is no a priori reason to expect the product of k of these numbers to have a simple expression, or to have magnitude greater than one. In fact, the product amounts to precisely 2^k for a period-k orbit.

Finding a conjugacy between a map under study and an easier map is a useful trick. Unfortunately, it is not always possible to find such a conjugacy. For the family $g_a(x) = ax(1 - x)$, no useful conjugacy exists for most values of the parameter a.

Remark 3.12 There is one more fact (already used in Chapter 1) that is a consequence of conjugacy. Because of the one-to-one correspondence C from $[0, 1]$ to itself, there is a direct correspondence between the itinerary subintervals of Figure 3.2 (for the tent map) and Figure 1.12 of Chapter 1 (for the logistic map). The logistic map subintervals are exactly the images of the tent map subintervals under the transformation C. If the tent map subintervals are placed along the x-axis of Figure 3.4, vertical lines are drawn to the curve C and then extended horizontally to the y-axis, the logistic map subintervals will be produced.

The length of each tent map subinterval of level k (represented by a sequence $S_1 \ldots S_k$ of R's and L's) is 2^{-k}. If one of these subintervals is denoted by $[x_1, x_2]$, then the length of $[C(x_1), C(x_2)]$ is

$$C(x_2) - C(x_1) = \int_{x_1}^{x_2} C'(x)\,dx = \int_{x_1}^{x_2} \frac{\pi}{2}\sin \pi x\,dx$$

$$\leq \frac{\pi}{2}\int_{x_1}^{x_2} dx = \frac{\pi}{2}(x_2 - x_1) = \frac{\pi}{2^{k+1}}. \tag{3.4}$$

This establishes the upper bound $\pi/2^{k+1}$ for the level-k subintervals of the logistic map G.

To finish this section, we'll use the tools we developed in this chapter to compute the Lyapunov exponent of orbits of the logistic map G. In particular, we'll further exploit the correspondence between orbits of the logistic map and the tent map T.

Consider an orbit $\{x_i\}$ of T that does not contain the point 0. The conjugacy provides a corresponding orbit $C(x_i)$ of the logistic map, as pictured in Figure 3.5. We'll use the chain rule (3.3) to get information about the derivatives $G'(C(x_i))$, and then find the Lyapunov exponent. Equation (3.3) implies that

$$T'(x_k) \cdots T'(x_2)T'(x_1)$$
$$= \frac{G'(C(x_k))C'(x_k)}{C'(x_{k+1})} \cdots \frac{G'(C(x_2))C'(x_2)}{C'(x_3)}\frac{G'(C(x_1))C'(x_1)}{C'(x_2)}$$
$$= G'(C(x_k)) \cdots G'(C(x_1))\frac{C'(x_1)}{C'(x_{k+1})}. \tag{3.5}$$

Then

$$\ln|T'(x_k) \cdots T'(x_1)| = \sum_{i=1}^{k} \ln|T'(x_i)|$$

$$= \ln|C'(x_1)| - \ln|C'(x_{k+1})| + \sum_{i=1}^{k} \ln|G'(C(x_i))|.$$

Now we need to divide by k and take the limit. Note that $\ln C'(x_1)/k \to 0$ as $k \to \infty$ since the numerator is constant. Suppose that for the orbit we're considering, the condition

$$\frac{\ln |C'(x_{k+1})|}{k} \to 0 \text{ as } k \to \infty \qquad (3.6)$$

is satisfied. Under this condition,

$$\lim_{k \to \infty} \frac{1}{k} \sum_{i=1}^{k} \ln |T'(x_i)| = \lim_{k \to \infty} \frac{1}{k} \sum_{i=1}^{k} \ln |G'(C(x_i))|,$$

so that the Lyapunov exponents of the corresponding orbits of T and G are identical.

Condition (3.6) holds, in particular, if the orbit of x never has the sequence LL in its symbol sequence. For then the orbit never enters the intervals $[0, 1/4]$ or $[7/8, 1]$. Since $C'(x) = (\pi/2) \sin \pi x$, we have

$$\frac{\pi}{2} \sin \frac{7\pi}{8} \le |C'(x)| \le \frac{\pi}{2}$$

for any x in the orbit of x_1 under T. The natural log of the left-hand side is therefore a lower bound (away from $-\infty$) for $\ln |C'(x_{k+1})|$, and condition (3.6) is satisfied.

More generally, if the orbit of T never has a sequence of m consecutive L's in its symbol sequence, then the orbit never enters the intervals $[0, 2^{-m}]$ or $[1 - 2^{-m-1}, 1]$, and the Lyapunov exponent of the corresponding orbit of G will be again $\ln 2$.

In order to prove that many orbits of G are not periodic or eventually periodic, we use the fact that the set of eventually periodic orbits of G is a "countable" set, while the set of all orbits of G is an uncountable set. Readers unfamiliar with these concepts may want to consult the first section of Chapter 4, where the concept of countability is developed.

Theorem 3.13 *The logistic map G has chaotic orbits.*

Proof: Through the conjugacy with T, we can determine the periodic points of G; this set is countable. We showed above that all periodic points of G are sources, and therefore no orbits besides periodic orbits—and eventually periodic orbits, another countable set—can be asymptotically periodic. Then any orbit whose corresponding symbol sequence is not eventually periodic, and which never contains the sequence LL, has Lyapunov exponent $\ln 2$ and is chaotic. □

How does what we have proved correspond with what we observe on a computer screen? Suppose we begin with a typical number between 0 and 1, and

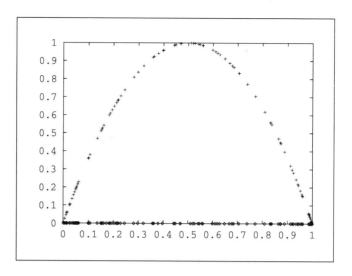

Figure 3.6 The logistic map.
One hundred iterates are plotted on the x-axis; their images are plotted on the vertical axis.

compute the resulting orbit of G. If you try this, you will see the iterates fill up the unit interval. Figure 3.6 shows the first 100 iterates of a typical orbit.

We have developed enough mathematics by now to actually prove that there is a chaotic orbit that travels throughout the entire unit interval. The chaotic orbits of the proof, for example, never enter the intervals $[0, 1/4]$ or $[7/8, 1]$. With just a little extra work we can exhibit an orbit that visits every neighborhood, no matter how small, of the unit interval $[0, 1]$.

There is a term for a subset A, such as an orbit, that visits every vicinity of a set B. We make the following definition.

Definition 3.14 Let A be a subset of B. The set A is **dense** in B if arbitrarily close to each point in B there is a point of A. In other words, for each point x in B and each $\epsilon > 0$, the neighborhood $N_\epsilon(x)$ contains a point in A.

The set of rational numbers is dense in the set of real numbers. To see this, let x be any real number. We assume that x is a number between 0 and 1, and leave the general case to the reader. For a given $\epsilon > 0$, choose n sufficiently large that $10^{-n} < \epsilon$. Let $a_1, a_2, \ldots, a_n$ be the first n digits of x. Then $|x - .a_1 a_2 \ldots a_n| < \epsilon$. The rational number $.a_1 a_2 \ldots a_n$ is in $N_\epsilon(x)$.

Orbits are countable by their nature, since the points are in one-to-one correspondence with the natural numbers. Suppose we plot an orbit on a computer screen, and observe it filling up the interval $[0, 1]$. Of course, speaking precisely, it is impossible for one orbit to be the entire interval—the orbit is countable and the interval is uncountable. But an orbit that leaves no gap unfilled in the interval, that eventually comes arbitrarily close to every point of the interval, is an orbit that comes as close as is possible to filling up the interval. Such an orbit is dense in the interval. Can we find a dense chaotic orbit for the logistic map G in $[0, 1]$?

Consider the candidate orbit whose itinerary begins with R and L, followed by all possible pairs of R, L, followed by all triples of R, L, and so on:

$$.R\ L\ RR\ RL\ LR\ LL\ RRR\ RRL\ \ldots$$

This orbit is not eventually periodic. If it were, its itinerary would be eventually periodic. Furthermore, given any interval of length 2^{-n} represented by a symbol sequence of n symbols, the orbit, after a sufficient wait, enters that interval. This is a dense orbit.

Now check the Lyapunov exponent of this dense orbit. If it is positive, the orbit is a chaotic orbit that is dense in the unit interval. A little checking verifies that no sequence of m consecutive L's occurs before the 2^mth symbol of the sequence. Therefore x_k does not visit $[0, 2^{-m}]$ or $[1 - 2^{-m-1}, 1]$ until $k > 2^m$. If m is a positive integer, then for $k < 2^m$,

$$\frac{\pi}{2} \sin \frac{\pi}{2^{m+1}} \leq |C'(x_k)| \leq \frac{\pi}{2}.$$

Taking logs and dividing by k preserves the directions of the inequality. For $k < 2^m$ we have

$$\frac{\ln \frac{\pi}{2} + \ln \sin \frac{\pi}{2^{m+1}}}{2^m} \leq \frac{\ln |C'(x_k)|}{k} \leq \frac{\ln \frac{\pi}{2}}{k}.$$

As $k \to \infty$, both the far left and far right quantities approach zero (use L'Hospital's rule for the left side). Thus condition (3.6) is satisfied, and the Lyapunov exponent of the orbit is $\ln 2$.

We have shown that this orbit is not eventually periodic. It is not asymptotically periodic because G has no periodic sinks. Since the Lyapunov exponent of the orbit is $\ln 2 > 0$, it is a dense chaotic orbit.

EXAMPLE 3.15

While on the subject of conjugacy, there is another example that we have been using implicitly all along. We will demonstrate using the tent map T. If $S_0 S_1 S_2 \cdots$ is the itinerary of an initial condition x_1, then $S_1 S_2 S_3 \cdots$ is the itinerary of $T(x_1)$. Because of this, we can define a map on the set of symbol sequences

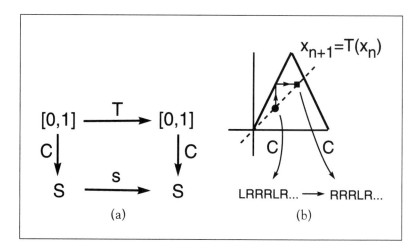

Figure 3.7 Tent map conjugacy.
The map T is conjugate to the shift map on the two symbols L and R. (a) The conjugacy betwen T and the shift s. (b) Schematic view of the action of the conjugacy map C.

whose dynamical properties mirror those of T. The translation between them can be thought of as a conjugacy.

Definition 3.16 The set **S** of all infinite itineraries of a map is called the **symbol space** for the map. The **shift map** s is defined on the symbol space **S** as follows:

$$s(S_0S_1S_2\ldots) = S_1S_2S_3\ldots.$$

The shift map chops the leftmost symbol, which is the analogue on the itinerary of iterating the map on the point.

Figure 3.7 is an analogue of Figure 3.5, but for the conjugacy of the tent map with the shift map. The conjugacy C moves a point in $[0, 1]$ to its itinerary. An orbit of T in $[0, 1]$ has a corresponding "orbit" in symbol space where the dynamics consist of chopping off one symbol from the left end of the infinite sequence each iterate.

3.4 TRANSITION GRAPHS AND FIXED POINTS

Imagine that a 12-inch plastic ruler is melted and stretched lengthwise, beyond its true length, and then laid down to completely cover a true 12-inch ruler. We'll allow the stretched ruler to be laid in the same orientation as the true ruler, or in the reversed orientation. In either case there must be a real number between 0 and 12 for which the rulers line up exactly. This fact is expressed in Theorem 3.17.

Theorem 3.17 *[Fixed-Point Theorem] Let f be a continuous map of the real line, and let $I = [a, b]$ be an interval such that $f(I) \supseteq I$. Then f has a fixed point in I.*

Proof: Since $f(I)$ contains numbers as large as b and as small as a, there is a point in I for which the function $f(x) - x \geq 0$, and a point in I for which $f(x) - x \leq 0$. By the Intermediate Value Theorem, there is a point c in I such that $f(c) - c = 0$. $\qquad\qquad \Box$

This Fixed-Point Theorem says that if the image of an interval I covers the interval I itself, then I contains a fixed point. Since periodic points of a map are fixed points of higher iterates of the map, the same theorem can be exploited to prove the existence of periodic points. Assume that $I_1, \ldots, I_n$ are closed intervals, and that $f(I_1) \supseteq I_2, f(I_2) \supseteq I_3, \ldots, f(I_{n-1}) \supseteq I_n$, and that $f(I_n) \supseteq I_1$. In that case we can conclude that $f^n(I_1) \supseteq I_1$, and that f^n has a fixed point in I_1. It corresponds to a periodic orbit of f (of period n or possibly less), that moves through the intervals I_i in succession before returning to I_1.

Our goal in this section is to use this theorem in conjunction with the itineraries developed in the previous section to establish the existence of large quantities of periodic orbits. Recall the definition of transition graphs in Chapter 1. We begin with a **partition** of the interval I of interest, which is a collection of

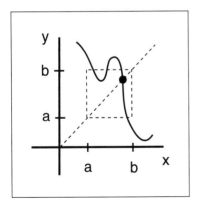

Figure 3.8 The content of Theorem 3.17:
If a function f maps the interval $[a, b]$ across itself, as in the picture, then f must have a fixed point in that interval.

subintervals that are pairwise disjoint except at the endpoints, whose union is I. The arrows in a transition graph are drawn subject to the following rule.

COVERING RULE FOR TRANSITION GRAPHS

An arrow is drawn from **A** to **B** in a transition graph if and only if the image $f(\mathbf{A})$ contains the subinterval **B**.

Figure 3.2(a) shows a transition graph for the tent map. The set $\{\mathbf{L}, \mathbf{R}\}$ is an example of a partition of the unit interval.

The covering rule has a very interesting consequence for paths in the graph that are allowed by the arrows. First note that set containment is preserved by a map. That is, if $\mathbf{A} \subset \mathbf{B}$, then $f(\mathbf{A}) \subset f(\mathbf{B})$. The covering rule means that if $\mathbf{A} \to \mathbf{B} \to \mathbf{C}$ is an allowed path in the transition graph, then $f(\mathbf{A}) \supset \mathbf{B}$ and $f(\mathbf{B}) \supset \mathbf{C}$. Since $f^2(\mathbf{A}) \supset f(\mathbf{B})$, it follows that $f^2(\mathbf{A})$ covers $\mathbf{C}$. More generally, if the sequence $\mathbf{S_1} \ldots \mathbf{S_{k+1}}$ is an allowable path, then $f^k(\mathbf{S_1}) \supset \mathbf{S_{k+1}}$.

Now we can see how Theorem 3.17, the Fixed-Point Theorem, can be used to establish the existence of periodic orbits. Assume that $\{\mathbf{S_1}, \ldots, \mathbf{S_n}\}$ is a partition. If the transition graph of f allows a sequence of symbols that returns to the same symbol, such as $\mathbf{S_1} \ldots \mathbf{S_k S_1}$, then $f^k(\mathbf{S_1}) \supseteq \mathbf{S_1}$, so that f^k has a fixed point lying in $\mathbf{S_1}$. Of course, the periodic orbit we have found may not be a period-k orbit of f—its period could be an integer that divides evenly into k. We can sum this up as a consequence of the Fixed-Point Theorem.

Corollary 3.18 Assume that $\mathbf{S_1} \ldots \mathbf{S_k S_1}$ is a path in the transition graph of a map f. Then the subinterval denoted by $\mathbf{S_1} \cdots \mathbf{S_k S_1}$ contains a fixed point of f^k.

For the logistic map $G(x) = 4x(1 - x)$, Corollary 3.18 can be used as another way to prove the existence of periodic orbits of every period. For example, the transition graph of G in Chapter 1 shows that there is an orbit with sequence beginning **LRLRRLL**. We can immediately conclude that there is a fixed point of G^6 belonging to the subinterval **L**. What can we say about the corresponding periodic orbit of G? From what we know so far, it could be an orbit of period 1, 2, or 6. But we know a little more from the itinerary. The orbit cannot be a fixed point of G; according to the itinerary, it moves between **L** and **R**, and the only point in both L and R is $x = 1/2$, which does not have this itinerary. If the corresponding orbit of G were a period-two orbit, then the symbols would be forced to repeat with period two (for example, as LRLRLRL). So the orbit is not a period-two orbit of G, nor period-3, by similar reasoning. So there is a period-6 orbit for G in L. An argument of this type can be used to prove the existence of a periodic orbit for any period.

✎ **EXERCISE T3.9**

(a) Find a scheme to provide, for any positive integer n, a sequence $S_1 \ldots S_{n+1}$ of the symbols L and R such that $S_1 = S_{n+1}$, and such that the sequence $S_1 \ldots S_n$ is not the juxtaposition of identical subsequences of shorter length. (b) Prove that the logistic map G has a periodic orbit for each integer period.

✎ **EXERCISE T3.10**

Period-three implies all periods! Consider the period-three map of Chapter 1, shown in Figure 1.14. (a) Find itineraries that obey the transition graph of the period-three map for any period, which are not periodic for any lower period. (b) Prove that the period-three map has periodic orbits for each positive integer period.

Exercise T3.10 applies only to the particular map of the type drawn in Figure 1.14 of Chapter 1. This is because we verified the transition graph only for this particular case. However, the fact is that the existence of a period-three orbit for a continuous map implies the existence of points of every period. An even more general result, called Sharkovskii's Theorem, is the subject of Challenge 3 at the end of this chapter.

EXAMPLE 3.19

Consider the map f graphed in Figure 3.9. The three subintervals I, J, and K form a partition. The transition graph is shown in Figure 3.9(b). From the transition graph we can write down some valid symbol sequences. For example, $\bar{J}$, meaning an infinite sequence of the symbol J, is valid according to the figure. The symbol sequences $\bar{I}$ and $\bar{K}$ are not valid. In fact, we can conclude from the figure that $\bar{J}$, $\overline{IK}$, $\overline{KI}$, and each of the latter two preceded by a finite number of J's, are the only valid sequences.

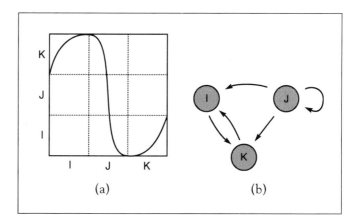

Figure 3.9 Example 3.19.
(a) The subintervals I, J, and K form a partition for the map shown. (b) Transition graph.

So far we are guaranteed a fixed point (because JJ is a path in the transition graph) and a period-two orbit (because IKI is legal, and because the corresponding fixed point of f^2 clearly cannot be a fixed point of f).

✎ **EXERCISE T3.11**

(a) Prove that f in Figure 3.9(a) has a fixed point and no other periodic points whose period is an odd number. (b) Prove that there are only two periodic orbits whose periods are even, both of period two. Where are they?

EXAMPLE 3.20

Consider the map f graphed in Figure 3.10(a). There are four subintervals I, J, K, and L that form a partition. The transition graph is shown in Figure 3.10(b). This time, notice that the sequence JKLJ is possible so the map f has a period-three orbit. Note also that JKL . . . LJ is possible. This implies that f has periodic orbits of all periods.

✎ **EXERCISE T3.12**

List all periodic sequences for periodic orbits of f in Figure 3.10 of period less than or equal to 5. Note: $\overline{JKL}$ and $\overline{KLJ}$ are not distinct since they represent the same orbit.

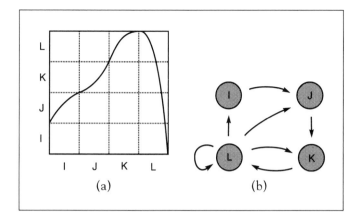

Figure 3.10   A map with a four-piece partition.
(a) The subintervals I, J, K, L form a covering partition for the map shown. (b) Transition graph for I, J, K, L.

3.5 BASINS OF ATTRACTION

The concept of "stable manifold" was introduced in Chapter 2 to refer to the set of points whose forward orbits converge to a fixed or periodic point. In this section we investigate more closely the set of points whose orbits converge to an attracting fixed point or periodic point, called the "basin of attraction" or just "basin" of the sink, and prove a useful theorem about attractors for one-dimensional maps.

Definition 3.21   Let **f** be a map on $\mathbb{R}^n$ and let **p** be an attracting fixed point or periodic point for **f**. The **basin of attraction** of **p**, or just **basin** of **p**, is the set of points **x** such that $| \mathbf{f}^k(\mathbf{x}) - \mathbf{f}^k(\mathbf{p}) | \to 0$, as $k \to \infty$.

EXAMPLE 3.22

For the map $f(x) = ax$ on $\mathbb{R}^1$ with $|a| < 1$, zero is a fixed point sink whose basin is the entire real line. More generally, if **f** is a linear map on $\mathbb{R}^n$ whose matrix representation has distinct eigenvalues that are less than one in magnitude, then the origin is a fixed sink whose basin is $\mathbb{R}^n$.

Theorem 3.23 is useful for finding basins for sinks of some simple maps on $\mathbb{R}^1$.

Theorem 3.23   *Let f be a continuous map on $\mathbb{R}^1$.*
(1) If $f(b) = b$ and $x < f(x) < b$ for all x in $[a, b)$, then $f^k(a) \to b$.
(2) If $f(b) = b$ and $b < f(x) < x$ for all x in $(b, c]$, then $f^k(c) \to b$.

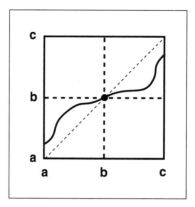

Figure 3.11   Illustration of Theorem 3.23.
The sink shown at $x = b$ attracts all initial values in the interval $[a, c]$.

The content of this theorem is expressed in Figure 3.11, which shows the graph of a map f on $\mathbb{R}^1$. Part (1) of the theorem says that the basin of b contains all initial values to the left of b, and part (2) says it contains all initial values to the right of b.

Proof: We establish (1), and leave (2) to the reader. Let $x_0 = a$, $x_{i+1} = f(x_i)$ for $i \geq 0$. If $x \in [a, b)$, then $f(x) \in [a, b)$. In fact, $a \leq x < f(x) < b$. Thus all $x_i \in [a, b)$. Further, the x_i are strictly increasing and bounded above by b. Since increasing bounded sequences must converge, $x_i \to x_*$ for some $x_* \in [a, b]$. Taking limits, we have

$$x_* = \lim_{i \to \infty} x_{i+1} = \lim_{i \to \infty} f(x_i) = f(x_*),$$

by the continuity of f. Since b is the only fixed point in $[a, b]$, $x_* = b$.

EXAMPLE 3.24

Consider the map $f(x) = (4/\pi) \arctan x$ on $\mathbb{R}^1$. See Figure 3.12. This map has three fixed points: $-1, 0, 1$. Using Theorem 1.9 of Chapter 1, it is easy to check that -1 and 1 are sinks, and that 0 is a source. It follows from Theorem 3.23 that the basin of the fixed point 1 is the set of all positive numbers. The basin of 1 is colored gray in Figure 3.12. Likewise, the basin of -1 is the set of all negative numbers, and is shown in black.

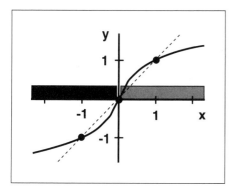

Figure 3.12 The map $y = f(x) = (4/\pi)\arctan x$.
The basin of the sink -1 is shown in black, and the basin of 1 is in gray.

EXAMPLE 3.25

Consider the logistic map $f(x) = ax(1 - x)$, shown in Figure 1.5 of Chapter 1. If $0 < a < 1$, there is a single attracting fixed point $x = 0$. Theorem 3.23 says that the interval $((a - 1)/a, 1]$ lies in the basin of $x = 0$. From graphical representation of orbits, it is clear that in addition, the interval $[1, 1/a)$ is contained in the basin of 0. It is also clear that the intervals $(-\infty, (a - 1)/a)$ and $(1/a, \infty)$ consist of initial conditions that diverge to infinity. We could say that these points belong to the "basin of infinity".

If $1 < a < 2$, the sink has moved to the right from 0 to $(a - 1)/a$. Theorem 3.23 and some graphical analysis implies that the basin of the sink $(a - 1)/a$ is $(0, 1)$.

EXAMPLE 3.26

Consider the map of the plane defined by

$$f(r, \theta) = (r^2, \theta - \sin \theta),$$

where $r \geq 0$ and $0 \leq \theta < 2\pi$ are polar coordinates. There are three fixed points (the origin, $(r, \theta) = (1, 0)$ and $(1, \pi)$). The origin and infinity are attractors. Every initial condition inside the unit circle tends toward the origin upon iteration, and every point outside the unit circle tends toward infinity. The basins of these two attractors are shown in gray and white, respectively, in Figure 3.13.

The dynamics on the dividing circle are also rather tame; there is a fixed point at $(r, \theta) = (1, 0)$ to which all points on the circle tend, except for the fixed point $(r, \theta) = (1, \pi)$. The basin boundary itself is unstable, in the sense that points near it are repelled, except for points precisely on the boundary.

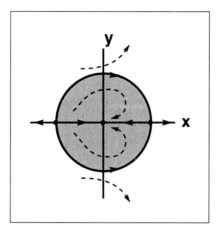

Figure 3.13 The map of Example 3.26.
The gray region is the basin of the origin. The white region is the basin of infinity.

One of the most interesting features of nonlinear maps is their ability to have more than one attractor. The arctan map of Example 3.24 has coexisting sinks. However, the diagram of attractors of the logistic family of maps in Figure 1.6 of Chapter 1 appears not to have coexisting attractors—for each value of the parameter a, only one attractor is visible. This section is devoted to proving that this observation is correct. It is useful to introduce the Schwarzian derivative for this purpose. Theorem 3.29 states that maps like the logistic map, with negative Schwarzian and finite basins, can have at most one attracting periodic orbit. The remainder of this section pertains only to one-dimensional maps.

Definition 3.27   Let f be a smooth map on $\mathbb{R}^1$. The **Schwarzian derivative** of f is defined by

$$S(f)(x) = \frac{f'''(x)}{f'(x)} - \frac{3}{2}\left(\frac{f''(x)}{f'(x)}\right)^2.$$

We will say that a map has **negative Schwarzian** if $S(f)(x)$ is negative whenever $f'(x) \neq 0$.

EXAMPLE 3.28

Check that $g(x) = ax(1 - x)$ has Schwarzian derivative

$$S(g)(x) = -\frac{3}{2}\left(\frac{-2a}{a - 2ax}\right)^2,$$

and therefore has negative Schwarzian.

✎ **EXERCISE T3.13**

Show that if f and g have negative Schwarzian, then $f \circ g$ has negative Schwarzian. Therefore if f has negative Schwarzian, so does each iterate f^k of f.

In the following theorem, we say that a fixed point or periodic point p has an "infinite basin" if the basin of p contains an interval of infinite length. Recall that a "critical point" of a map f is a point c such that $f'(c) = 0$.

Theorem 3.29 *If the map f on $\mathbb{R}^1$ has negative Schwarzian, and if p is a fixed point or a periodic point for f, then either:*

1. *p has an infinite basin; or*
2. *there is a critical point of f in the basin of p; or*
3. *p is a source.*

Proof: We will assume that p is not a source nor a sink with an infinite basin, and prove that there is a critical point c of f in the basin of p.

First consider the simpler case $f(p) = p, f'(p) \geq 0$. If p is itself a critical point of f, then we are done. Otherwise, $0 < f'(p) \leq 1$, since p is not a source. Note that $f'(x)$ cannot be constant in a neighborhood of p, since on that interval $f'' = f''' = 0$ would imply $S(f) = 0$.

It is clear from Theorem 3.23 that since p does not have an infinite basin, we can conclude that either f has a critical point in the basin of p, in which case we are done, or there exists an interval (a,b) containing p such that $f'(a) \geq 1$ and $f'(b) \geq 1$. Since $f'(p) \leq 1$, there is a local minimum m for f' in the basin of p. Note that $f''(m) = 0$ and $f'''(m) > 0$, so that negative Schwarzian implies $f'(m) < 0$. By the Intermediate Value Theorem there is a number c between p and m such that $f'(c) = 0$. Since the interval (a, b) is contained in the basin of p and $a < c < b$, we have found a critical point in the basin of p.

We can now describe the general case, in which p is a periodic point of period k. Since p is not a source orbit nor a sink orbit with infinite basin for the map f, the same is true for the fixed point p of the map f^{2k}. Since $(f^{2k})'(p) = (f^k)'(p)^2$, we know that $0 \leq (f^{2k})'(p) \leq 1$. If $(f^{2k})'(p) = 0$, then there is a critical point in the orbit of p (check this), and therefore in the basin, and we are done. We are left with the case $0 < (f^{2k})'(p) \leq 1$, and we can apply the above argument to f^{2k}, since by Exercise T3.13, it also has negative Schwarzian. We conclude that f^{2k} has a critical point in the basin of p, and so f has also.

Corollary 3.30   The logistic map $g(x) = ax(1 - x)$, where $0 \le a \le 4$, has at most one periodic sink.

Proof: The map g has negative Schwarzian, so Theorem 3.29 applies. All orbits that begin outside $[0, 1]$ tend toward $-\infty$, so no points in $[0, 1]$ have an infinite basin. Since the only critical point of g is $x = 1/2$, there can be at most one attracting periodic orbit. $\square$

We found earlier in this chapter that for $a = 4$, all periodic orbits are sources. Here is another way to see that fact, since when $a = 4$, the orbit with initial value $x = 1/2$ maps in two iterates to the fixed source 0.

Sharkovskii's Theorem

A CONTINUOUS MAP of the unit interval $[0, 1]$ may have one fixed point and no other periodic orbits (for example, $f(x) = x/2$). There may be fixed points, period-two orbits, and no other periodic orbits (for example, $f(x) = 1 - x$). (Recall that f has a point of period p if $f^p(x) = x$ and $f^k(x) \neq x$ for $1 \leq k < p$.)

As we saw in Exercise T3.10, however, the existence of a periodic orbit of period three, in addition to implying sensitive dependence on initial conditions (Challenge 1, Chapter 1), implied the existence of orbits of all periods. We found that this fact was a consequence of our symbolic description of itineraries using transition graphs.

If we follow the logic used in the period-three case a little further, we can prove a more general theorem about the existence of periodic points for a map on a one-dimensional interval. For example, although the existence of a period-5 orbit may not imply the existence of a period-3 orbit, it does imply orbits of all other periods.

Sharkovskii's Theorem gives a scheme for ordering the natural numbers in an unusual way so that for each natural number n, the existence of a period-n point implies the existence of periodic orbits of all the periods higher in the ordering than n. Here is Sharkovskii's ordering:

$$3 < 5 < 7 < 9 < \ldots < 2 \cdot 3 < 2 \cdot 5 < \ldots < 2^2 \cdot 3 < 2^2 \cdot 5 < \ldots$$

$$\ldots < 2^3 \cdot 3 < 2^3 \cdot 5 < \ldots < 2^4 \cdot 3 < 2^4 \cdot 5 < \ldots < 2^3 < 2^2 < 2 < 1.$$

Theorem 3.31 *Assume that f is a continuous map on an interval and has a period p orbit. If $p < q$, then f has a period-q orbit.*

Thus, the existence of a period-eight orbit implies the existence of at least one period-four orbit, at least one period-two orbit, and at least one fixed point. The existence of a periodic orbit whose period is not a power of two implies the existence of orbits of all periods that are powers of two. Since three is the "smallest" natural number in the Sharkovskii ordering, the existence of a period-three orbit implies the existence of all orbits of all other periods.

The simplest fact expressed by this ordering is that if f has a period-two orbit, then f has a period-one orbit. We will run through the reason for this fact, as it will be the prototype for the arguments needed to prove Sharkovskii's Theorem.

Let x_1 and $x_2 = f(x_1)$ be the two points of the period-two orbit. Since $f(x_1) = x_2$ and $f(x_2) = x_1$, the continuity of f implies that the set $f([x_1, x_2])$ contains $[x_1, x_2]$. (Sketch a rough graph of f to confirm this.) By Theorem 3.17, the map f has a fixed point in $[x_1, x_2]$.

The proof of Sharkovskii's theorem follows in outline form. We adopt the general line of reasoning of Block et al. 1979. In each part, you are expected to fill in an explanation. Your goal is to prove as many of the propositions as possible.

Assume f has a period p orbit for $p \geq 3$. This means that there is an x_1 such that $f^n(x_1) = x_1$ holds for $n = p$ but not for any other n smaller than p. Let $x_1 < \cdots < x_p$ be the periodic orbit points. Then $f(x_1)$ is one of the x_i, but we do not know which one. We only know that the map f permutes the x_i. In turn, the x_i divide the interval $[a, b] = [x_1, x_p]$ into $p - 1$ subintervals $[x_1, x_2], [x_2, x_3], \ldots, [x_{p-1}, x_p]$. Note that the image of each of these subintervals contains others of the subintervals. We can form a transition graph with these $p - 1$ subintervals, and form itineraries using $p - 1$ symbols.

Let A_1 be the rightmost subinterval whose left endpoint maps to the right of itself. Then $f(A_1)$ contains A_1 (see Figure 3.14 for an illustration of the $p = 9$ case).

Step 1   Recall that the image of an interval under a continuous map is an interval. Use the fact that $A \subseteq B \Rightarrow f(A) \subseteq f(B)$ to show that

$$A_1 \subseteq f(A_1) \subseteq f^2(A_1) \subseteq \ldots.$$

(We will say that the subintervals $A_1, f(A_1), f^2(A_1), \ldots$ form an increasing "chain" of subintervals.)

Step 2   Show that the number of orbit points x_i lying in $f^j(A_1)$ is stictly increasing with j until all p points are contained in $f^k(A_1)$ for a certain k. Explain why $f^k(A_1)$ contains $[x_1, x_p]$. Use the important facts that the endpoints of each

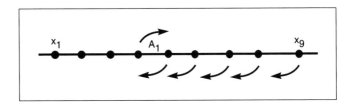

Figure 3.14   Definition of A_1.
A_1 is chosen to be the rightmost subinterval whose left-hand endpoint maps to the right under f.

subinterval are obliged to map to the subinterval endpoints from the partition, and that each endpoint must traverse the entire period-p orbit under f.

As a consequence of Step 2, the endpoints of A_1 cannot simply map among themselves under f—there must be a new orbit point included in $f(A_1)$, since $p \neq 2$. So at least one endpoint maps away from the boundary of A_1, implying that $f(A_1)$ contains not only A_1 but another subinterval, which we could call A_2.

Step 3 Prove that either (1) there is another subinterval (besides A_1) whose image contains A_1, or (2) p is even and f has a period-two orbit. [Hints: By definition of period, there are no periodic orbits of period less than p among the x_i. Therefore, if p is odd, the x_i on the "odd" side of A_1 cannot map entirely amongst themselves, and cannot simply exchange points with the "even" side of A_1 for arithmetic reasons. So for some subinterval other than A_1, one endpoint must be mapped to the odd side of A_1 and the other to the even side. If p is even, the same argument shows that either there is another subinterval whose image contains A_1, or else the x_i on the left of A_1 map entirely to the x_i on the right of A_1, and vice versa. In this case the interval consisting of all points to the left of A_1 maps to itself under f^2.]

Step 4 Prove that either (1) f has a periodic orbit of period $p - 2$ in $[x_1, x_p]$, (2) p is even and f has a period-two orbit, or (3) $k = p - 2$. Alternative (3) means that $f(A_1)$ contains A_1 and one other interval from the partition called A_2, $f^2(A_1)$ contains those two and precisely one more interval called A_3, and so on. [Hint: If $k \leq p - 3$, use Step 3 and the Fixed-Point Theorem (Theorem 3.17) to show that there is a length $p - 2$ orbit beginning in A_1.]

Now assume that p is the smallest odd period greater than one for which f has a periodic orbit. Steps 5 and 6 treat the concrete example case $p = 9$.

Step 5 Beginning with Figure 3.14, show that the endpoints of subintervals $A_1, \ldots, A_8$ map as in Figure 3.15, or as its mirror image. Conclude that $A_1 \subseteq f(A_8)$, and that the transition graph is as shown in Figure 3.16 for the itineraries of f. In particular, A_8 maps over A_i for all odd i.

Step 6 Using symbol sequences constructed from Figure 3.16, prove the existence of periodic points of the following periods:

(a) Even numbers less than 9;
(b) All numbers greater than 9;
(c) Period 1.

This proves Sharkovskii's Theorem for maps where 9 is the smallest odd period.

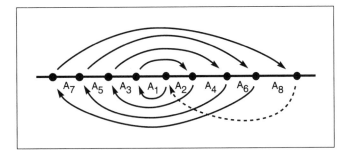

Figure 3.15 A map with a period-nine orbit and no period-three, -five, or -seven orbits.

It must map subintervals as shown, or as the mirror image of this picture.

Step 7   In Steps 5 and 6, we assumed that the smallest odd period was 9. Explain how to generalize the proof from 9 to any odd number greater than 1. Note that Step 4 is not required for the $p = 3$ case.

Step 8   Prove that if f has a periodic orbit of even period, then f has a periodic orbit of period-two. [Hint: Let p be the smallest even period of f. Either Step 3 gives a period-two orbit immediately, or Step 4 applies, in which case Steps 5 and 6 can be redone with p even to get a period-two orbit.]

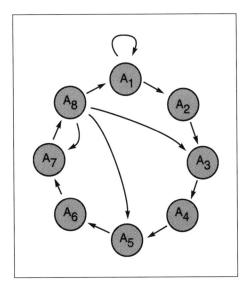

Figure 3.16 Transition graph for period-nine map.

The existence of orbits of periods 1, 2, 4, 6, 8, and all numbers greater than 9 is implied by this graph.

Step 9   Prove that if f has a periodic orbit of period 2^k, then f has periodic orbits of periods $2^{k-1}, \ldots, 4, 2, 1$. (Since $f^{2^{k-2}}$ has a period-four point, it has a period-two point, by Step 8. Ascertain the period of this orbit as an orbit of f.)

Step 10   Assume $p = 2^k q$ is the leftmost number on the list for which f has a period-p point, where q is an odd number greater than 1. The integers to the right of p in the list are of two types: either $2^k r$, where r is greater than q, or an integer power of 2. Since f^{2^k} has a period-q orbit, Step 7 implies that f^{2^k} has orbits of every period r greater than q. Our choice of p implies that these orbits are not orbits of f whose period divides evenly into p. Use these periodic orbits and Step 9 to complete the proof of Sharkovskii's Theorem.

EXERCISES

3.1.   Let $f_a(x) = a - x^2$, where a is a constant.

    (a) Find a value a_1 of the parameter a for which f_a has exactly one fixed point.

    (b) Describe the limit of all orbits of f_a for $a < a_1$.

    (c) The map f_a has an attracting fixed point for a in the open interval (a_1, a_2). Find a_2.

    (d) The map f_a has an attracting period-two point for a in the open interval (a_2, a_3). Find a_3.

    (e) Describe the dynamics of f_a for $a = 2$.

3.2.   Decide on a partition for the map $f(x) = 2x \pmod 1$ on $[0, 1]$, and draw its transition graph and schematic itineraries as in Figure 3.2(a)(b). How do they differ from those for the tent map?

3.3.   (a) Find a conjugacy C between $G(x) = 4x(1 - x)$ and $g(x) = 2 - x^2$.

    (b) Show that $g(x)$ has chaotic orbits.

3.4.   Show that $g(x) = 2.5x(1 - x)$ has no chaotic orbits.

3.5.   (a) Sketch a graph of the map $f(x) = -2x^2 + 8x - 5$.

    (b) Find a set of two subintervals that form a partition.

    (c) Draw the transition graph for f. What are the possible periods for periodic orbits?

3.6.   Repeat the previous exercise with $f(x) = 2x(x - 1)$.

3.7.   Let T be the tent map. Prove that the periodic points of T are dense in I.

3.8.   Assume that $\{x_1, x_2, \ldots, x_9\}$ is a periodic orbit for a continuous map f on the real line, with $x_1 < \cdots < x_9$. Assume $f(x_i) = x_{i+1}$ if $i < 9$ and $f(x_9) = x_1$.

    (a) What periods does Sharkovskii's Theorem guarantee on the basis of the period-nine orbit?

    (b) Draw the transition graph for this map. Which periods are certain to exist?

3.9.   Assume that $x_1 < x_2 < x_3 < x_4$ are points on the real line, and that f is a continuous map satisfying $f(x_1) = x_2, f(x_2) = x_4, f(x_3) = x_1$ and $f(x_4) = x_3$. For simplicity, assume that f is monotonic (increasing or decreasing) except possibly at the four points mentioned.

(a) Sketch a graph of f.

(b) Draw the transition graph for f.

(c) What periods must exist?

3.10. Assume that $x_1 < \cdots < x_5$ are points on the real line, and that f is a continuous map satisfying $f(x_1) = x_3, f(x_2) = x_5, f(x_3) = x_4, f(x_4) = x_2$ and $f(x_5) = x_1$. Assume that f is monotonic (increasing or decreasing) except possibly at the five points mentioned.

(a) Sketch a graph of f.

(b) Draw the transition graph for f.

(c) What periods must exist?

3.11. Let f be a continuous map from the unit interval *onto* itself (that is, such that $f([0, 1]) = [0, 1]$).

(a) Prove that f must have at least one fixed point.

(b) Prove that f^2 must have at least two fixed points. (Hint: Explain why either f or f^2 must have points $0 \leq x_1 < x_2 \leq 1$ such that x_1 maps to 0 and x_2 maps to 1.)

(c) If in addition 0 and 1 are not fixed points for f, show that f^2 must have at least 3 fixed points.

3.12. Let $f(x) = rx(1 - x), r > 2 + \sqrt{5}$. Show that the Lyapunov exponent of any orbit that remains in $[0, 1]$ is greater than zero, if it exists.

3.13. Let n be a positive integer, and $f(x) = nx \pmod 1$ on $[0, 1]$. Which points are periodic, eventually periodic, and asymptotically periodic? Which orbits of f are chaotic orbits?

3.14. (From Seth Patinkin) Let I be an interval in the real line and f a continuous function on I. The goal of this exercise is to prove the following theorem:

> Assume there is an initial value whose orbit is dense in I and assume f has two distinct fixed points. Then there must be a point of period 3.

The following two facts are to be used as help in proving the theorem.

(i) Suppose there is a point a in I such that $a < f(a)$ and there is a $b < a$ in I such that $f(b) \leq b$ and there is a $k > 1$ such that $f^k(a) \leq b$. Then there is a point of period 3. (Of course it is possible to reverse the inequalities, $f(a) < a < b \leq f(b)$, and $b \leq f^k(a)$, to again get a true statement.)

(ii) Assume there is no period 3 point and that there is some orbit dense in I. Then there is a unique fixed point c and $x - c$ and $f(x) - x$ always have opposite signs for every $x \neq c$ in I. Let $A = \{x : f(x) > x\}$ and $B = \{x : f(x) < x\}$. Use (i) to show that A and B must each be an interval.

Discussion: In Chapters 1 and 3 we discussed that implications of having a period 3 orbit. It implies the existence of periodic orbits of all other periods and it implies sensitivity to initial data. This exercise provides a partial converse. The assumption that some orbit is dense in an interval is quite reasonable. It can be shown that for any piecewise expanding map F (see Chapter 6), the orbit of almost every initial point is dense in the union of a finite number of intervals, and for some k the piecewise expanding map F^k has the property that the orbit of almost every point in dense in an interval. The proof of these results is beyond the scope of this book. We could choose such an interval to be I and $f = F^k$ to be our function. It might or might not have two fixed points.

3.15. (Party trick.) (a) A **perfect shuffle** is performed by dividing a 52-card deck in half, and interleaving the halves, so that the cards from the top half alternate with the cards from the bottom half. The top card stays on top, and so it and the bottom card are fixed by this operation. Show that 8 perfect shuffles return the deck to its original order. [Hint: Number the original card order from 0 to 51. Then a perfect shuffle can be expressed as the map

$$f(n) = \begin{cases} 2n & \text{if } 0 \le n \le 25 \\ 2n - 51 & \text{if } 26 \le n \le 51 \end{cases}$$

The goal is to show that all integers are fixed points under f^8. First show that $f^8(n) = 2^8 n - 51k$ for some integer k, where k may be different for different n.] Caution: when demonstrating at actual parties, be sure to remove the jokers first! If the deck consists of 54 cards, then 52 perfect shuffles are required.

(b) If the bottom card 51 is ignored (it is fixed by the map anyway), the above map is $f(x) = 2x \pmod{51}$, where we now consider x to be a real number. Nonperiodic orbits have Lyapunov number equal to 2, yet every integer point is periodic with period a divisor of 8. Sharkovskii's Theorem shows that it is typical for chaotic maps to contain many periodic orbits. Find all possible periods for periodic orbits for this map on the interval $[0, 51]$.

Periodicity and Chaos in a Chemical Reaction

IN ELEMENTARY chemistry classes, a great deal of emphasis is placed on finding the equilibrium state of a reaction. It turns out that equilibria present only one facet of possible behavior in a chemical reaction. Periodic oscillations and even more erratic behaviors are routinely observed in particular systems.

Studies on oscillating reactions originally focused on the Belousov-Zhabotinskii reaction, in which bromate ions are reduced by malonic acid in the presence of a catalyst. The mechanism of the reaction is complicated. More than 20 species can be identified at various stages of the reaction. Experiments on this reaction by a group of researchers at the University of Texas were conducted in a continuous-flow stirred tank reactor (CSTR), shown schematically in Figure 3.17. The solution is stirred at 1800 rpm by the impeller. There is a constant flow in and out of the tank, perfectly balanced so that the total fluid volume does not change as the reaction proceeds. The feed chemicals are fixed concentrations of malonic acid, potassium bromate, cerous sulfate, and sulfuric acid. The flow rate is maintained as a constant throughout the reaction, and the bromide concentration is measured with electrodes immersed in the reactor. The output of the experiment is monitored solely through the bromide measurement.

The constant flow rate can be treated as a system parameter, which can be changed from time to time to look for qualitative changes in system dynamics, or bifurcations. Figure 3.18 shows several different periodic behaviors of the bromide concentration, for different settings of the flow rate. The progression of flow rate values shown here is decreasing in the direction of the arrows and results in periodic behavior of periods 6, 5, 3, 5, 4, 6, and 5. Each oscillation (an up and down movement of the bromide concentration) takes around 2 minutes. Approximately one hour was allowed to pass between changes of the flow rate to allow the system to settle into its asymptotic behavior.

Roux, J.-C., Simoyi, R.H., Swinney, H.L., "Observation of a strange attractor". Physica D **8**, 257-266 (1983).

Coffman, K.G., McCormick, W.D., Noszticzius, Z., Simoyi, R.H., Swinney, H.L., "Universality, multiplicity, and the effect of iron impurities in the Belousov-Zhabotinskii reaction." J. Chemical Physics **86**, 119-129 (1987).

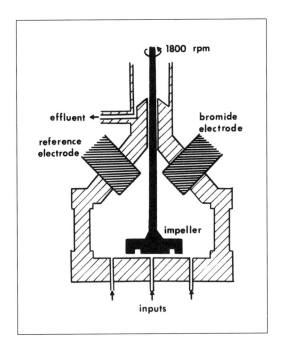

Figure 3.17 A schematic diagram of the chemical reactor used for the BZ reaction.

The volume of the cylindrically symmetric chamber is 33 cc.

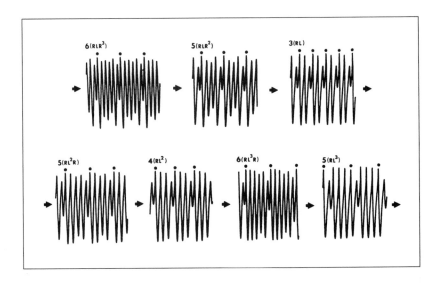

Figure 3.18 Periodic oscillations of the bromide concentration.

The horizontal axis is time, and the concentration is graphed vertically. Dots are shown to mark the period of the oscillation.

The time series of concentration in Figure 3.18 are quite compelling; certainly they reflect a periodic state in the chemistry. Chaotic states present a greater challenge to data analysis. How can we see deterministic structure, if it exists, in a single time series recording of bromide concentration? A solution to this problem is illustrated in Figure 3.19. The bromide concentration $B(t_i)$ is plotted against two delayed versions of itself, $B(t_i + T)$ and $B(t_i + 2T)$, in a 3-D plot. (The time unit is seconds; $T = 8.8$ sec.) This type of plot, called a delay coordinate plot, reveals the characteristic shape of a chaotic attractor. Delay coordinate plots are discussed in more detail in Chapter 13.

From the plot, a one-dimensional map can be constructed as a Poincaré return map. Using the plane fragment shown, the horizontal coordinates of successive intersections are recorded. If x denotes the bromide concentration at one intersection, then $f(x)$ is the concentration at the next intersection. The resulting pairs of points $(x, f(x))$ are plotted in Figure 3.20. Using standard approximation techniques, a function f was drawn through the points that best fit the experimental data.

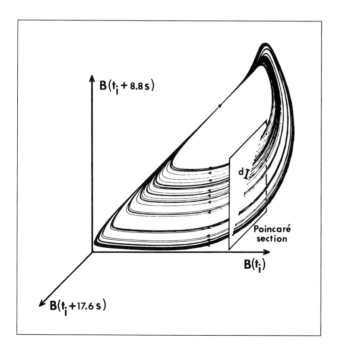

Figure 3.19   Reconstruction of dynamics from the bromide time series.
The three coordinates are the concentrations at three equally spaced time intervals. The Poincaré section, shown as a plane, intersects the data essentially in a curve, which allows a reduction to the one-dimensional map of Figure 3.20.

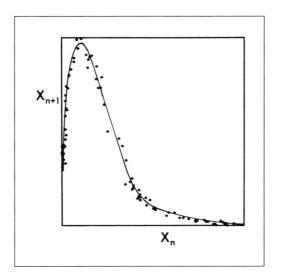

Figure 3.20 One-dimensional map reconstructed from the time series.
A spline fit was made to the data points from Poincaré map of Figure 3.19.

The one-dimensional map derived from this process is quite easy to analyze using the symbol sequences we have discussed in Chapter 3. The interval can be divided into three subintervals I, II, and III. The effect of one iteration of the map is shown in Figure 3.21. The subintervals I and II each stretch across subinterval III, and III stretches across both I and II. The result is a transition graph as in Figure 3.21. Using Figure 3.21 and Sharkovskii's Theorem, one can find the set

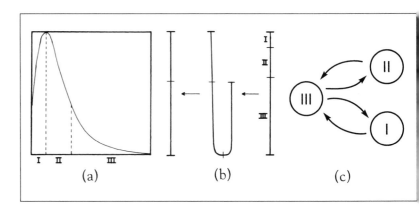

Figure 3.21 Symbol analysis of the one-dimensional map.
(a) The one-dimensional map fit from the data in Figure 3.20. (b) Schematic representation of the stretching and folding of the interval on the three subintervals I, II, and III. (c) Transition graph of the dynamics.

of periods possessed by periodic orbits of the system: all periods that are an even number of returns to the Poincaré surface of section shown in Figure 3.19.

An attempt was made to calculate the Lyapunov exponent from the one-dimensional map of Figure 3.20. It was found that the calculation of this number was very difficult, in the sense that small changes in the way the function approximation was made had large effects on the Lyapunov exponent that was derived. A careful estimate turned up $\lambda_1 \approx 0.3 \pm 0.1$.

Partway through the series of experiments by the Texas group, they had an opportunity to upgrade the laboratory apparatus, significantly improving the control of the conditions of the reaction. When they restarted the experiments, they found that the same experimental conditions gave different results. In particular, the one-dimensional map of Figure 3.20, although still chaotic, had significantly changed.

A detailed analysis showed that the difference arose not from changes in equipment but primarily from differences in the malonic acid reagent, which now came from a new supplier. The group eventually bought malonic acid of 99.5% purity from seven different vendors, and found that each yielded its own one-dimensional map. Results within any given sample of malonic acid were reproducible over long periods of time, even with modified experimental apparatus. Finally, the research group purified the different lots of malonic acid, and again found identical dynamics independent of the supplier as well as the particular purification procedure. As a result, there is great confidence in the reproducibility of the results. One of the several morals of this story is the extreme sensitivity of the dynamics with respect to system parameters.

CHAPTER FOUR

Fractals

A FRACTAL is a complicated geometric figure that, unlike a conventional complicated figure, does not simplify when it is magnified. In the way that Euclidean geometry has served as a descriptive language for the classical mechanics of motion, fractal geometry is being used for the patterns produced by chaos. Trajectories of the two-body problem, for example, consist of conic sections: ellipses, parabolas, and hyperbolas. Chaotic attractors, on the other hand, often have features repeated on many length or time scales.

Scientists know a fractal when they see one, but there is no universally accepted definition. The term "fractal" was coined in the 1960's by B. Mandelbrot, a mathematician at IBM. It is generally acknowledged that fractals have some or all of the following properties: complicated structure at a wide range of length scales, repetition of structures at different length scales (self-similarity), and a "fractal dimension" that is not an integer. We will exhibit and analyze these properties for several examples that are generally agreed to be fractals, and in so doing define fractal dimension. Perhaps the simplest geometric object that deserves to be called a fractal is a Cantor set.

4.1 CANTOR SETS

EXAMPLE 4.1

Begin with the unit interval $I = [0, 1]$ and make a new set according to the following instructions. (See Figure 4.1.) First remove the open interval $(1/3, 2/3)$, the middle third of I. The set of points that remain after this first step will be called K_1. The set K_1 is the union $[0, 1/3] \cup [2/3, 1]$. In the second step, remove the middle thirds of the two segments of K_1. That is, remove $(1/9, 2/9) \cup (7/9, 8/9)$ and set $K_2 = [0, 1/9] \cup [2/9, 3/9] \cup [6/9, 7/9] \cup [8/9, 1]$ to be what remains after the first two steps. Delete the middle thirds of the four remaining segments of K_2 to get K_3. Repeating this process, the limiting set $K = K_\infty$ is called the **middle-third Cantor set**. The set K is the set of points that belong to all of the K_n.

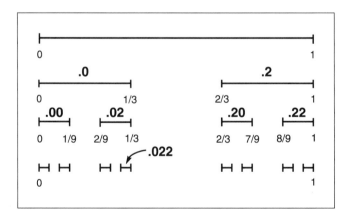

Figure 4.1 Construction of the middle-third Cantor set.
In step 1, the middle third of the unit interval is removed. In further steps, the middle third of every remaining subinterval is removed. Here three steps are shown. The points that are never removed make up the Cantor middle-third set. The set marked .02 consists of all numbers in the unit interval whose ternary expansion begins with .02.

What is the length of the set K? First of all, the set K is contained in K_n for each n. Just as K_1 consists of 2 intervals of length $1/3$, and K_2 consists of 4 intervals of length $1/9$, in general K_n consists of 2^n intervals, each of length $(1/3)^n$, so its total length is $(2/3)^n$. Hence K can be "covered" by a collection K_n of intervals whose total length can be made as small as you like. For example, K_{40} has length less than 10^{-6}. Since K_∞ is a subset of all of these sets, we say that K_∞ has length zero.

Although we doubt anyone disagrees on determining the length of an interval $[a, b]$, it can be a perplexing task to assign length to every possible set of points. There is a field of mathematics called measure theory that tries to solve this problem. We return to the concept of measure in Chapter 6. For now, it is sufficient to say that there is a definition of zero length, or measure zero, that is useful in practice.

A set S is said to have **measure zero** if it can be covered with intervals whose total length is arbitrarily small. In other words, for each predetermined $\epsilon > 0$, one can find a countable collection of intervals containing S whose total length is at most ϵ. The logic is that if the set in question can be viewed as a subset of sets with ordinary length as small as you want, then the set should be assigned a length of zero.

The set $\{1, 2, 3, \ldots, 10\}$ has measure zero, since for any predetermined $\epsilon > 0$, the set can be covered by 10 intervals of length $\epsilon/10$ centered at the 10 integers. Therefore it has a covering set of length ϵ, for ϵ as small as you want. To show that the set $\{1, 2, 3, \ldots\}$ of natural numbers has measure zero takes a little more work. Given the predetermined ϵ, consider the countable set of intervals of length $2^{-n}\epsilon$ centered at the natural number n. The sum total of the interval lengths is ϵ, as required. Finally, since the Cantor set K is covered by the set K_n of length $(2/3)^n$, which is arbitrarily small for large enough n, we can say that the Cantor set K has measure zero.

Although K has no length, it contains many points of the unit interval. Obviously, 0 and 1 belong to K. Similarly, both endpoints of any deleted middle third belong to K. After all, they will never end up in the middle third of a remaining subinterval at any stage of the construction.

Surprisingly, the endpoints make up only an insignificant portion of the points of K. For example, the number $1/4$, although never an endpoint of a subinterval of the construction, belongs to the middle-third Cantor set K. To see this, it is useful to express the numbers between 0 and 1 in base-3 representation. If $0 \le r \le 1$, then r can be written

$$r = a_1 \times 3^{-1} + a_2 \times 3^{-2} + \cdots + a_n \times 3^{-n} + \cdots$$

for numbers $a_k = 0, 1,$ or 2. The a_k's are the **ternary digits** of r.

To find the base 3 representation of a number r between 0 and 1, proceed as follows. Multiply r by 3, and take the integer part of the result (which is 0, 1, or 2) as a_1. Then take the fractional part and repeat this step to get a_2, etc. For example, the fraction 1/3 has a base-3 representation of .1, and 1/2 has a representation of $.\bar{1}$, that is, all digits are 1. This is analogous to the binary representations introduced in Chapter 3.

This representation for r is unique except for numbers $r = .a_1 a_2 \ldots a_n$ with a finite base-3 representation. By a finite base-3 representation, we mean that the ternary digit a_n is nonzero, and $0 = a_{n+1} = a_{n+2} = \ldots$. Then r is represented by exactly two base-3 expansions:

$$r = .a_1 a_2 \ldots a_n = .a_1 a_2 \ldots (a_n - 1) 222 \ldots .$$

The subinterval $[1/3, 2/3)$ consists of the points whose base-3 representations satisfy $a_1 = 1$. The number 1/3 can be expressed in two ways, as $.1 = .0\bar{2}$ in base 3. Therefore, the set $K_1 = [0, 1/3] \cup [2/3, 1]$ consists of all numbers in $[0, 1]$ that can be represented in base 3 with $a_1 = 0$ or 2. Similarly the set

$$K_2 = [0, 1/9] \cup [2/9, 1/3] \cup [2/3, 7/9] \cup [8/9, 1]$$

from the second step of the Cantor set construction is a set that consists of all numbers having representations with a_1 and a_2 each being either 0 or 2. We can ask what the analogous property is for K_n, and then ask what property a number must have if it is simultaneously in all of the K_n, that is, if it is in K_∞. From this reasoning follows a simple theorem.

Theorem 4.2 *The middle-third Cantor set K consists of all numbers in the interval $[0, 1]$ that can be represented in base 3 using only the digits 0 and 2.*

For example, the base-3 number $r = .\overline{02}$ belongs to K. Note that

$$r = 0 \times 3^{-1} + 2 \times 3^{-2} + 0 \times 3^{-3} + 2 \times 3^{-4} + \ldots$$
$$= \frac{2}{9}(1 + 3^{-2} + 3^{-4} + \ldots)$$
$$= \frac{2}{9}\left(\frac{1}{1 - 1/9}\right) = 1/4.$$

As mentioned above, some numbers have two base-3 representations: for example, one-third can expressed as either $.0\bar{2}$ or .1 in ternary expansion. However, each number in K has exactly one representation that includes no 1's.

We have been viewing the real numbers [0, 1] in terms of their base-3 expansions, but each number also has a binary expansion. That is, the number can be expressed using base-2 arithmetic as an infinite sequence of two symbols, 0 and 1. Therefore Theorem 4.2 shows that there is a one-to-one correspondence between the set [0, 1] and part of the Cantor set K, which is somewhat surprising in view of the fact that K has no length. In particular, K is a typical example of what is called an uncountable set, which we define next.

The idea of putting two sets in a one-to-one correspondence is the basis of counting. Saying that a particular deck contains 52 cards means that we can associate the cards with the set of numbers $\{1, 2, \ldots, 52\}$. Saying that the correspondence is "one-to-one" means that for each number there is exactly one card. No cards are counted twice, but all are counted. Another way of expressing a one-to-one correspondence with the set $\{1, \ldots, 52\}$ is to say that we can make a list with 52 entries. If there is no finite list containing all elements of the set, we call the set infinite.

Cantor took this idea of counting a step further. He called a set **countably infinite** if it can be put in a one-to-one correspondence with the natural numbers (positive integers). We will say that a set is **countable** if it is a finite set or a countably infinite set. Another way to say this is that a set is countable if its elements can be put in a finite or infinite list with each element listed exactly once. We call a set **uncountable** if it is not countable.

For example, the set of positive even integers is a countable set, as is the set of squares $\{1, 4, 9, \ldots\}$. A little more thought shows that the set of (positive and negative) integers is countable. Moreover, a subset of any countable set is countable. Thus the set of prime numbers is countable, even though no one knows an explicit formula for the nth prime number.

✎ **EXERCISE T4.1**

(a) Show that the union of two countable sets is countable. (b) Let $S_1, S_2, S_3, \ldots$ be a countable collection of countable sets. Show that the union of the S_i is countable.

The set of rational numbers (fractions m/n between 0 and 1, where m and n are nonzero integers) is countable. A scheme for counting the rational numbers in the interval $(0, 1]$ is illustrated in Figure 4.2. Notice that there are repetitions in the ordering; for example, $1/2, 2/4, 3/6$, etc., are all counted separately, so the figure shows that the rationals are a subset of a countable set, and therefore countable. Furthermore, once we know that the rationals in $(n, n + 1]$ form

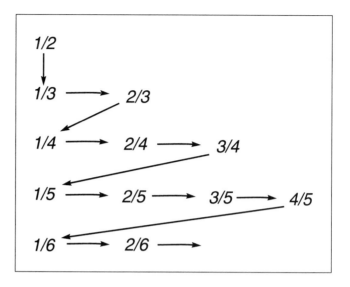

Figure 4.2 A scheme for counting the rationals.
The set of rational numbers is called countable because they can be put into a single (infinite) list.

a countable set for each integer n, then the entire set of rational numbers is a countable union of countable sets, which according to Exercise T4.1 is again a countable set.

Another example of a countable set is the set of all numbers in the middle-third Cantor set K having a finite number of ternary digits. Verify that these numbers are the right-hand endpoints of the removed open intervals of the Cantor set construction process. The correspondence could go as follows:

Natural number	Number in K
1	.2
2	.02
3	.22
4	.002
5	.022
6	.202
7	.222
$\vdots$	$\vdots$

Note that every point in the Cantor set with a terminating ternary expansion is identified with one of the natural numbers. By analogy, the subset of the Cantor set consisting of left-hand endpoints is also a countable set. Next we will show that the entire Cantor set is not a countable set.

✎ **EXERCISE T4.2**

Characterize the set of left-hand endpoints of the removed intervals for the Cantor set in terms of ternary expansions.

Cantor invented a technique for showing that certain sets are uncountable. A set is uncountable if every attempt to list all the elements must miss some of the set. We will demonstrate his technique by using it to show that the set K is uncountable (even though the subset of elements that have a terminating ternary expansion is countable).

Any list of numbers in K can be written

Integer	Number in K_∞
1	$r_1 = .a_{11}a_{12}a_{13} \cdots$
2	$r_2 = .a_{21}a_{22}a_{23} \cdots$
3	$r_3 = .a_{31}a_{32}a_{33} \cdots$
⋮	⋮
n	$r_n = .a_{n1}a_{n2}a_{n3} \cdots$
⋮	⋮

where each entry a_{ij} is a 0 or 2. Define the number $r = .b_1b_2b_3 \ldots$ in the following way. Choose its first digit b_1 to be 0 if a_{11} is 2 and 2 if a_{11} is 0. In general, choose its nth digit b_n to be a 0 or 2 but different from a_{nn}. Notice that the number r we have defined cannot be in the list—in particular, it can't be the seventh number on the list, for the seventh digit b_7 does not agree with a_{77}. Every element in the list has at least one digit that doesn't match. We know that each element of the Cantor set has exactly one representation using 0 and 2. We conclude that no single list can contain all the elements of K. Therefore the middle-third Cantor set is an uncountable set. With similar reasoning, using the symbols 0 and 1 instead of 0 and 2, we can show that the unit interval $[0, 1]$ is an uncountable set.

4.2 PROBABILISTIC CONSTRUCTIONS OF FRACTALS

In the first section we described a deterministic process which produced the middle-third Cantor set. Next we devise a probabilistic process which reproduces this set.

EXAMPLE 4.3

Play the following game. Start with any point in the unit interval $[0, 1]$ and flip a coin. If the coin comes up heads, move the point two-thirds of the way towards 1. If tails, move the point two-thirds of the way to 0. Plot the point that results. Then repeat the process. Flip the coin again and move two-thirds of the way from the new point to 1 (if heads) or to 0 (if tails). Plot the result and continue.

Aside from the first few points plotted, the points that are plotted appear to fill out a middle-third Cantor set. More precisely, the Cantor set is an attractor for the probabilistic process described, and the points plotted in the game approach the attractor at an exponential rate.

The reason for this is clear from the following thought experiment. Start with a homogeneous distribution, or cloud, of points along the interval $[0, 1]$. The cloud represents all potential initial points. Now consider the two possible outcomes of flipping the coin. Since there are two equally likely outcomes, it is fair to imagine half of the points of the cloud, randomly chosen, moving two thirds of the way to 0, and the other half moving two-thirds of the way to 1. After this, half of the cloud lies in the left one-third of the unit interval and the other half lies in the right one-third. After one hypothetical flip of the coin, there are two clouds of points covering $K_1 = [0, 1/3] \cup [2/3, 1]$ with a gap in between.

After the second flip, there are four clouds of points filling up K_2. Reasoning in this way, we see that an orbit that is generated randomly by coin flips will stay within the clouds we have described, and in the limit, approach the middle-third Cantor set. In fact, the reader should check that after k flips of the coin, the randomly-generated orbit must lie within $1/(3^k 6)$ of a point of the Cantor set. Verify further that the Cantor set is invariant under the game: that is, if a point in the Cantor set is moved two-thirds of the way either to 0 or 1, then the resulting point is still in the Cantor set.

The game we have described is an example of a more general concept we call an iterated function system.

Definition 4.4 An **iterated function system** on $\mathbb{R}^m$ is a collection $\{f_1, \ldots, f_r\}$ of maps on $\mathbb{R}^m$ together with positive numbers $p_1, \ldots, p_r$ (to be treated as probabilities) which add up to one.

Given an iterated function system, an orbit is generated as follows. Begin with a point in $\mathbb{R}^m$, and choose a map from the system with which to iterate the point. The map is chosen randomly according to the specified probabilities, thus map f_i is chosen with probability p_i. Use the randomly chosen map to iterate the point, and then repeat the process.

For certain kinds of maps f_i, the iterated function system will generate a fractal. For example, assume that each f_i is an **affine contraction map** on $\mathbb{R}^m$, defined to be the sum of a linear contraction and a constant. This means that $f_i(v) = L_i v + c_i$ where L_i is an $m \times m$ matrix with eigenvalues smaller than one in magnitude. Then for almost every initial condition, the orbit generated will converge exponentially fast to the same bounded set.

Example 4.3 is an iterated function system on $\mathbb{R}^1$. Let $f_1(x) = x/3$ and $f_2(x) = (2 + x)/3$, with associated probabilities $p_1 = p_2 = 1/2$. Check that both maps are affine contraction maps and have unique sinks at 0 and 1, respectively.

Recently, iterated function systems have proved to be a useful tools in data and image compression. See (Barnsley, 1988) for an introduction to this subject.

⇨ **COMPUTER EXPERIMENT 4.1**

The experiments in this chapter require the ability to plot graphics on a computer screen or printer. Define an iterated function system by

$$f_1(x, y) = (x/2, y/2), \quad f_2(x, y) = ((1 + x)/2, y/2), \quad f_3(x, y) = (x/2, (1 + y)/2)$$

with probabilities $p_1 = p_2 = p_3 = 1/3$. Begin with any point in the unit square $[0, 1] \times [0, 1]$ and use a random number generator or three-headed coin to generate an orbit. Plot the attractor of the iterated function system. This fractal is revisited in Exercise 4.9.

EXAMPLE 4.5

Consider the **skinny baker map** on $\mathbb{R}^2$ shown in Figure 4.3. This map exhibits the properties of stretching in one direction and shrinking in the other

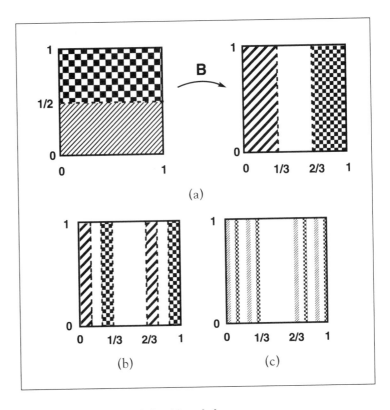

Figure 4.3 Definition of the skinny baker map.
(a) The top half maps to the right strip. The bottom half maps to the left strip.
(b) The second and (c) third iterate of the map. In the limit, the invariant set is a
Cantor middle-third set of vertical lines.

that are typical for chaotic two-dimensional maps. The equations of the map are

$$B(x, y) = \begin{cases} (\frac{1}{3}x, 2y) & \text{if } 0 \le y \le \frac{1}{2} \\ (\frac{1}{3}x + \frac{2}{3}, 2y - 1) & \text{if } \frac{1}{2} < y \le 1. \end{cases}$$

The map is discontinuous, since points (x, y) where y is less than $1/2$ are mapped
to the left side of the unit square, while points with y greater than $1/2$ are mapped
to the right. So there are pairs of points on either side of the line $y = 1/2$ that
are arbitrarily close but are mapped at least $1/3$ unit apart.

After one application of the map, the image of the unit square lies in the
left one-third and right one-third of the square. After two iterations, the image
of the unit square is the union of four strips, as shown in Figure 4.3(b). The set
to which all points of the unit square are eventually attracted is a Cantor set of
line segments. In fact, there is a close relationship between the skinny baker map
and the probabilistic game of Example 4.3. Sending the bottom half of the square

two-thirds of the way to the left and the top half two-thirds of the way to the right in the skinny baker map is analogous to the thought experiment described there.

EXAMPLE 4.6

The Cantor set construction of Example 4.1 can be altered to create interesting fractals in the plane. Start with a triangle T_0 with vertices A, B, and C, as in Figure 4.4. Delete the middle triangle from T_0, where by middle triangle we mean the one whose vertices are the midpoints of the sides of T_0. The new shape T_1 is the union of 3 subtriangles. Repeat the process indefinitely by deleting the middle triangles of the remaining 3 subtriangles of T_1, and so on. The points that remain make up the **Sierpinski gasket**.

EXAMPLE 4.7

As with the Cantor set, there is a probabilistic game that leads to the Sierpinski gasket. This game was introduced in Computer Experiment 4.1, although the triangle differs from Figure 4.4. Let A, B, and C be the vertices of a triangle. Start with a random point in the plane, and move the point one-half of the distance to one of the vertices. Choose to move with equal likelihood toward each of A, B, and C. From the new point, randomly choose one of the vertices and repeat.

The attractor for this process is the Sierpinski gasket. Initial points asymptotically approach the attractor at an exponential rate. Except for the first few points, the picture is largely independent of the initial point chosen (as it is for the Cantor set game). This is another example of an iterated function system.

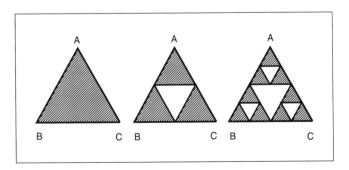

Figure 4.4 Construction of the Sierpinski gasket.
Start with a triangle, remove the central triangle, and repeat with each remaining triangle.

✏️ **EXERCISE T4.3**

(a) Find the maps f_1, f_2, f_3 to express Example 4.6 as an iterated function system. Assume that the triangle is equilateral with side-length one. (b) Find the exponential convergence rate of an orbit to the attractor. That is, if x_0 is a point inside the triangle with vertices A, B, and C, find an upper limit for the distance from the point x_k of the orbit of the iterated function system to the Sierpinski gasket.

EXAMPLE 4.8

A similar construction begins with a unit square. Delete a symmetric cross from the middle, leaving 4 corner squares with side-length 1/3, as in Figure 4.5. Repeat this step with each remaining square, and iterate. The limiting set of this process is called the **Sierpinski carpet**.

✏️ **EXERCISE T4.4**

Repeat Exercise 4.3 using the Sierpinski carpet instead of the gasket. Find four maps and the exponential convergence rate.

⇨ **COMPUTER EXPERIMENT 4.2**

Plot the Sierpinski gasket and carpet in the plane using the iterated function systems developed in Exercises 4.3 and 4.4.

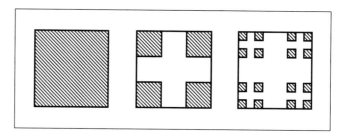

Figure 4.5 Construction of the Sierpinski carpet.
Remove a central cross from the square, and repeat for each remaining square.

4.3 FRACTALS FROM DETERMINISTIC SYSTEMS

In this section we see how the complexities of the previous abstract examples occur in familiar maps. One of the most basic maps on the real line is the tent map, which we studied in Chapter 3. We define a slightly more general version here. For $a > 0$, the tent map with slope a is given by

$$T_a(x) = \begin{cases} ax & \text{if } x \leq 1/2 \\ a(1 - x) & \text{if } 1/2 \leq x \end{cases}.$$

This map is continuous, but not smooth because of the corner at $x = 1/2$.

In the case of the slope-2 tent map ($a = 2$), the unit interval $I = [0, 1]$ is mapped onto itself by $T_2(x)$. The itineraries and transition graph were developed in Chapter 3, and it was noted that T_2 exhibits sensitive dependence on initial conditions.

In the case $0 < a < 1$, the tent map has a single fixed point at 0, which is attracting. All initial conditions are attracted to 0. For $a > 1$, the complement of I maps to the complement of I. Therefore if a point x is mapped outside of I, further iterations will not return it to I. For $1 \leq a \leq 2$, the points of I stay within I. For $a > 2$ most of the points of the unit interval eventually leave the interval on iteration, never to return.

✎ **EXERCISE T4.5**

For the tent map, define the set L of all points x in $[0, 1]$ such that $f^n(x) < 0$ for some n. (a) Prove that if $f^n(x) < 0$, then $f^k(x) < 0$ for all $k \geq n$. (b) Prove that for any $a > 2$, the length of L is 1 (the complement of L has measure zero). (c) Prove that for $0 \leq a \leq 2$, $L = \varnothing$, the empty set.

EXAMPLE 4.9

Consider the particular tent map T_3, which is sketched in Figure 4.6. The slope-3 tent map has dynamical properties that differ from those of the slope-2 tent map and its conjugate, the logistic map. For the slope-3 tent map, the basin of infinity is interesting. From the graph of T_3 in Figure 4.6, it is clear that initial conditions in the intervals $(-\infty, 0)$ and $(1, \infty)$ converge to $-\infty$ upon iteration by T_3. The same is true for initial conditions in $(1/3, 2/3)$, since that interval maps into $(1, \infty)$.

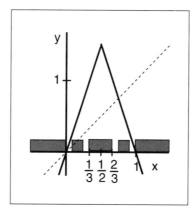

Figure 4.6 The tent map with slope 3.
The shaded points map out of the unit interval after two or fewer iterations. There is another region of points (not shown) that maps out after three iterations, and so on. All points not lying in the middle-third Cantor set are attracted to $-\infty$.

Let B be the basin of infinity, the set of points whose orbits diverge to $-\infty$. So far we have noted that B contains $(-\infty, 0) \cup (1, \infty)$ as well as $(1/3, 2/3)$, since $T_3(1/3, 2/3) \subset (1, \infty)$. Further, B contains $(1/9, 2/9)$, since $T_3^2(1/9, 2/9) \subset (1, \infty)$. The same goes for $(7/9, 8/9)$. The pattern that is emerging here is summarized in the next exercise.

✎ **EXERCISE T4.6**

Show that B, the basin of infinity of T_3 is the complement in $\mathbb{R}^1$ of the middle-third Cantor set C.

The long-term behavior of points of $\mathbb{R}^1$ under f can now be completely understood. The points of B tend to $-\infty$. The points of C, the remainder, bounce around within C in a way that can be described efficiently using the base-3 representation of C.

Up to this point, the examples of this chapter are almost pathologically well-organized. They are either the result of highly patterned constructions or, in the case of iterated function systems, achieved by repeated applications of affine contraction mappings. Sets such as these that repeat patterns on smaller scales are called **self-similar**. For example, if we magnify by a factor of three the portion of the middle-third Cantor set that is in the subinterval $[0, 1/3]$, we recover the entire Cantor set. Next we will see that this type of regularity can develop in more general nonlinear systems.

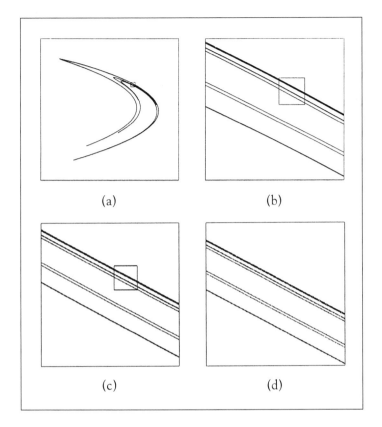

Figure 4.7   Self-similarity of the Hénon attractor.
(a) An attracting orbit of (4.1). Parts (b),(c),(d) are successive magnifications, showing the striated structure repeated on smaller and smaller scales. This sequence is zooming in on a fixed point. (a) $[-2.5, 2.5] \times [-2.5, 2.5]$. (b) $[0.78, 0.94] \times [0.78, 0.94]$. (c) $[0.865, 0.895] \times [0.865, 0.895]$. (d) $[0.881, 0.886] \times [0.881, 0.886]$.

EXAMPLE 4.10

Figure 4.7(a) depicts an orbit of the Hénon map of the plane

$$f(x, y) = (1.4 - x^2 + 0.3y, x). \tag{4.1}$$

After ten million iterations, the orbit remains in the region shown but (apparently) does not converge to a periodic orbit. M. Hénon proposed this map as an example of a dynamical system with a fractal attractor. The orbit of almost any initial point in this region will converge to this attractor. In Figure 4.7(b)(c)(d), the attractor is shown on progressively smaller scales.

EXAMPLE 4.11

Often, fractal structure is revealed indirectly, as in the case of basin boundaries. Consider, for example, the plane Hénon map with different parameter values:

$$f(x, y) = (1.39 - x^2 - 0.3y, x). \tag{4.2}$$

Figure 4.8(a) shows in black all initial conditions in a rectangular region whose orbits diverge to infinity, that is, the basin of infinity. The orbits of the initial conditions colored white stay within a bounded region. In Figure 4.8(b)(c) successive blow-ups reveal the fractal nature of the boundary between the black and white regions.

4.4 FRACTAL BASIN BOUNDARIES

Figure 2.3 of Chapter 2 shows basins of a period-two sink of the Hénon map. In part (a) of the figure, the basin boundary is a smooth curve. Part (b) of the figure shows a much more complicated basin. In fact, the boundary between two basins for a map of the plane can be far more complicated than a simple curve. In order to understand how complicated boundaries can develop, we investigate a simple model.

Consider a square R whose image is an S-shaped configuration, mapping across itself in three strips, as in Figure 4.9(a). The map also has two attracting fixed points outside the square, and we might assume that all points in the square that are mapped outside the square to the left eventually go to the sink A_1 on the left. Similarly, assume all points in this square that are mapped to the right of the square will eventually go to the sink A_2 on the right. We will see that this innocuous set of assumptions already implies a fractal basin boundary.

In Figure 4.9(b) the vertical strips shaded light grey represent points that map out of the square to the left in one iterate, while points in the strips shaded dark grey map out to the right in one iterate. The points in the three strips that are not shaded stay in the square for one iterate. Each of these strips maps horizontally across the square and contains two substrips which, on the next iteration, will map to the left and two that map to the right. We could therefore think of continuing the shading in this figure so that each white strip contains two light grey vertical substrips alternating with two dark grey vertical substrips. The white substrips in between will be further subdivided and points will be shaded or not, depending

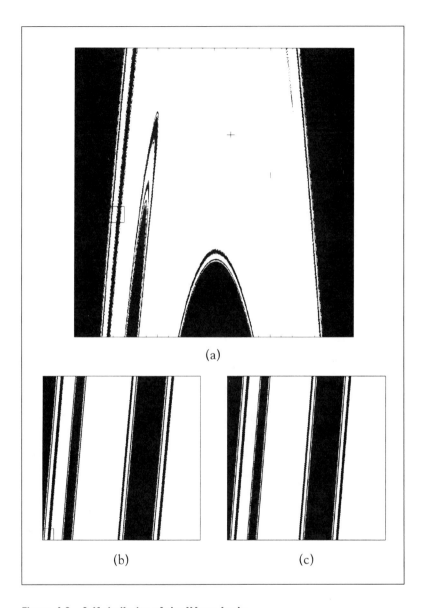

Figure 4.8 Self-similarity of the Hénon basin.
The points in white are attracted to the period-two attractor $\{(1, 0.3), (0.3, 1)\}$
of (4.2), marked with crosses. The points in black are attracted to infinity
with iteration. (a) The region $[-2.5, 2.5] \times [-2.5, 2.5]$. (b) The subregion
$[-1.88, -1.6] \times [-0.52, -0.24]$, which is the box in part (a). (c) The subregion
$[-1.88, -1.86] \times [-0.52, -0.5]$, the box in the lower left corner of part (b).

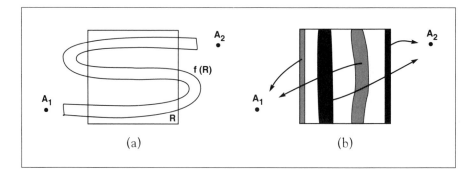

Figure 4.9 Construction of a fractal basin boundary.
(a) The image of the rectangle R is an S-shaped strip. Points that map outside and to the left of R are attracted to the sink A_1, and points that map outside and to the right of R are attracted to the sink A_2. (b) The shaded regions are mapped out of the rectangle in one iteration. Each of the three remaining vertical strips will lose 4 shaded substrips on the next iteration, and so on. The points remaining inside forever form a Cantor set.

on whether the third iterate goes to the left, goes to the right, or stays in the square.

This analysis implies that the subset of R that remains inside the square for n iterates consists of 3^n vertical strips. Each of these segments has three substrips that will remain inside R for a total of $n + 1$ iterates. If we have set up the map in a reasonable manner, the width of the vertical strips at the nth stage will shrink geometrically to zero as n tends to infinity. Thus we have a Cantor set construction; that is, there is a Cantor set of vertical curves whose images will remain inside R for all future times. Each of these vertical curves stretches from the top of the square to the bottom, and each point in the union of the curves has nearby points which go to A_1 and nearby points which go to A_2. Therefore, the Cantor set of vertical curves is the boundary between the basin of A_1 and the basin of A_2.

EXAMPLE 4.12

(Julia sets.) We return to the quadratic map, but with a difference: We now view it as a function of one complex variable. Let $P_c(z) = z^2 + c$, where z is a complex variable and $c = a + bi$ is a complex constant. Notice that P_c is a planar map; one complex variable $z = x + yi$ is composed of two real variables, x and y.

Multiplication of two complex numbers follows the rule

$$(u + vi)(x + yi) = ux - vy + (uy + vx)i$$

where $i^2 = -1$. In terms of real variables,

$$P_c(x + yi) = (x + yi)^2 + a + bi = x^2 - y^2 + a + (2xy + b)i.$$

We begin by considering the dynamics of this complex map when the parameter $c = 0$. Then the map $P_0(z) = z^2$ has an attracting fixed point at $z = 0$ whose basin of attraction is $\{z : |z| < 1\}$, the interior of the unit disk. A point on the unit circle

$$S = \{z : |z| = 1\}$$

maps to another point on the unit circle (the angle is doubled). The orbit of any point in $\{z : |z| > 1\}$, the exterior of the unit disk, diverges to infinity. The circle S forms the boundary between the basin of $z = 0$ and basin of infinity. Notice that points in the invariant set S are not in either basin.

For different settings of the constant c, $z = 0$ will no longer be a fixed point. Since we are considering all complex numbers, the equation $z^2 + c = z$ will have roots. Therefore P_c has fixed points. In fact, there is an easy way of finding all the attracting fixed and periodic points of P_c, due to a theorem of Fatou: Every attracting cycle for a polynomial map P attracts at least one critical point of P. Actually, Fatou proved the result for all rational functions (functions that are quotients of polynomials). Compare this statement with Theorem 3.29 of Chapter 3. Since our function P_c has only one critical point ($z = 0$), it can have at most one attracting periodic orbit.

Sometimes P_c has no attractors. Consider, for example, $P_{-i}(z) = z^2 - i$. Then $P^2(0) = -1 - i$, which is a *repelling* period-two point. We need look no further for an attractor.

Recognizing the important role that the orbit of 0 plays in the dynamics of P_c, we define the **Mandelbrot set** as follows:

$$M = \{c : 0 \text{ is not in the basin of infinity for the map } P_c(z) = z^2 + c\}.$$

We have seen that $c = 0$ and $c = -i$ are in the Mandelbrot set. Check that $c = 1$ is not. Figure 4.10 shows the set in white, where the number $c = a + bi$ is plotted in the plane as the point (a, b). See Color Plates 11–12 for color versions of the Mandelbrot set.

For each c in the Mandelbrot set, there are orbits of P_c that remain bounded and orbits that do not. Therefore, the boundary of the basin of infinity is non-empty. This boundary is called the **Julia set** of P_c, after the French mathematician G. Julia. Technically, the Julia set is defined as the set of repelling fixed and

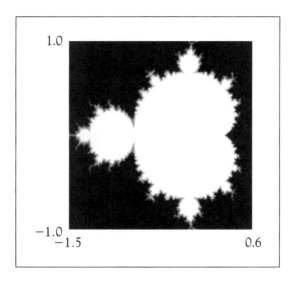

1.0

−1.0
−1.5   0.6

Figure 4.10 The Mandelbrot set.
A constant c is colored white if the orbit of $z^2 + c$ with initial value 0 does not diverge to infinity.

periodic points together with the limit points of this set. In the case of polynomials, however, the two definitions coincide.

In Figure 4.11 we show Julia sets for values of c inside the Mandelbrot set. For $c = -0.17 + 0.78i$, there is a period-three sink, whose basin is shown in Figure 4.11(a). Each point in white is attracted to the period-three sink. The points in black diverge to infinity. The boundary between these two basins is the Julia set. This picture shows many "rabbits" and other interesting shapes. The rabbit at the top of the picture is magnified in part (b). Part (c) shows a Julia set which is the boundary of the basin of a period-five sink.

The constants c used to make Figures 4.11(a) and (c) lie in distinctive places in the Mandelbrot set. Each c chosen from the small lobe at the top of the Mandelbrot set, such as $c = -0.17 + 0.78i$, creates a white basin of a period-three attractor, as in Figure 4.11(a). The period-five lobe lies at about 2 o'clock on the Mandelbrot set; it contains $c = 0.38 + 0.32i$, whose period-five sink is shown in Figure 4.11(c). Part (d) of the figure has a period-11 sink. The value of c used lies in the period-11 lobe, which is almost invisible in Figure 4.10. For more on Julia sets and the Mandelbrot set, consult (Devaney, 1986), (Devaney and Keen, 1989), or (Peitgen and Richter, 1986).

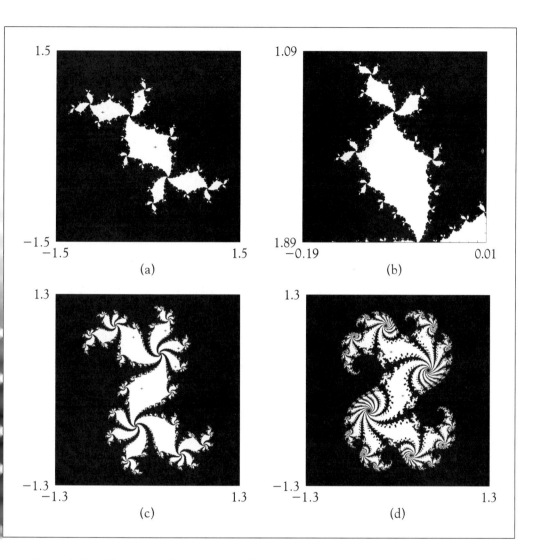

Figure 4.11 Julia sets for the map $f(z) = z^2 + c$.
(a) The constant is set at $c = -0.17 + 0.78i$. The white points are the basin of a period-three sink, marked with crosses, while the black points are the basin of infinity. The fractal basin boundary between black and white is the Julia set. (b) The uppermost rabbit of (a) is magnified by a factor of 15. (c) $c = 0.38 + 0.32i$. The white points are the basin of a period-five sink, marked with crosses. (d) $c = 0.32 + 0.043i$. The white points are the basin of a period-11 sink.

⇨   C O M P U T E R   E X P E R I M E N T   4 . 3

Draw the Julia set for $f(z) = z^2 + c$ with $c = 0.29 + 0.54i$. Plot the basin of infinity in black. Divide the square $[-1.3, 1.3] \times [-1.3, 1.3]$ into an $N \times N$ grid for $N = 100$. For each of the N^2 small boxes, iterate $f(z)$ using the center of the box as initial value. Plot a black point at the box (or better, fill in the box) if the iteration diverges to infinity; plot a different color if the iteration stays bounded. Your picture should bear some resemblance to the examples in Figure 4.11. What is the bounded attracting periodic sink for this Julia set? Locate c in the Mandelbrot set. Increase N for better resolution. Further work: Can you find a constant c such that $f(z) = z^2 + c$ has a period-six sink?

EXAMPLE 4.13

(Riddled basins.) In the examples we have studied so far, a basin of attraction is an open set. In particular, it contains entire disk neighborhoods. In this example, each disk, no matter how small, contains a nonzero area of points whose orbits move to different attractors. A basin that is shot through with holes in this sense is called a **riddled** basin.

Define the map

$$f(z) = z^2 - (1 + ai)\bar{z}, \tag{4.3}$$

where $z = x + iy$ is a complex variable, and $\bar{z} = x - iy$ denotes the complex conjugate of z. The number a is a parameter that we will specify later. The map can be rewritten in terms of the real and imaginary parts, as

$$f(x, y) = (x^2 - y^2 - x - ay, 2xy - ax + y).$$

The line $N_1 = \{z = x + iy : y = a/2\}$ is invariant under f, meaning that any point on that line maps back on the line:

$$f(x + ia/2) = (x + ia/2)^2 - (1 + ai)(x - ia/2)$$
$$= x^2 - 3a^2/4 - x + i(a/2). \tag{4.4}$$

If we consider the map restricted to that line, the above formula shows that it is the one-dimensional quadratic map, $g(x) = x^2 - x - 3a^2/4$.

✎ **EXERCISE T4.7**

Show that the one-dimensional map $g(x) = x^2 - x - 3a^2/4$ is conjugate to the map $h(w) = w^2 - c$, where the constant $c = 3(1 + a^2)/4$.

Define the lines N_2 and N_3 to be the result of rotating N_1 clockwise through $120°$ and $240°$, respectively. It can be checked that the map on N_2 is the same as the map on N_1, except that the images lie on N_3, and vice versa for the map on N_3. Thus N_2 and N_3 map to one another with quadratic one-dimensional dynamics. The second iterate f^2 on N_2 is exactly g^2, and the same for N_3.

In Figure 4.12 the three lines are shown in white along with three basins of attraction, for a particular parameter value $a = 1.0287137\ldots$, which is a value of a such that c satisfies the equality $c^3 - 2c^2 + 2c - 2 = 0$. This value is chosen so that the quadratic map $h(x) = w^2 - c$, to which g is conjugate, has the property that $h^3(0)$ is a fixed point. The reason for requiring this is hard to explain for now,

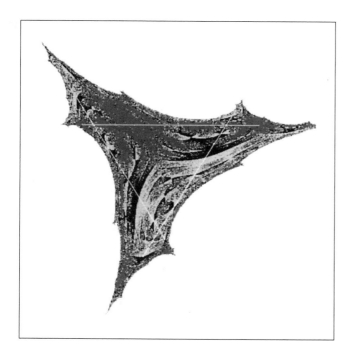

Figure 4.12   The riddled basin.
There are three attractors in the shaded region. Any disk of nonzero radius in this region, no matter how small, contains points from all 3 basins. Color Plate 2 is a color version of this figure.

but we will see in Chapter 6 that when the critical point maps in three iterates to a fixed point for a one-dimensional map, a continuous natural measure is created, which turns out to be desirable for this application.

The three attractors for this system are contained in the union of the three lines. The first, A_1, is the union of two subintervals of the line N_1, and the second, A_2, is the union of two slanted intervals that intersect A_1. The third, A_3, is an "X" at the intersection of the lines N_2 and N_3.

Figure 4.12 shows the basin of infinity in white, and the basins of A_1, A_2, and A_3 in dark gray, light gray, and black. The basins of all three attractors have nonzero area, and are riddled. This means that any disk of nonzero radius in the shaded region, no matter how small, has points from all 3 basins. Proving this fact is beyond the scope of this book. Color Plate 2 is a color version of this figure, which shows more of the detail.

The message of this example is that prediction can be difficult. If we want to start with an initial condition and predict the asymptotic behavior of the orbit, there is no limit to the accuracy with which we need to know the initial condition. This problem is addressed in Challenge 4 in a simpler context: When a basin boundary is fractal, the behavior of orbits near the boundary is hard to predict. A riddled basin is the extreme case when essentially the entire basin is made up of boundary.

4.5 FRACTAL DIMENSION

Our operational definition of fractal was that it has a level of complication that does not simplify upon magnification. We explore this idea by imagining the fractal lying on a grid of equal spacing, and checking the number of grid boxes necessary for covering it. Then we see how this number varies as the grid size is made smaller.

Consider a grid of step-size $1/n$ on the unit interval $[0, 1]$. That is, there are grid points at $0, 1/n, 2/n, \ldots, (n-1)/n, 1$. How does the number of grid boxes (one-dimensional boxes, or subintervals) depend on the step-size of the grid? The answer, of course, is that there are n boxes of grid size $1/n$. The situation changes slightly if we consider the interval $[0, 8]$. Then we need $8n$ boxes of size $1/n$. The common property for one-dimensional intervals is that the number of boxes of size ϵ required to cover an interval is no more than $C(1/\epsilon)$, where C is a constant depending on the length of the interval. This proportionality is often expressed by saying that the number of boxes of size ϵ *scales* as $1/\epsilon$, meaning that the number of boxes is between C_1/ϵ and C_2/ϵ, where C_1 and C_2 are fixed constants not depending on ϵ.

The square $\{(x, y) : 0 \leq x, y \leq 1\}$ of side-length one in the plane can be covered by n^2 boxes of side-length $1/n$. It is the exponent 2 that differentiates this two-dimensional example from the previous one. Any two-dimensional rectangle in $\mathbb{R}^2$ can be covered by $C(1/\epsilon)^2$ boxes of size ϵ. Similarly, a d-dimensional region requires $C(1/\epsilon)^d$ boxes of size ϵ.

The constant C depends on the rectangle. If we consider a square of side-length 2 in the plane, and cover by boxes of side-length $\epsilon = 1/n$, then $4(1/\epsilon)^2$ boxes are required, so $C = 4$. The constant C can be chosen as large as needed, as long as the scaling $C(1/\epsilon)^2$ holds as ϵ goes to 0.

We are asking the following question. Given an object in m-dimensional space, how many m-dimensional boxes of side-length ϵ does it take to cover the object? For example, we cover objects in the plane with $\epsilon \times \epsilon$ squares. For objects in three-dimensional space, we cover with cubes of side ϵ. The number of boxes, in cases we have looked at, comes out to $C(1/\epsilon)^d$, where d is the number we would assign to be the dimension of the object. Our goal is to extend this idea to more complicated sets, like fractals, and use this "scaling relation" to *define* the dimension d of the object in cases where we don't start out knowing the answer.

Notice that an interval of length one, when viewed as a subset of the plane, requires $1/\epsilon$ two-dimensional boxes of size ϵ to be covered. This is the same scaling that we found for the unit interval considered as a subset of the line, and matches what we would find for a unit interval inside $\mathbb{R}^m$ for any integer m. This scaling is therefore intrinsic to the unit interval, and independent of the space in which it lies. We will denote by $N(\epsilon)$ the number of boxes of side-length ϵ needed to cover a given set. In general, if S is a set in $\mathbb{R}^m$, we would like to say that S is a d-dimensional set when it can be covered by

$$N(\epsilon) = C(1/\epsilon)^d$$

boxes of side-length ϵ, for small ϵ. Stated in this way, it is not required that the exponent d be an integer.

Let S be a bounded set in $\mathbb{R}^m$. To measure the dimension of S, we lay a grid of m-dimensional boxes of side-length ϵ over S. (See Figure 4.13.) Set $N(\epsilon)$ equal to the number of boxes of the grid that intersect S. Solving the scaling law for the

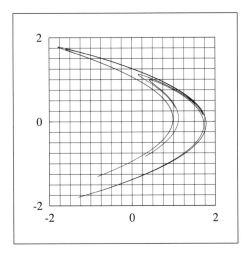

Figure 4.13   Grid of boxes for dimension measurement.
The Hénon attractor of Example 4.10 is shown beneath a grid of boxes with side-length $\epsilon = 1/4$. Of the 256 boxes shown, 76 contain a piece of the attractor.

dimension d gives us

$$d = \frac{\ln N(\epsilon) - \ln C}{\ln(1/\epsilon)}.$$

If C is constant for all small ϵ, the contribution of the second term in the numerator of this formula will be negligible for small ϵ. This justifies the following:

Definition 4.14   A bounded set S in $\mathbb{R}^n$ has **box-counting dimension**

$$\text{boxdim}(S) = \lim_{\epsilon \to 0} \frac{\ln N(\epsilon)}{\ln(1/\epsilon)},$$

when the limit exists.

We can check that this definition of dimension gives the correct answer for a line segment in the plane. Let S be a line segment of length L. The number of boxes intersected by S will depend on how it is situated in the plane, but roughly speaking, will be at least L/ϵ (if it lies along the vertical or the horizontal) and no more than $2L/\epsilon$ (if it lies diagonally with respect to the grid, and straddles pairs of neighboring boxes). As we expect, $N(\epsilon)$ scales as $1/\epsilon$ for this one-dimensional set. In fact, $N(\epsilon)$ is between L times $1/\epsilon$ and $2L$ times $1/\epsilon$. This remains true for infinitesimally small ϵ. Then Definition 4.14 gives $d = 1$.

✎ EXERCISE T4.8

Show that the box-counting dimension of a disk (a circle together with its interior) is 2.

Three simplifications will be introduced that make Definition 4.14 easier to use. First, the limit as $\epsilon \to 0$ need only be checked at a discrete sequence of box sizes, for example $\epsilon = 2^{-n}$ or $\epsilon = n^{-2}$, $n = 1, 2, 3, \ldots$. Second, the boxes need not be cemented into a grid—they can be moved around to more easily fit the set at hand. This approach may slightly decrease the number of boxes needed, without changing the dimension. Third, boxes don't need to be square boxes: they could be spheres or tetrahedra, for example.

Simplification 1. It is sufficient to check $\epsilon = b_n$, where $\lim_{n \to \infty} b_n = 0$ and $\lim_{n \to \infty} \frac{\ln b_{n+1}}{\ln b_n} = 1$.

We begin by describing the first simplification for a set in $\mathbb{R}^2$, and then generalize to $\mathbb{R}^m$. For each ϵ smaller than 1, there is some $n \geq 0$ such that ϵ lies between b_{n+1} and b_n. In $\mathbb{R}^2$, any box of side b_{n+1} is covered by 4 or fewer boxes of the ϵ-grid, as shown in Figure 4.14. It follows that if $N(b_{n+1})$ boxes of side-length b_{n+1} cover a set S, then the number of ϵ-boxes that will cover S satisfies $N(\epsilon) \leq 4N(b_{n+1})$. By the same token, any ϵ-box is covered by 4 or fewer boxes of the b_n-grid, so $N(b_n) \leq 4N(\epsilon)$. Therefore we have the inequality

$$\frac{N(b_n)}{4} \leq N(\epsilon) \leq 4N(b_{n+1}).$$

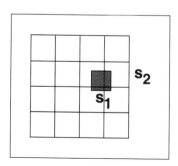

Figure 4.14 A box within a larger grid.
If $s_1 < s_2$, then four two-dimensional boxes of side s_2 are sufficient to cover a box of side s_1. In $\mathbb{R}^m$, 2^m boxes are sufficient. This fact is the key to Simplification 1.

For the case of $\mathbb{R}^m$, the 4 is replaced by 2^m, so that

$$2^{-m}N(b_n) \leq N(\epsilon) \leq 2^m N(b_{n+1}). \tag{4.5}$$

It follows from $b_{n+1} \leq \epsilon < b_n$ that

$$-\ln b_{n+1} \geq -\ln \epsilon > -\ln b_n. \tag{4.6}$$

Putting (4.5) and (4.6) together,

$$\frac{\ln b_n}{\ln b_{n+1}} \frac{-m\ln 2 + \ln N(b_n)}{-\ln b_n} \leq \frac{\ln N(\epsilon)}{\ln(1/\epsilon)} \leq \frac{m\ln 2 + \ln N(b_{n+1})}{-\ln b_{n+1}} \frac{\ln b_{n+1}}{\ln b_n}. \tag{4.7}$$

Thus, if $\lim\limits_{n\to\infty} \dfrac{\ln N(b_n)}{\ln(1/b_n)} = d$, then the terms on the left and right tend to the limit d as $n \to \infty$, and likewise the middle term must approach d as $\epsilon \to 0$.

Theorem 4.15   *Assume that $b_1 > b_2 > \ldots$, $\lim_{n\to\infty} b_n = 0$, and $\lim_{n\to\infty} \dfrac{\ln b_{n+1}}{\ln b_n} = 1$. If*

$$\lim_{n\to\infty} \frac{\ln N(b_n)}{\ln(1/b_n)} = d,$$

then $\lim\limits_{\epsilon\to 0} \dfrac{\ln N(\epsilon)}{\ln(1/\epsilon)} = d$, and therefore the box-counting dimension is d.

Simplification 2. Boxes can be moved to improve efficiency of the cover.

The second simplification refers to the fact that we could have alternatively defined the box-counting dimension by replacing $N(\epsilon)$, the number of grid boxes hit by the set, with the smallest possible number $N_0(\epsilon)$ of ϵ-boxes (not necessarily from any particular grid) that cover the set S. The boxes must be translates of grid boxes. In this formulation, rotations of the boxes are not allowed. We might be able to cover S more efficiently this way—by nudging some of the grid boxes to more convenient locations. Clearly $N_0(\epsilon)$ is at most $N(\epsilon)$, and could be less.

For all the simplicity of this alternate definition, it may be very difficult in practice to determine $N_0(\epsilon)$. On the other hand, the calculation of $N(\epsilon)$ is always accessible—one can lay a grid over the set and count boxes (this assumes we have complete knowledge of the set). As we show next, both definitions result in the same box-counting dimension.

By definition, $N_0(\epsilon) \leq N(\epsilon)$. We also know, as above, that any ϵ-box whatsoever is covered by 2^m or fewer ϵ-grid-boxes, so that $N(\epsilon) \leq 2^m N_0(\epsilon)$. The string of inequalities

$$N_0(\epsilon) \leq N(\epsilon) \leq 2^m N_0(\epsilon) \tag{4.8}$$

shows as in (4.7) that the grid definition (using $N(\epsilon)$) and the gridless definition (using $N_0(\epsilon)$) are equivalent.

Simplification 3. Other sets can be used in place of boxes.

We could have defined $N_0(\epsilon)$ as the smallest number of ϵ-disks (disks of radius ϵ) that cover the set S. The reasoning above goes through with little change. Other shapes, such as triangles or tetrahedra, could be used. We use triangles to determine the dimension of the Sierpinski gasket below.

4.6 COMPUTING THE BOX-COUNTING DIMENSION

We are now ready to compute the dimension of the middle-third Cantor set. Recall that the Cantor set K is contained in K_n, which consists of 2^n intervals, each of length $1/3^n$. Further, we know that K contains the endpoints of all 2^n intervals, and that each pair of endpoints lie 3^{-n} apart. Therefore the smallest number of 3^{-n}-boxes covering K is $N_0(3^{-n}) = 2^n$. We compute the box-counting dimension of K as

$$\text{boxdim}(K) = \lim_{n \to \infty} \frac{\ln 2^n}{\ln 3^n} = \lim_{n \to \infty} \frac{n \ln 2}{n \ln 3} = \frac{\ln 2}{\ln 3}.$$

We can compute the dimension of the Sierpinski gasket by exploiting the second and third simplifications above. We will use equilateral triangles of side-length $(1/2)^n$. After step n of the construction of Example 4.6, there remain 3^n equilateral triangles of side 2^{-n}. This is the smallest number of triangles of this size that contains the completed fractal, since all edges of the removed triangles lie in the fractal and must be covered. Therefore $N_0(2^{-n}) = 3^n$, and the box-counting dimension works out to $\ln 3 / \ln 2$.

✎ **EXERCISE T4.9**

Find the box-counting dimension of the invariant set of the skinny baker map of Example 4.5.

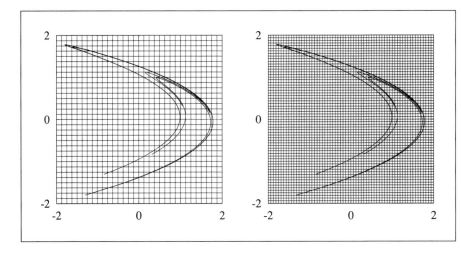

Figure 4.15 Finding the box-counting dimension of the Hénon attractor.
Two grids are shown, with gridsize $\epsilon = 1/8$ and $1/16$ respectively.

For a more complicated example such as the Hénon attractor of Example 4.10, no exact formula can be found for the box-counting dimension. We are stuck with drawing pictures and counting boxes, and using the results to form an estimate of the dimension. In Figure 4.15, we extend the grid of Figure 4.13 to smaller boxes. The side-length ϵ of the boxes is $1/8$ and $1/16$, respectively, in the two plots. A careful count reveals a total of 177 boxes hit by the attractor for $\epsilon = 1/8$, 433 for $\epsilon = 1/16$, 1037 for $\epsilon = 1/32$, 2467 for $\epsilon = 1/64$, and 5763 for $\epsilon = 1/128$.

In Figure 4.16 we graph the results of the box count. We graph the quantity $\log_2 N(\epsilon)$ versus $\log_2 (1/\epsilon)$ because its ratio is the same as $\ln N(\epsilon)/\ln(1/\epsilon)$, which defines box-counting dimension in the limit as $\epsilon \to 0$. We used box sizes $\epsilon = 2^{-2}$ through 2^{-7}, and take $\log_2$ of the box counts given above. The box-counting dimension corresponds to the slope in the graph. Ideally, Figure 4.16 would be extended as far as possible to the right, in order to make the best approximation possible to the limit. The slope in the picture gives a value for the box-counting dimension approximately equal to 1.27.

⇨ **COMPUTER EXPERIMENT 4.4**

Write a program for calculating box-counting dimension of planar sets. Test the program by applying it to a long trajectory of the iterated function system

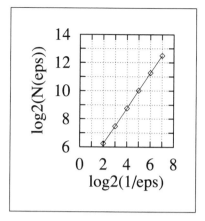

Figure 4.16 Box-counting dimension of the Hénon attractor.
A graphical report of the results of the box counts in Figures 4.13 and 4.15. The box-counting dimension is the limit of $\log_2 N(\epsilon)/\log_2(1/\epsilon)$. The dimension corresponds to the limiting slope of the line shown, as $\epsilon \to 0$, which is toward the right in this graph. The line shown has slope ≈ 1.27.

for the Sierpinski gasket. To how many correct digits can you match the correct value $\ln 3/\ln 2$?

In the remainder of this section we investigate whether there is a relationship between box-counting dimension and the property of measure zero. By constructing a Cantor set with a different ratio of removed intervals—say, $1/2$ instead of $1/3$—we can find a measure-zero set that has a different box-counting dimension. What about Cantor sets in the unit interval for which the lengths of the removed intervals sum to a number less than one? Such sets, called "fat" fractals, are not measure-zero sets.

✎ **EXERCISE T4.10**

Consider the Cantor set D formed by deleting the middle subinterval of length 4^{-k} from each remaining interval at step k. (a) Prove that the length of D is $1/2$. Thus D is a fat fractal. (b) What is the box–counting dimension of D? (c) Let f be the function on $[0, 1]$ which is equal to 1 on D and 0 elsewhere. It is the limit of functions that are Riemann integrable. Note that f is not Riemann integrable. What is the value of any lower Riemann sum for f?

Theorem 4.16 *Let A be a bounded subset of $\mathbb{R}^m$ with boxdim(A) $= d < m$. Then A is a measure zero set.*

Proof: The set A is contained in the union of $N(\epsilon)$ boxes of side ϵ. Then

$$\lim_{\epsilon \to 0} \frac{\ln \epsilon^m N(\epsilon)}{\ln \epsilon} = \lim_{\epsilon \to 0} \frac{m \ln \epsilon + \ln N(\epsilon)}{\ln \epsilon} = m - d > 0.$$

Since $\ln \epsilon \to -\infty$, we conclude that $\ln \epsilon^m N(\epsilon) \to -\infty$, so that $\epsilon^m N(\epsilon) \to 0$. We have found ϵ-boxes covering **A** whose total volume $\epsilon^m N(\epsilon)$ is as small as desired. This is the definition of measure zero. □

The converse of Theorem 4.16 does not hold; there are subsets of the unit interval (in fact, countable subsets) with box-counting dimension one and measure zero.

✎ **EXERCISE T4.11**

(a) Find the box-counting dimension of the set of integers $\{0, \ldots, 100\}$.

(b) Find the box-counting dimension of the set of rational numbers in $[0, 1]$.

4.7 CORRELATION DIMENSION

Box-counting dimension is one of several definitions of fractal dimension that have been proposed. They do not all give the same number. Some are easier to compute than others. It is not easy to declare a single one to be the obvious choice to characterize whatever fractal dimension means.

Correlation dimension is an alternate definition that is popular because of its simplicity and lenient computer storage requirements. It is different from box-counting dimension because it is defined for an orbit of a dynamical system, not for a general set. More generally, it can be defined for an invariant measure, which we describe in Chapter 6.

Let $S = \{v_0, v_1, \ldots\}$ be an orbit of the map f on $\mathbb{R}^n$. For each $r > 0$, define the **correlation function** $C(r)$ to be the proportion of pairs of orbit points within r units of one another. To be more precise, let S_N denote the first N points of the orbit S. Then

$$C(r) = \lim_{N \to \infty} \frac{\#\{\text{pairs } \{w_1, w_2\} : w_1, w_2 \text{ in } S_N, |w_1 - w_2| < r\}}{\#\{\text{pairs } \{w_1, w_2\} : w_1, w_2 \text{ in } S_N\}} \tag{4.9}$$

The correlation function $C(r)$ increases from 0 to 1 as r increases from 0 to ∞. If $C(r) \approx r^d$ for small r, we say that the **correlation dimension** of the orbit S is d. More precisely:

$$\text{cordim}(S) = \lim_{r \to 0} \frac{\log C(r)}{\log(r)}, \qquad (4.10)$$

if the limit exists.

Figure 4.17 shows an attempt to measure the correlation dimension of the orbit of the Hénon attractor shown in Figure 4.13. An orbit of length $N = 1000$ was generated, and of the $(1000)(999)/2$ possible pairs, the proportion that lie within r was counted for $r = 2^{-2}, \ldots, 2^{-8}$. According to the definition (4.10), we should graph $\log C(r)$ versus $\log r$ and try to estimate the slope as $r \to 0$. This estimate gives $\text{cordim}(S) \approx 1.23$ for the Hénon attractor, slightly less than the box-counting dimension estimate.

For dimension measurements in high-dimensional spaces, correlation dimension can be quite practical when compared with counting boxes. The number of ϵ-boxes in the unit "cube" of $\mathbb{R}^n$ is ϵ^{-n}. If $\epsilon = 0.01$ and $n = 10$, there are potentially 10^{20} boxes that need to be tracked, leading to a significant data structures problem. Because no boxes are necessary to compute correlation dimension, this problem doesn't arise.

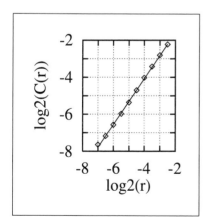

Figure 4.17 Correlation dimension of the Hénon attractor.
A graphical report of the results of the correlation dimension estimate for the Hénon attractor. The correlation dimension is the limit of $\log_2 C(r)/\log_2(r)$. The dimension corresponds to the limiting slope of the line shown, as $r \to 0$, which is toward the left. The line shown has slope ≈ 1.23.

The use of correlation dimension as a way to characterize chaotic attractors was suggested by theoretical physicists (Grassberger and Procaccia, 1983). Since then it has become a common tool for the analysis of reconstructed attractors from experimental data, such as the example of Lab Visit 3. Lab Visit 4 shows two illustrative applications of this method. More of the theory and practical issues of correlation dimension and other dimensions can be found in (Ott, Sauer, and Yorke, 1994).

☞ CHALLENGE 4

Fractal Basin Boundaries and the Uncertainty Exponent

WHY IS THE fractal dimension of a set important? In a broad sense, of course, it tells us something about the geometry of the set through its scaling behavior. But how do we use such information? In this challenge, we explore how the complexity of a fractal can influence final state determination within a dynamical system and see what the dimension of the fractal says about the resulting uncertainty.

Our model will be a one-dimensional, piecewise linear map F, which is illustrated in Figure 4.18. Under this map, almost all initial conditions will have orbits that tend to one or the other of two final states. Specifically, F is given by the following formula:

$$F(x) = \begin{cases} 5x + 4 & \text{if } x \leq -0.4 \\ -5x & \text{if } -0.4 \leq x \leq 0.4 \\ 5x - 4 & \text{if } x \geq 0.4 \end{cases}$$

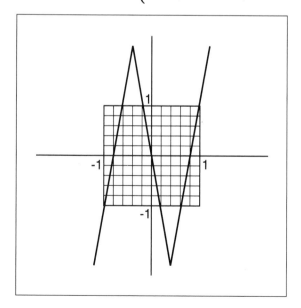

Figure 4.18 A piecewise linear map with fractal basin boundaries.
The orbits of almost every initial value tend to either $+\infty$ or $-\infty$; the boundary between the two basins is fractal.

Notice that any initial condition $x_0 > 1$ generates an orbit that tends to $+\infty$. Similarly, the orbit of any $x_0 < -1$ goes to $-\infty$. There are no fixed-point sinks or periodic sinks. In fact, the orbits of most points, even those in the interval $[-1, 1]$, become unbounded under iteration by F. Normally, we group together points whose orbits diverge to infinity (here either plus or minus infinity) and call the set the "basin of infinity". For the purposes of this challenge, however, we distinguish points whose orbits go to $+\infty$ from those that go to $-\infty$. Let $B_{+\infty}$ be the basin of $+\infty$, let $B_{-\infty}$ be the basin of $-\infty$, and let J be the set of points whose orbits stay in $[-1, 1]$ for all iterates of F.

Step 1   Describe $B_{+\infty}$, $B_{-\infty}$, and J. [Hint: Start by dividing the interval $[-1, 1]$ into five equal parts and deciding where each part goes under the map.]

Step 2   Show: $\text{boxdim}(J) = \dfrac{\ln 3}{\ln 5}$.

Step 3   Show that for every x in $B_{+\infty}$ there exists $\epsilon > 0$ such that $N_\epsilon(x) \subset B_{+\infty}$. (The analogous statement for $B_{-\infty}$ also holds.)

Step 4   Show that the following characterization holds for each y in J: For every $\epsilon > 0$, $N_\epsilon(y)$ contains points of both $B_{+\infty}$ and $B_{-\infty}$. In other words, the basin is an open set.

By definition, the orbit of an initial point x in $B_{+\infty}$ or $B_{-\infty}$, will tend to $+\infty$ or $-\infty$. If x is near a boundary point (a point in J), however, lack of measurement accuracy can make final state prediction impossible. For example, if accuracy is specified to within $\epsilon = 0.1$, then there is no way to decide whether a point in the intervals $(-1, -0.5)$, $(-0.3, 0.3)$, or $(0.5, 1.0)$ is in $B_{+\infty}$ or $B_{-\infty}$. The problem is that these points are all within 0.1 of a boundary point, and therefore all within 0.1 of points of both basins. Points within distance $\epsilon > 0$ of a basin boundary point are called ϵ-**uncertain** points. The complete set of 0.1-uncertain points between -1 and $+1$ is the union of the three above open intervals. The total length of these intervals is 1.6, or 80% of the interval $[-1, 1]$.

More generally, suppose that J is the fractal boundary of two basins in $\mathbb{R}$. In Steps 5 and 6, show there exists a number $p > 0$, which depends on $\text{boxdim}(J)$, such that the total length of ϵ-uncertain points (as $\epsilon \to 0$) is proportional to ϵ^p. The number p is called the **uncertainty exponent**.

Step 5   Let $L(\epsilon)$ be the total length of ϵ-uncertain points. Show that $\epsilon N(\epsilon) \leq L(\epsilon) \leq 3\epsilon N(\epsilon)$.

Step 6   Let $p = 1 - \text{boxdim}(J)$. Show: $1 \leq \lim\limits_{\epsilon \to 0} \dfrac{L(\epsilon)}{\epsilon^p} \leq 3$, if the limit exists.

Step 7   The previous step says that $L(\epsilon)$ is proportional to ϵ^p for small ϵ. This quantifies the fact that the larger the fractal dimension of J, the greater the uncertainty in prediction. For example, suppose boxdim $(J) = 0.9$, so that $p = 0.1$. How much of a reduction in measurement accuracy ϵ is necessary to reduce the total length of ϵ-uncertain points by a factor of two?

Step 8   Prove a two-dimensional version. Assume that J is a fractal basin boundary in $\mathbb{R}^2$, and that each point in J is in the boundary of at least two basins. Show that the total area $A(\epsilon)$ of ϵ-uncertain points satisfies the formula

$$A(\epsilon) \leq C\epsilon^p,$$

where $p = 2 - \text{boxdim}(J)$, and C is a constant.

Postscript.   The subject of this Challenge is the damaging effect on final state prediction due to fractal basin boundaries in one-dimensional maps. The characterization of J in Step 4 means that each point in J is in the boundary of each of the two basins. Basin boundaries of two-dimensional maps are even more intricate and interesting—they form the subject of Challenge 10. There we investigate points that are simultaneously in the boundary of *three* basins.

EXERCISES

4.1.   (a) Establish a one-to-one correspondence between the portion of the middle-third Cantor set in $[0, 1/3]$ and the entire Cantor set.

(b) Repeat part (a), replacing $[0, 1/3]$ by $[19/27, 20/27]$.

4.2.   Consider the **middle-half Cantor set** $K(4)$ formed by deleting the middle half of each subinterval instead of the middle third.

(a) What is the total length of the subintervals removed from $[0,1]$?

(b) What numbers in $[0, 1]$ belong to $K(4)$?

(c) Show that $1/5$ is in $K(4)$. What about $17/21$?

(d) Find the length and box-counting dimension of the middle-half Cantor set.

(e) Let S be the set of initial conditions that never map outside of the unit interval for the tent map T_a, as in Exercise T4.6. For what value of the parameter a is $S = K(4)$?

4.3.   Let T be the tent map with $a = 2$.

(a) Prove that rational numbers map to rational numbers, and irrational numbers map to irrationals.

(b) Prove that all periodic points are rational.

(c) Prove that there are uncountably many nonperiodic points.

4.4.   Show that the set of rational numbers has measure zero.

4.5.   Let $\hat{K}$ be the subset of the Cantor set K whose ternary expansion does not end in a repeating 2. Show that there is a one-to-one correspondence between $\hat{K}$ and $[0, 1]$. Thus K is uncountable because it contains the uncountable set $\hat{K}$.

4.6.   Let K be the middle-third Cantor set.

(a) A point x is a limit point of a set S if every neighborhood of x contains a point of S aside from x. A set is closed if it contains all of its limit points. Show that K is a closed set.

(b) A set S is **perfect** if it is closed and if each point of S is a limit point of S. Show that K is a perfect set. A perfect subset of $\mathbb{R}$ that contains no intervals of positive length is called a Cantor set.

(c) Let S be an arbitrary Cantor set in $\mathbb{R}$. Show that there is a one-to-one correspondence between S and $K(3)$. In fact a correspondence can be chosen so that if $s, t \in S$ are associated with $s', t' \in K(3)$, respectively, then $s < t$ implies $s' < t'$; that is, the correspondence preserves order.

4.7.   Find the box-counting dimension of:

(a) The middle $1/5$ Cantor set.

(b) The Sierpinski carpet.

4.8. Find the box-counting dimension of the set of endpoints of removed intervals for the middle-third Cantor set K.

4.9. Another way to find the attracting set of the Sierpinski gasket of Computer Experiment 4.1 is to iterate the following process. Remove the upper right one-quarter of the unit square. For each of the remaining three quarters, remove the upper right quarter; for each of the remaining 9 squares, remove the upper right quarter, and so on. (a) Find the box-counting dimension of the resulting gasket. (b) Show that the gasket consists of all pairs (x, y) such that for each n, the nth bit of the binary representations of x and y are not both 1.

4.10. Let A and B be bounded subsets of $\mathbb{R}$. Let $A \times B$ be the set of points (x, y) in the plane such that x is in A and y is in B. Show that $\mathrm{boxdim}(A \times B) = \mathrm{boxdim}(A) + \mathrm{boxdim}(B)$.

4.11. (a) True or false: The box-counting dimension of a finite union of sets is the maximum of the box-counting dimensions of the sets. Justify your answer.

(b) Same question for countable unions, assuming that it is bounded.

4.12. This problem shows that countably infinite sets S can have either zero or nonzero box-counting dimension.

(a) Let $S = \{0\} \cup \{1, 1/2, 1/3, 1/4, \ldots\}$. Show that $\mathrm{boxdim}(S) = 0.5$.

(b) Let $S = \{0\} \cup \{1, 1/2, 1/4, 1/8, \ldots\}$. Show that $\mathrm{boxdim}(S) = 0$.

4.13. Generalize the previous problem by finding a formula for $\mathrm{boxdim}(S)$:

(a) $S = \{0\} \cup \{n^{-p} : n = 1, 2, \ldots\}$, where $0 < p$.

(b) $S = \{0\} \cup \{p^{-n} : n = 0, 1, 2, \ldots\}$, where $1 < p$.

4.14. The time-2π map of the forced damped pendulum was introduced in Chapter 2. The following table was made by counting boxes needed to cover the chaotic orbit of the pendulum, using various box sizes. Two hundred million iterations of the time-2π map were made. Find an estimate of the boxdim of this set. If you know some statistics, say what you can about the accuracy of your estimate.

Box size	Boxes hit
2^{-2}	111
2^{-3}	327
2^{-4}	939
2^{-5}	2702
2^{-6}	7839
2^{-7}	22229
2^{-8}	62566
2^{-9}	178040

☞ LAB VISIT 4

Fractal Dimension in Experiments

FINDING THE fractal dimension of an attractor measured in the laboratory requires careful experimental technique. Fractal structure by its nature covers many length scales, so the construction of a "clean" experiment, where only the phenomenon under investigation will be gathered as data, is important. This ideal can never be achieved exactly in a laboratory. For this reason, computational techniques which "filter" unwanted noise from the measured data can sometimes help.

A consortium of experts in laboratory physics and data filtering, from Germany and Switzerland, combined forces to produce careful estimates of fractal dimension for two experimental chaotic attractors. The first experiment is the hydrodynamic characteristics of a fluid caught between rotating cylinders, called the Taylor-Couette flow. The second is an NMR laser which is being driven by a sinusoidal input.

The Taylor-Couette apparatus is shown in Figure 4.19. The outside glass cylinder, which has a diameter of about 2 inches, is fixed, and the inner steel

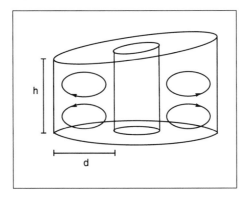

Figure 4.19 Setup of the Taylor-Couette flow experiment.
As the inner steel cylinder rotates, the viscous fluid between the cylinders undergoes complicated but organized motion.

Kantz, H., Schreiber, T., Hoffman, I., Buzug, T., Pfister, G., Flepp, L. G., Simonet, J., Badii, R., Brun, E., 1993. Nonlinear noise reduction: A case study on experimental data. Physical Review E **48**:1529-1538.

cylinder, which has a diameter of 1 inch, rotates at a constant rate. Silicon oil, a viscous fluid, fills the area between cylinders. Care was taken to regulate the temperature of the oil, so that it was constant to within 0.01° C. The speed of the rotating inner cylinder could be controlled to within one part in 10,000. As shown in the figure, the top of the chamber is not flat, but set at a tiny angle, to destroy unwanted effects due to symmetries of the boundary conditions. The top surface is movable so that the "aspect ratio" h/d can be varied between 0 and

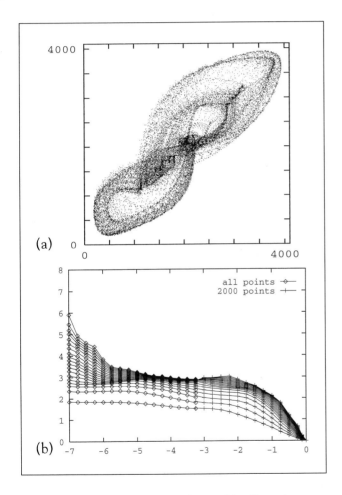

Figure 4.20   The Taylor-Couette data and its dimension.
(a) A two-dimensional projection of the reconstructed attractor obtained after filtering the experimental data. The time delay is $T = 5$. (b) Slope estimates of the correlation sum. There are 19 curves, each corresponding to correlation dimension estimates for an embedding dimension between 2 and 20. The 32,000 points plotted in Figure 4.20(a) were used in the left half, and only 2000 points in the right half, for comparison. The conclusion is that the correlation dimension is approximately 3.

more than 10, although an aspect ratio of 0.374 was used for the plots shown here.

The measured quantity is the velocity of the fluid at a fixed location in the chamber. Light-scattering particles were mixed into the silicon oil fluid, and their velocities could be tracked with high precision using a laser Doppler anemometer. The velocity was sampled every 0.02 seconds.

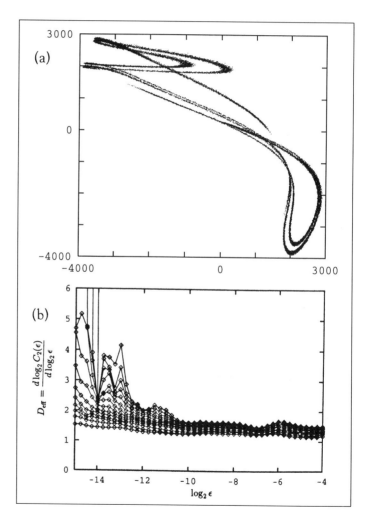

Figure 4.21 The NMR laser data and its estimated dimension.
(a) Time-delay plot of a Poincaré section of the NMR laser data. About 39,000 returns to the Poincaré surface were counted and plotted. (b) Estimated slopes from the correlation sum from the data in (a). The researchers report a correlation dimension estimate of 1.5.

Figure 4.20(a) shows a time-delay plot of the velocity of the Taylor-Couette flow experiment. We saw a first example of a time-delay plot in Figure 3.19; we will give a more comprehensive treatment in Chapter 13. For now, we recall that a time-delay reconstruction of a data series (Y_t), which in this case consists of the Taylor-Couette velocity readings, is constructed by plotting vectors of form $(y_t, y_{t+T}, \ldots, y_{t+(m-1)T})$ in $\mathbb{R}^m$. The delay used in Figure 4.20(a) is $T = 0.1$ seconds, or 5 sampling units. The dimensionality of the time-delay plot, m, is called the *embedding dimension*. In Figure 4.20(a), $m = 2$.

Once the set of vectors is plotted in $\mathbb{R}^m$, its properties can be analyzed as if it were an orbit of a dynamical system. In fact, it turns out that many of the dynamical properties of the orbit of the Taylor-Couette system that is producing the measurements are passed along to the set of time-delay vectors, as explained in Chapter 13. In particular, as first suggested in (Grassberger and Procaccia, 1983), the correlation dimension can be found by computing the correlation function of (4.9) using the time-delay vectors, and approximating the slope of a line as in Figure 4.17.

The resulting slope approximations for the line $\log C(\epsilon)/\log \epsilon$, which are the correlation dimension estimates as $\epsilon \to 0$, are graphed in Figure 4.20(b) as a function of $\log \epsilon$. The several curves correspond to computations of the correlation functions in embedding dimensions 2, 3, ..., 20. Here it is evident how evaluation of a limit as $\epsilon \to 0$ can be a challenge when experimental data is involved. We want to see the trend toward smaller ϵ, the left end of Figure 4.20(b). However, the effects of experimental uncertainties such as measurement noise are most strongly felt at very small scales. The range $2^{-5} < \epsilon < 2^{-2}$ shows agreement on a dimension of about 3.

The ingredients of a laser are the radiating particles (atoms, electrons, nuclei, etc.) and the electromagnetic field which they create. An external energy source causes a "population inversion" of the particles, meaning that the higher-energy states of the particles are more heavily populated than the lower ones. The laser cavity acts as a resonant chamber, which provides feedback for the laser field, causing coherence in the excitation of the particles. For the ruby NMR laser used in this experiment, the signal is a essentially a voltage measurement across a capacitor in the laser cavity.

The laser output was sampled at a rate of 1365 Hz (1365 times per second). A Poincaré section was taken from this data, reducing the effective sampling rate to 91 Hz. The time-delay plot of the resulting 39,000 intersections with the Poincaré surface is shown in Figure 4.21(a). The researchers estimate the noise level to be about 1.1%. After filtering, the best estimate for the correlation dimension is about 1.5, as shown in Figure 4.21(b).

CHAPTER FIVE

Chaos in Two-Dimensional Maps

THE CONCEPTS of Lyapunov numbers and Lyapunov exponents can be extended to maps on $\mathbb{R}^m$ for $m \geq 1$. In the one-dimensional case, the idea is to measure separation rates of nearby points along the real line. In higher dimensions, the local behavior of the dynamics may vary with the direction. Nearby points may be moving apart along one direction, and moving together along another.

In this chapter we will explain the definition of Lyapunov numbers and Lyapunov exponents in higher dimensions, in order to develop a definition of chaos in terms of them. Following that, we will extend the Fixed-Point Theorems of the one-dimensional case to higher dimensions, and investigate prototypical examples of chaotic systems such as the baker map and the Smale horseshoe.

5.1 Lyapunov Exponents

For a map on $\mathbb{R}^m$, each orbit has m Lyapunov numbers, which measure the rates of separation from the current orbit point along m orthogonal directions. These directions are determined by the dynamics of the map. The first will be the direction along which the separation between nearby points is the greatest (or which is least contracting, if the map is contracting in all directions). The second will be the direction of greatest separation, chosen from all directions perpendicular to the first. The third will have the most stretching of all directions perpendicular to the first two directions, and so on. The stretching factors in each of these chosen directions are the Lyapunov numbers of the orbit.

Figures 5.1 and 5.2 illustrate this concept pictorially. Consider a sphere of small radius centered on the first point v_0 of the orbit. If we examine the image $f(S)$ of the small sphere under one iteration of the map, we see an approximately ellipsoidal shape, with long axes along expanding directions for f and short axes along contracting directions.

After n iterates of the map f, the small sphere will have evolved into a longer and thinner ellipsoid-like object. The per-iterate changes of the axes of this image "ellipsoid" are the Lyapunov numbers. They quantify the amount of stretching and shrinking due to the dynamics near the orbit beginning at v_0. The natural logarithm of each Lyapunov number is a Lyapunov exponent.

For the formal definition, replace the small sphere about v_0 and the map f by the unit sphere U and the first derivative matrix $\mathbf{Df}(v_0)$, since we are interested in the infinitesimal behavior near v_0. Let $J_n = \mathbf{Df}^n(v_0)$ denote the first derivative

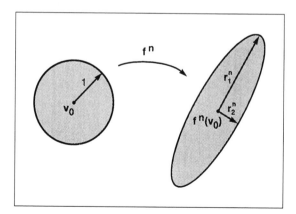

Figure 5.1   Evolution of an initial infinitesimal disk.
After n iterates of a two-dimensional map, the disk is mapped into an ellipse.

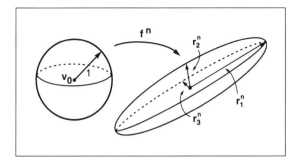

Figure 5.2   A three-dimensional version of Figure 5.1.
A small ball and its image after n iterates, for a three-dimensional map.

matrix of the nth iterate of $\mathbf{f}$. Then $J_n U$ will be an ellipsoid with m orthogonal axes. This is because for every matrix A, the image AU is necessarily an ellipsoid, as we saw in Chapter 2. The axes will be longer than 1 in expanding directions of $\mathbf{f}^n(\mathbf{v}_0)$, and shorter than 1 in contracting directions, as shown in Figure 5.1. The m average multiplicative expansion rates of the m orthogonal axes are the Lyapunov numbers.

Definition 5.1   Let $\mathbf{f}$ be a smooth map on $\mathbb{R}^m$, let $J_n = \mathbf{Df}^n(\mathbf{v}_0)$, and for $k = 1, \ldots, m$, let r_k^n be the length of the kth longest orthogonal axis of the ellipsoid $J_n U$ for an orbit with initial point $\mathbf{v}_0$. Then r_k^n measures the contraction or expansion near the orbit of $\mathbf{v}_0$ during the first n iterations. The kth **Lyapunov number** of $\mathbf{v}_0$ is defined by

$$L_k = \lim_{n \to \infty} (r_k^n)^{1/n},$$

if this limit exists. The kth **Lyapunov exponent** of $\mathbf{v}_0$ is $h_k = \ln L_k$. Notice that we have built into the definition the property that $L_1 \geq L_2 \geq \cdots \geq L_m$ and $h_1 \geq h_2 \geq \cdots \geq h_m$.

If U is the unit sphere in $\mathbb{R}^m$ and A is an $m \times m$ matrix, then the orthogonal axes of the ellipsoid AU can be computed in a straightforward way, as we showed in Theorem 2.24 of Chapter 2. The lengths of the axes are the square roots of the m eigenvalues of the matrix AA^T, and the axis directions are given by the m corresponding orthonormal eigenvectors.

Using the concept of Lyapunov exponent, we can extend the definition of chaotic orbit to orbits of higher dimensional maps. For technical reasons which we describe later (Example 5.5), we will require that no Lyapunov exponent is exactly zero for a chaotic orbit.

Definition 5.2 Let $\mathbf{f}$ be a map of $\mathbb{R}^m$, $m \geq 1$, and let $\{v_0, v_1, v_2, \ldots\}$ be a bounded orbit of $\mathbf{f}$. The orbit is **chaotic** if

1. it is not asymptotically periodic,
2. no Lyapunov number is exactly one, and
3. $L_1(v_0) > 1$.

In terms of Lyapunov exponents, part 3 of Definition 5.2 is equivalent to $h_1(v_0) > 0$.

EXAMPLE 5.3

The skinny baker map was defined in Chapter 4. It is a map of the unit square S in the plane which exhibits many typical properties of chaotic maps. See Figure 4.3 of Chapter 4 to recall the geometric behavior. The equations are written there, but we write them again in a slightly different form:

$$B(x_1, x_2) = \begin{cases} \begin{pmatrix} \frac{1}{3} & 0 \\ 0 & 2 \end{pmatrix} \begin{pmatrix} x_1 \\ x_2 \end{pmatrix} & \text{if } 0 \leq x_2 \leq \frac{1}{2} \\[2em] \begin{pmatrix} \frac{1}{3} & 0 \\ 0 & 2 \end{pmatrix} \begin{pmatrix} x_1 \\ x_2 \end{pmatrix} + \begin{pmatrix} \frac{2}{3} \\ -1 \end{pmatrix} & \text{if } \frac{1}{2} < x_2 \leq 1. \end{cases} \tag{5.1}$$

The middle third of the rectangles that remain are mapped out on each iteration. Define the **invariant set** A of $\mathbf{B}$ to be the points that lie in $\mathbf{B}^n(S)$ for every positive and negative integer n. The invariant set is a middle-third Cantor set of vertical lines. We call the set A invariant because it has the property that $\mathbf{B}^{-1}(A) = A$.

✎ **EXERCISE T5.1**

Find the area of $\mathbf{B}^n(S)$. Note that $\mathbf{B}$ is an area-contracting map, hence its name.

We are interested in the following questions about the skinny baker map. First, what are the Lyapunov exponents for orbits? Second, are there periodic orbits, and what do they look like? Finally, are there chaotic orbits? We begin with the first question, which is relatively easy to answer. We will defer the discussion of the periodic orbits until later in the chapter, where it will motivate the development of a fixed-point theorem and covering partitions in a direct analogy with those concepts in Chapter 3.

The Jacobian matrix is constant for each point in the unit square for which it is defined:

$$\mathbf{DB(v)} = \begin{pmatrix} \frac{1}{3} & 0 \\ 0 & 2 \end{pmatrix}$$

for all $\mathbf{v}$ except along the discontinuity line $x_2 = 1/2$. Consider a small circle of radius r centered at a point in the unit square. After one iteration of the skinny baker map, it is transformed into an ellipse with axes of length $\frac{1}{3}r$ in the horizontal direction and $2r$ in the vertical. After n iterates, the ellipse is $(\frac{1}{3})^n r$ by $2^n r$. (All of this is true provided that the ellipses never map across the line $x_2 = \frac{1}{2}$, in which case they get chopped in two on the next iterate. To avoid this, consider an even smaller initial circle.) We conclude that the Lyapunov numbers of $\mathbf{B}$ are $\frac{1}{3}$ and 2, or equivalently, that the Lyapunov exponents are $-\ln 3$ and $\ln 2$ for every orbit. Since $\ln 2 > 0$, every orbit that is not asymptotically periodic is chaotic. Determining orbits that are not asymptotically periodic will be left for later in the chapter.

✎ **EXERCISE T5.2**

Find a general formula for (area-contracting) skinny baker maps, where the strips have width w instead of $1/3$. Find the area contraction factor (per iteration) and the Lyapunov exponents.

EXAMPLE 5.4

The cat map, introduced in Challenge 2 (Chapter 2), is also fairly simple to analyze for Lyapunov exponents. The map on the unit square in $\mathbb{R}^2$ is defined by

$$\mathbf{f}\begin{pmatrix} x_1 \\ x_2 \end{pmatrix} = \begin{pmatrix} 2 & 1 \\ 1 & 1 \end{pmatrix}\begin{pmatrix} x_1 \\ x_2 \end{pmatrix} \pmod{1} = \mathbf{Av} \pmod{1}.$$

The Jacobian matrix $\mathbf{Df(v)}$ is the constant matrix $\mathbf{A}$ for any point $\mathbf{v}$. Unlike the skinny baker map case, it is not diagonal, so we turn to Theorem 2.25 of Chapter 2 in order to calculate the ellipses formed by applying $\mathbf{f}$ to an infinitesimal disk.

The theorem says that the axes of the ellipse formed by applying $\mathbf{f}^n$ will be the square roots of the eigenvalues of $\mathbf{A}^n(\mathbf{A}^n)^T$. Since $\mathbf{A}$ is a symmetric matrix ($\mathbf{A} = \mathbf{A}^T$), these are the same as the eigenvalues of $\mathbf{A}^n$, which can be found as e_i^n, where e_i are the eigenvalues of $\mathbf{A}$. We calculate $e_1 = (3 + \sqrt{5})/2 \approx 2.618$ and $e_2 = (3 - \sqrt{5})/2 \approx 0.382$. The product of the two eigenvalues is $1 = \det(\mathbf{A})$.

This tells us that a disk of radius r is transformed into an ellipse with axes e_1 and e_2 by the cat map, and into an ellipse with axes e_1^n and e_2^n by n applications of the cat map. Since $e_1 > 1$ and $e_2 < 1$, the fixed point $0 = (0, 0)$ is a saddle. The Lyapunov numbers L_1 and L_2 for the fixed point 0 and for any other orbit of the cat map are e_1 and e_2, and the Lyapunov exponents are $h_1 = \ln e_1 \approx 0.962$ and $h_2 = \ln e_2 \approx -0.962$.

The cat map is unusual in that the Jacobian matrix $\mathbf{Df}$ is independent of $\mathbf{v}$. Ordinarily, unless the orbit is periodic, computing eigenvalues of individual Jacobians tells nothing about Lyapunov numbers.

Since every orbit of the cat map has a positive Lyapunov exponent (and no Lyapunov exponent equal to 0), any orbit that is not asymptotically periodic will be chaotic. In the remainder of this example, we argue that a large number of orbits are not asymptotically periodic.

By Step 4 of Challenge 2, the periodic points of the cat map are those points in the unit square with both coordinates rational. This set is countable. Since all periodic orbits are saddles, the only asymptotically periodic orbits are these saddles and the orbits of points on their stable manifolds. As described in Chapter 2, the stable manifold of a saddle in the plane is a one-dimensional curve containing the saddle point. In the case of the cat map, a stable manifold emanates from a saddle at an irrational angle (from the horizontal). In order to understand the behavior of a stable manifold globally, we need to think of the cat map as defined on the torus (as described in Challenge 2), where it is continuous. Any one stable manifold will then wind around the torus without end. Viewing the torus as the unit square with left and right sides glued and top and bottom glued, we can see how this curve repeatedly crosses the line segment $I = \{(x, 0) : 0 \le x < 1\}$, a cross-sectional circle on the torus.

Following one branch of the stable manifold as it emanates from a saddle $\mathbf{p}$, we count successive crossings of the manifold with I. Since the slope is irrational, there will be infinitely many of these crossings; however, the set of all crossings will be countable. (This can be seen by the fact that there is a definite ordering of the crossings as we follow the stable manifold from $\mathbf{p}$.) Taking the union of all crossings for all stable manifolds, we have a countable union of countable sets, which is again countable. Since I contains an uncountable number of points, there must be points in I that are not on any stable manifold. Therefore, the cat map has chaotic orbits.

In Definition 5.2 the condition that 0 not be a Lyapunov exponent (of a chaotic orbit) is included to rule out cases of "quasiperiodicity", as illustrated by the next example.

EXAMPLE 5.5

Let

$$\mathbf{f}(r, \theta) = (r^2, \theta + q),$$

where r and θ are polar coordinates, and q is an angle irrationally related to 2π. The orbits of all points inside the unit circle converge to the origin, which is a fixed point. The unit circle is invariant under $\mathbf{f}$, as points on it rotate through the angle q. Outside the unit circle, orbits are unbounded. The quasiperiodic orbits at $r = 1$ have Lyapunov exponents $\ln 2$ and 0. Although they are not asymptotically periodic and they do exhibit sensitive dependence (due to a positive Lyapunov exponent in the r direction), we do not want the definition of "chaotic orbit" to apply to these orbits since their motion is quite predictable.

⇨ COMPUTER EXPERIMENT 5.1

Plot the orbit of the two-dimensional Tinkerbell map

$$\mathbf{f}(x, y) = (x^2 - y^2 + c_1 x + c_2 y, 2xy + c_3 x + c_4 y)$$

where $c_1 = -0.3$, $c_2 = -0.6$, $c_3 = 2$, $c_4 = 0.5$, with initial value $(x, y) = (0.1, 0.1)$. The orbit tends toward an oval-shaped quasiperiodic attractor. Find out what replaces the quasiperiodic attractor when the parameter c_4 is decreased or increased.

5.2 NUMERICAL CALCULATION OF LYAPUNOV EXPONENTS

For most interesting maps, there is no direct way to determine Lyapunov exponents from knowledge of the map and its Jacobian matrices. The skinny baker map and the cat map of the previous section are exceptions to this rule. Normally, the matrix $J_n = \mathbf{Df}^n(v_0)$ is difficult to determine exactly for large n, and we must resort to the approximation of the image ellipsoid $J_n U$ of the unit sphere by computational algorithms.

The ellipsoid $J_n U$ has semi-major axes of length s_i in the directions u_i. The direct approach to calculating the Lyapunov exponents would be to explicitly form $J_n J_n^T$ and find its eigenvalues s_i^2. In the case that the ellipsoid has stretching and shrinking directions, it will be very long and very thin for large n. The eigenvalues of $J_n J_n^T$ will include both very large and very small numbers. Because of the limited number of digits allowed for each stored number, computer calculations become difficult when numbers of vastly different sizes are involved in the same calculation. The problem of computing the s_i gets worse as n increases. For this reason, the direct calculation of the ellipsoid $J_n U$ is usually avoided.

The indirect approach that works better in numerical calculations involves following the ellipsoid as it grows. Since

$$J_n U = \mathbf{Df}(v_{n-1}) \cdots \mathbf{Df}(v_0) U, \tag{5.2}$$

we can compute one iterate at a time. Start with an orthonormal basis $\{w_1^0, \ldots, w_m^0\}$ for $\mathbb{R}^m$, and compute the vectors $z_1, \ldots, z_m$:

$$z_1 = \mathbf{Df}(v_0) w_1^0, \ldots, z_m = \mathbf{Df}(v_0) w_m^0. \tag{5.3}$$

These vectors lie on the new ellipse $\mathbf{Df}(v_0) U$, but they are not necessarily orthogonal. We will remedy this situation by creating a new set of orthogonal vectors $\{w_1^1, \ldots, w_m^1\}$ which generate an ellipsoid with the same volume as $\mathbf{Df}(v_0) U$. Use the Gram-Schmidt orthogonalization procedure, which defines

$$y_1 = z_1$$

$$y_2 = z_2 - \frac{z_2 \cdot y_1}{\|y_1\|^2} y_1$$

$$y_3 = z_3 - \frac{z_3 \cdot y_1}{\|y_1\|^2} y_1 - \frac{z_3 \cdot y_2}{\|y_2\|^2} y_2$$

$$\vdots$$

$$y_m = z_m - \frac{z_m \cdot y_1}{\|y_1\|^2} y_1 - \cdots - \frac{z_m \cdot y_{m-1}}{\|y_{m-1}\|^2} y_{m-1}, \tag{5.4}$$

where $\cdot$ denotes the dot or scalar product and $\| \cdot \|$ denotes Euclidean length.

Notice what the equations do: First z_1 is declared to be kept as is in the new set. The part of z_2 which is perpendicular to z_1 is retained as y_2; the term being subtracted away is just the projection of z_2 to the z_1 direction. The vector y_3 is defined to be the part of z_3 perpendicular to the plane spanned by z_1 and z_2, and so on.

✎ EXERCISE T5.3

Write the result of the Gram-Schmidt orthogonalization procedure as an $m \times m$ matrix equation $Z = YR$, where the columns of Y are $\mathbf{y}_1, \ldots, \mathbf{y}_m$, the columns of Z are $\mathbf{z}_1, \ldots, \mathbf{z}_m$, and R is an upper triangular matrix. Show that $\det(R) = 1$, so that $\det(Y) = \det(Z)$. Explain why this implies that the ellipsoids YU and ZU have equal m-dimensional volume. Depending on your matrix algebra skills, you may want to start with the case $m = 2$.

According to Exercise T5.3, if we set $\mathbf{w}_1^1 = \mathbf{y}_1, \ldots, \mathbf{w}_m^1 = \mathbf{y}_m$ for the new orthogonal basis, they will span an ellipsoid of the same volume as $\mathbf{Df}(\mathbf{v}_0)U$.

Next apply the Jacobian $\mathbf{Df}(\mathbf{v}_1)$ at the next orbit point, and reorthogonalize the set

$$\mathbf{Df}(\mathbf{v}_1)\mathbf{w}_1^1, \ldots, \mathbf{Df}(\mathbf{v}_1)\mathbf{w}_m^1 \tag{5.5}$$

to produce a new orthogonal set $\{\mathbf{w}_1^2, \ldots, \mathbf{w}_m^2\}$. Repeating this step n times gives a final set $\{\mathbf{w}_1^n, \ldots, \mathbf{w}_m^n\}$ of vectors which approximate the semi-major axes of the ellipsoid $J_n U$.

The total expansion r_i^n in the ith direction after n iterations, referred to in Definition 5.1, is approximated by the length of the vector $\mathbf{w}_i^n$. Thus $||\mathbf{w}_i^n||^{1/n}$ is the approximation to the ith largest Lyapunov number after n steps.

To eliminate the problem of extremely large and small numbers, this algorithm should be amended to normalize the orthogonal basis at each step. Denote the $\mathbf{y}$ vectors recovered from the application of Gram-Schmidt orthogonalization to

$$\mathbf{Df}(\mathbf{v}_j)\mathbf{w}_1^j, \ldots, \mathbf{Df}(\mathbf{v}_j)\mathbf{w}_m^j$$

by $\mathbf{y}_1^{j+1}, \ldots, \mathbf{y}_m^{j+1}$. Set $\mathbf{w}_i^{j+1} = \mathbf{y}_i^{j+1}/||\mathbf{y}_i^{j+1}||$, making the $\mathbf{w}_i^{j+1}$ unit vectors. Then $||\mathbf{y}_i^{j+1}||$ measures the one-step growth in direction i, and since $r_i^n \approx ||\mathbf{y}_i^n|| \cdots ||\mathbf{y}_i^1||$, the expression

$$\frac{\ln ||\mathbf{y}_i^n|| + \cdots + \ln ||\mathbf{y}_i^1||}{n}$$

is a convenient estimate for the ith largest Lyapunov exponent after n steps.

EXAMPLE 5.6

Consider the Hénon map introduced in Chapter 2 with parameters $a = 1.4$ and $b = 0.3$. Typical initial conditions have Lyapunov exponents which can be approximated by the method described above. Reasonably accurate approximations are $h_1 = 0.42$ and $h_2 = -1.62$.

Figure 5.3 The Ikeda attractor of Example 5.7.
The attractor has fractal structure and a largest Lyapunov exponent of approximately 0.51.

EXAMPLE 5.7

The Ikeda map is given by

$$F(x, y) = (R + C_2(x \cos \tau - y \sin \tau), C_2(x \sin \tau + y \cos \tau)), \qquad (5.6)$$

where $\tau = C_1 - C_3/(1 + x^2 + y^2)$, and R, C_1, C_2, and C_3 are real parameters. This map was proposed as a model, under some simplifying assumptions, of the type of cell that might be used in a optical computer. The map is invertible with Jacobian determinant C_2^2 for each (x, y), therefore $L_1 L_2 = C_2^2$. In certain ranges of the parameters, there are two fixed point sinks, corresponding to two stable light frequencies in the cell. Setting $R = 1, C_1 = 0.4, C_2 = 0.9$, and $C_3 = 6$, one of these sinks has developed into what is numerically observed to be a chaotic attractor, with Lyapunov numbers $L_1 = 1.66$ and $L_2 = 0.487$ (Lyapunov exponents 0.51 and -0.72). The orbit of one initial condition is shown in Figure 5.3. The orbits of most initial points chosen in a neighborhood of the attractor appear to converge to the same limit set.

⇨   **C O M P U T E R E X P E R I M E N T 5 . 2**

Write a program to measure Lyapunov exponents. Check the program by comparing your approximation for the Hénon or Ikeda map with what is given in the text. Calculate the Lyapunov exponents of the Tinkerbell map quasiperiodic orbit from Computer Exercise 5.1 (one should be zero). Finally, change the first parameter of Tinkerbell to $c_1 = 0.9$ and repeat. Plot the orbit to see the graceful-looking chaotic attractor which gives the map its name.

5.3 LYAPUNOV DIMENSION

There is a relationship between the Lyapunov exponents and the fractal dimension of a typical chaotic attractor. A definition of dimension that acknowledges this relationship has been proposed, called the Lyapunov dimension. In general, this dimension gives a different number than the box-counting dimension, although they are usually not far apart. The appealing feature of this dimension is that it is easy to calculate, if the Lyapunov exponents are known. No boxes need to be counted. We will begin with the definition, which is fairly simple to state, and then explain where it comes from.

Definition 5.8 Let **f** be a map on $\mathbb{R}^m$. Consider an orbit with Lyapunov exponents $h_1 \geq \cdots \geq h_m$, and let p denote the largest integer such that

$$\sum_{i=1}^{p} h_i \geq 0. \tag{5.7}$$

Define the **Lyapunov dimension** D_L of the orbit by

$$D_L = \begin{cases} 0 & \text{if no such } p \text{ exists} \\ p + \dfrac{1}{|h_{p+1}|} \displaystyle\sum_{i=1}^{p} h_i & \text{if } p < m \\ m & \text{if } p = m \end{cases} \tag{5.8}$$

In the case of a two-dimensional map with $h_1 > 0 > h_2$ and $h_1 + h_2 < 0$ (for example, the Hénon map and the skinny baker map), (5.8) yields

$$D_L = 1 + \frac{h_1}{|h_2|}. \tag{5.9}$$

Inserting the Lyapunov exponents for the skinny baker map into the Lyapunov dimension formula yields

$$D_L = 1 + \frac{\ln 2}{\ln 3}. \tag{5.10}$$

This number exactly agrees with the box-counting dimension of the skinny baker invariant set (see Exercise T4.9). To be precise, the definition of Lyapunov dimension applies to an orbit. In dealing with an arbitrary invariant set, such as a chaotic attractor, we are making the assumption that the orbit whose dimension we have calculated is in some way representative of the invariant set. For example, we have seen that the invariant set of the logistic map (the unit interval) contains a dense orbit. (Recall that a "dense" orbit is one that comes arbitrarily close to each point in the invariant set.) The idea of chaotic attractors containing representative orbits (and, in particular, dense orbits) is developed further in Chapter 6.

All that remains is to explain the reasoning behind the definition. We start with an invariant set of a map on $\mathbb{R}^2$, such as that of the skinny baker map. We assume that the Lyapunov exponents satisfy $h_1 \geq 0 \geq h_2$ and $h_1 + h_2 < 0$. There is one direction with a stretching factor of $e^{h_1} > 1$ per iterate (on average), and a perpendicular direction with shrinking factor $0 < e^{h_2} < 1$. In the vicinity of the orbit, areas change at a rate proportional to $e^{h_1} e^{h_2} = e^{h_1 + h_2} < 1$ per iterate, which means that they decrease toward zero.

Figure 5.4 illustrates the situation. A small square of side d becomes (approximately) a rectangle with sides de^{kh_1} and de^{kh_2} after k iterates of the map $\mathbf{f}$. The area is $d^2(e^{h_1 + h_2})^k$, which tends to zero, but the dimension of the resulting invariant set is at least 1, due to the expansion in one direction. Assuming that the invariant set of the map is bounded, the box shrinks down into a long one-dimensional curve that eventually doubles back on itself; this is repeated indefinitely. We expect the dimension of the invariant set to be one plus a fractional amount due to fractal structure perpendicular to the expanding direction.

With this picture in mind, we can count boxes as we did in Chapter 4, the difference being that instead of having an actual set to cover, all we have is Figure 5.4. We can cover the image of the initial box under $\mathbf{f}^k$ by de^{kh_1}/de^{kh_2} boxes of

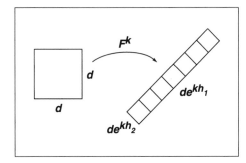

Figure 5.4   Image of a square under a plane map.
A square of side d is mapped approximately into a rectangle by $\mathbf{f}^k$.

side $\epsilon = de^{kh_2}$. Proceeding as in the case of box dimension, we find that for an invariant set which can be covered by $N(\epsilon)$ boxes of side length ϵ

$$\frac{\ln N(\epsilon)}{-\ln \epsilon} \approx -\frac{\ln N(d)e^{k(h_1 - h_2)}}{\ln de^{kh_2}} = -\frac{k(h_1 - h_2) + \ln N(d)}{kh_2 + \ln d} \quad (5.11)$$

$$= \frac{h_2 - h_1 + [\ln N(d)]/k}{h_2 - (\ln d)/k}.$$

If we let the box size ϵ go to zero by letting k approach infinity, we find the value $1 - h_1/h_2$ for the dimension. We call the result of this heuristic argument the Lyapunov dimension $D_L = 1 + h_1/|h_2|$, which agrees with our general definition above.

EXAMPLE 5.9

The Lyapunov dimension of the invariant set of the Hénon map discussed above is $1 + 0.39/1.59 \approx 1.25$.

EXERCISE T5.4

Find the Lyapunov dimensions of the invariant sets of the cat and Ikeda maps described above.

For completeness, note that the case $0 > h_1 \geq h_2$ is trivial, since it corresponds to an invariant set that is discrete (such as a periodic sink). The formula (5.8) gives 0 in this case. If $0 < h_2 \leq h_1$, area is expanding on the bounded invariant set, and the formula gives a dimension of 2.

To complete this section we will extend our heuristic argument to higher dimensions, in order to understand the definition of Lyapunov dimension for maps on $\mathbb{R}^m$. To decide on the size of boxes that are relevant, we need to find the largest integer p for which p-dimensional volume is not shrinking under the application of the map $\mathbf{f}$. For example, for the area-contracting Hénon map of Example 5.6, $p = 1$, since area is contracting by a factor of 0.3 per iteration, while the map is typically expanding in one direction. Then we visualize the invariant set as p-dimensional tubes winding around one another, giving a dimension somewhat larger than p. It is clear that p is the largest integer such that $e^{h_1 + \cdots + h_p} \geq 1$, or equivalently, such that $h_1 + \cdots + h_p \geq 0$.

We will illustrate a couple of interesting cases in $\mathbb{R}^3$, assuming that there is a single positive Lyapunov exponent. With the assumption $h_1 > 0 > h_2 \geq h_3$, (5.8) distinguishes two cases, depending on the sign of $h_1 + h_2$. Thus arises the spaghetti-lasagna dichotomy.

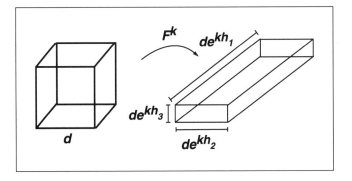

Figure 5.5   Three-dimensional version of Figure 5.4.
A small cube of side d maps into box.

Case 1. $h_1 + h_2 < 0.$

In this case $p = 1$, meaning that the box in Figure 5.5 shrinks toward a one-dimensional curve which winds around itself as in Figure 5.6(a). The curves have lengths and widths regulated by h_1 and h_2, while the perpendicular direction associated to h_3 is irrelevant, since the invariant set contracts to a plane (locally) in this direction. Ignoring this direction, we cover the invariant set by 2-dimensional boxes of side de^{kh_2} as in Figure 5.4, and get $D_L = 1 + h_1/|h_2|$.

Case 2. $h_1 + h_2 \geq 0.$

Now $p = 2$, so that the invariant set is the limiting case of two-dimensional "ribbons" winding throughout a bounded region of $\mathbb{R}^3$, as in Figure 5.6(b). The fractal width of this limit will be decided by the h_3 direction. Imagine cutting up the rectangles with thickness in this figure into three-dimensional cubes of side de^{kh_3}. The number we need is the product of e^{kh_1}/e^{kh_3} along one direction and e^{kh_2}/e^{kh_3} along the other. In total, it takes $e^{k(h_1-h_3)} \times e^{k(h_2-h_3)}$ of these boxes to cover, resulting in

$$
\frac{\ln N(\epsilon)}{-\ln \epsilon} \approx -\frac{\ln N(d)e^{k(h_1-h_3)+k(h_2-h_3)}}{\ln de^{kh_3}}
$$

$$
= -\frac{k(h_1 + h_2 - 2h_3) + \ln N(d)}{kh_3 + \ln d}
$$

$$
= \frac{2h_3 - h_1 - h_2 + [\ln N(d)]/k}{h_3 - (\ln d)/k}, \tag{5.12}
$$

which approaches $D_L = 2 + (h_1 + h_2)/|h_3|$ in the limit as $k \to \infty$, since $h_3 < 0$.

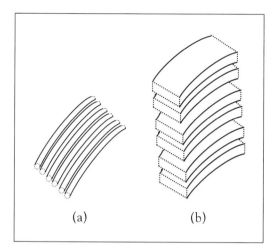

(a)   (b)

Figure 5.6 The spaghetti-lasagna dichotomy.
(a) Illustration of Case 1. Since $h_1 + h_2 < 0$, a cube eventually shrinks into strands
of spaghetti. (b) In Case 2, where $h_1 + h_2 \geq 0$, you get lasagna.

The general case is an extension of this reasoning. Assuming that $h_1 + \cdots +$
$h_p \geq 0$ and $h_1 + \cdots + h_p + h_{p+1} < 0$, we can use $(p + 1)$-dimensional boxes of side
$de^{kh_{p+1}}$. The number of boxes needed is $e^{k(h_1 - h_{p+1})} \times e^{k(h_2 - h_{p+1})} \times \cdots \times e^{k(h_p - h_{p+1})}$,
resulting in a Lyapunov dimension of

$$\frac{\ln N(\epsilon)}{-\ln \epsilon} \approx -\frac{k(h_1 + \ldots + h_p - ph_{p+1})}{kh_{p+1} + \ln d} \to p + \frac{h_1 + \ldots + h_p}{|h_{p+1}|}. \qquad (5.13)$$

5.4 A Two-Dimensional Fixed-Point Theorem

In Section 3.4 of Chapter 3, we developed a fixed-point theorem in conjunction with itineraries for the purpose of determining the periodic points of one-
dimensional maps. A two-dimensional analogue is shown in Figure 5.7. Start with
a geographical map of a rectangular area, say the state of Colorado. Take a second
identical map, shrink it in one direction, stretch it in the other, and lay it across
the first. Then there must be a point in Colorado that lies exactly over itself in
the two maps. The exact orientation of the maps is not important; if the top map
is moved a little from the position shown in Figure 5.7, the fixed point may be
moved slightly, but there will still be one. The principal hypothesis is that the
horizontal sides of the top map are both outside the horizontal sides of the lower

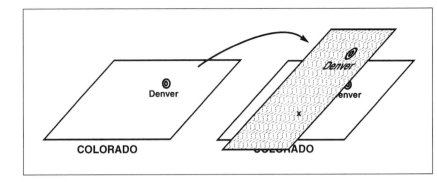

Figure 5.7   The Colorado Corollary.
If a map of Colorado is overlaid with a stretched version of the same map, then there must be a place in the state lying over itself on the two maps.

map, and the vertical sides of the top one are both inside the vertical sides of the lower one.

We want to extract the fundamental elements of this interesting fact and explain why it works. Let **f** be a continuous map on the plane and S be a rectangular region. Imagine the image $\mathbf{f}(S)$ as some deformed version of a rectangle (as in Figure 5.8). Now consider restricting your view of the map to the boundary of S only. In the figure, vectors are drawn connecting each point **v** on the boundary to its image point $\mathbf{f}(\mathbf{v})$. If you begin at one point on the rectangle boundary and make one trip around, returning where you started, the (directions of the) vectors $\mathbf{V}(\mathbf{x}) = \mathbf{f}(\mathbf{x}) - \mathbf{x}$ will make an integer number of turns. The vectors will travel through a cumulative $n \cdot 360°$ for some integer n. Here we are considering the net effect of going around the entire boundary; the vectors could go several turns in one direction, and then unwind to end up with zero net turns by the time you return to the starting point. The net rotations must be an integer, since you return to the vector $\mathbf{V}(\mathbf{x})$ where you started. In Figure 5.8, part (a) shows a net rotation of 1 turn on the boundary, while part (c) shows a net rotation of 0 turns. The next theorem guarantees that (a) implies the existence of a fixed point; (c) does not.

Theorem 5.10   *Let **f** be a continuous map on $\mathbb{R}^2$, and S a rectangular region such that as the boundary of S is traversed, the net rotation of the vectors $\mathbf{f}(\mathbf{x}) - \mathbf{x}$ is nonzero. Then **f** has a fixed point in S.*

Proof: There is nothing to do if the center **c** of the rectangle is already a fixed point. If it isn't, we slowly shrink the rectangle down from its original size to the point **c**, and find a fixed point along the way. As the rectangle shrinks,

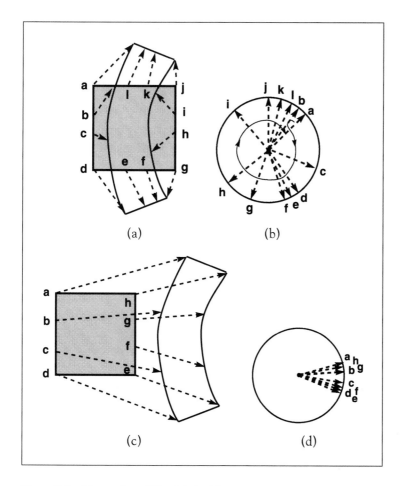

Figure 5.8   Illustration of Fixed-Point Theorem.
A direction $V(x) = f(x) - x$ is defined for each x on the rectangle's boundary, assuming that $f(x) - x$ is never zero on the boundary. (a) As x is followed counterclockwise around the rectangle, the directions $V(x)$ make one complete clockwise turn when $f(S)$ lies across S. (b) The direction vectors from (a) are normalized to unit length and moved to a unit circle. A net rotation of one can be seen. (c) A rectangle on which the map does not have a fixed point. (d) The vector directions have a net rotation of zero turns.

the image shrinks, and the vectors $V(x)$ defined along the boundary of S change continuously, but they must continue to make at least one full turn, unless there is some point x where the direction $V(x)$ fails to make sense because $V(x) = \vec{0}$. (The net rotation is an integer and cannot jump from one integer to another while the vector directions are changing continuously.) Of course, $V(x) = \vec{0}$ implies a fixed point $x = f(x)$.

But this failure must occur, because when the shrinking rectangle gets small enough, its image must move completely away from it, as in Figure 5.8(c), since $\mathbf{f}(\mathbf{c})$ is a nonzero distance from $\mathbf{c}$, and $\mathbf{f}$ is continuous. At this point the net rotation of $\mathbf{V}(\mathbf{x})$ is zero, so that such a failure must have occurred sometime during the shrinking process.   □

What is really happening here is that as we shrink the rectangle, we will in fact develop a fixed point on the boundary of the rectangle, so the net rotation is not defined. At this point the net rotation can jump discontinuously. As the rectangle gets very small, all the vectors on its boundary nearly point in the same direction and the net rotation must be zero.

In fact, this technique could be used to locate a fixed point. Consider a computer program that computes the net rotation for a rectangle; then when the rectangle has a nonzero rotation number, the program shrinks the rectangle slightly. The rotation number will be unchanged. The program keeps shrinking it until the rotation number suddenly jumps. This locates a rectangle with a fixed point on it. The program could then search the rectangle's boundary to locate the fixed point with high accuracy.

More about fixed point theorems and the use of net rotation, or winding number, to prove them can be found in texts on elementary topology (Chinn and Steenrod, 1966).

Remark 5.11   The Fixed-Point Theorem is a property of the topology of the map alone, and doesn't depend on starting with a perfect rectangle, as long as the region has no holes.

Figure 5.9(a) shows a general rectangular set S, and part (b) shows an image $f(S)$ lying across S, similar to the situation with the map of Colorado. Note that $f(S_L)$ lies entirely to the right of S_L, $f(S_R)$ lies entirely to the left of S_R, $f(S_T)$ lies entirely above S_T, and $f(S_B)$ lies entirely below S_B. These facts imply that the net rotation is one counterclockwise turn. The theorem says that there must be a fixed point lying in S.

✎   **EXERCISE T5.5**

Prove the Brouwer Fixed-Point Theorem in two dimensions: If $\mathbf{f}$ is a one-to-one map, S is a square and $\mathbf{f}(S)$ is contained in S, then there is a fixed point in S.

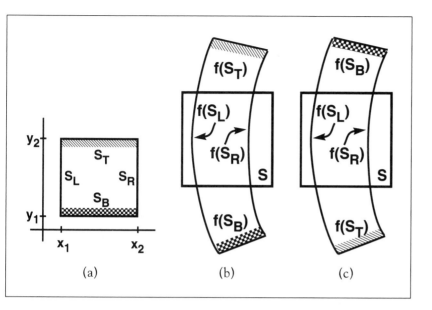

Figure 5.9 The setting for Corollary 5.12.
(a) The domain rectangle S. (b) The image f(S) lies across S consistent with the hypotheses of Corollary 5.12. There must be a fixed point of f inside S. (c) Another way the hypotheses can be satisfied.

We summarize these ideas in the next corollary. Check that the nonzero rotation hypothesis of Theorem 5.10 is satisfied by the hypothesis of Corollary 5.12.

Corollary 5.12 (The Colorado Corollary.) Let f be a continuous map on $\mathbb{R}^2$, and let S be a rectangle in $\mathbb{R}^2$ with vertical sides s_L and s_R, and horizontal sides s_T and s_B. Assume that $f(s_L)$ and $f(s_R)$ are surrounded by s_L and s_R (in terms of x-coordinates) and that $f(s_T)$ and $f(s_B)$ surround s_T and s_B (in terms of y-coordinates). Then f has a fixed point in S.

Referring to Figure 5.10, we will say that "$f(A)$ lies across B vertically" if the images of the right and left sides of A lie inside the right and left sides of B (in terms of x-coordinates), and the images of the top and bottom sides lie outside B, one on or above the top of B and one on or below the bottom of B. There is a similar definition where $f(A)$ lies across B horizontally. In fact, if f is invertible and $f(A)$ stretches across B, then $f^{-1}(B)$ stretches across A.

Note that this "lying across" property is transitive. That is, if $f(A)$ lies across B and if $f(B)$ lies across C with the same orientation, then some curvilinear subrectangle of $f^2(A)$ lies across C with this orientation. (See Figure 5.10.) If

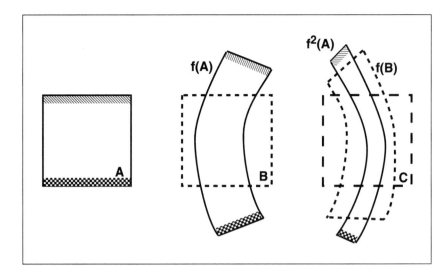

Figure 5.10   Transitivity of "lying across".
If $\mathbf{f}(A)$ lies across B, and $\mathbf{f}(B)$ lies across C, then $\mathbf{f}^2(A)$ lies across C.

there is a sequence $S_1 \dots S_k S_1$ such that $\mathbf{f}(S_1)$ lies across S_2, $\mathbf{f}(S_2)$ lies across S_3, etc., and $\mathbf{f}(S_k)$ lies across S_1, then we conclude that $\mathbf{f}^k(S_1)$ lies across S_1, and by Theorem 5.10, there is a fixed point of $\mathbf{f}^k$ in S_1. The orbit of this point travels through all of the S_i, so there is a fixed point of $\mathbf{f}^k$ in each of the S_i.

Corollary 5.13   Let $\mathbf{f}$ be a map and $\{S_1, \dots, S_k\}$ be rectangular sets in $\mathbb{R}^2$ such that $\mathbf{f}(S_i)$ lies across S_{i+1} for $1 \le i \le k - 1$ and $\mathbf{f}(S_k)$ lies across S_1, all with the same orientation. Then $\mathbf{f}^k$ has a fixed point in S_1.

5.5 MARKOV PARTITIONS

We are now ready to update our definition of covering partition from Chapter 3 to two-dimensional maps. Let $\mathbf{f}$ be a one-to-one map of $\mathbb{R}^2$.

Definition 5.14   Assume that $S_1, \dots, S_r$ are rectangular subsets of a rectangle S whose interiors do not overlap. For simplicity we will assume that the rectangles are formed from segments parallel to the coordinate axes and that the map $\mathbf{f}$ stretches the rectangles in the direction of one axis and contracts in the direction of the other. Assume that whenever $\mathbf{f}(S_i)$ intersects S_j in a set of nonzero area, $\mathbf{f}(S_i)$ "lies across S_j" so that stretching directions are mapped to stretching directions, and shrinking directions to shrinking directions. Then we say that $\{S_1, \dots, S_r\}$ is a **Markov partition** of S for $\mathbf{f}$.

Corollary 5.13 gives a way of constructing fixed points from symbol sequences of Markov partitions. We can define itineraries of an orbit by assigning a symbol to each subset in the Markov partition and tracing the appearances of the orbit in these subsets. To illustrate this idea, we return to the skinny baker map.

EXAMPLE 5.15

The skinny baker map $\mathbf{B}$ in (5.1) has a Markov partition of the unit square consisting of two rectangles L and R. It is clear from Figure 5.11 that L maps across L and R vertically under the map $\mathbf{B}$, and similarly for R.

We will construct itineraries for the points in the unit square that belong to the invariant set of $\mathbf{B}$; that is, those points whose forward iterates and inverse images all lie in the unit square. In Example 5.3 we denoted the invariant set by A.

In contrast to one-dimensional Markov partitions, the itineraries for the skinny baker map $\mathbf{B}$ are "bi-infinite"; that is, they are defined for $-\infty < i < \infty$. The itinerary of a point $\mathbf{v} = (x_1, x_2)$ is a string of subsets from the Markov partition

$$\cdots S_{-2}S_{-1}S_0 \bullet S_1 S_2 S_3 \cdots ,$$

where the symbol $\mathsf{S_i}$ is defined by $\mathbf{B}^i(\mathbf{v}) \in \mathsf{S_i}$. Note that $\mathbf{B}$ is one-to-one, so that $\mathbf{B}^{-1}$ is defined on the image of $\mathbf{B}$, and in particular for the invariant set A of the square S. We will form itineraries only for points that lie in the invariant set A

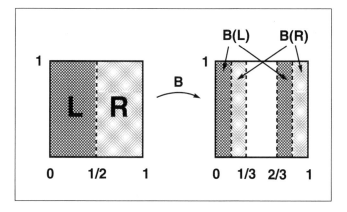

Figure 5.11 A Markov partition of the skinny baker map.
The partition consists of the left and right halves of the unit square. The center third of the square has no pre-images.

of the map **B**; these points have inverse images $\mathbf{B}^{-i}(\mathbf{v})$ for all i. Note also that $\mathbf{B}^{-i}(\mathbf{v})$ is in S_j if and only if $\mathbf{v}$ is in $B^i(\mathsf{S}_j)$.

To see what itineraries look like, we first investigate the right-hand side of the decimal point. Figure 5.12(a) shows the subsets of the unit square that begin with **.L** and **.R**, respectively. The lower half is denoted **.L** because after one iteration of B, those points all land in the left half of the square. The **.L** rectangle is divided into **.LL** and **.LR** in Figure 5.12(b), separating the points in the lower half of **.L**, which map to the left side after two iterations from the points of the upper half of **.L**, which map to the right after two iterations. Further subdivisions, which track the third iterate of **B**, are displayed in Figure 5.12(c).

We can already see the use of Markov partitions for locating periodic points of **B**. For example, the strip marked **.LL** in Figure 5.12 must contain a fixed point, as does **.RR**, because of the repeated symbol. (The fixed points are $(0, 0)$ and $(1, 1)$, respectively.)

✎ EXERCISE T5.6

Find the (x_1, x_2)-coordinates of the period-two orbits in the strips **.LRL** and **.RLR** of Figure 5.12. Are there any other period-two orbits?

Figure 5.13 shows the forward images of the unit square under the map **B**. Figures 5.13(a)-(c) show the images of **B**, $\mathbf{B}^2$, and $\mathbf{B}^3$, respectively. The part of the

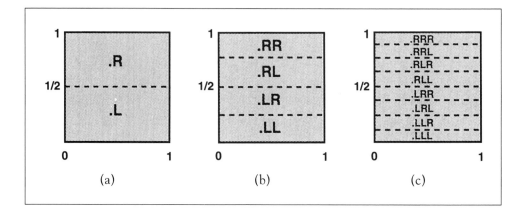

Figure 5.12   Forward itineraries of points of the unit square under the skinny baker map.

The unit squares in (a), (b), and (c) show successive refinements of regions defined by itineraries. The points in **.RLR**, for example, map to the right half under **B**, the left half under $\mathbf{B}^2$, and the right half under $\mathbf{B}^3$.

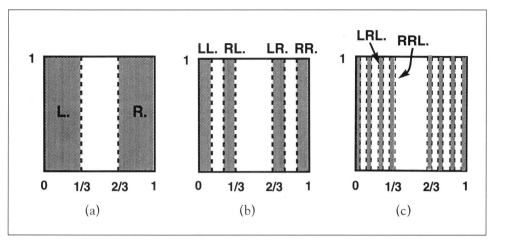

Figure 5.13   Backward itineraries of points of the unit square under the skinny baker map.
Each point **p** in **RRL.**, for example, satisfies **p** is in **L**, **p** is in $B(\mathbf{R})$, and **p** is in $B^2(\mathbf{R})$.

itinerary to the left of the decimal point shows the inverse image history of that point. For example, consider a point lying in the vertical strip denoted by **RRL.**. Reading symbols from right to left, the point lies in **L**, it is the iterate of a point $\mathbf{p}_1$ in **R**, and it is the second iterate of a point $\mathbf{p}_2$ in **R**. (The reader is encouraged to choose a point in **RRL.** and determine $\mathbf{p}_1$ and $\mathbf{p}_2$ visually.)

It is clear from Figure 5.12 that the horizontal strip represented by $.\mathbf{S}_1 \cdots \mathbf{S}_k$ has width 2^{-k}. An infinite sequence $.\mathbf{S}_1\mathbf{S}_2 \cdots$ corresponds to a single horizontal line. Similarly, Figure 5.13 shows that the points with symbol sequence $\mathbf{S}_{-k} \cdots \mathbf{S}_{-1}\mathbf{S}_0.$ form a vertical strip of width $3^{-(k+1)}$, and any particular infinite sequence extending to the left corresponds to a single vertical line of points.

✎ EXERCISE T5.7

Construct the periodic table for the skinny baker map, for periods up to 6.

Putting these facts together, we see that each bi-infinite sequence of symbols is owned by a single point in the unit square. So the itinerary of a point in the invariant set A gives us a good address to identify that point. The correspondence between bi-infinite sequences and points in A is not quite one-to-one, since some sequences identify the same point. For example, $\overline{\mathbf{L}}.\mathbf{R}\overline{\mathbf{L}}$ and $\overline{\mathbf{L}}.\mathbf{L}\overline{\mathbf{R}}$ both identify the point $(0, 1/2)$.

Consider a point with itinerary $\cdots S_{-k} \cdots S_{-1} S_0 . S_1 S_2 \cdots S_k \cdots$. The image of this point under the map $\mathbf{B}$ has the itinerary $\cdots S_{-k} \cdots S_{-1} S_0 S_1 . S_2 \cdots S_k \cdots$. That is, the effect of the map $\mathbf{B}$ on itineraries is to shift the symbols one to the left, with respect to the decimal point.

Definition 5.16 The **shift map** s is defined on the space of two-sided itineraries by:

$$s(\ldots S_{-2} S_{-1} S_0 . S_1 S_2 S_3 \ldots) = \ldots S_{-2} S_{-1} S_0 S_1 . S_2 S_3 \ldots .$$

Secondly, as we mentioned above, if the itinerary consists of a repeated sequence of k symbols, the point is contained in a periodic orbit of period k (or smaller). The orbit is asymptotically periodic if and only if the itinerary is eventually periodic toward the right.

This gives us a good way to identify chaotic orbits. Any itinerary that is not periodic toward the right is not asymptotically periodic. Recall that the Lyapunov exponents of every orbit of the baker map are $-\ln 3$ and $\ln 2$, the latter being positive. So any itinerary that does not eventually repeat a single finite symbol sequence toward the right is a chaotic orbit.

Theorem 5.17 *The skinny baker map has chaotic orbits.*

5.6 THE HORSESHOE MAP

The skinny baker map has a regular construction that makes the itineraries fairly easy to organize. Its discontinuity at the line $y = 1/2$ makes it less than ideal as a model for continuous natural processes. The horseshoe map, on the other hand, is a model that can easily be identified in continuous nonlinear maps like the Hénon map.

For example, Figure 5.14 shows a quadrilateral, which we shall refer to loosely as a rectangle, and its image under the Hénon map (2.8) of Chapter 2. Notice that the image of the rectangle takes a shape vaguely reminiscent of a horseshoe, and lies across the original rectangle. The dynamics that result in the invariant set are in many ways similar to those of the baker map.

The **horseshoe map** is a creation of S. Smale. Define a continuous one-to-one map $\mathbf{h}$ on $\mathbb{R}^2$ as follows: Map the square $W = ABCD$ to the overlapping horseshoe image, as shown in Figure 5.15, with $\mathbf{h}(A) = A^*$, $\mathbf{h}(B) = B^*$, $\mathbf{h}(C) = C^*$, and $\mathbf{h}(D) = D^*$. We assume that in W the map uniformly contracts distances horizontally and expands distances vertically. To be definite, we could assume

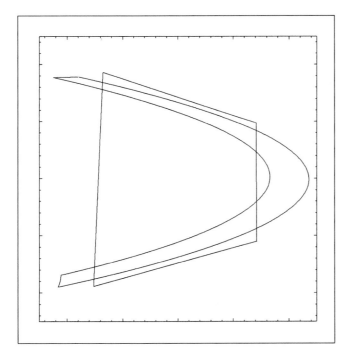

Figure 5.14   A horseshoe in the Hénon map.
A quadrilateral and its horseshoe-shaped image are shown. Parameter values are
$a = 4.0$ and $b = -0.3$.

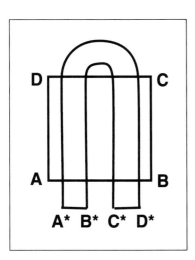

Figure 5.15   The horseshoe map.
The map sends the square $W = ABCD$ to its image $A^*B^*C^*D^*$, which is shaped
like a horseshoe.

points are stretched vertically by a factor of 4, and squeezed horizontally by a factor of 4. Outside W, the only restriction we make on $\mathbf{h}$ is that it be continuous and one-to-one.

In our analysis of the horseshoe map, we focus on the points in the plane that remain in the rectangle W for all forward and backward iterates. These points form the invariant set H of $\mathbf{h}$.

Notice that any point in H must be in either the left leg V_L or the right leg V_R of $W \cap \mathbf{h}(W)$. We can assign itineraries to these points that depend on the location, V_L or V_R, of each iterate. As with the baker map, the itinerary that gives the forward iterates does not uniquely define a point, and both the past and future of a point are needed in order to identify it. For a point $\mathbf{v}$ in H, we define the itinerary $\overline{\mathbf{v}}$ of $\mathbf{v}$ to be $\ldots S_{-3}S_{-2}S_{-1}S_0 \bullet S_1 S_2 \ldots$, as follows:

1. If $\mathbf{h}^i(\mathbf{v})$ lies in V_L, set $S_i = \mathsf{L}$.
2. If $\mathbf{h}^i(\mathbf{v})$ lies in V_R, set $S_i = \mathsf{R}$.

Which points in W map into V_L and which map into V_R? Imagine unfolding the horseshoe image, stretching it in one direction and shrinking it in another, so that square $A^*B^*C^*D^*$ can be placed exactly on square $ABCD$. Then the points corresponding to V_L and V_R form two horizontal strips, labeled $\bullet \mathsf{L}$ and $\bullet \mathsf{R}$ in Figure 5.16. The points in $\bullet \mathsf{L}$ (respectively, $\bullet \mathsf{R}$) are those that map into V_L (respectively, V_R) and for which S_1 is L (respectively, R).

In order to assign coordinate S_2, "unfold" $\mathbf{h}^2(W)$, as shown in Figure 5.17. The four sets of points whose itineraries begin $S_1 S_2$ form four horizontal strips, nested in pairs inside the strips $\bullet \mathsf{L}$ and $\bullet \mathsf{R}$. Again, each time we specify an

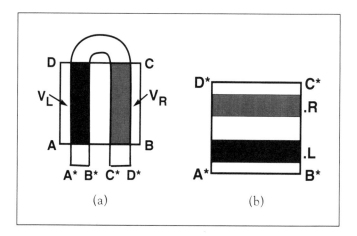

(a)   (b)

Figure 5.16   Unfolding of the horseshoe map.
The grey (resp., black) region in (b) maps to the grey (resp., black) region in (a).

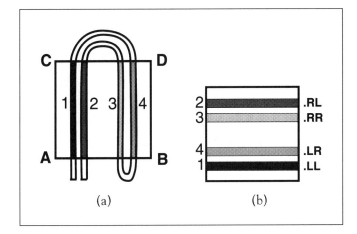

Figure 5.17 Unfolding $h^2(W)$, the second iterate of the horseshoe map.
Forward itineraries of points in the unit square are shown.

additional coordinate, we double the number of horizontal strips. When the entire sequence $.S_1S_2S_3 \ldots$ is specified, the set of points represented is a horizontal line segment. The collection of these segments, for all possible such sequences, forms a Cantor set of horizontal line segments.

Now we turn to the inverse iterates. Which points have $\mathbf{h}^{-1}$ in V_L and which in V_R? To begin with, if the inverse iterate of a point $\mathbf{v}$ in W is to remain in W, $\mathbf{v}$ must be in V_L or V_R. If we do the inverse procedure of $\mathbf{h}$, i.e., unfolding $\mathbf{h}(W)$, then, as we saw previously, the inverse images of V_L and V_R are the horizontal strips $.\mathbf{L}$ and $.\mathbf{R}$. The set of points within $.\mathbf{L}$ that are in V_L are labeled $\mathbf{L}.\mathbf{L}$ in Figure 5.18(a); those in V_R are labeled $\mathbf{R}.\mathbf{L}$. Figure 5.18(b) shows the points $\mathbf{v}$ in W for which $\bar{\mathbf{v}}$ has $S_{-1}S_{0\bullet}$ equal to each of the four partial sequences: $\mathbf{LL}_\bullet$, $\mathbf{RL}_\bullet$, $\mathbf{RR}_\bullet$, and $\mathbf{LR}_\bullet$.

By specifying the entire one-sided sequence $\ldots S_{-2}S_{-1}S_{0\bullet}$, we represent a vertical line segment of points. The collection of all such sequences is a Cantor set of vertical line segments. To specify exactly one point in H, we intersect the vertical line segment represented by $\ldots S_{-2}S_{-1}S_{0\bullet}$ and the horizontal line segment represented by $.S_1S_2S_3 \ldots$. Thus each two-sided itinerary $\ldots S_{-2}S_{-1}S_{0\bullet}S_1S_2S_3 \ldots$ corresponds to exactly one point in H. Notice that an assumption of uniform stretching in one direction and contraction in the other is necessary for the width of the vertical strips and the height of the horizontal strips to go to zero, ensuring the one-to-one correspondence.

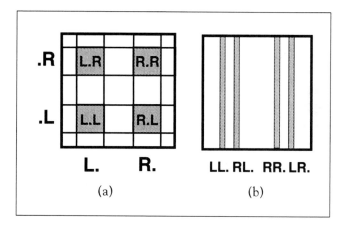

Figure 5.18 Itineraries of the horseshoe map.
(a) Two-sided itineraries are intersections of vertical and horizontal strips. (b) Backward itineraries correspond to vertical strips.

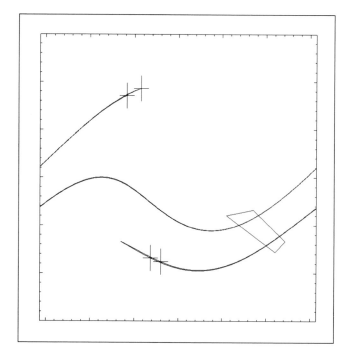

Figure 5.19 Horseshoe in the forced damped pendulum.
The rectangular-shaped region is shown along with its first image under the time-2π map. The image is stretched across the original shape, and is so thin that it looks like a curve, but it does have width. The crosses show the image of the corner points of the domain rectangle.

✎ **EXERCISE T5.8**

Construct the periodic table for the horseshoe map, for periods up to 6.

We can also define a shift map, identical to Definition 5.16 for the baker map. Notice that in the case of two-sided itineraries the shift map is invertible. As with the baker map, the one-to one correspondence says in the context of this definition that $h(v) = w$ if and only if $s(\bar{v}) = \bar{w}$. In particular, there are exactly two fixed points in W, whose symbol sequences are ...RR•RR... and ...LL•LL.....

If we assume that the horseshoe map stretches and contracts uniformly at points in the invariant set and that stretching directions map to stretching directions and contracting directions map to contracting directions, then each orbit in the invariant set has a positive Lyapunov exponent. For example, assume that the square shaped into a horseshoe in Figure 5.15 has width $1/4$ of the width of the original square ABCD, and length three times the original. Then the Lyapunov exponents are $\ln 3$ and $-\ln 4$. With this assumption, we can say the following.

Theorem 5.18 *The horseshoe map has chaotic orbits.*

The forced damped pendulum (2.10) of Chapter 2 also has a horseshoe. Figure 5.19 shows a "rectangle" of initial conditions $(\theta, \dot{\theta})$. The long thin image set represents the images of those states 2π time units later. The corners of the image rectangle are marked with crosses. Recall that θ, the horizontal variable, is periodic with period 2π, so that the image set is a connected set.

The horseshoe map of Theorem 5.18 is a prototype. To the extent that a system such as the forced damped pendulum maps a rectangle across itself in the shape of a horseshoe, and satisfies appropriate stretching and shrinking conditions, the rich itinerary structure of the prototype horseshoe of this section will exist. One of the most profound facts of chaotic dynamics is Theorem 10.7, due to S. Smale, that shows that horseshoes must exist in the vicinity of any transversal homoclinic intersection of stable and unstable manifolds.

We have seen that the existence of a horseshoe in a map forces a great deal of complexity (a two-sided shift map) in the system dynamics. Theoretical discovery of the horseshoe was followed by the identification of its presence in real systems. Color Plates 21–22 show a horseshoe in the dynamics of a laboratory mixing apparatus along with the intricate patterns that follow.

☞ **CHALLENGE 5**

Computer Calculations and Shadowing

ARE COMPUTER calculations reliable? We have shown the results of computer calculations in this book to demonstrate mathematical concepts. Some of these calculations involve many thousands of iterations of a map. Because floating-point computers have a finite degree of precision, there will be small errors made on essentially every operation the computer carries out. Should the pictures we make, and conclusions we reach on the basis of computer calculation, be trusted?

This is a deep question with no definitive answer, but we will begin with some simple examples, and in Challenge 5 work our way to a surprisingly strong positive conclusion to the question. We start by considering fixed points, and establish that it is certainly possible for a computer to produce a misleading result. Let

$$f(x, y) = (x + d, y + d), \tag{5.14}$$

where $d = .000001$. Consider the initial condition $\mathbf{x}_0 = (0, 0)$, and suppose the computer makes an error of exactly $-d$ in each coordinate when f is computed. Then the computer will calculate the incorrect $\hat{f}(0, 0) = (0, 0)$, instead of the correct $f(0, 0) = (d, d)$. The computer says there is a fixed point but it is wrong. The true map has no fixed points or periodic points of any period.

It seems extremely easy to make such a mistake with this map—to find a fixed point when there isn't one. From this example one might infer that using a computer to make mathematical conclusions about interesting maps means trading the world of mathematical truth for "close enough".

The problem is compounded for longer-term simulations. If $k = 10^6$, the correct $f^k(0, 0) = (1, 1)$ has been turned into $\hat{f}^k(0, 0) = (0, 0)$ by the computer. Many small errors have added up to a significant error.

Add to this the consideration that the above map is not chaotic. Suppose we are using a computer to simulate the iteration of a map with sensitive dependence on initial conditions. A digital computer makes small errors in floating-point calculations because its memory represents each number by a finite number of binary digits (bits). What happens when a small rounding error is made by the computer? Essentially, the computer has moved from the true orbit it was supposed to follow to another nearby orbit. As we know, under a chaotic map, the nearby orbit will diverge exponentially fast from the true orbit. To make matters worse, these small rounding errors are being made each iteration! From this point of view, it sounds as if the computer-generated orbit is garbage, and that simulating a chaotic system is hopeless.

The goal of Challenge 5 is to have you take a closer look, before you melt your chips down for scrap. It is true that the orbit the computer generates will not be the actual orbit starting at the initial value the computer was given. As discussed above, sensitive dependence makes it impossible to do otherwise. However, it is possible that the computer-generated orbit closely approximates a true orbit starting from another initial condition, very close to the given initial condition. Often it is close enough so that the computed orbit found is acceptable for the intended purpose.

You will start with the skinny baker map **B**. The first goal is to prove the fact that if the image of $\mathbf{B}(\mathbf{x}_0)$ lies within a small distance of $\mathbf{x}_0$, then there *must* be a fixed point near that pair of points. In other words, the baker map is immune to the problem we saw above for map (5.14).

Step 1 Assume that $\mathbf{B}(\mathbf{x}_0)$ differs from $\mathbf{x}_0$ by less than d in each coordinate. In Figure 5.20 we draw a rectangle centered at $\mathbf{x}_0$ with dimensions $3d$ in the horizontal direction and $2d$ in the vertical direction. Assume that the rectangle lies on one side or the other of the line $y = 1/2$, so that it is not chopped in two by

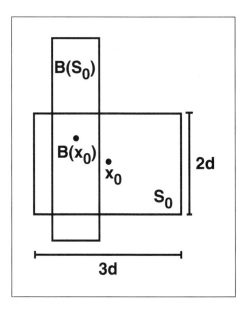

Figure 5.20 Image of a small square under the baker map.
The wide rectangle S_0 maps across itself, implying that a fixed point exists in the overlap.

the map. Then its image is the rectangle shown; the center of the rectangle is of course $\mathbf{B}(\mathbf{x}_0)$. Show that the image of the rectangle is guaranteed to "map across" the original rectangle. Explain why there is a fixed point of $\mathbf{B}$ in the rectangle, within $2d$ of $\mathbf{x}_0$.

Step 2 Now suppose our computer makes mistakes in evaluating $\mathbf{B}$ of size at most 10^{-6}, and it tells us that $\mathbf{B}(\mathbf{x}_0)$ and $\mathbf{x}_0$ are equal within 10^{-6}. Prove that $\mathbf{B}$ has a fixed point within 10^{-5} of $\mathbf{x}_0$.

So far these results are less than astonishing. In order for $\mathbf{B}(\mathbf{x}_0)$ to map near $\mathbf{x}_0$ under the baker map, $\mathbf{x}_0$ must be either near the lower left or upper right corners of the square. Since the two fixed points are $(0,0)$ and $(1,1)$, it's no surprise that $\mathbf{x}_0$ is near a fixed point. The next fact, however, we find amazing.

Step 3 Prove Theorem 5.19.

Theorem 5.19 *Let* $\mathbf{B}$ *denote the skinny baker map, and let* $d > 0$. *Assume that there is a set of points* $\{\mathbf{x}_0, \mathbf{x}_1, \ldots, \mathbf{x}_{k-1}, \mathbf{x}_k = \mathbf{x}_0\}$ *such that each coordinate of* $\mathbf{B}(\mathbf{x}_i)$ *and* $\mathbf{x}_{i+1}$ *differ by less than* d *for* $i = 0, 1, \ldots, k - 1$. *Then there is a periodic orbit* $\{\mathbf{z}_0, \mathbf{z}_1, \ldots, \mathbf{z}_{k-1}\}$ *such that* $|\mathbf{x}_i - \mathbf{z}_i| < 2d$ *for* $i = 0, \ldots, k - 1$.

[*Hint:* Draw a $3d \times 2d$ rectangle S_i centered at each $\mathbf{x}_i$ as in Figure 5.21. Show that $\mathbf{B}(S_i)$ lies across S_{i+1} by drawing a variant of Figure 5.20, and use Corollary 5.13.]

Therefore, if you are computing with the skinny baker map on a computer with rounding errors of $d = .000001$ and find an orbit that matches up within the first seven digits, then you know there is a *true* periodic orbit within at

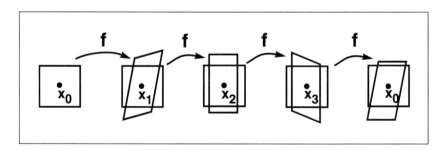

Figure 5.21   Transitivity of lying across.
For the rectangle S_0 on the left and repeated on the right, $\mathbf{f}^4(S_0)$ lies across S_0. This implies that $\mathbf{f}^4$ has a fixed point.

most $2d = .000002$. (The 2 can be replaced by the slightly smaller $\sqrt{13}/2$.) The striking fact about the theorem is that the length of the orbit is nowhere relevant. The theorem can be used to verify the existence of a period-one-million orbit as well as a fixed point. The true orbit that slinks along very close to a computer-generated "approximate orbit" is called a **shadowing orbit**. Moreover, the phenomenon holds even more generally than for periodic orbits.

Step 4   Let **f** be any continuous map, and assume that there is a set of rectangles $S_0, \ldots, S_k$ such that $\mathbf{f}(S_i)$ lies across S_{i+1} for $i = 0, \ldots, k-1$, each with the same orientation. Prove that there is a point $\mathbf{x}_0$ in S_0 such that $\mathbf{f}^i(\mathbf{x}_0)$ lies in the rectangle S_i for all $0 \leq i \leq k$. By the way, does k have to be finite?

Step 5   Let **B** denote the baker map and let $d > 0$. Prove the following: If $\{\mathbf{x}_0, \mathbf{x}_1, \ldots, \mathbf{x}_k\}$ is a set of points such that each coordinate of $\mathbf{B}(\mathbf{x}_i)$ and $\mathbf{x}_{i+1}$ differ by less than d for $i = 0, 1, \ldots, k-1$, then there is a true orbit within $2d$ of the $\mathbf{x}_i$, that is, there exists an orbit $\{\mathbf{z}_0, \mathbf{z}_1, \ldots, \mathbf{z}_k\}$ of **B** such that $|\mathbf{x}_i - \mathbf{z}_i| < 2d$ for $i = 0, \ldots, k$.

Step 6   To what extent can the results for the baker map be reproduced in other maps? What properties are important? Show that the same steps can be carried out for the cat map, for example, by replacing the $3d \times 2d$ rectangle appropriately.

Step 7   Assume that a plot of a length one million orbit of the cat map is made on a computer screen, and that the computer is capable of calculating an iteration of the cat map accurately within 10^{-6}. Do you believe that the dots plotted represent a true orbit of the map (to within the pixels of the screen)?

Step 8   Decide what property is lacking in map (5.14) that allows incorrect conclusions to be made from the computation.

Postscript.   Computer-aided techniques based on Challenge 5 have been used to verify the existence of true orbits near computer-simulation orbits for many different systems, including the logistic map, the Hénon map, the Ikeda map, and the forced damped pendulum (Grebogi, Hammel, Yorke, Sauer, 1990). For these systems, double-precision computations of orbits of several million iterates in length have been shown to be within 10^{-6} of a true orbit, point by point. Therefore, the computer pictures shown in this book represent real orbits, at least within the resolution available to the printer which produced your copy. Astronomers have begun to apply shadowing ideas to simulations of celestial mechanics in order to investigate the accuracy of very long orbits of n-body problems (Quinn and Tremaine, 1992).

EXERCISES

5.1.   Assume that the map $\mathbf{f}$ on $\mathbb{R}^m$ has constant Jacobian determinant, say det $\mathbf{Df(x)} = D$ for each $\mathbf{x}$. Explain why the product of all m Lyapunov numbers is D (equivalently, the sum of the Lyapunov exponents is $\ln D$).

5.2.   Let

$$\mathbf{f}\begin{pmatrix} x_1 \\ x_2 \end{pmatrix} = \begin{pmatrix} 1 & 1 \\ 1 & 0 \end{pmatrix}\begin{pmatrix} x_1 \\ x_2 \end{pmatrix} \ (\mathrm{mod} \ 1).$$

Then $\mathbf{f}$ is defined on the unit square in $\mathbb{R}^2$ (or on the torus). Find the Lyapunov exponents of any orbit of the map. Notice that these numbers are exactly half those of the cat map of Example 5.4. Why?

5.3.   Show that the cat map of Example 5.4 is one-to-one.

5.4.   Draw the transition graphs for the Markov partitions of the skinny baker map and the horseshoe map.

5.5.   Show that the set of chaotic orbits of the horseshoe map is uncountable.

5.6.   Show that the invariant set of the horseshoe map contains a dense chaotic orbit.

5.7.   Consider the map $f(z) = z^2$, where z represents complex numbers.

(a) Use Euler's formula from complex arithmetic to show that f corresponds to the map $p(r, \theta) = (r^2, 2\theta)$ in polar coordinates.

(b) Find all initial points whose trajectories are bounded and do not converge to the origin.

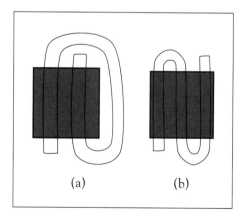

(a)   (b)

Figure 5.22   Images of the unit square.
The maps used in Exercise 5.8. (a) The image of the unit square intersects itself in two pieces, and (b) in three pieces.

(c) Find the Lyapunov exponents of all bounded orbits.

(d) Show that f has chaotic orbits.

5.8. Explore itineraries and periodic points for the two maps defined on the unit square shown in Figure 5.22. For each map, decide in what ways it is similar to, and different from, the horseshoe map. Draw and label with symbols the strips of points which stay in the square for two forward iterates. Repeat for backward iterates. Find the transition graph for each map, and show examples of periodic orbits.

5.9. Let f be a continuous map on $\mathbb{R}^2$, and S a rectangle in $\mathbb{R}^2$ with vertical sides s_L and s_R, and horizontal sides s_T and s_B. Assume that $f(s_L)$ and $f(s_R)$ surround s_L and s_R (in terms of x-coordinates) and that $f(s_T)$ and $f(s_B)$ surround s_T and s_B (in terms of y-coordinates). Show that f has a fixed point in S.

5.10. (from Clark Robinson) Let $f(x, y) = (5 - x^2 - 0.3y, x)$, a version of the Hénon map. Define the rectangles $V_L = [-3, -1] \times [-3, 3]$ and $V_R = [1, 3] \times [-3, 3]$. (a) Show that the net rotation of f on the boundary of the rectangle is nonzero, for both V_L and V_R. (b) Show that $\{V_L, V_R\}$ is a Markov partition for f. (c) What periods are present in the periodic orbits of f?

☞ LAB VISIT 5

Chaos in Simple Mechanical Devices

ONE OF THE most appealing properties of chaos is the fact that it is exhibited by systems on a wide variety of size scales, including scales within the first-hand experience of human beings. One of the most familiar physical systems, to anyone who has used a swing as a child or watched a grandfather clock, is the pendulum. We have already used a mathematical model of the pendulum to illustrate concepts. The equation for the forced, damped pendulum apparently has chaotic trajectories, as shown by a plot of its time-2π map in Figure 2.7 of Chapter 2.

Here we present two experiments which were carried out jointly by the Daimler-Benz Corporation, which has a long-term interest in mechanical systems, and a nearby university in Frankfurt, Germany. Both start with a simple mechanical system, and apply periodic forcing. These nonchaotic elements combine to produce chaos. The researchers built a mechanical pendulum that could be forced externally by torque at its pivot, and a metal ribbon whose oscillations can be forced by placing it in an alternating magnetic field. Both exhibit chaotic orbits in a two-dimensional Poincaré map.

Figure 5.23(a) shows a schematic picture of the pendulum constucted by the group. The damping is due to mechanical friction, and the forcing is applied through an electric motor at the pivot point of the pendulum. The motor applies force sinusoidally, so that the torque alternates in the clockwise/counterclockwise direction, exactly as in Equation (2.10) of Chapter 2.

An experimental time-T map of the pendulum is shown in Figure 5.23(b). The time T is taken to be the forcing period, which is 1.2 seconds. The plot of an orbit of the map bears a strong resemblance to the theoretical pendulum attractor of Figure 2.7 of Chapter 2. There are differences due to discrepancies in parameter settings of the computer-generated system as compared to the experimental

Hübinger, B., Doerner, R., Martienssen, W., Herdering, M., Pitka, R., Dressler, U. 1994. Controlling chaos experimentally in systems exhibiting large effective Lyapunov exponents. Physical Review E **50**:932–948.

Dressler, U., Ritz, T., Schenck zu Schweinsberg, A., Doerner, R., Hübinger, B., Martienssen, W. 1995. Tracking unstable periodic orbits in a bronze ribbon experiment. Physical Review E **51**:1845–8.

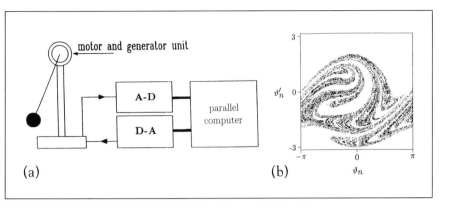

Figure 5.23 The experimental forced damped pendulum.
(a) The experimental setup shows an electric motor at the pivot, controlled by the computer via the digital-analog converter, provides the periodic forcing. (b) A plot of $(\theta, \dot{\theta})$ done in time increments of length 1.2 seconds, the period of the forcing torque at the pivot.

system. The fractal dimension of the time-T map in Figure 5.23(b) was estimated to be 1.8. Since each point represents an entire loop of the trajectory, the entire attractor should have dimension 2.8.

The second experimental setup is an elastic bronze ribbon, shown in Figure 5.24(a). Two small permanent magnets have been attached to the free end of the

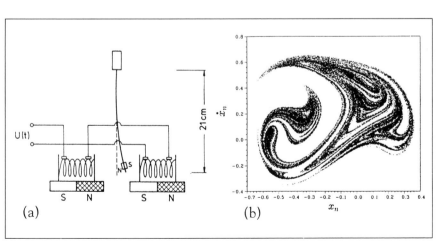

Figure 5.24 The bronze ribbon experiment.
(a) Two small magnets are attached to the free end of the ribbon. Two large permanenet magnets provide an inhomogeneous magnetic field, and an alternating current through the two coils provide a periodic forcing to the ribbon. (b) An orbit of the time-2π map. One hundred thousand $(x, \dot{x})$ pairs are plotted.

ribbon. Define the variable x to be the distance of the tip from its equilibrium position with no bend. Two larger magnets set up a magnetic field such that the original equilibrium $x = 0$ is no longer stable. Instead, there are two stable equilibrium positions symmetrically situated about $x = 0$. When the oscillating magnetic field set up by the coils is added, there are no stable equilibria, and the ribbon wobbles chaotically.

The period of the oscillatory voltage in the coil is 1 second in the experiment reported here. Figure 5.24(b) shows a plot of the time-1 map for the bronze ribbon. The fractal structure of the chaotic attractor is clearly visible.

CHAPTER SIX

Chaotic Attractors

AN IMPORTANT aspect of explaining dynamical phenomena is the description of attractors. Newton knew of two types of attracting motion that systems settle into: the apple sitting on the ground is in equilibrium, and the planets in the solar system are undergoing periodic, or more properly quasiperiodic motion, at least to good approximation. For the next 300 years, these were the only kinds of motion known for simple dynamical systems. Maxwell and Poincaré were among a small number of scientists who were not content with this view. The small number grew, but it was not until the widespread availability of desktop computers in the last quarter of the 20th century that the third type of motion, chaos, became generally recognized.

In previous chapters, we have developed the concept of a chaotic orbit. We have seen they occur in certain one-dimensional quadratic maps and in two-dimensional maps with horseshoes. But can chaotic motion be attracting?

If chaotic motion is to be observed in the motion of a physical system, it must be because the set on which the chaotic motion is occurring attracts a significant portion of initial conditions. If an experimentalist observes a chaotic motion, he or she has chosen (often randomly) an initial condition whose trajectory has converged to a chaotic attractor. This motion could perhaps be described as "stable in the large" (it attracts a large set of initial conditions) while "locally unstable" (it is a chaotic orbit).

Figure 6.1 shows two numerically observed chaotic attractors. The black set in each picture was obtained by plotting a finite trajectory that appears to be chaotic. Throughout the calculation, the orbit appears to have a positive Lyapunov exponent and it fails to approach periodic behavior. Figure 6.1(a) shows a chaotic orbit (in black) of the Hénon map:

$$\mathbf{f}(x, y) = (a - x^2 + by, x), \tag{6.1}$$

where $a = 1.4$ and $b = 0.3$. The gray set in Figure 6.1(a) is the basin of the chaotic orbit, the set of initial values whose orbits converge to the black set. The iteration of any point chosen randomly in the gray region would produce essentially the same black set—that is, after throwing out the initial segment of the trajectory

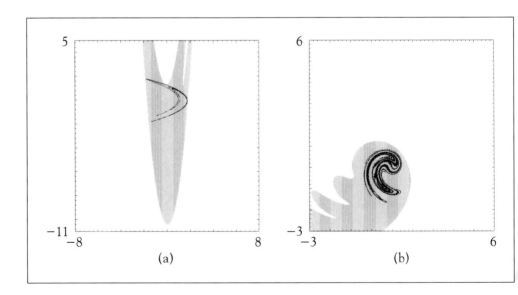

Figure 6.1 Chaotic attractors of plane maps.
Each part shows a chaotic attractor in black inside a gray basin. (a) The chaotic attractor of the Hénon map. White points are attracted to infinity. (b) The chaotic attractor of the Ikeda map. Gray points approach the chaotic attractor; white points approach the sink in the upper right corner.

during the approach to the attractor. The white set consists of initial values whose orbits diverge to infinity.

Figure 6.1(b) shows a chaotic orbit of the Ikeda map, whose governing equation is:

$$f(x, y) = (R + C_2(x \cos \tau - y \sin \tau), C_2(x \sin \tau + y \cos \tau)), \qquad (6.2)$$

where $\tau = C_1 - C_3/(1 + x^2 + y^2)$, and R, C_1, C_2, and C_3 are fixed parameters. For this picture, the settings $R = 0.9, C_1 = 0.4, C_2 = 0.9$, and $C_3 = 6$ were used. This map was proposed as a model of the type of cell that might be used in an optical computer. The chaotic orbit is shown in black; the gray set consists of initial values whose orbits are attracted to the chaotic orbit. The white set consists of initial values whose orbits converge to the sink located approximately at $(3.0026, 3.8945)$.

The two most important properties of a chaotic attractor are demonstrated in Figure 6.1. A chaotic attractor: (1) contains a chaotic orbit, and (2) attracts a set of initial values that has nonzero area in the plane. Both of these properties are fairly intuitive, but both need further elaboration to make them more precise.

6.1 FORWARD LIMIT SETS

The idea of a chaotic orbit is familiar by now. By definition, it is not periodic or asymptotically periodic, and it has at least one positive Lyapunov exponent. Now we introduce the idea of "chaotic attractor", keeping in mind as our model the black limit sets in the previous figures. First of all, a chaotic attractor is a "forward limit set", which we define formally below. It is in some sense what remains after throwing away the first one thousand, or one million, or any large initial number of points of a chaotic orbit. That means that the orbit continually returns to the vicinity of these points far into the future.

Figure 6.2(a) shows the first 1000 points of two different orbits of the Ikeda map. The two initial conditions are marked by crosses. The two initial values lie in separate basins (see Figure 6.1(b)), so their orbits have different asymptotic behavior. Figure 6.2(b) shows iterates 1,000,001 to 2,000,000 of the same two orbits.

For one orbit, the 10^6 plotted points lie on the chaotic Ikeda attractor shown in Figure 6.1(b). We predict you will see a similar picture no matter how many iterates are thrown away (and no matter which initial condition is chosen within the basin). This set of points that won't go away is the forward limit set of the orbit. In the other case, the 10^6 points lie on top of the sink at

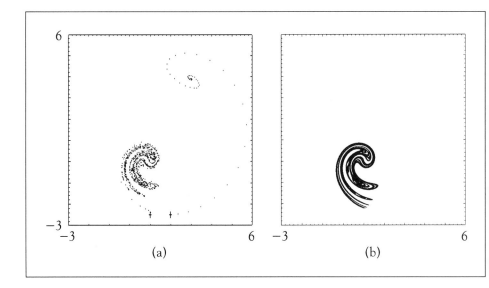

Figure 6.2   Two orbits of the Ikeda map.
(a) Two initial values are denoted by crosses. The next 1000 iterates of each initial value are plotted. One orbit approaches the chaotic attractor, the other approaches a sink. (b) Points 1,000,001–2,000,000 of each orbit are plotted. Part (b) shows some points of the forward limit sets of the two orbits of part (a).

$\approx (3.0026, 3.8945)$. Higher iterates of the orbit do not return to the initial point since the orbit becomes trapped by the sink. The forward limit set in this case is the set consisting of the sink alone. It is an attractor but it is not chaotic.

The forward limit set of an orbit $\{f^n(x_0)\}$ is the set of points x to which the orbit forever makes occasional neighborhood visits. No matter how small $\epsilon > 0$ is chosen, and no matter how long N we wait, there is an $n > N$ with $|f^n(x_0) - x| < \epsilon$. For example, if $f^n(x_0)$ tends toward a period-two orbit, then the forward limit set is comprised of the two points of the periodic orbit. If x is one of these two points, then $|f^n(x_0) - x|$ would be small only for large odd values of n or large even values.

Definition 6.1   Let f be a map and let x_0 be an initial condition. The **forward limit set** of the orbit $\{f^n(x_0)\}$ is the set

$$\omega(x_0) = \{x : \text{for all } N \text{ and } \epsilon \text{ there exists } n > N \text{ such that } |f^n(x_0) - x| < \epsilon\}.$$

This set is sometimes called the ω-**limit set** of the orbit; ω is the Greek letter omega, the last letter of the Greek alphabet. If $\omega(x_0)$ is a forward limit set of some orbit and x_1 is another initial condition, then we will say that the orbit $\{f^n(x_1)\}$ (or the point x_1) is **attracted** to $\omega(x_0)$ if $\omega(x_1)$ is contained in $\omega(x_0)$.

The definition of forward limit set of an orbit is the set of points to which the orbit returns arbitrarily close, infinitely often. Points in an orbit may or may not be contained in its forward limit set. The forward limit set may have no points in common with the orbit, as is the case with the forward limit set of an orbit converging to a sink. In this case the forward limit set is one point, the sink, which is approached by the orbit as closely as you specify and as far forward in time as you want to require. The orbit is attracted to the sink.

Fixed-points are forward limit sets because $x_0 = \omega(x_0)$. So are periodic orbits, for the same reason: The forward limit set of the orbit is the orbit itself. These are examples of finite sets, but larger sets can also be forward limit sets. In Chapter 3, we found a chaotic orbit for the logistic map $G = g_4$ that is dense in the unit interval $[0, 1]$. This orbit returns arbitrarily close infinitely often to every number between 0 and 1. Therefore the forward limit set of this orbit is $[0, 1]$. The fact that the chaotic orbit is dense in $[0, 1]$ is an important one. It means that no smaller set could be its forward limit set.

✎ **EXERCISE T6.1**

Show that if x_0 is in a forward limit set, say $\omega(y)$, then the entire orbit $\{f^n(x_0) : n \geq 0\}$ is in $\omega(y)$.

✎ **EXERCISE T6.2**

Show that a forward limit set cannot contain a fixed point sink unless it is itself a fixed point sink. Can a forward limit set contain any fixed points and be more than a single point?

Definition 6.2 Let $\{f^n(x_0)\}$ be a chaotic orbit. If x_0 is in $\omega(x_0)$, then $\omega(x_0)$ is called a **chaotic set**. In other words, a chaotic set is the forward limit set of a chaotic orbit which itself is contained in its forward limit set. An **attractor** is a forward limit set which attracts a set of initial values that has nonzero measure (nonzero length, area, or volume, depending on whether the dimension of the map's domain is one, two, or higher). This set of initial conditions is called the basin of attraction (or just basin), of the attractor. A **chaotic attractor** is a chaotic set that is also an attractor.

The requirement that the defining chaotic orbit be in its own forward limit set ensures that a chaotic set has a dense orbit (why?), and thus provides that the set be irreducible. We stress that not all chaotic sets are chaotic attractors, as Exercise T6.3 shows.

✎   **EXERCISE T6.3**

(a) Show that the logistic map $g_a(x) = ax(1 - x)$, for $a > 4$, has a chaotic set, which is not a chaotic attractor.

(b) Show that the hyperbolic horseshoe (described in Chapter 5) contains a chaotic set.

Figure 6.3 shows the observable attractors for the one-parameter family of one-dimensional maps $g_a(x) = ax(1 - x)$ over a values in the range 2 to 4. Each vertical slice of the figure is the attractor for a fixed value of a. The figure was made as follows: for each value of a, a random initial value in $[0, 1]$ is chosen; the orbit of this point is calculated and the first 100 iterates are discarded. During this time, the initial condition converges toward the attractor. The next 10,000 iterates are plotted and should reflect an orbit on (more precisely, infinitesimally close to) the attractor.

The attractors vary widely for different values of the parameter a. When for a given a value the chosen initial point is in the basin of a fixed point attractor, only one point will appear in the vertical slice at that a value. This behavior characterizes a values below 3. When the initial value is in the basin of a periodic attractor, isolated points in the attracting orbit will appear in the horizontal slices. This behavior occurs for subintervals of a values throughout the diagram. Beginning at $a = 3$, for example, there is a period-two attracting orbit, which becomes a period-four attractor, then a period-eight attractor, etc., as the parameter is changed. An interval of parameter values for which the only attractor is a periodic orbit is called a **periodic window** in the bifurcation diagram.

For certain a values (larger than $a = 3.57$), one observes an entire interval or intervals of plotted points. (Recall this is *one* orbit plotted.) Calculation of Lyapunov exponents at these parameter values indicates that the orbits are indeed chaotic. In addition, virtually every choice of initial value in $(0, 1)$ yields the same diagram, indicating a large basin of attraction for these sets. The last a value with a visible limit set is $a = 4$. For a values larger than 4, the orbits of almost all points in $[0, 1]$ leave the interval and become unbounded. Although there appear to be entire intervals of a values with chaotic attractors, looks can be deceiving, as the Computer Experiment 6.1 shows.

➭   **COMPUTER EXPERIMENT 6.1**

Let $g_a(x) = ax(1 - x)$. Choose a subinterval of length 0.01 in the parameter a which appears to contain only chaotic attractors (for example, a set within

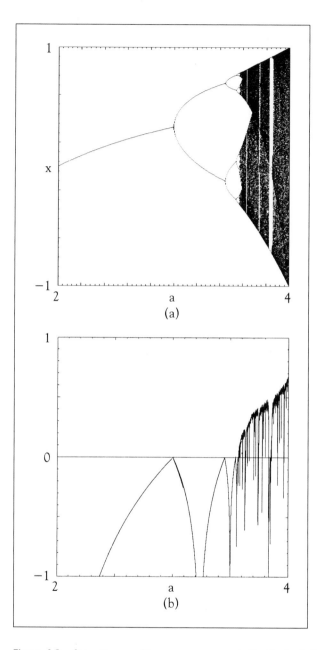

Figure 6.3 Attractors and Lyapunov exponents for the logistic family of maps.
(a) Attractors of g_a are plotted on the vertical scale. The horizontal axis is the parameter a. (b) Lyapunov exponents of the attracting orbit of g_a versus parameter a. The exponent rises and hits zero at period doublings and becomes positive in the chaotic region. It drops below 0 when there is a periodic attractor.

[3.9, 4.0], according to Figure 6.3(a)). Magnify the region until a periodic window is located.

Disclaimer 6.3　This is a good place to remark on the difficulty of proving rigorously that orbits are chaotic, even for simple systems. For a particular vertical slice in Figure 6.3, even if the best computer approximation can indicate a positive Lyapunov exponent and a nonperiodic orbit, this is not a mathematical proof. If the orbit is periodic with period longer than the number of atoms in the universe, no simple computer iteration scheme will tell us. The same caution is due for the two attractors in Figure 6.1. Although they appear to be chaotic orbits from computer experiments, there is at the time of this writing no rigorous proof of that fact.

⇨　**COMPUTER EXPERIMENT 6.2**

Choose a point p on the attractor of g_a for $a = 3.9$. (Find this point by taking the 1000th point of a trajectory in the basin.) Choose another point x in the basin, and let m_n be the minimum distance between p and the first n iterates of x. Print out values of n and m_n each time m_n changes. Does $m_n \to 0$ as $n \to \infty$? Can you quantify the rate at which it goes to zero? (In other words, what is the functional relation of m_n and n?)

Figure 6.4 shows two other examples of probable chaotic attractors of plane maps. In each case, the chaotic attractor was plotted by choosing an initial point in the gray region and plotting trajectory points $1001, 1002, \ldots, 1,000,000$. Experience has shown that almost any other point chosen from the gray would have yielded the same picture. Of course, we can subvert the process by choosing a nonattracting fixed point for the initial point, on or off the attractor. Then the plot would only show a single point, assuming that small computer roundoff errors did not push us away.

6.2 CHAOTIC ATTRACTORS

A fixed-point sink is easily seen to satisfy the definition of attractor, since it attracts an entire ϵ-neighborhood. (The forward limit set of each point in the neighborhood is the sink.) Periodic sinks are also attractors.

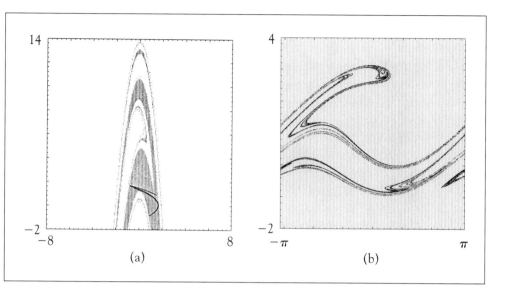

Figure 6.4 More chaotic attractors of plane maps.
The attractors are shown in black inside their basins, which are shown in gray.
(a) A chaotic attractor of the Hénon map (6.1) with $a = 2.1, b = -0.3$. The basin
boundary, as well as the attractor, is fractal. White points are attracted to infinity.
(b) The chaotic attractor of the time–2π map of the forced damped pendulum. The
basin of this attractor, shown in gray, consists of virtually all initial conditions.

The unit interval is the forward limit set of a chaotic orbit of the logistic
map $G = g_4$ and it attracts an interval of initial conditions (itself), so it is a
chaotic attractor. Similarly, the unit interval is a chaotic attractor for the tent
map $T_2(x)$.

EXAMPLE 6.4

Let $f(x) = 2x \pmod 1$. As in Exercise 3 of Chapter 4, we can associate
with each point in $[0, 1]$ an itinerary (an infinite sequence of 0's and 1's) on
which the shift map represents the action of f. All points in the subinterval
$[(k - 1)2^{-n}, k2^{-n}], 1 \leq k \leq 2^n$, are represented by a finite sequence of n symbols.
By constructing a symbol sequence that contains every possible finite sequence,
a point with an orbit that is dense in $[0, 1]$ can be shown to exist.

✐ EXERCISE T6.4

Show that $[0, 1]$ is a chaotic attractor for the $2x \pmod 1$ map of Example
6.4.

WHAT IS AN ATTRACTOR?

The term attractor is used for the forward-time limit of an orbit that attracts a significant portion of initial conditions. A sink is an example, since it attracts at least a small neighborhood of initial values.

An attractor should be irreducible in the sense that it includes only what is necessary. The set consisting of the sink together with one of the orbits approaching the sink is also a set that attracts initial conditions, but for the reason that it contains the sink. Only the sink is actually needed. Irreducibility is guaranteed by requiring that the attractor contain a dense orbit, an orbit that comes arbitrarily close to each point in the attractor.

Besides irreducibility, the attractor must have the property that a point chosen at random should have a greater-than-zero probability of converging to the set. A saddle fixed point is irreducible in the above sense and does attract orbits: for example, the one whose initial condition is the fixed point itself. However, this initial condition is very special; the definition requires that an attractor must attract a set of initial values of nonzero state space volume.

Chaos introduces a new twist. Chaotic orbits can be attracting, as shown in this chapter. If the forward limit set of such a chaotic orbit contains the orbit itself (and therefore contains a dense orbit), then the attractor is a chaotic attractor.

EXAMPLE 6.5

The map of Example 6.4, although not continuous as a map of the interval, is continuous when viewed as the map $f(\theta) = 2\theta$ of the circle. Using polar coordinates, we can embed f into a map of the plane

$$f(r, \theta) = (r^{1/2}, 2\theta).$$

See Figure 6.5. There is a repelling fixed point at the origin, which is the forward limit set $\omega(0)$. The circle $r = 1$ is also the forward limit set of an orbit with Lyapunov exponent $\ln 2$, which we can conclude from Example 6.4. This means

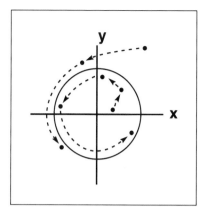

Figure 6.5 Map for which the unit circle is a chaotic attractor.
Two orbits are shown for the map $f(r, \theta) = (r^{1/2}, 2\theta)$ given in polar coordinates. All orbits except the origin converge to the unit circle $r = 1$, which has a dense chaotic orbit.

the circle is a chaotic set. All points other than the origin are attracted to this circle, making the circle a chaotic attractor.

EXAMPLE 6.6

The piecewise linear map on the unit interval illustrated in Figure 6.6 is called the W-map. If we restrict our attention to the subinterval $S = [1/4, 3/4]$,

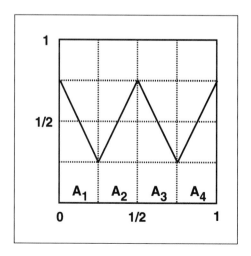

Figure 6.6 The W-map of the unit interval.
The set $[1/4, 3/4]$ is a chaotic attractor.

the map operates like the tent map. Points from the remainder of the unit interval are immediately mapped into S. Because of the analogy with the tent map, we know that S contains a dense orbit with Lyapunov exponent $\ln 2$. That makes S a chaotic set. Further, the entire unit interval is attracted to S, making S a chaotic attractor.

EXAMPLE 6.7

Hénon observed the attractor shown in Figure 6.7 for the map (6.1) with $a = 1.4$ and $b = 0.3$. In addition, he found a quadrilateral in the plane, shown in Figure 6.7(b), which maps into itself when the function is applied to it once.

A bounded neighborhood that the map sends into itself, such as the quadrilateral, is called a **trapping region**. A quadrilateral trapping region, together with its forward image, is shown in Figure 6.7(c). The Jacobian determinant of H is $J = -0.3$ (minus the coefficient of y) at all points in the plane. Thus the image under H of a region of the plane is decreased by the factor $|J| = 0.3$. Because H

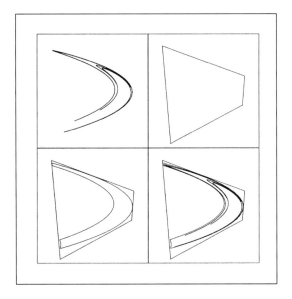

Figure 6.7   The Hénon attractor and a trapping region.
The attractor of the map $f(x, y) = (1.4 - x^2 + 0.3y, x)$ shown in (a) is contained in a quadrilateral trapping region shown in (b). The trapping region maps into itself and contracts in area by the factor 0.3. The quadrilateral and its image under f are shown in (c). Since f is area-contracting, forward iterates of the trapping region shrink down to a limit set (the attractor) that has zero area. The figures in (a) and (c) are superimposed in (d).

is area-contracting, forward iterates of the quadrilateral shrink to zero area in the limit.

What are the implications of a trapping region for an area-contracting map such as the Hénon map? One iterate maps the region inside itself to the shape in Figure 6.7(c), which is 30% of the original size. Two iterates map the quadrilateral to 9% of its original size, to a shape folded further within itself. The limit of this process is the chaotic attractor shown in Figure 6.7(d). Numerical approximation of the Lyapunov exponents on the attractor yield .39 and -1.59, so that this is a chaotic attractor—assuming that it is not just a long periodic orbit. That fact is surprisingly hard to prove and as of this writing, it has not been established for these parameter values of the map. It is likely that there is a chaotic attractor here but it may never be proved. One difficulty is that arbitrarily small changes in the parameters (0.3 and 1.4) can be made so that the system has an attracting periodic orbit.

EXAMPLE 6.8

The circle $r = 1$ in Example 6.5 is a good prototype of a chaotic attractor. It is properly contained in a basin of attraction (unlike the logistic map g_4), and the map is smooth, at least around the attractor (unlike the W-map). It is not, however, an invertible map. By increasing the dimension of the underlying space, we can define a similar example, called the **solenoid**, in which the map is one-to-one. The underlying set of this chaotic attractor was described first by topologists (see, for example, (Hocking and Young, 1961)) and then as a dynamical system by (Smale, 1967).

We define the map $\mathbf{f}$ on the solid torus (a subset of $\mathbb{R}^3$), which we think of as a circle of two-dimensional disks. The disk D in $\mathbb{R}^2$ can be defined with one complex coordinate z, as $D = \{z : |z| \le 1\}$. Then points in the solid torus T can be described by two coordinates:

$$T = \{(t, z) : t \in [0, 1) \text{ and } |z| \le 1, z \in \mathbb{C}\}.$$

The map $\mathbf{f} : T \to T$ is defined as follows:

$$\mathbf{f}(t, z) = \left(2t \,(\text{mod } 1), \frac{1}{4}z + \frac{1}{2}e^{2\pi it}\right),$$

where $e^{ix} = \cos x + i \sin x$.

In order to understand this map geometrically, we refer to the picture of T and $\mathbf{f}(T)$ in Figure 6.8(a). Think of the solid torus T as being stretched to a longer, thinner loop. Then it is twisted, doubled over, and placed back into T. The image

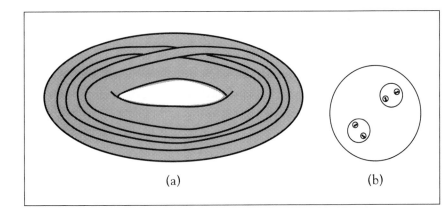

Figure 6.8 Schematic picture of the solenoid map.
(a) The solid torus T together with the image $\mathbf{f}(T)$ is shown. (b) A cross-sectional disk D_t in the solid torus T is shown together with the intersection of $\mathbf{f}(T)$, $\mathbf{f}^2(T)$, and $\mathbf{f}^3(T)$ with D_t.

of T now intersects each cross-sectional disk D_t in two smaller disks, D_0 and D_1, each $\frac{1}{4}$ the diameter of the original. The second iterate $\mathbf{f}^2(T)$ intersects each D_t in four smaller disks, etc. Figure 6.8(b) shows a cross-sectional disk D_t of T together with intersections of $\mathbf{f}(T)$, $\mathbf{f}^2(T)$, and $\mathbf{f}^3(T)$ with D_t. Any two points in T whose t-coordinates differ by $\frac{1}{2}$ map to the same cross-sectional disk.

✎ **EXERCISE T6.5**

(a) Identify the pre-images of D_0 and D_1, the intersection of $\mathbf{f}(T)$ with the cross-sectional disk D_t. (b) Show that D_0 and D_1 are symmetric with respect to the origin in D_t.

The map $\mathbf{f}$ is uniformly stretching in the t direction: Each orbit has a positive Lyapunov exponent of $\ln 2$. The attractor $A = \bigcap_{n\geq0} \mathbf{f}^n(T)$ intersects each D_t in a Cantor set, and A has zero volume inside T, the basin of attraction. The solenoid has many interesting topological and dynamical properties that we do not pursue here. For a detailed discussion, see (Robinson, 1995).

We leave the task of identifying a dense orbit for A to the following exercise.

✎ **EXERCISE T6.6**

Find a way to assign bi-infinite symbol sequences to points in A, as done earlier for the horseshoe map. Determine an orbit that is dense in A.

6.3 CHAOTIC ATTRACTORS OF EXPANDING INTERVAL MAPS

In this section, we take a closer look at a class of maps of the unit interval that have chaotic attractors. Let $p_0 < p_1 < \cdots < p_k$ be points on the real line and let I be the interval $[p_0, p_k]$. Define the closed intervals $A_i = [p_{i-1}, p_i]$. Let $f : I \rightarrow I$ be a map whose derivative satisfies $|f'(x)| \geq \alpha > 1$ except possibly at the p_i (where f may have a corner, or in some other way not be differentiable). We will call such a map a **piecewise expanding map** with **stretching factor** α. We say that $\{p_0, p_1, \ldots, p_k\}$ is a **stretching partition** for the piecewise expanding map **f** if, for each $i, f(A_i)$ is exactly the union of some of the intervals $A_1, \ldots, A_k$. A stretching partition satisfies the covering rule of Chapter 3, which allows the construction of transition graphs for the partition intervals A_i. (It is also the one-dimensional analogue of the concept of "Markov partition," defined in Chapter 5.) An example of a piecewise expanding map is the W-map of Figure 6.6.

We found in Chapter 3 that when there is a partition that satifies the covering rule, symbol sequences can be used to keep track of the itinerary of an orbit of f. For example, if the orbit begins in interval A_1, then maps into A_2, and then to A_3, its itinerary would begin: $.A_1 A_2 A_3 \ldots$. Of course, there is a countable set of orbits that eventually land precisely on one of the points p_i. We ignore these special orbits in this discussion and concentrate on the remaining uncountably many orbits.

Not all sequences of the symbols $A_1, \ldots, A_k$ may be achievable by orbits of f. For example, let f be the W-map as shown in Figure 6.6. Any orbit that begins in subinterval $A_1 = [0, 1/4]$ maps to the subinterval $[1/4, 3/4] = A_2 \cup A_3$; but $f(A_2) = f(A_3) = A_2 \cup A_3$, meaning that the orbit will never return to A_1. That is, $.A_1 A_2 \ldots$ is an allowable symbol sequence for f, but $.A_2 A_1 \ldots$ is not. In general, if B and C are two subintervals for a stretching partition, C is allowed to follow B in the symbol sequence of a point of the interval I if and only if $f(B) = C \cup$ (other subintervals).

The fact that a continuous map f is stretching by at least the factor $\alpha > 1$ causes the dynamics of f to be well organized. Let L be the length of the entire interval I. For an allowable sequence $.B_1 \ldots B_n$ of n symbols (repetitions allowed), there is a subinterval of length at most $\frac{L}{\alpha^{n-1}}$, which we call an **order n subinterval**, whose points follow that itinerary. For each n, the order n subintervals virtually fill up I: every point of I either lies in some order n subinterval or is on the boundary of an order n subinterval. An infinite allowable sequence represents one point in I, since $L/\alpha^{n-1} \rightarrow 0$ as $n \rightarrow \infty$.

The derivative $(f^n)'(x_0) = f'(x_{n-1}) \cdots f'(x_0)$ is greater than α^n for each x_0, so

$$|(f^n)'(x_0)|^{1/n} \geq \alpha. \tag{6.3}$$

The Lyapunov number for the orbit starting at x_0 is therefore at least α, if the limit exists. Although our formal definition of chaotic orbit in Chapter 3 requires the Lyapunov number to be greater than 1, we will relax the requirement that the limit of (6.3) exists for the statement of Theorem 6.11. For readers with an advanced calculus background, we point out that whether the limit exists or not,

$$\liminf |(f^n)'(x_0)|^{1/n} \geq \alpha > 1, \tag{6.4}$$

which is a more inclusive definition of chaos.

EXAMPLE 6.9

Let f be the tent map on the unit interval $[0, 1]$. We studied the itineraries of this map in Chapter 3 as an example of chaotic orbits. It is clear that $\{0, \frac{1}{2}, 1\}$ is a stretching partition for the tent map, with $\alpha = 2$.

EXAMPLE 6.10

Let f be the W-map of Figure 6.6. The stretching partition is $0 < \frac{1}{4} < \frac{1}{2} < \frac{3}{4} < 1$, and $\alpha = 2$. Allowable symbol sequences are arbitrary sequences of A_1, A_2, A_3, A_4 with the restriction that every symbol must be followed by either A_2 or A_3. Thus A_1 and A_4 can appear only as the leftmost symbol. This corresponds to the fact that the intervals $[0, 1/4]$ and $[3/4, 1]$ map into the interval $[1/4, 3/4]$ and the points never return. The transition graph for this partition is shown in Figure 6.9.

The facts we have developed in this section are summarized in the following theorem. The proof of the last property is left to the reader.

Theorem 6.11 *Let f be a continuous piecewise expanding map on an interval I of length L with stretching factor α, and let $p_0 < \cdots < p_k$ be a stretching partition for f.*

1. *Each allowable finite symbol sequence $.A_1 \cdots A_n$ corresponds to a subinterval of length at most $\frac{L}{\alpha^{n-1}}$.*
2. *Each allowable infinite symbol sequence $.A_1 A_2 A_3 \cdots$ corresponds to a single point x of I such that $f^i(x) \in A_{i+1}$ for $i \geq 0$, and if the symbol sequence is not periodic or eventually periodic, then x generates a chaotic orbit.*

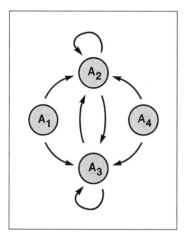

Figure 6.9 Transition graph for the W-map.
The unit interval is partitioned into four subsets, A_1, A_2, A_3, and A_4, on each of which the map is stretching. The ω-limit set of any orbit is $A_2 \cup A_3$.

3. *If, in addition, each pair of symbols* **B** *and* **C** *(possibly* **B** = **C***) can be connected by allowable finite symbol sequences* **B** $\cdots$ **C** *and* **C** $\cdots$ **B***, then* f *has a dense chaotic orbit on* I, *and* I *is a chaotic attractor.*

✎ **EXERCISE T6.7**

Assuming the hypothesis of Theorem 6.11 and, in addition, that each pair of symbols can be connected by an allowable symbol sequence, show that f has a chaotic orbit that is dense in I.

Theorem 6.11 is extremely useful for proving the existence of chaotic attractors on the real line. Consider the tent map, with stretching partition $0 < \frac{1}{2} < 1$ and symbols A_1 and A_2 corresponding to the two subintervals. Since $f(A_1) = A_1 \cup A_2$ and $f(A_2) = A_1 \cup A_2$, the pair of symbols A_1 and A_2 can occur in either order, and part 3 of Theorem 6.11 applies. The unit interval is a chaotic attractor for the tent map.

The W-map of Example 6.10 has a stretching partition $0 < \frac{1}{4} < \frac{1}{2} < \frac{3}{4} < 1$. Because A_2A_1 cannot occur, part 3 of Theorem 6.11 does not apply. However, if we restrict the map to $[\frac{1}{4}, \frac{3}{4}]$, part 3 again applies to show that $[\frac{1}{4}, \frac{3}{4}]$ is a chaotic attractor. In fact this attractor is essentially the tent map attractor.

EXAMPLE 6.12

(Critical point is period three.) Define

$$f(x) = \begin{cases} \frac{1}{a+1} + ax & \text{if } 0 \leq x \leq \frac{1}{a+1} \\ a - ax & \text{if } \frac{1}{a+1} \leq x \leq 1 \end{cases},$$

where $a = \frac{\sqrt{5}+1}{2}$ is the **golden mean**. The map is sketched in Figure 6.10. Note that a is a root of the equation $a^2 = a + 1$. Because of the careful choice of a, we have $f(\frac{1}{a+1}) = 1$, and since also $f(1) = 0$ and $f(0) = \frac{1}{a+1}$, the peak of the function f occurs at $c = \frac{1}{a+1} \approx 0.382$, which is a period-three point for f.

The partition $0 < c < 1$ is a stretching partition for f. Define the subintervals $A = [0, c]$ and $B = [c, 1]$. Notice that $f(A) = B$ and $f(B) = A \cup B$. Allowable symbol sequences for orbits of f consist of strings of **A** and **B** such that **A** cannot be followed by **A**. The stretching factor under f is the golden mean, which is approximately 1.618.

Can part 3 of Theorem 6.11 be applied? Symbol sequences **AB**, **BA**, and **BB** can obviously occur, and since **ABA** is also permitted, all possible pairs can be connected. Therefore $[0, 1]$ is a chaotic attractor for the map f.

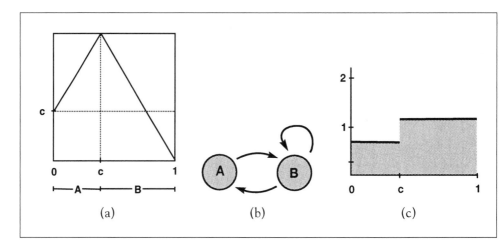

Figure 6.10 Map for which the critical point is period three.
(a) Sketch of map. (b) Transition graph. (c) Invariant measure for (a); see Section 6.4.

6.4 MEASURE

In our quest to identify and describe fixed-point, periodic, and chaotic attractors, we have considered them as sets of points. Although this does a good job of characterizing the attractor when the point set is finite, the case of chaotic attractors presents an extra challenge, since they must contain infinitely many points.

When studying chaotic attractors we must give up on the idea of keeping track of individual points, and instead do our bookkeeping over regions. The term "measure", to which this section is devoted, refers to a way of specifying how much of the attractor is in each conceivable region.

For example, imagine a box drawn in the plane that contains part of the Ikeda attractor, such as box 1 in Figure 6.11. How could we measure the proportion of points on the attractor that are contained in the box? We will use the "rain

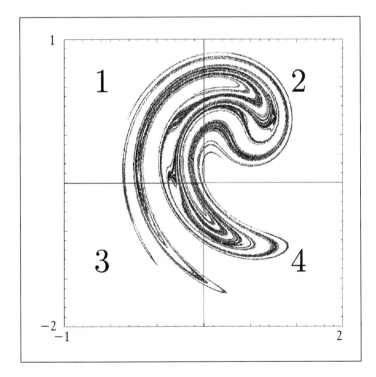

Figure 6.11   100,000 points on the Ikeda attractor.
The proportion landing in each of the 4 boxes is an approximation to the natural measure of the Ikeda attractor for that box. Boxes 1, 2, 3, and 4 contain about 30%, 36%, 17%, and 17%, respectively.

gauge" technique and measure the proportion of points that fall into box 1 from a typical orbit.

Choose an initial point at random to start an orbit. At each iteration, we record whether the new orbit point fell into the box or not. We keep this up for a long time, and when we stop we divide the number that landed in the box by the total number of iterates. The result would be a number between 0 and 1 that we could call the "Ikeda measure" of the box. In Figure 6.11, 29,798 of the 100,000 points shown lie in box 1, so that approximately 30% of the points fall into the box.

This number could be regarded as the probability that a point of the Ikeda attractor lies in the box. A box located away from the attractor would have Ikeda measure zero. In fact, on the basis of the 100,000 points shown, the Ikeda measure of box 1 is 0.29798. The measures of boxes 2, 3, and 4 are 0.35857, 0.17342, and 0.17003, respectively. These four numbers add up to 1 because each of the 100,000 points landed in one of the four boxes.

Now we will make our ideas on measure a little more precise. The two most important properties of the rain-gauge method are nonnegativity and additivity, namely:

> (a) The measure of any set is a nonnegative number, and
> (b) The measure of a disjoint union of a finite or countably infinite number of sets is equal to the sum of the measures of the individual sets.

If A is a subset of B, we will denote by $B \setminus A$ the **complement** of A in B, the set of points in B that do not belong to A. It follows from (b) that if A is a subset of B, then the measure of the complement $B \setminus A$ is the difference of the measures of B and A.

A method of assigning a number to each closed set so that (a) and (b) are satisfied is called a **measure**. (See Section 8.2 for an introduction to closed sets.) The ordinary measures we are used to, such as length on $\mathbb{R}^1$, area on $\mathbb{R}^2$, and volume on $\mathbb{R}^3$ are called **Lebesgue measures** when extended to apply to all closed sets. Although fact (b) is true when the union is a countably infinite union, we can't expect (b) to hold for uncountable disjoint unions. The length of a single point is zero, and an uncountable union of them makes a unit interval with length one.

Once a measure is defined for all closed sets, we can extend the definition to many other sets with the help of properties (a) and (b). For example, using Lebesgue measure (length) m on the real line gives the length of the unit interval to be one, $m([0, 1]) = 1$. Since we know that the length of the single point set

$\{0\}$ is zero, we can find the measure of the half-open interval $(0, 1]$. Rule (b) gives $m(\{0\}) + m((0, 1]) = m([0, 1])$, so that we find $m((0, 1]) = 1$. In this way we can define the measure of many nonclosed sets. A **Borel set** is a set whose measure can be determined by the knowledge of the measure of closed sets and a chain of applications of rules (a) and (b). The set $(0, 1)$ is a Borel set because rule (b) allows definition of its measure from other sets: $m((0, 1)) + m(\{1\}) = m((0, 1])$, from which we conclude that $m((0, 1)) = 1 - 0 = 1$. To summarize, the Borel sets are those that have a well-defined number assigned to them by the measure.

It is hard to conceive of a set of real numbers that is not a Borel set, but they do exist. See any book on measure theory for examples. Strictly speaking, our definition of measure, with its scope limited to closed sets and those that can be formed from them using (b), is often called a **Borel measure**. Less restrictive definitions of measure exist but will not be considered in this book.

✎ **EXERCISE T6.8**

Show that if μ is a measure, A and B are Borel sets, $\mu(B) = 0$, and $A \subset B$, then $\mu(A) = 0$.

The rain-gauge method of generating a measure has two further properties that are important. First, since it measures the proportion of points falling into a set, it satisfies the following property:

> (c) The measure of the entire space equals 1.

A measure that satisfies rule (c) is called a **probability measure**.

Finally, notice that the amount of rain-gauge measure in a box using points generated by the map f is the same as the measure of the pre-image of the box under the map f. The reason is that all of the pre-iterates of B will end up in B on their next iteration. A measure μ that satisfies the property

> (d) $\mu(f^{-1}(S)) = \mu(S)$ for each closed set S

is called an **f-invariant** measure. Ikeda measure satisfies both (c) and (d), so it is an f-invariant probability measure.

Lebesgue measure is generated by the rain-gauge technique by replacing the Ikeda attractor with a random number generator. Assume that we have a method for producing pairs (x, y) of real numbers that randomly fill the square $[0, 1] \times [0, 1]$ in a uniform way. In theory, if the random number generator is truly random and uniform, we could use this device to find area: We could call it "area measure", by analogy with Ikeda measure. A box with dimensions $a \times b$ lying in the square $[0, 1] \times [0, 1]$ will get a proportion of randomly generated points that converges to ab with time.

A random number generator on a computer is pseudo-random, meaning that it is a deterministic algorithm designed to imitate a random process as closely as possible. If the random number program is run twice with the same starting conditions, it will of course yield the same sequence of random numbers, barring machine failure. That is the meaning of "deterministic". This is sometimes useful, as when a simulation needs to be repeated a second time with the same random inputs. However, it is important to be able to produce a completely different random sequence. For this reason most random number generators allow the user to set the seed, or initial condition, of the program. If the algorithm is well-designed, it will measure the area of an $a \times b$ rectangle to be ab, no matter which seed is used to start the program.

The relationship of an invariant measure to a chaotic attractor is the same as the relationship of standard area to a uniform random number generator. The important concept is that the percentages of points in a given rectangle in Ikeda measure are independent of the initial value of the orbit used in the rain-gauge technique. If this is true, then Ikeda measure of a rectangle has a well-defined meaning. In the same way, we know that an $a \times b$ rectangle in the plane has area ab, even though there may be no uniform random number generator nearby to check with.

Both chaotic attractors and random number generators could fail to give the correct measure. For all we know, our random number generator might produce the infinite sequence of points $(1/2, 1/2), (1/2, 1/2), \ldots$. In fact, this sequence of numbers is as likely to occur as any other sequence. Obviously this output of random numbers will not correctly measure sets. It would imply that the set $[0, 1/4] \times [0, 1/4]$ has measure zero. The corresponding problem with an attractor occurs, for example, if the initial value used to generate the points happens to be a fixed point. The orbit generated will not generate Ikeda measure; it will generate a measure that is 1 if the box contains the fixed point, and 0 if not.

To have a good measure, we need to require that almost every initial value produces an orbit that in the limit measures every set identically. That is, if we ignore a set of initial values that is a measure zero set, then the limit of the proportion of points that fall into each set is independent of initial value. A measure with this property will be called a natural measure.

6.5 NATURAL MEASURE

Now we are ready to make the rain-gauge measure into a formal definition. We need to introduce some fine print to the definition of rain-gauge measure that is important for borderline cases. First, define the **fraction** of iterates of the orbit $\{f^n(x_0)\}$ lying in a set S by

$$F(x_0, S) = \lim_{n \to \infty} \frac{\#\{f^i(x_0) \text{ in } S : 1 \leq i \leq n\}}{n}. \tag{6.5}$$

EXAMPLE 6.13

Let $f(x) = x/2$. All orbits are attracted to the sink $x = 0$. First define $S_1 = [-1, 1]$. For any initial point x_0, the fraction $F(x_0, S_1) = 1$. Even if the orbit doesn't start out inside S_1, it eventually moves inside S_1 and never leaves. The ratio in the definition (6.5) tends to one in the limit as $n \to \infty$. If we define S_2 to be a closed interval that does not contain 0, then the orbit would eventually leave the interval and never return, so that $F(x_0, S_2) = 0$ for any x_0. For these sets S_1 and S_2, the rain-gauge technique gives the correct answer, since the measure should be concentrated at $x = 0$.

Now let $S_3 = [0, \infty)$ and compute the fraction $F(x_0, S_3)$ for various x_0. If $x_0 < 0$, no iterates of the orbit lie in S_3, and $F(x_0, S_3) = 0$. If $x_0 \geq 0$, then all iterates of the orbit lie in S_3, and $F(x_0, S_3) = 1$.

Because one of the boundaries of S_3 lies at the sink, this is a borderline case. The rain-gauge technique is too unrefined to work here. When we ask what proportion of a typical orbit lies in S_3, we get conflicting answers, depending on x_0. A similar problem occurs for an orbit starting in the basin of, say, a chaotic Hénon attractor S. If v_0 is attracted to S but not in S, then $\mathbf{f}^n(v_0)$ gets very close to S, but is never in S. Hence the fraction $F(v_0, S) = 0$. On the other hand, if we fatten the attractor by a tiny amount, the problem is fixed. In other words, if $r > 0$, then $F(v_0, N(r, S)) = 1$, where we have used the notation

$$N(r, S) = \{x : \text{dist}(x, S) \leq r\}, \tag{6.6}$$

to denote the set of points within r of the set S.

The difficulty in assigning the correct value to the fraction of the orbit that lands inside a given set is solved by the following definition of natural measure. Assume that f is a map and that S is a closed set.

Definition 6.14 The **natural measure generated by the map** f, also called f-**measure**, is defined by

$$\mu_f(S) = \lim_{r \to 0} F(x_0, N(r, S)) \tag{6.7}$$

for each closed set S, as long as all x_0 except for a measure zero set give the same answer.

Quite often we will want to apply this definition to f limited to a subset of its domain. It is common for this subset to be a basin of attraction. For example, the logistic map $G(x) = 4x(1 - x)$ on the unit interval $[0, 1]$ has a natural measure which is investigated in detail in Challenge 6. Notice that even in this case, we must allow for a measure zero set of x_0 which do not go along with the crowd. Not every orbit $\{f^n(x_0)\}$ can be used to evaluate $\mu_f(S)$. For example, we know that G has a two fixed-point sources in $[0, 1]$. Neither of these orbits can be used; nor can any of the periodic orbits of G. Together they make up a countable set, which has measure zero. On the other hand, the map G on the entire real line does not have a natural measure, since initial values outside of $[0, 1]$ are attracted to ∞, and so the condition that the exceptions be a measure zero set is not satisfied.

✎ **EXERCISE T6.9**

Show that properties (a), (b), (c), and (d) in the definition of measure hold for a natural measure.

With this more sophisticated version of natural measure, the difficulty of determining the fraction of an orbit lying within a set disappears. For $f(x) = x/2$ and $S = [0, \infty)$, the set $N(r, S) = [-r, \infty)$, and $F(x_0, N(r, S)) = 1$ for any x_0, as long as $r > 0$. Therefore $\mu_f(S)$ is the limit as $r \to 0$, which is 1. (The fact that $F(x_0, N(0, S)) = 0$ if $x_0 < 0$ is now irrelevant.)

Any interval $[a, b]$ with $a < 0 < b$ will have f-measure equal to 1. Even if some of the original iterates miss the interval, eventually they will stay within a neighborhood of 0 so small that it is contained entirely within $[a, b]$. The limiting ratio of orbit points in the interval to the total approaches 1 in the limit.

Now consider an interval with one endpoint equal to 0. This is another borderline case. An orbit converging to 0 eventually moves into the set $N(r, S)$ for any $r > 0$. Therefore $\mu_f([a, 0]) = \mu_f([0, b]) = 1$.

A singleton set $\{x\}$ has f-measure zero if $x \neq 0$ because it is contained in an interval with f-measure zero, by Exercise T6.8. On the other hand, $\mu(\{0\}) = 1$. The measure of other sets can be found using the above facts in conjunction with Property (b). For example, $\mu((0, 1]) = \mu([0, 1]) - \mu(\{0\}) = 1 - 1 = 0$.

WHY MEASURE?

Why do we study measure in connection with chaotic attractors? At the very least, we must know that the natural measure of a map is not atomic if there is a chaotic attractor. More importantly, the existence of a natural measure allows us to calculate quantities that are sampled and averaged over the basin of attraction and have these quantities be well-defined. Perhaps the most important such quantity for the purposes of this book is the Lyapunov exponent. For an orbit of a one-dimensional map f, the Lyapunov exponent is $\ln |f'|$ averaged over the entire orbit. In order to know that the average really tells us something about the attractor (in this case, that orbits on the attractor separate exponentially), we must know that we will obtain the same average no matter which orbit we choose. We must be guaranteed that an orbit chosen at random spends the same portion of its iterates in a given region as any other such orbit would. That is precisely what a natural measure guarantees.

Recent progress has been made in the mathematical verification of the existence of chaotic attractors for the Hénon family $f_{a,b}$. (Benedicks and Carleson, 1991) have shown that for small negative Jacobian (small fixed b), there is a set of parameter a values with positive Lebesgue measure such that the attracting set for $f_{a,b}$ is a chaotic attractor. Interestingly, the particular a values with chaotic attractors cannot be specified. Also, (Benedicks and Young, 1996) have shown that there is a natural measure associated with these attractors. See (Palis and Takens, 1993) for more details.

A measure is **atomic** if all of the measure is contained in a finite or countably infinite set of points. To summarize our conclusions for Example 6.13, the natural measure for f is the atomic measure located at the sink $x = 0$. In general, a map for which almost every orbit is attracted to a fixed-point sink will, by the same reasoning, have an atomic natural measure located at the sink.

Ikeda measure is an example of a measure generated by a map. A different measure would be generated by a chaotic Hénon attractor as in Figure 6.1(a) or 6.4(a). Other map-generated measures are extremely simple and look nothing like a random number generator.

✎ **EXERCISE T6.10**

Let f be a map of the interval I for which almost every orbit is attracted by a period-k sink. Find the natural measure of f and justify your answer.

6.6 INVARIANT MEASURE FOR ONE-DIMENSIONAL MAPS

In this section we show how to find invariant measures, and in some cases natural measures, for a class of one-dimensional maps. We will call a map f on $[0, 1]$ **piecewise smooth** if $f(x), f'(x)$, and $f''(x)$ are continuous and bounded except possibly at a finite number of points. Recall that a map is **piecewise expanding** if furthermore there is some constant $\alpha > 1$ such that $|f'(x)| \geq \alpha$ except at a finite number of points in $[0, 1]$. Example 6.12 and the W-map of Example 6.10 satisfy these conditions.

✎ **EXERCISE T6.11**

Verify that the following maps on $[0, 1]$ are piecewise expanding. (a) The piecewise linear map $f(x) = a + bx$ mod 1, where $a \geq 0$ and $b > 1$. (b) The tent map $T_b(x)$ with slopes $\pm b$, where $1 < b \leq 2$.

Theorem 6.15 *Assume that the map f on $[0, 1]$ is piecewise smooth and piecewise expanding. Then f has an invariant measure μ. Furthermore the density is bounded, meaning that there is a constant c such that $\mu([a, b]) \leq c|b - a|$ for every $0 \leq a < b \leq 1$.*

The proof of this theorem is beyond the scope of this book (see (Lasota and Yorke, 1973) and (Li and Yorke, 1978)). Next we give some particularly nice examples for which μ can be exactly determined. It is possible for piecewise expanding maps to have more than one attractor, in which case each attractor will have a natural measure. We will see that for some choices of $[a, b]$, it is possible that $\mu([a, b]) = 0$.

In the examples we discuss, invariant measures have a simple description as the integral of a piecewise constant nonnegative function. That means that the measure of a subset S of I will be given by

$$\mu(S) = \int_S p(x)\, dx$$

for some

$$p(x) = \begin{cases} p_1 & \text{if } x \in A_1 \\ \vdots & \\ p_n & \text{if } x \in A_n \end{cases},$$

where the $p_i \geq 0$ and $\sum_{i=1}^{n} p_i \text{length}(A_i) = 1$. When the measure of a set is given by the integral of a function over the the set, such as the relationship between the measure μ and the function p in this case, then the function p is called the **density** of the measure.

For example, consider the W-map, shown in Figure 6.12 along with the density that defines its invariant measure:

$$p(x) = \begin{cases} p_1 = 0 & \text{if } 0 \leq x < 1/4 \\ p_2 = 2 & \text{if } 1/4 \leq x < 1/2 \\ p_3 = 2 & \text{if } 1/2 \leq x < 3/4 \\ p_4 = 0 & \text{if } 3/4 \leq x \leq 1 \end{cases}. \qquad (6.8)$$

Notice that the function $p(x)$ is the density of a probability measure, since $\int_0^1 p(x) \, dx = 1$. This measure is invariant under the W-map f because

$$\mu(S) = \int_S p(x) \, dx = \int_{f^{-1}(S)} p(x) \, dx = \mu(f^{-1}(S))$$

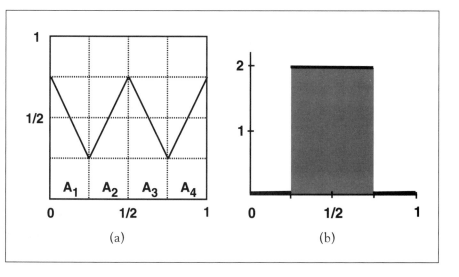

Figure 6.12 The W-map.
(a) The map is linear on each of four subintervals. (b) The density $p(x)$ that defines the invariant measure of (a). According to this graph, the measure of an interval inside $[1/4, 3/4]$ is twice its length.

for each set S. Let $S = [1/4, 3/8]$ for purposes of illustration. Then $f^{-1}(S) = [3/16, 5/16] \cup [11/16, 13/16]$, as can be checked using Figure 6.12. The μ-measure of S is

$$\mu(S) = \int_S p(x)\, dx = \int_{1/4}^{3/8} 2\, dx = 2(3/8 - 1/4) = 1/4,$$

which is the same as

$$\mu(f^{-1}(S)) = \int_{f^{-1}(S)} p(x)\, dx$$

$$= \int_{3/16}^{4/16} 0\, dx + \int_{4/16}^{5/16} 2\, dx + \int_{11/16}^{12/16} 2\, dx + \int_{12/16}^{13/16} 0\, dx$$

$$= 0 + 1/8 + 1/8 + 0. \tag{6.9}$$

As another example, assume instead that S is completely contained in $[0, 1/4]$. No points map to it, so $f^{-1}(S)$ is the empty set, which must have measure zero as required by the definition of measure. We can state this as a general fact.

✎ **EXERCISE T6.12**

Let f be a map on the interval I and let μ be an invariant measure for f. Show that if S is a subset of I that is outside the range of f, then $\mu(S) = 0$.

✎ **EXERCISE T6.13**

Prove that $\mu(S) = \mu(f^{-1}(S))$ for the invariant measure μ of the W-map, for any subset S of the unit interval.

Next we show how to calculate the invariant measures for expanding piecewise-linear maps of the interval. In general we assume that the interval is the union of subintervals $A_1, \ldots, A_k$, and that the map f has constant slope s_i on subinterval A_i. We also assume that the image of each A_i under f is exactly a union of some of the subintervals. Define the **Z-matrix** of f to be a $k \times k$ matrix whose (i, j) entry is the reciprocal of the absolute value of the slope of the map from A_j to A_i, or zero if A_j does not map across A_i. For the W-map, defined in four subintervals A_1, A_2, A_3, A_4, the Z-matrix is

$$Z = \begin{pmatrix} 0 & 0 & 0 & 0 \\ 1/2 & 1/2 & 1/2 & 1/2 \\ 1/2 & 1/2 & 1/2 & 1/2 \\ 0 & 0 & 0 & 0 \end{pmatrix}. \tag{6.10}$$

As we explain below, a Z-matrix has no eigenvalues greater than 1 in magnitude, and 1 is always an eigenvalue. Any eigenvector $\mathbf{x} = (x_1, \ldots, x_k)$ associated to the eigenvalue 1 corresponds to an invariant measure of f, when properly normalized so that the total measure is 1. The normalization is done by dividing $\mathbf{x}$ by $x_1 L_1 + \cdots + x_k L_k$, where L_i is the length of A_i. The ith component of the resulting eigenvector gives the height of the density p_i on A_i.

For example, the Z-matrix (6.10) for the W-map has characteristic polynomial $P(\lambda) = \lambda^3(\lambda - 1)$ and eigenvalues 1,0,0,0. All eigenvectors associated with the eigenvalue 1 are scalar multiples of $\mathbf{x} = (0, 1, 1, 0)$. Normalization entails dividing $\mathbf{x}$ by $0 \cdot 1/4 + 1 \cdot 1/4 + 1 \cdot 1/4 + 0 \cdot 1/4 = 1/2$, yielding the density $p = (0, 2, 2, 0)$ of Equation (6.8). Because the other eigenvalues are 0, the measure defined by this density is a natural measure for the W-map.

There may be several different invariant measures for f if the space of eigenvectors associated to the eigenvalue 1 has dimension greater than one. An eigenvalue is called **simple** if this is not the case, if there is only a one-dimensional space of eigenvectors. If 1 is a simple eigenvalue and if all other eigenvalues of the Z-matrix are strictly smaller than 1 in magnitude, then the resulting measure is a natural measure for f, meaning that almost every point in the interval will generate this measure.

Next we explain why the Z-matrix procedure works. Let p_i be the total density of the invariant measure on A_i, assumed to be constant on the subinterval. The amount of measure contained in A_i is $L_i p_i$. Since the slope of f is constant on A_i, one iteration of the map distributes this measure evenly over other subintervals. The image of A_i has length $s_i L_i$, where s_i is the absolute value of the slope of the map on A_i. Therefore the proportion of A_i's measure deposited into A_j is $L_j/(L_i s_i)$. (For the W-map, the s_i are all 2.) Since the p_i represent the density of an invariant measure by assumption, the total measure mapped into A_j from all A_i's must total up to $L_j p_j$. For the W-map this can be summarized in the matrix equation

$$
\begin{pmatrix}
0 & 0 & 0 & 0 \\
\frac{L_2}{L_1 s_1} & \frac{L_2}{L_2 s_2} & \frac{L_2}{L_3 s_3} & \frac{L_2}{L_4 s_4} \\
\frac{L_3}{L_1 s_1} & \frac{L_3}{L_2 s_2} & \frac{L_3}{L_3 s_3} & \frac{L_3}{L_4 s_4} \\
0 & 0 & 0 & 0
\end{pmatrix}
\begin{pmatrix}
L_1 p_1 \\
L_2 p_2 \\
L_3 p_3 \\
L_4 p_4
\end{pmatrix}
=
\begin{pmatrix}
L_1 p_1 \\
L_2 p_2 \\
L_3 p_3 \\
L_4 p_4
\end{pmatrix}.
\tag{6.11}
$$

Since the L_i are known, finding the invariant measure $\{p_i\}$ is equivalent to finding an eigenvector of the matrix in (6.11) with eigenvalue 1. This matrix turns out to have some interesting properties. For example, notice that each column adds up to exactly 1. That is because the length of the image $f(A_i)$ is $s_i L_i$, so the numerators in each column must add up to $s_i L_i$.

Definition 6.16 A square matrix with nonnegative entries, whose columns each add up to 1, is called a **Markov matrix**.

A Markov matrix always has the eigenvalue $\lambda = 1$, and its corresponding eigenvector has nonnegative entries. If the remaining eigenvalues are smaller than 1 in magnitude, then all vectors except linear combinations of eigenvectors of the other eigenvalues tend to the dominant eigenvector upon repeated multiplication by the matrix. See, for example, (Strang, 1988). In our application, when this eigenvector is normalized so that the sum of its entries is 1, its ith entry is the amount of measure $L_i p_i$ in A_i.

We can simplify (6.11) quite a bit by defining

$$
D = \begin{pmatrix} L_1 & & \\ & \ddots & \\ & & L_k \end{pmatrix}, \quad p = \begin{pmatrix} p_1 \\ \vdots \\ p_k \end{pmatrix}.
$$

It is a fact that multiplying a matrix by a diagonal matrix $\mathrm{diag}(L_1, \ldots, L_k)$ on the left results in multiplying the ith row by L_i, while multiplying by a diagonal matrix on the right multiplies the ith column by L_i. Then (6.11) can be written $D Z D^{-1} D p = D p$, which simplifies to $Z p = p$ by multiplying both sides by D^{-1} on the left. This concludes our explanation since it shows that solving for an eigenvector of Z with eigenvalue 1 will give an invariant measure.

✎ **EXERCISE T6.14**

Show that the Z-matrix for the tent map is $\begin{pmatrix} \frac{1}{2} & \frac{1}{2} \\ \frac{1}{2} & \frac{1}{2} \end{pmatrix}$, and that the density which defines the natural measure is the constant $p(x) = 1$ on $[0, 1]$. Thus the natural measure for this map is ordinary Lebesgue (length) measure.

EXAMPLE 6.17

(Critical point is period-three.) Recall the map from Example 6.12, shown in Figure 6.10(a). The Z-matrix is

$$
\begin{pmatrix} 0 & \frac{1}{a} \\ \frac{1}{a} & \frac{1}{a} \end{pmatrix}, \tag{6.12}
$$

where $a = (\sqrt{5} + 1)/2$ satisfies $a^2 = a + 1$. The characteristic equation is $\lambda^2 - (1/a)\lambda - 1/a^2 = 0$. We know $(\lambda - 1)$ is a factor, and so the factorization $(\lambda - 1)(\lambda + c) = 0$ follows, where $c = 1/a^2 \approx 0.382$. The eigenvalues are 1 and $-c$. Since the latter is smaller than 1 in absolute value, there is a natural measure. An eigenvector associated to 1 is $(x_1, x_2) = (1, a)$. The normalization involves dividing this eigenvector by $x_1 L_1 + x_2 L_2 = 1 \cdot c + a \cdot (1 - c)$. The result is that the natural measure for f is $\mu(S) = \int_S p(x)\, dx$, where

$$p(x) = \begin{cases} 1 - 1/(1 + a^2) & \text{if } 0 \le x \le c \\ a - a/(1 + a^2) & \text{if } c \le x \le 1 \end{cases}.$$

The measure is illustrated in Figure 6.10(c).

EXAMPLE 6.18

Let

$$f(x) = \begin{cases} \frac{1}{2} - 2x & \text{if } 0 \le x \le \frac{1}{4} \\ -\frac{1}{2} + 2x & \text{if } \frac{1}{4} \le x \le \frac{3}{4} \\ \frac{5}{2} - 2x & \text{if } \frac{3}{4} \le x \le 1 \end{cases},$$

as shown in Figure 6.13. The transition graph is shown in Figure 6.13(b).

There is a stretching partition $0 < 1/4 < 1/2 < 3/4 < 1$, but no dense orbit, because A_1 and A_4 cannot be connected by an itinerary. There are, however,

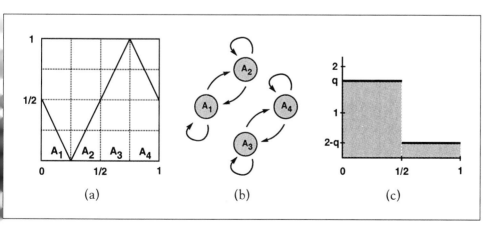

(a)   (b)   (c)

Figure 6.13 A piecewise linear map with no dense orbit.
(a) Sketch of map. (b) Transition graph. (c) One invariant measure for (a). In this case, there is no unique invariant measure—there are infinitely many.

dense orbits for the map restricted to $[0, 1/2]$ and to $[1/2, 1]$. The Z-matrix is

$$Z = \begin{pmatrix} 1/2 & 1/2 & 0 & 0 \\ 1/2 & 1/2 & 0 & 0 \\ 0 & 0 & 1/2 & 1/2 \\ 0 & 0 & 1/2 & 1/2 \end{pmatrix}.$$

The eigenvalues of Z are $1, 1, 0, 0$. Both $(2, 2, 0, 0)$ and $(0, 0, 2, 2)$, as well as linear combinations of them, are eigenvectors associated to 1. As a result, there are many invariant measures for f: for any $0 < q < 1$, the measure defined by

$$p(x) = \begin{cases} q & \text{if } 0 \le x \le 1/2 \\ 2 - q & \text{if } 1/2 \le x \le 1 \end{cases}.$$

is invariant. The map has no natural measure.

EXAMPLE 6.19

(Critical point eventually maps onto fixed point.) Let

$$f(x) = \begin{cases} 2 + \sqrt{2}(x - 1) & \text{if } 0 \le x \le c \\ \sqrt{2}(1 - x) & \text{if } c \le x \le 1 \end{cases},$$

where $c = \frac{2 - \sqrt{2}}{2}$ and $d = 2 - \sqrt{2}$ is a fixed point of f. See Figure 6.14. Check that the slopes of the map are $\sqrt{2}$ and $-\sqrt{2}$. The stretching factor for f is $\alpha = \sqrt{2}$,

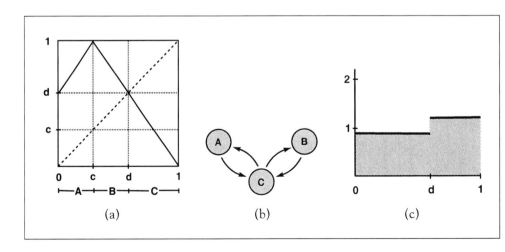

(a)   (b)   (c)

Figure 6.14   Map in which the critical point is eventually periodic.
(a) Sketch of map. (b) Transition graph. (c) Invariant measure for (a).

and the Lyapunov exponent is $\ln \sqrt{2}$ for each orbit for which it is defined. The transition graph is shown in Figure 6.14(b).

Notice that $f^3(c) = d$, so that the critical point is an eventually fixed point. The partition $0 < c < d < 1$ is a stretching partition for f. Define the subintervals $A = (0, c)$, $B = (c, d)$, and $C = (d, 1)$. Then $f(A) = C, f(B) = C$, and $f(C) = A \cup B$. Check that all pairs of symbols can be connected. According to Theorem 6.11, f has a dense chaotic orbit on I, and I is a chaotic attractor for f. The Z-matrix is

$$
Z = \begin{pmatrix} 0 & 0 & 1/\alpha \\ 0 & 0 & 1/\alpha \\ 1/\alpha & 1/\alpha & 0 \end{pmatrix},
$$

where the stretching factor for f is $\alpha = \sqrt{2}$. The eigenvectors of this matrix are $1, -1$, and 0. The invariant measure associated to the eigenvalue 1 turns out to be a natural measure. Check that the measure is defined by

$$
p(x) = \begin{cases} 1/(2d) & \text{if } 0 \le x \le d \\ 1/(\sqrt{2}d) & \text{if } d \le x \le 1 \end{cases} . \tag{6.13}
$$

⇨ COMPUTER EXPERIMENT 6.3

Write a computer program to approximate the invariant measure of an interval map. Divide the interval into equally-spaced bins, and count the number of points that fall into each bin from a given orbit. Calculate the invariant measure for Example 6.19 and compare with (6.13). Then calculate the invariant measure for the logistic map and compare to the answer in Challenge 6.

☞ CHALLENGE 6

Invariant Measure for the Logistic Map

WE HAVE SEEN how to construct invariant measures for simple piecewise linear maps of the interval. More complicated maps can be significantly more difficult. In Challenge 6 you will work out the invariant measure of the logistic map $G(x) = 4x(1 - x)$.

In Chapter 3 we found a conjugacy $C(x) = (1 - \cos \pi x)/2$ between the tent map $T = T_2$ and the logistic map, satisfying $CT = GC$. Exercise T6.14 shows the invariant measure of the tent map to be m_1, ordinary length measure on the unit interval. Some elementary calculus will allow the transfer of the invariant measure of the tent map to one for the logistic map.

Step 1   If S is a subset of the unit interval, prove that the sets $T^{-1}(S)$ and $C^{-1}G^{-1}C(S)$ are identical. (Remember that T and G are not invertible; by $T^{-1}(S)$ we mean the points x such that $T(x)$ lies in S.)

Step 2   Use the fact that m_1 is invariant for the tent map to prove that $m_1(S) = m_1(C^{-1}G^{-1}C(S))$ for any subset S.

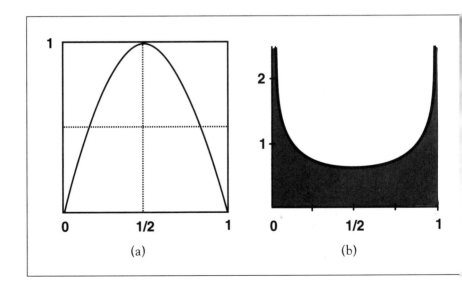

(a)

(b)

Figure 6.15   The logistic map G.
(a) Sketch of map. (b) The invariant measure for G.

Step 3 Prove that the the definition $\mu(S) = m_1(C^{-1}(S))$ results in a probability measure μ on the unit interval.

Step 4 Use Step 2 to show that μ is an invariant measure for G.

Step 5 It remains to compute the density $p(x)$. So far we know

$$\mu(S) = \int_{C^{-1}(S)} 1 \, dx.$$

Using the change of variable $y = (1 - \cos \pi x)/2$, rewrite $\mu(S)$ as an integral

$$\int_S p(y) \, dy$$

in y over the set S, and find the density $p(y)$. See Figure 6.15 for a graph of $p(x)$. (Answer: $p(x) = 1/\pi\sqrt{x - x^2}$)

EXERCISES

6.1. Show that a chaotic attractor cannot contain a sink.

6.2. For each of the piecewise linear maps shown in Figures 6.10(a), 6.13(a), and 6.14(a), find all periods for which there are periodic points.

6.3. Let $f(x, y) = (\frac{4}{\pi} \arctan x, y/2)$. Find all forward limit sets, attractors, and basins for each attractor. What is the basin boundary, and where do these points go under iteration?

6.4. Let $f(x) = -x/2$. The attractor is the set $\{0\}$. Find the fractions $F(x_0, \{0\})$ and $F(x_0, N(r, \{0\}))$ for all x_0 and $r > 0$. Compute the fraction $F(x_0, [0, \infty))$ for all x_0. Find the natural measure μ_f.

6.5. Let $g(x) = 2x(1 - x)$, which as a map of the interval $[0, 1]$ has a source at $x = 0$ and a sink at $x = 1/2$. What natural measure does g generate?

6.6. Let f be the map of Example 6.19, shown in Figure 6.14(a). Let $g = f^2$ be the second iterate of f. Sketch g, and explain why it doesn't have a natural measure. What are the invariant measures of g?

6.7. Find an invariant measure for the map $f(x) = 2 - x^2$ on the interval $[-2, 2]$.

6.8. Define the interval map

$$f(x) = \begin{cases} 3/4 - 2x & \text{if } 0 \le x \le 1/4 \\ 3x/2 - 1/8 & \text{if } 1/4 \le x \le 3/4 \\ 4 - 4x & \text{if } 3/4 \le x \le 1 \end{cases}$$

(a) Sketch the graph of f.

(b) Find a partition and transition graph for f.

(c) For what periods does f have a periodic orbit?

(d) Find the minimum stretching factor α. Is the interval $I = [0, 1]$ a chaotic attractor for f?

(e) Find the natural measure of f.

6.9. Let p be a point in $\mathbb{R}^n$. Prove that the assignment of 0 to each set not containing p and 1 to each set containing p is a probability measure on $\mathbb{R}^n$. This is the atomic measure located at p.

☞ **LAB VISIT 6**

Fractal Scum

A PLANE MAP was studied in an innovative experiment involving the hydrodynamics of a viscous fluid. A tank was filled with a sucrose solution, 20% denser than water, at a temperature of 32°C (think corn syrup). The plane on which the dynamics was observed was the surface of the syrupy fluid at the top of the tank.

The two-dimensional map on the fluid surface was defined as follows. Starting with the fluid at rest, a pump located below the surface was turned on for a fixed time period, and then turned off. The experimental configuration is shown in Figure 6.16. The pump draws off some of the solution from the middle of the

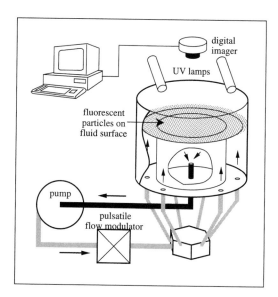

Figure 6.16   Schematic view of the tank.
Sucrose solution is intermittently pumped out of the tank from the bottom and returns through outlets in the bottom. Floating flourescent tracer particles (scum) on the surface of the fluid are excited by ultraviolet lamps and photographed by a digital camera.

Sommerer, J. 1994. Fractal tracer distributions in complicated surface flows: an application of random maps to fluid dynamics. Physica D **76**:85-98.

tank, and reinjects it at the bottom of the tank. There is a resulting mixing effect on the fluid surface. After the pump is turned off, the fluid is allowed to come to rest. The rest state of the fluid surface is by definition the result of one iteration of the plane map.

In order to follow an individual trajectory, tiny tracer particles were distributed on the fluid surface. The particles were fluorescent plastic spheres, each

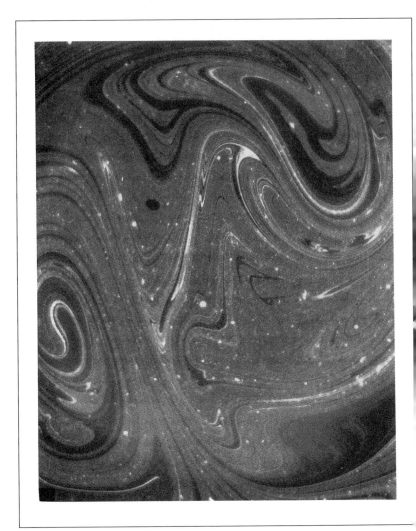

Figure 6.17 A 6-inch × 6-inch snapshot of the tracer distribution for the pulsed flow.
Bright areas are due to a dense collection of the tracer material, which correspond to a buildup of invariant measure. The distribution has a fractal appearance. This exposure took several seconds.

one being four one-thousandths of a millimeter in diameter. When ultraviolet lamps were turned on, the tracer particles floating on the surface were illuminated and photographed by the camera looking down on the surface. A typical photograph is shown in Figure 6.17. The variations in the relative illumination are caused by variations in the density of the tracer particles, which in turn approximate the natural measure of the attractor. Notice the pronounced fractal structure caused by the mixing effects of the map.

This experiment should be modeled, strictly speaking, as a map with some randomness. Due to the difficulty with exact control of the experimental conditions, each time the pump is run, slightly different effects occur; the map applied is not exactly the same each iteration, but an essentially random choice from a family of quite similar maps. For such maps it is important to focus not so much on individual trajectories but on collective, average behavior.

In order to estimate the Lyapunov exponents of this two-dimensional map, a different arrangement of tracer particles was used. A droplet of tracer was applied to the surface with a pipette, forming a small disk as shown in Figure 6.18(a). After the map is iterated once by running the pump as described, the disk becomes an approximate ellipse as shown in Figure 6.18(b).

Determining the Lyapunov exponent from the definition would mean starting with a tiny disk and observing the evolution of the ellipse for many iterations. The stretching factor along the longest ellipse axis, when expressed on a per-step basis, would give the Lyapunov number. However, this approach is impractical

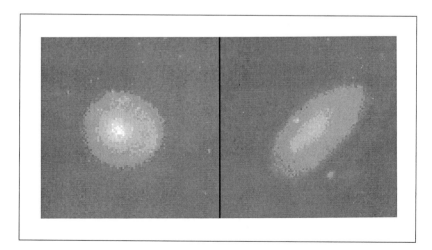

Figure 6.18 Evolution of a disk under the map.
On the left is a circular tracer particle patch on the surface of the fluid. After a few iterations of the map (pumping oscillations), the disk is transformed into an ellipse.

in this experiment. The ellipse cannot have an infinitesimal size, so it becomes folded and distorted after just a few iterates.

This problem was solved by starting a new disk after each iteration. After the first iteration, the ellipse of tracer particles is carefully removed by suction. A new circular patch was placed at the point of removal, and a new iteration commenced. Several dozen iterations were done in this way, and the ellipses recorded. For each ellipse, the (geometric) average radius among all directions of the ellipse was calculated. Then the geometric average of these radii was declared the dominant Lyapunov number. This is a reasonable approximation for the Lyapunov number as long as the stretching directions of the various ellipses are evenly distributed in two dimensions.

For each ellipse, the ratio of the ellipse area to the original disk area gives an estimate for the Jacobian determinant of the map. Therefore the smaller of the two Lyapunov numbers can also be found. The estimates for the larger Lyapunov number 1.68 and the Jacobian determinant $J = 0.83$ are shown in Figure 6.19, converging to their apparent limiting values after 100 iterations. Since we know that the Lyapunov exponents satisfy $\lambda_1 + \lambda_2 = \ln J$, the Lyapunov exponents are

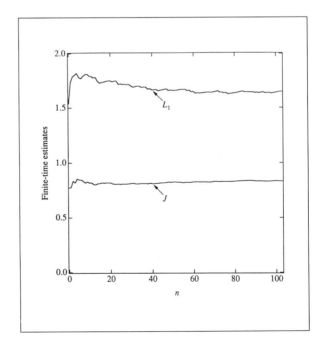

Figure 6.19   Convergence of the Lyapunov number estimate.
Experimental estimates of the largest Lyapunov number L_1 and area-contraction J.

ln 1.68 = 0.52 and −0.71. The Lyapunov dimension of the syrup attractor shown in Figure 6.16 is therefore $1 + 0.52/0.71 \approx 1.73$.

This experiment was carried out and the results interpreted with considerably more care than our description shows. In particular, great pains were taken to correctly interpret the results in light of the random influences on the dynamics. In addition, the experiment was repeated several times at different temperatures, where the viscosity of the solution is different. At 38°C, the syrup is runnier, and the Lyapunov dimension is about 1.6. At 27°C, the Lyapunov dimension is estimated to be 1.9. Consult the original source for more details.

CHAPTER SEVEN

Differential Equations

IN THE FIRST six chapters, we modeled physical processes with maps. One of the most important uses of maps in scientific applications is to assist in the study of a differential equation model. We found in Chapter 2 that the time-T map of a differential equation may capture the interesting dynamics of the process while affording substantial simplification from the original differential equation.

A map describes the time evolution of a system by expressing its state as a function of its previous state. Iterating the map corresponds to the system moving through time in discrete updates. Instead of expressing the current state as a function of the previous state, a differential equation expresses the *rate of change* of the current state as a function of the current state.

A simple illustration of this type of dependence is Newton's law of cooling, which we discussed in Chapter 2. Consider the state x consisting of the difference between the temperature of a warm object and the temperature of its surroundings. The rate of change of this temperature difference is negatively proportional to the temperature difference itself:

$$\dot{x} = ax, \tag{7.1}$$

where $a < 0$. Here we have used the notation $\dot{x}$ to represent the derivative of the function x. The solution of this equation is $x(t) = x(0)e^{at}$, meaning that the temperature difference x decays exponentially in time. This is a linear differential equation, since the terms involving the state x and its derivatives are linear terms.

Another familiar example from Chapter 2, which yields a nonlinear differential equation, is that of the pendulum. The pendulum bob hangs from a pivot, which constrains it to move along a circle, as shown in Figure 2.4 of Chapter 2. The acceleration of the pendulum bob in the tangential direction is proportional to the component of the gravitational downward force in the tangential direction, which in turn depends on the current position of the pendulum. This relation of the second derivative of the angular position with the angular position itself is one of the most fundamental equations in science:

$$\ddot{x} = -\sin x. \tag{7.2}$$

The pendulum is an example of a nonlinear oscillator. Other nonlinear oscillators that satisfy the same general type of differential equation include electric circuits, feedback systems, and many models of biological activity.

Most physical laws that have been successful in the study of dynamically changing quantities are expressed in the form of differential equations. The prime example is Newton's law of motion $F = ma$. The acceleration a is the second derivative of the position of the object being acted upon by the force F. Newton and Leibniz developed the calculus in the seventeenth century to express the fact that a relationship between x and its derivatives $\dot{x}$, $\ddot{x}$ and so on, can determine the motion in the past and future, given a specified present (initial condition). Since then, calculus and differential equations have become essential tools in the sciences and engineering.

Ordinary differential equations are differential equations whose solutions are functions of one independent variable, which we usually denote by t. The variable t often stands for time, and the solution we are looking for, $x(t)$, usually stands for some physical quantity that changes with time. Therefore we consider x as a dependent variable. Ordinary differential equations come in two types:

- **Autonomous**, for example

$$\dot{x} = ax, \tag{7.3}$$

  in which the time variable t does not explicitly appear, and
- **Nonautonomous**, as in the forced damped pendulum equation

$$\ddot{x} = -c\dot{x} - \sin x + \rho \sin t, \tag{7.4}$$

  for which t appears explicitly in the differential equation.

Autonomous differential equations are the ones that directly capture the spirit of a deterministic dynamical system, in which the law for future states is written only in terms of the present state x. However, the distinction is somewhat artificial: Any nonautonomous equation can be written as an autonomous system by defining a new dependent variable y equal to t; then for example we could write (7.4) as

$$\ddot{x} = -c\dot{x} - \sin x + \rho \sin y$$

$$\dot{y} = 1. \tag{7.5}$$

The system of equations (7.5) is autonomous because t does not appear on the right-hand side. In effect, t has been turned into one of the dependent variables by renaming it y. Because autonomous equations are the more general form, we will restrict our attention to them in this chapter.

The **order** of an equation is the highest derivative that occurs in the equation. We will begin by discussing first-order equations, in which only first derivatives of the dependent variable occur. The equations may be linear or nonlinear, and there may be one or more dependent variables. We will discuss several cases, in order of increasing complexity.

7.1 ONE-DIMENSIONAL LINEAR DIFFERENTIAL EQUATIONS

First, let us explain the title of this section. The dimension refers to the number of dependent variables in the equation. In this section, there is one (the variable x), which is a function of the independent variable t. The differential equation will express $\dot{x}$, the instantaneous rate of change of x with respect to t, in terms of the current state x of the system. If the expression for $\dot{x}$ is linear in x, we say that it is a linear differential equation.

Let

$$\dot{x} \equiv \frac{dx}{dt} = ax, \tag{7.6}$$

where x is a scalar function of t, a is a real constant, and $\dot{x}$ denotes the instantaneous rate of change with respect to time. For $a > 0$, (7.6) is a simple model of population growth when the population is small. The rate dx/dt at which the population grows is proportional to the size x of the population. Solutions of (7.6) with $a > 0$ are shown in Figure 7.1(a). (Compare these with the population model $x_{n+1} = ax_n$ in Chapter 1. In that case, the size of the new population is proportional to the previous population. These are different models.)

The differential equation (7.6) has infinitely many solutions, each of form $x(t) = ce^{at}$, for a constant real number c. By substituting $t = 0$, it follows that $x(0) = c$. The number $x_0 = x(0)$ is called the **initial value** of the function x. A problem is usually stated in the form of an **initial value problem**, which consists of a differential equation together with enough initial values (one, in this case) to specify a single solution. Using this terminology, we say that the solution of the initial value problem

$$\dot{x} = ax$$

$$x(0) = x_0 \tag{7.7}$$

is

$$x(t) = x_0 e^{at}. \tag{7.8}$$

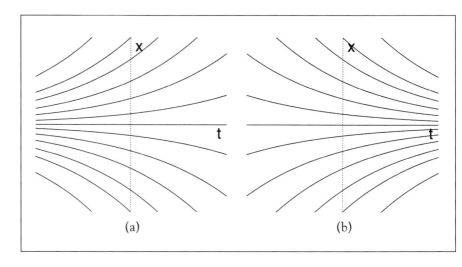

Figure 7.1 The family of solutions of $\dot{x} = ax$.
(a) $a > 0$: exponential growth (b) $a < 0$: exponential decay.

Figure 7.1(a) shows the family of all solutions of a differential equation, for various initial conditions x_0. Each choice of initial value x_0 puts us on one of the solution curves. This is a picture of the so-called flow of the differential equation.

Definition 7.1 The **flow** F of an autonomous differential equation is the function of time t and initial value x_0 which represents the set of solutions. Thus $F(t, x_0)$ is the value at time t of the solution with initial value x_0. We will often use the slightly different notation $F_t(x_0)$ to mean the same thing.

The reason for two different notations is that the latter will be used when we want to think of the flow as the time-t map of the differential equation. If we imagine a fixed $t = T$, then $F_T(x)$ is the map which for each initial value produces the solution value T time units later. For Newton's law of cooling, the time-10 map has the current temperature as input and the temperature 10 time units later as the output. The definition of a time-T map allows us to instantly apply all that we have learned about maps in the previous six chapters to differential equations.

Figures 7.1 (a) and (b) show the family of solutions (depending on x_0) for $a > 0$ and for $a < 0$, respectively. For (7.6), the flow is the function of two variables $F(t, x) = xe^{at}$. Certain solutions of (7.6) stand out from the others. For example, if $x_0 = 0$, then the solution is the constant function $x(t) = 0$, denoted $x \equiv 0$.

Definition 7.2 A constant solution of the autonomous differential equation $\dot{x} = f(x)$ is called an **equilibrium** of the equation.

An equilibrium solution x necessarily satisfies $f(x) = 0$. For example, $x \equiv 0$ is an equilibrium solution of (7.6). For all other solutions of (7.6) with $a > 0$, $\lim_{t \to \infty} |x(t)| = \infty$, as shown in Figure 7.1(a). An equilbrium like $x_0 = 0$ is a fixed point of the time-T map for each T.

For some purposes, too much information is shown in Figure 7.1. If we were solely interested in where solutions curves end up in the limit as $t \to \infty$, we might eliminate the t-axis, and simply show on the x-axis where solution trajectories are headed. For example, Figure 7.2 (a) shows that the x-values diverge from 0 as t increases. This figure, which suppresses the t-axis, is a simple version of a phase portrait, which we describe at length later. The idea of the phase portrait is to compress information. The arrows indicate the direction of solutions (toward or away from equilibria) without graphing specific values of t. As with maps, we are often primarily interested in understanding qualitative aspects of final state

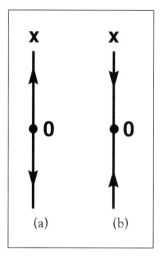

Figure 7.2 Phase portraits of $\dot{x} = ax$.
Since x is a scalar function, the phase space is the real line $\mathbb{R}$. (a) The direction of solutions is away from the equilibrium for $a > 0$. (b) The direction of solutions is toward the equilibrium for $a < 0$.

behavior. Other details, such as the rate at which orbits approach equilibria is lost in a phase portrait.

For $a < 0$, (7.6) models exponential decay, as shown in Figure 7.2(b), and $x \equiv 0$ is still an equilibrium solution. Examples include Newton's law of cooling, which was described above, and radioactive decay, where the mass of a radioactive isotope decreases according to the exponential decay law. No matter how we choose the initial condition x_0, we see that $x(t)$ tends toward 0 as time progresses; that is, $\lim_{t \to \infty} x(t) = \lim_{t \to \infty} x_0 e^{at} = 0$. The flow $F(t, x_0)$ of solutions is shown in Figure 7.1 (b). In the phase portrait of Figure 7.2(b) we suppress t and depict the motion of the solutions toward 0.

7.2 ONE-DIMENSIONAL NONLINEAR DIFFERENTIAL EQUATIONS

EXAMPLE 7.3

Equation (7.6) ceases to be an appropriate model for large populations x because it ignores the effects of overcrowding, which are modeled by non-linear terms. It is perhaps more accurate for the rate of change of population to be a logistic function of the population. The differential equation

$$\dot{x} = ax(1 - x), \tag{7.9}$$

where a is a positive constant, is called the **logistic differential equation**. As x gets close to 1, the limiting population, the rate of increase of population decreases to zero.

This equation is nonlinear because of the term $-ax^2$ on the right side. Although the equation can be solved analytically by separating variables (see Exercise 7.4), in this section we will describe alternative methods that are geometric in nature and that can quickly reveal some important qualitative properties of solutions. We might ask, for example, which initial conditions yield increasing solutions? Which yield decreasing solutions? Given an initial population, to what final state will the population evolve? The qualitative methods developed here are of critical importance because most nonlinear equations are impossible to solve analytically in closed form.

We begin by finding constant solutions of (7.9). Setting $\dot{x}$ equal to 0 and solving for x yields two equilibrium solutions: namely $x \equiv 0$ and $x \equiv 1$. Figure 7.3 shows the family of solutions, or flow, of (7.9). For each initial condition x_0 there is a single solution curve that we denote by $F(t, x_0)$ which satisfies (1) the

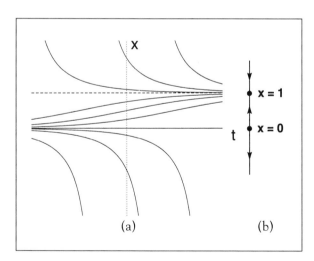

Figure 7.3 Solutions of the logistic differential equation.
(a) Solutions of the equation $\dot{x} = x(1 - x)$. Solution curves with positive initial conditions tend toward $x = 1$ as t increases. Curves with negative initial conditions diverge to $-\infty$. (b) The phase portrait provides a qualitative summary of the information contained in (a).

differential equation, and (2) the initial condition x_0. The solution curve $F(t, x_0)$ may or may not be defined for all future t. The curve $F(t, 1/2)$, shown in Figure 7.3, is asymptotic to the equilibrium solution $x \equiv 1$, and is defined for all t. On the other hand, the curves $F(t, x_0)$ shown in the figure with $x_0 < 0$ "blow up in finite time"; that is, they have a vertical asymptote for some finite t. (Since negative populations have no meaning, this is no great worry for the use of the logistic equation as a population model.) See Exercise 7.4 to work out the details. Another example of this blow-up phenomenon is given in Example 7.4.

Since we have not explicitly solved (7.9), how were we able to graph its solutions? We rely on three concepts:

1. **Existence:** Each point in the (t, x)-plane has a solution passing through it. The solution has slope given by the differential equation at that point.
2. **Uniqueness:** Only one solution passes through any particular (t, x).
3. **Continuous dependence:** Solutions through nearby initial conditions remain close over short time intervals. In other words, the flow $F(t, x_0)$ is a continuous function of x_0 as well as t.

Using the first concept, we can draw a **slope field** in the (t, x)-plane by evaluating $\dot{x} = ax(1 - x)$ at several points and putting a short line segment with the evaluated slope at each point, as in Figure 7.4. Recall that for an autonomous

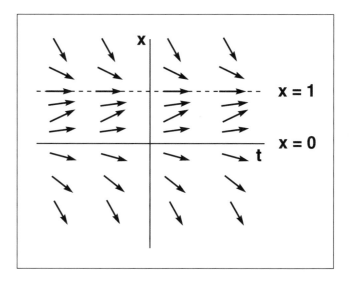

Figure 7.4 Slope field of the logistic differential equation.
At each point (t, x), the a small arrow with slope $ax(1 - x)$ is plotted. Any solution must follow the arrows at all times. Compare the solutions in Figure 7.3.

equation such as this, the slope for a given x-value is independent of the t-value, since t does not appear in the right-hand side of (7.9).

Curves can now be constructed that follow the slope field and which, by the second concept, do not cross. The third concept is referred to as "continuous dependence of solutions on initial conditions". All solutions of sufficiently smooth differential equations exhibit the property of continuous dependence, which is a consequence of continuity of the slope field. Theorems pertaining to existence, uniqueness, and continuous dependence are discussed in Section 7.4.

The concept of continuous dependence should not be confused with "sensitive dependence" on initial conditions. This latter property describes the behavior of unstable orbits over longer time intervals. Solutions may obey continuous dependence on short time intervals and also exhibit sensitive dependence, and diverge from one another on longer time intervals. This is the characteristic of a chaotic differential equation.

The phase portrait of the logistic differential equation (7.9) is shown in Figure 7.3(b). Note how the phase portrait summarizes the information about the asymptotic behavior of all solutions. It does this by suppressing the t-coordinate. The arrows on the phase portrait show the sign of the derivative (positive or negative) for points between, greater than, or less than the equilibrium values. For one-dimensional autonomous differential equations, the phase portrait gives almost all the important information about solutions.

When an interval of nonequilibrium points in the phase portrait is bounded by (finite) equilibria, then solutions for initial conditions in the interval converge to one equilibrium for positive time (as $t \to \infty$) and to the other for negative time (as $t \to -\infty$). The latter corresponds to following all arrows in the slope field backwards. This fact is illustrated in Figure 7.3, and is the subject of Theorem 8.3 of Chapter 8.

As in the case of iterated maps, an equilibrium solution is called **attracting** or a **sink** if the trajectories of nearby initial conditions converge to it. It is called **repelling** or a **source** if the solutions through nearby initial conditions diverge from it. Thus in (7.6), $x \equiv 0$ is attracting in the case $a < 0$ and repelling in the case $a > 0$. In (7.9), $x \equiv 0$ is repelling and $x \equiv 1$ is attracting.

✎ **EXERCISE T7.1**

Draw the slope field and phase portrait for $\dot{x} = x^3 - x$. Sketch the resulting family of solutions. Which initial conditions lead to bounded solutions?

Next we explore the phenomenon of "blow up in finite time". As mentioned above, all initial conditions that lie between two equilibria on the one-dimensional phase portrait will move toward one of the equilibria, unless it is an equilibrium itself. In particular, the trajectory of the initial condition will exist for all time. The following example shows that for initial conditions that are not bounded by equilibria, solutions do not necessarily exist for all time.

EXAMPLE 7.4

Consider the initial value problem

$$\dot{x} = x^2$$

$$x(t_0) = x_0. \tag{7.10}$$

We solve this problem by a method called **separation of variables**. First, divide by x^2, then integrate both sides of the equation

$$\frac{1}{x^2}\frac{dx}{dt} = 1$$

with respect to time from t_0 to t:

$$\int_{t_0}^{t} \frac{1}{x^2}\frac{dx}{dt}dt = \int_{t_0}^{t} dt = t - t_0. \tag{7.11}$$

Making the change of variables $x = x(t)$ means that dx replaces $\frac{dx}{dt}dt$, yielding

$$t - t_0 = \int_{x(t_0)}^{x(t)} \frac{dx}{x^2} = -\frac{1}{x}\bigg|_{x(t_0)}^{x(t)} = -\frac{1}{x(t)} + \frac{1}{x(t_0)}. \tag{7.12}$$

Solving for $x(t)$, we obtain

$$x(t) = \frac{1}{\frac{1}{x_0} + t_0 - t} = \frac{x_0}{1 + x_0(t_0 - t)}. \tag{7.13}$$

The result is valid if x is nonzero between time t_0 and t. Thus $x(t)$ must have the same sign (positive or negative) as $x(t_0)$.

When $x_0 = 0$ (or when $x = 0$ at any time), the unique solution is the equilibrium $x \equiv 0$, which is defined for all t. When $x_0 \neq 0$, the solution is not defined if the denominator $1 + x_0(t_0 - t)$ is 0. In this case, let $t_\infty = t_0 + 1/x_0$; t_∞ is the solution of $1 + x_0(t_0 - t) = 0$. Then $\lim_{t \to t_\infty} x(t) = \infty$. For $x_0 \neq 0$, therefore, the solution $x(t)$ exists only for t in the interval $(-\infty, t_\infty)$ or (t_∞, ∞), whichever contains t_0.

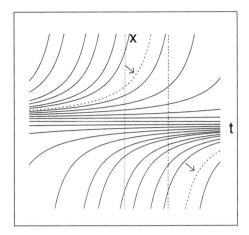

Figure 7.5 Solutions that blow up in finite time.
Curves shown are solutions of the equation $\dot{x} = x^2$. The dashed curve in the upper left is the solution with initial value $x(0) = 1$. This solution is $x(t) = 1/(1 - t)$, which has a vertical asymptote at $x = 1$, shown as a dashed vertical line on the right. The dashed curve at lower right is also a branch of $x(t) = 1/(1 - t)$, one that cannot be reached from initial condition $x(0) = 1$.

The solution curves of (7.10) are shown in Figure 7.5. The constant $x \equiv 0$ is an equilibrium. All curves above the horizontal t-axis blow up at a finite value of t; all curves below the axis approach the equilibrium solution $x \equiv 0$. For example, if $t_0 = 0$ and $x_0 = 1$, then $t_\infty = 1$, and the solution $x(t) = 1/(1 - t)$ exists on $(-\infty, 1)$. This solution is the dashed curve in the upper left of Figure 7.5. Equation (7.13) seems to suggest that the solution continues to be defined on $(1, \infty)$, but as a practical matter, the solution of the initial value problem with $x(0) = x_0$ no longer exists for t beyond the point where $x(t)$ goes to ∞.

In general, notice that the solution (7.13) has 2 branches. In Figure 7.5, arrows point to the 2 dashed branches of $x(t) = 1/(1 - t)$. Only the branch with $x(t)$ and x_0 having the same sign is valid for the initial value problem with $x_0 = 1$. The other branch is spurious as far as the initial value problem is concerned, since (7.12) fails to make sense when $x(t_0)$ and $x(t)$ have opposite signs.

Solutions that blow up in finite time are of considerable interest in mathematical modeling. When a solution reaches a vertical asymptote, the time interval over which the differential equation is valid has reached an end. Either the model needs to be refined to better reflect the properties of the system being modeled, or the system itself has a serious problem!

↻   COMPUTER EXPERIMENT 7.1

To do the Computer Experiments in the next few chapters, you may need to read Appendix B, an introduction to numerical solution methods for differential equations. A good practice problem is to choose some version of the Runge-Kutta algorithm described there and plot solutions of the differential equations of Figures 7.1, 7.3 and 7.5.

7.3 LINEAR DIFFERENTIAL EQUATIONS IN MORE THAN ONE DIMENSION

So far we have discussed first-order equations with one dependent variable. Now we move on to more dependent variables, beginning with a system of two linear equations. The solution of the system

$$\dot{x} = 2x$$

$$\dot{y} = -3y \tag{7.14}$$

can be determined by solving each equation separately. For an initial point (x_0, y_0) at time $t = 0$, the solution at time t is the vector $(x_0 e^{2t}, y_0 e^{-3t})$.

Figure 7.6 shows a graphical representation of the vector field of (7.14). A **vector field** on $\mathbb{R}^n$ is a function that assigns to each point in $\mathbb{R}^n$ an n-dimensional vector. In the case of a differential equation, the coordinates of the vector assigned to point (x, y) are determined by evaluating the right side of the equation at (x, y). For (7.14), the vector $(2x, -3y)$ is placed at the point (x, y), as seen in Figure 7.6(a). As in the case of slope fields, these vectors are tangent to the solutions. Thus the vector $(2x, -3y)$ gives the direction the solution moves when it is at the point (x, y). The length $|(2x, -3y)|$ of the vector is the speed of the solution as it moves through the point.

Since (7.14) is autonomous, (the variable t does not appear explicitly on the right side of (7.14)), the vector assigned to (x, y) is independent of time. A solution passing through (x, y) at one time t will not go a different direction or speed when passing through (x, y) at a different time. Autonomous equations have a particularly useful property that enables us to draw phase portraits for these systems: graphs of solutions $(x(t), y(t))$ drawn in the xy-plane, ignoring the t-axis, do not cross. (This property holds in addition to the fact that they do not cross in (x, y, t)–space, which is a consequence of uniqueness of solutions.) Figure 7.6(b)

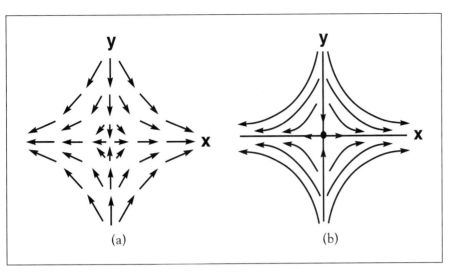

Figure 7.6 Vector field and phase plane for a saddle equilibrium.
(a) The vector field shows the vector $(\dot{x}, \dot{y})$ at each point (x, y) for (7.14). (b) The phase portrait, or phase plane, shows the behavior of solutions. The equilibrium $(x, y) \equiv (0, 0)$ is a saddle. The time coordinate is suppressed in a phase portrait.

shows the phase portrait of solutions of (7.14). A phase portrait in two dimensions is often called a **phase plane**, and in higher dimensions it is called **phase space**. The dimension of the phase space is the number of dependent variables. Arrows on the solution curves in phase space indicate the direction of increasing time.

The system (7.14) is uncoupled, meaning that neither of the dependent variables appear in the other's equation. More generally, the equation has the form

$$\dot{\mathbf{v}} = \mathbf{A}\mathbf{v}, \tag{7.15}$$

where $\mathbf{v}$ is a vector of variables, and $\mathbf{A}$ is a square matrix. Since the right-hand side of (7.15) is the zero vector when $\mathbf{v} = \mathbf{0}$, the origin is an equilibrium for (7.15). The stability of the equilibrium can be determined by the eigenvalues of the matrix of coefficients.

EXAMPLE 7.5

We show how eigenvalues are used to find the solution $\mathbf{v} = (x, y)$ of

$$\dot{x} = -4x - 3y$$

$$\dot{y} = 2x + 3y \tag{7.16}$$

with initial value $\mathbf{v}(0) = (1, 1)$. In vector form, (7.16) is given by (7.15) where

$$\mathbf{v} = \begin{pmatrix} x \\ y \end{pmatrix} \text{ and } \mathbf{A} = \begin{pmatrix} -4 & -3 \\ 2 & 3 \end{pmatrix}.$$

The eigenvalues of the matrix $\mathbf{A}$ are 2 and -3. Corresponding eigenvectors of $\mathbf{A}$ are $(1, -2)$ and $(3, -1)$, respectively. We refer the reader to Appendix A for fundamentals of eigenvalues and eigenvectors.

For a real eigenvalue λ of $\mathbf{A}$, an associated eigenvector $\mathbf{u}$ has the property that $\mathbf{Au} = \lambda\mathbf{u}$. Set $\mathbf{v}(t) = e^{\lambda t}\mathbf{u}$. By definition of eigenvector, $\mathbf{Av}(t) = e^{\lambda t}\mathbf{Au} = e^{\lambda t}(\lambda\mathbf{u})$. Since $\mathbf{u}$ is a fixed vector, $\dot{\mathbf{v}}(t) = \lambda e^{\lambda t}\mathbf{u}$ as well. Each eigenvalue-eigenvector pair of $\mathbf{A}$ leads to a solution of $\dot{\mathbf{v}} = \mathbf{Av}$. As the phase plane in Figure 7.7 shows, any vector along the line determined by an eigenvector $\mathbf{u}$ is stretched or contracted as t increases, depending on whether the corresponding eigenvalue λ is positive or negative. The phase plane of (7.16) is similar to that of (7.14) except that lines in the direction of the eigenvectors $\mathbf{u}_1 = (1, -2)$ and $\mathbf{u}_2 = (3, -1)$ take the place of the x and y axes, respectively. Both $\mathbf{v}(t) = e^{\lambda_1 t}\mathbf{u}_1$ and $\mathbf{v}(t) = e^{\lambda_2 t}\mathbf{u}_2$ are solutions.

If this vector argument is not clear, write the vectors in coordinates. Since $\mathbf{u}_1 = (1, -2)$, the corresponding solution $\mathbf{v}(t) = (x(t), y(t))$ is $(1e^{\lambda_1 t}, -2e^{\lambda_1 t})$. Differentiating each coordinate separately gives $(1\lambda_1 e^{\lambda_1 t}, -2\lambda_1 e^{\lambda_1 t})$, which is $\lambda_1\mathbf{v}(t)$, as needed. The same argument works for $\mathbf{u}_2$. When $\lambda_1 \neq \lambda_2$, then $\mathbf{u}_1$ and $\mathbf{u}_2$ are linearly independent, and the general solution of $\dot{\mathbf{v}} = \mathbf{Av}$ is given by

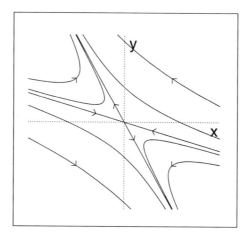

Figure 7.7 Phase plane for a saddle equilibrium.
For (7.16), the origin is an equilibrium. Except for two solutions that approach the origin along the direction of the vector $(3, -1)$, solutions diverge toward infinity, although not in finite time.

$\mathbf{v}(t) = c_1 e^{\lambda_1 t} \mathbf{u}_1 + c_2 e^{\lambda_2 t} \mathbf{u}_2$, for real constants c_1 and c_2. The constants need to be chosen to fit the initial values.

For example, suppose we want the solution with initial condition $\mathbf{v}(0) = (1, 1)$ at $t = 0$. Setting t equal to 0 in the general solution gives $\mathbf{v}(0) = c_1 \mathbf{u}_1 + c_2 \mathbf{u}_2$. Written in coordinates, we have $c_1(1, -2) + c_2(3, -1) = (1, 1)$. Solving gives $c_1 = -4/5$ and $c_2 = 3/5$. Thus the specific solution to this initial value problem is $\mathbf{v}(t) = -4/5 e^{2t} \mathbf{u}_1 + 3/5 e^{-3t} \mathbf{u}_2$. This solution corresponds to the uppermost curve in Figure 7.7. As t becomes large, the $\mathbf{u}_2$ term becomes negligible, so that this curve asymptotically approaches the line L_1 defined by the direction $\mathbf{u}_1$.

The eigenvectors corresponding to a particular eigenvalue λ together with the origin form a linear subspace, called the eigenspace corresponding to λ. In Figure 7.7 the straight lines are the eigenspaces; the steeper line L_1 is all multiples of $(1, -2)$, and the other line L_2 consists of all multiples of $(3, -1)$. The two eigenspaces play important roles in the long-time dynamics of (7.16). All initial values $\mathbf{v}(0)$ lying on L_2 approach the origin as t increases. All initial values except those on L_2 approach L_1 in the limit as $t \to \infty$.

In this example, the eigenvectors together span the entire phase space $\mathbb{R}^2$. In other cases, it is possible that the sum of the dimensions of all the eigenspaces does not equal the dimension of the phase space. In the following example, there is a single eigenvalue and a single linearly independent eigenvector, even though the phase space is two dimensional.

EXAMPLE 7.6

Let

$$\dot{x} = 3x + y$$

$$\dot{y} = 3y \tag{7.17}$$

The coefficient matrix

$$\mathbf{A} = \begin{pmatrix} 3 & 1 \\ 0 & 3 \end{pmatrix}$$

has only one eigenvalue $\lambda = 3$, and $(1, 0)$ is the only eigenvector up to scalar multiple. The phase plane for this system is shown in Figure 7.8. The x-axis is the eigenspace; it contains all positive and negative scalar multiples of $(1, 0)$.

For simplicity we sometimes refer to the eigenvectors and eigenvalues of a linear differential equation like (7.17) when we actually mean the corresponding matrix A. We would say for example that $\lambda = 3$ is an eigenvalue of (7.17).

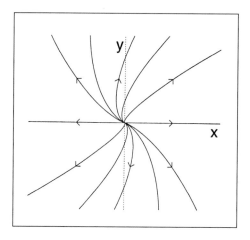

Figure 7.8   Phase plane for Equation (7.17).
The coefficient matrix $\mathbf{A}$ for this system has only one eigenvector, which lies along the x-axis. All solutions except for the equilibrium diverge to infinity.

✎   EXERCISE T7.2

(a) Verify the statements made in Example 7.6. (b) Find all solution curves of (7.17). Solve for $y(t)$ first, then try to guess the form of a solution for $x(t)$.

When the eigenvalues are complex, there are no corresponding real eigenvectors. We give two examples:

EXAMPLE 7.7

Let

$$\dot{x} = y$$

$$\dot{y} = -x. \tag{7.18}$$

Verify that the eigenvalues are $\pm i$. Solutions of this system are $x(t) = c_1 \cos t + c_2 \sin t$ and $y(t) = c_2 \cos t - c_1 \sin t$, where c_1 and c_2 are any real constants. The phase plane is shown in Figure 7.9. We will show that each solution remains a constant distance from the origin. If $\mathbf{v}(t) = (x(t), y(t))$ is a solution, then the distance squared is $|\mathbf{v}(t)|^2 = x^2 + y^2$. Differentiating this expression gives $2x\dot{x} + 2y\dot{y}$, which, from (7.18) is $2xy - 2yx = 0$. Thus the rate of change of the

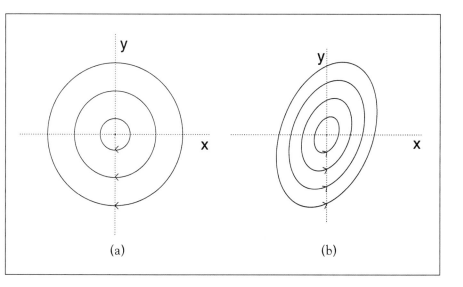

Figure 7.9 Phase planes for pure imaginary eigenvalues.
(a) In (7.18), the eigenvalues are $\pm i$. All solutions are circles around the origin, which is an equilibrium. (b) In (7.23), the eigenvalues are again pure imaginary. Solutions are elliptical. Note that for this equilibrium, some points initially move farther away, but not too far away. The origin is (Lyapunov) stable but not attracting.

distance from the origin is 0, meaning that $|\mathbf{v}(t)|^2$ and $|\mathbf{v}(t)|$ are constant. For those familiar with inner products, another way to see that solutions lie on circles about the origin is to note that at any point (x, y), the velocity vector $(y, -x)$ is perpendicular to the position vector (x, y). Hence the instantaneous motion is neither toward nor away from $(0, 0)$. Since this property holds at all points, $|\mathbf{v}(t)|$ never changes.

✎ **EXERCISE T7.3**

Explain why the trajectories in the phase planes of Figure 7.9(a) circle the origin clockwise.

EXAMPLE 7.8

A more general version of the above example is

$$\dot{x} = ax - by$$

$$\dot{y} = bx + ay. \tag{7.19}$$

Verify that the eigenvalues are $a \pm bi$. Solutions of this system are

$$x(t) = e^{at}(c_1 \cos bt + c_2 \sin bt)$$

$$y(t) = e^{at}(-c_2 \cos bt + c_1 \sin bt), \tag{7.20}$$

where c_1 and c_2 are any real constants. Check this solution by differentiating.

EXAMPLE 7.9

Here are two particular cases of Example 7.8. Let

$$\dot{x} = -x - 10y$$

$$\dot{y} = 10x - y. \tag{7.21}$$

Verify that the eigenvalues are $-1 \pm 10i$. Solutions of this system are $x(t) = e^{-t}(c_1 \cos 10t + c_2 \sin 10t)$ and $y(t) = e^{-t}(-c_2 \cos 10t + c_1 \sin 10t)$, where c_1 and c_2 are any real constants. The constants c_1 and c_2 are determined by matching initial conditions. All solutions spiral in toward the origin.

The slightly different system

$$\dot{x} = x - 10y$$

$$\dot{y} = 10x + y \tag{7.22}$$

has eigenvalues $1 \pm 10i$, and the solutions have form $x(t) = e^t(c_1 \cos 10t + c_2 \sin 10t)$ and $y(t) = e^t(-c_2 \cos 10t + c_1 \sin 10t)$. Solutions of this system spiral out from the origin.

See Figure 7.10 for a sketch of the phase planes of these two systems. The difference between them is that the origin is attracting when the eigenvalues of the right-hand side matrix have negative real part, and repelling when they have positive real part.

Definition 7.10 An equilibrium point $\bar{v}$ is called **stable** or **Lyapunov stable** if every initial point v_0 that is chosen very close to $\bar{v}$ has the property that the solution $F(t, v_0)$ stays close to $\bar{v}$ for $t \geq 0$. More formally, for any neighborhood N of $\bar{v}$ there exists a neighborhood N_1 of $\bar{v}$, contained in N, such that for each initial point v_0 in N_1, the solution $F(t, v_0)$ is in N for all $t \geq 0$. An equilibrium is called **asymptotically stable** if it is both stable and attracting. An equilibrium is called **unstable** if it is not stable. Finally, an equilibrium is **globally asymptotically stable** if it is asymptotically stable and all initial values converge to the equilibrium.

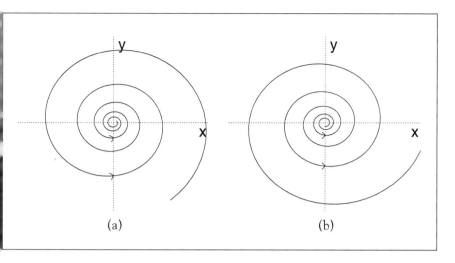

Figure 7.10   Phase planes for complex eigenvalues with nonzero real part.
(a) Under (7.21), trajectories spiral in to a sink at the origin. The eigenvalues of the coefficient matrix A have negative real part. (b) For (7.22), the trajectories spiral out from a source at the origin.

The two concepts of stability are independent; that is, there are examples of equilibria that are attracting but not stable and equilibria that are stable but not attracting. The equilibrium $\bar{v} = 0$ in (7.21) is asymptotically stable, therefore stable in both senses. In (7.17) and (7.22), the origin is unstable and not attracting. In Example 7.7, the origin is stable (take $N_1 = N$) but not attracting. For a linear system, the stability of the equilibrium at the origin is determined by the eigenvalues of the matrix A. If A has at least one eigenvalue with positive real part, at least one with negative real part, and no eigenvalues with real part zero, then 0 is called a **saddle**. In (7.14) and (7.16), the origin is a saddle. Note that saddles are unstable.

EXAMPLE 7.11

The origin of the system

$$\dot{x} = x - 2y$$

$$\dot{y} = 5x - y \qquad (7.23)$$

is a stable equilibrium with eigenvalues $\pm 3i$. The solutions are ellipses centered at the origin, as shown in Figure 7.9(b). This example shows why the definition of Lyapunov stability needs two neighborhoods: In order to have solutions stay within a neighborhood whose radius is the larger axis of an ellipse, initial conditions must be restricted to a neighborhood whose radius is no larger than the smaller axis of the solution.

✎ EXERCISE T7.4

Find the equations of the ellipses in Figure 7.9(b). They are given in parametric form by the solutions $x(t)$ and $y(t)$ of (7.23), which are linear combinations of $\cos 3t$ and $\sin 3t$.

Criteria for stability of a linear system are given by Theorem 7.12.

Theorem 7.12 *Let* A *be an* $n \times n$ *matrix, and consider the equation* $\dot{v} = Av$. *If the real parts of all eigenvalues of* A *are negative, then the equilibrium* $\bar{v} = 0$ *is globally asymptotically stable. If* A *has* n *distinct eigenvalues and if the real parts of all eigenvalues of* A *are nonpositive, then* $\bar{v} = 0$ *is stable.*

✎ EXERCISE T7.5

Let

$$\dot{x} = y$$

$$\dot{y} = 0. \tag{7.24}$$

Show that $(x(t), y(t)) = (at, a)$, $a \neq 0$, is an unbounded solution of (7.24). Therefore $\bar{v} = 0$ is an unstable equilibrium. Explain why this example does not contradict Theorem 7.12.

✎ EXERCISE T7.6

Determine the possible phase plane diagrams for two-dimensional linear systems with at least one eigenvalue equal to 0.

Thus far, we have only shown figures of one- and two-dimensional phase planes. The phase portraits of higher dimensional linear systems can be obtained by determining on which subspaces the equilibrium 0 is stable, asymptotically stable, or unstable.

EXAMPLE 7.13

Let

$$\dot{x} = 5x + y + z$$

$$\dot{y} = -2y - 3z$$

$$\dot{z} = 3y - 2z. \tag{7.25}$$

Verify that the eigenvalues are 5 and $-2 \pm 3i$. The eigenspace for the eigenvalue 5 is the x-axis. The phase space for this system is shown in Figure 7.11. The (y, z) coordinates of trajectories move toward $(y, z) = (0, 0)$. All trajectories except for those in the plane shown move away from origin $\mathbf{0} = (0, 0, 0)$ along the x-axis, and satisfy $|x(t)| \to \infty$ while $y(t), z(t) \to 0$. Trajectories in the plane with normal vector $(1, 5/29, 2/29)$ spiral in to the origin. This "eigenplane" is the stable manifold of the saddle; the x-axis is the unstable manifold. On the other hand, if the 5 is replaced by a negative number in (7.25), the origin $\mathbf{0}$ would be globally asymptotically stable.

If the number of linearly independent eigenvectors associated with a given eigenvalue λ is fewer than the multiplicity of λ as a root of the characteristic equation, then determination of the "generalized eigenspace" associated with λ is somewhat more complicated. We refer the reader to (Hirsch and Smale, 1974) for a complete treatment of this subject.

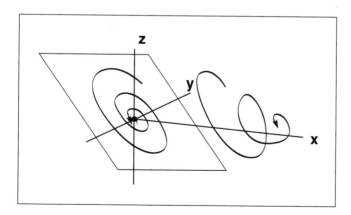

Figure 7.11   A three-dimensional phase portrait.
In Example 7.13, the origin $(0, 0, 0)$ is a saddle equilibrium. Trajectories whose initial values lie in the plane move toward the origin, and all others spiral away along the x-axis.

⇨ **COMPUTER EXPERIMENT 7.2**

Although linear systems can be solved with linear algebra, most nonlinear systems do not yield to analytic methods. For example, for the equation

$$\dot{x} = y - x^2$$

$$\dot{y} = -x, \tag{7.26}$$

obtained by adding a single nonlinear term to (7.18), we are reduced to numerical approximation methods, summarized in Appendix B. Plot solution curves of (7.26) near the origin. According to this evidence, do you expect that the origin is asymptotically stable? Lyapunov stable?

7.4 NONLINEAR SYSTEMS

In the first portion of this book, we used maps to model deterministic physical processes. We specified a map and the present state, and then as long as the map was well-defined, it told us unequivocally what happens for all future time. Now we want to use differential equations for the same purpose.

With differential equations, there are a few technicalities we need to consider. First, we have seen already that solutions to an initial value problem may blow up in finite time, and therefore not exist for all time. This happens for (7.10), for example. It is possible, both for differential equations and maps, for solutions to tend to infinity, but exist for all time in the process. But blow-up in finite time is different, and there is no analogue of this behavior for continuous maps.

Second, without any restrictions on the equations, an initial value problem may have more than one solution. This goes against the spirit of determinism. A good model should specify the future unambiguously, given the rule and the present state. But the initial value problem

$$\dot{x} = \sqrt{x}$$

$$x(0) = 0 \tag{7.27}$$

has two solutions, $x(t) = 0$ and $x(t) = t^2/4$. This does not make for a good model of a dynamical process.

Third, the utility of a model to give information about the dynamical process depends on the fact that the solution of the initial value problem does not depend too sensitively on the initial condition, at least at short time scales. In particular, for a fixed differential equation and two different initial values, we would like to know that the closer the two initial values are, the closer the solutions are for small t. This is called continuous dependence on initial conditions. For large t, we can't expect them to stay close—they may diverge toward opposite corners of phase space. Sensitivity at large t is called sensitive dependence on initial conditions.

Except for blow-up in finite time, these problems disappear under mild restrictions on the differential equation. We now present theorems on existence and uniqueness (Theorem 7.14) and continuous dependence on initial conditions (Theorem 7.16). Proofs of these theorems can be found in standard differential equations texts. Figure 7.12 shows two types of solution behavior which are ruled out.

Consider the first-order system

$$\dot{x}_1 = f_1(x_1, \ldots, x_n)$$

$$\vdots$$

$$\dot{x}_n = f_n(x_1, \ldots, x_n). \tag{7.28}$$

We denote this n-dimensional system of first-order ordinary differential equations by

$$\dot{\mathbf{v}} = \mathbf{f}(\mathbf{v}), \tag{7.29}$$

where $\mathbf{v} = (x_1, \ldots, x_n)$ is a vector.

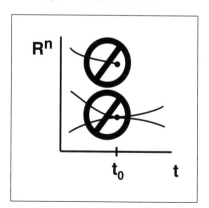

Figure 7.12 Solutions that are outlawed by the existence and uniqueness theorem.
Solutions cannot suddenly stop at t_0, and there cannot be two solutions through a single initial condition.

Theorem 7.14 **Existence and Uniqueness.** *Consider the first-order differential equation (7.29) where both* **f** *and its first partial derivatives with respect to* **v** *are continuous on an open set* U. *Then for any real number* t_0 *and real vector* $\mathbf{v}_0$, *there is an open interval containing* t_0, *on which there exists a solution satisfying the initial condition* $\mathbf{v}(t_0) = \mathbf{v}_0$, *and this solution is unique.*

Definition 7.15 Let U be an open set in $\mathbb{R}^n$. A function **f** on $\mathbb{R}^n$ is said to be **Lipschitz** on U if there exists a constant L such that

$$|\mathbf{f}(\mathbf{v}) - \mathbf{f}(\mathbf{w})| \leq L|\mathbf{v} - \mathbf{w}|,$$

for all $\mathbf{v}, \mathbf{w}$ in U. The constant L is called a **Lipschitz constant** for **f**.

If **f** has bounded first partial derivatives in U, then **f** is Lipschitz. For example, for the one-dimensional case, $f(x) = \sin x$ has Lipschitz constant $L = 1$. This follows from the Mean Value Theorem and the fact that $f'(x) = \cos x$.

✎ **EXERCISE T7.7**

The general two-dimensional linear equation is $\dot{\mathbf{v}} = \mathbf{A}\mathbf{v}$ where

$$\mathbf{A} = \begin{pmatrix} a & b \\ c & d \end{pmatrix}.$$

Find a Lipschitz constant for the function $\mathbf{A}\mathbf{v}$ on $\mathbb{R}^2$ in terms of a, b, c, and d.

Two neighboring solutions to the same differential equation can separate from each other at a rate no greater than e^{Lt}, where L is the Lipschitz constant of the differential equation. The Gronwall inequality, illustrated in Figure 7.13, is the basis of continuity of the flow as a function of the initial condition.

Theorem 7.16 **Continuous dependence on initial conditions.** *Let* **f** *be defined on the open set* U *in* $\mathbb{R}^n$, *and assume that* **f** *has Lipschitz constant* L *in the variables* **v** *on* U. *Let* $\mathbf{v}(t)$ *and* $\mathbf{w}(t)$ *be solutions of (7.29), and let* $[t_0, t_1]$ *be a subset of the domains of both solutions. Then*

$$|\mathbf{v}(t) - \mathbf{w}(t)| \leq |\mathbf{v}(t_0) - \mathbf{w}(t_0)|e^{L(t-t_0)},$$

for all t *in* $[t_0, t_1]$.

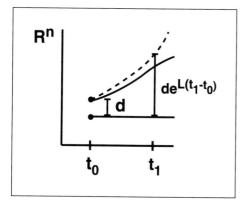

Figure 7.13 The Gronwall inequality.
Nearby solutions can diverge no faster than an exponential rate determined by the
Lipschitz constant of the differential equation.

Higher-order equations often can be transformed into a first-order system
of form (7.28). We illustrate the process with the nth-order differential equation

$$x^{(n)} = f(x, \dot{x}, \ddot{x}, \ldots, x^{(n-1)}),$$

where $x^{(n)}$ denotes the nth derivative of the function $x(t)$ with respect to t. Define
a new set of variables by

$$x_1 = x$$
$$x_2 = \dot{x}$$
$$x_3 = \ddot{x}$$
$$\vdots$$
$$x_n = x^{(n-1)}$$

These new variables satisfy the first-order autonomous system

$$\dot{x}_1 = x_2$$
$$\dot{x}_2 = x_3$$
$$\vdots$$
$$\dot{x}_{n-1} = x_n$$
$$\dot{x}_n = f(x_1, x_2, \ldots, x_n)$$

which is of form (7.28).

Unlike linear systems, most nonlinear systems of ordinary differential equations cannot be solved explicitly, meaning that the solutions cannot be found through an analytic calculation. Fortunately, much of the stability analysis for linear systems carries over to the study of equilibria of nonlinear systems.

Recall that an equilibrium of (7.29) is a vector $\bar{\mathbf{v}}$ in $\mathbb{R}^n$ such that $f_i(\bar{\mathbf{v}}) = 0$ for $i = 1, \ldots, n$. Again, for a given initial value $\mathbf{v}_0$, we denote by $\mathbf{F}(t, \mathbf{v}_0)$ the solution of (7.29) at time t. While a linear system either has only one equilibrium ($\bar{\mathbf{v}} = 0$) or has an entire line (or higher-dimensional subspace) of equilibria (see Exercise T7.6), a nonlinear system can have many isolated equilibria, as the following exercise illustrates.

✎ **EXERCISE T7.8**

Verify that the equilibria of the system

$$\dot{x} = y$$

$$\dot{y} = \sin x - y \qquad (7.30)$$

are $(n\pi, 0)$ for all integers n.

In order to determine the stability of an equilibrium $\bar{\mathbf{v}}$ of (7.29), we use the linear map that best approximates $\mathbf{f}$ at $\bar{\mathbf{v}}$—namely, the Jacobian matrix $\mathbf{Df}(\bar{\mathbf{v}})$ of partial derivatives evaluated at $\bar{\mathbf{v}}$. The Jacobian matrix $\mathbf{Df}(\bar{\mathbf{v}})$ can be expected to approximate $\mathbf{f}$ only in a small neighborhood of $\bar{\mathbf{v}}$. Since $\mathbf{f}(\bar{\mathbf{v}}) = 0$, the approximation $\mathbf{f}(\bar{\mathbf{v}} + \boldsymbol{\epsilon}) \approx \mathbf{Df}(\bar{\mathbf{v}})\boldsymbol{\epsilon}$ holds for small $\boldsymbol{\epsilon}$. Start a solution $\mathbf{w}(t) = \mathbf{F}(t, \bar{\mathbf{v}} + \boldsymbol{\epsilon})$ from an initial value $\bar{\mathbf{v}} + \boldsymbol{\epsilon}$ close to the equilibrium. Then $\mathbf{u}(t) = \mathbf{w}(t) - \bar{\mathbf{v}}$ satisfies

$$\dot{\mathbf{u}} = \dot{\mathbf{w}} = \mathbf{f}(\mathbf{w}(t)) = \mathbf{f}(\mathbf{u}(t) + \bar{\mathbf{v}}) \approx \mathbf{Df}(\bar{\mathbf{v}})\mathbf{u}(t),$$

at least for short times. The solutions of (7.29) near $\bar{\mathbf{v}}$ move toward or away from the equilibrium like the solutions of $\dot{\mathbf{u}} = \mathbf{Df}(\bar{\mathbf{v}})\mathbf{u}$. The behavior of the latter equation, which is linear, depends on the eigenvalues of $\mathbf{Df}(\bar{\mathbf{v}})$.

Definition 7.17 An equilibrium $\bar{\mathbf{v}}$ of $\dot{\mathbf{v}} = \mathbf{f}(\mathbf{v})$ is called **hyperbolic** if none of the eigenvalues of $\mathbf{Df}(\bar{\mathbf{v}})$ has real part 0.

When $\bar{\mathbf{v}}$ is hyperbolic, the linear part $\mathbf{Df}(\bar{\mathbf{v}})$ completely determines the stability of $\bar{\mathbf{v}}$. When all eigenvalues of $\mathbf{Df}(\bar{\mathbf{v}})$ have nonpositive real parts and there is at least one eigenvalue with zero real part, then higher-order terms must be taken into account to determine the stability; that is, it is not sufficient to know

$\mathbf{Df}(\bar{\mathbf{v}})$. Hyperbolic or not, $\bar{\mathbf{v}}$ is unstable if the real part of at least one eigenvalue of $\mathbf{Df}(\bar{\mathbf{v}})$ is strictly positive.

Theorem 7.18 *Let $\bar{\mathbf{v}}$ be an equilibrium of $\dot{\mathbf{v}} = \mathbf{f}(\mathbf{v})$. If the real part of each eigenvalue of $\mathbf{Df}(\bar{\mathbf{v}})$ is strictly negative, then $\bar{\mathbf{v}}$ is asymptotically stable. If the real part of at least one eigenvalue is strictly positive, then $\bar{\mathbf{v}}$ is unstable.*

EXAMPLE 7.19

We apply Theorem 7.18 to the following system:

$$\dot{x} = x^2 - y^2$$

$$\dot{y} = xy - 4. \tag{7.31}$$

There are two equilibria, $(2, 2)$ and $(-2, -2)$. The Jacobian matrix is

$$\mathbf{Df} = \begin{pmatrix} 2x & -2y \\ y & x \end{pmatrix}.$$

Evaluated at $(2, 2)$, $\mathbf{Df}$ has eigenvalues $3 \pm \sqrt{7}i$. Since the real part of the eigenvalues is $+3$, $(2, 2)$ is unstable. Evaluated at $(-2, -2)$, $\mathbf{Df}$ has eigenvalues $-3 \pm \sqrt{7}i$. Therefore, $(-2, -2)$ is asymptotically stable. A solution with initial condition (x_0, y_0) sufficiently close to the equilibrium $(-2, -2)$ will tend to $(-2, -2)$ as $t \to \infty$. Note that Theorem 7.18 does not tell us how close (x_0, y_0) must be to $(-2, -2)$ for the solution to converge to this equilibrium.

EXAMPLE 7.20

The one-dimensional equation

$$\dot{x} = -x^3 \tag{7.32}$$

has an equilibrium at $x = 0$. Since $f(x) = -x^3$ is a function of one variable, $\mathbf{Df}$ is the derivative $f'(x)$. Since $f'(0) = 0$, we cannot use Theorem 7.18 to determine the stability of $x = 0$.

EXERCISE T7.9

Decide whether $x = 0$ is an asymptotically stable equilibrium of (7.32). Solve by separating variables, as in Example 7.4. Does this equation have unique solutions? Find all solutions that satisfy $x(0) = 1$.

7.5 MOTION IN A POTENTIAL FIELD

Perhaps the most familiar system that illustrates the concepts of kinetic and potential energy is the pendulum equation

$$\ddot{x} + k \sin x = 0, \qquad (7.33)$$

where x is the angle of the pendulum rod from the vertical and k is the positive constant $k = g/l$, where l is the length of the pendulum rod and g is the acceleration of gravity. We will set $k = 1$ to simplify our analysis.

We are assuming no damping and no external forces aside from gravity. This equation can be rewritten as a first-order system with dependent variables x and y by setting $y = \dot{x}$. Then $\dot{y} = \ddot{x} = -\sin x$, and (7.33) becomes

$$\dot{x} = y$$

$$\dot{y} = -\sin x. \qquad (7.34)$$

✎ **EXERCISE T7.10**

(a) Show that the equilibria of (7.34) are $\{(n\pi, 0) : n = 0, \pm1, \pm2, \ldots\}$.
(b) Show that Theorem 7.18 identifies $(n\pi, 0)$ as an unstable (saddle) equilibrium if n is odd, but tells us nothing if n is even.

Taking a cue from mechanics, we use the principle of Conservation of Energy: In the absence of damping or any external forces, the system neither gains nor loses energy. Given an initial condition (x_0, y_0), the energy function E remains constant on the orbit $\mathbf{F}(t, (x_0, y_0))$ for all time t:

$$\frac{dE}{dt}(\mathbf{F}(t, (x_0, y_0))) = 0. \qquad (7.35)$$

Total energy is the sum of kinetic plus potential energies, which for the pendulum is given by

$$E(x, y) = (1/2)y^2 + 1 - \cos x. \qquad (7.36)$$

The potential energy is minus the integral of the force, $\int_0^x \sin t \, dt$ (the work required to raise the pendulum from angle 0 to angle x; we have set the mass of the pendulum bob to 1). To verify (7.35), write $(x(t), y(t))$ for a solution and compute $\frac{d}{dt}E(x(t), y(t))$ as

$$\frac{dE}{dt} = \frac{\partial E}{\partial x}\frac{dx}{dt} + \frac{\partial E}{\partial y}\frac{dy}{dt}$$

$$= (\sin x)\dot{x} + y\dot{y}$$

$$= (\sin x)y + y(-\sin x)$$

$$= 0. \tag{7.37}$$

Notice an interesting aspect of this analysis: We have completed this calculation without knowing the solutions $(x(t), y(t))$. Equation (7.37) says that $\frac{d}{dt}E(x(t), y(t)) = 0$. We conclude in this case that for each solution $E(x(t), y(t))$ remains constant as t varies. The function E provides a useful partition of the points (x, y) in the phase plane into individual solution trajectories.

Definition 7.21 Given a real number c and a function $E : \mathbb{R}^2 \longrightarrow \mathbb{R}$, the set $E_c = \{(x, y) : E(x, y) = c\}$ is called a **level curve** of the function E.

Notice that the minimum value of the energy function E is 0; hence, E_c is empty for $c < 0$. Some of the level curves of E are sketched in Figure 7.14(a).

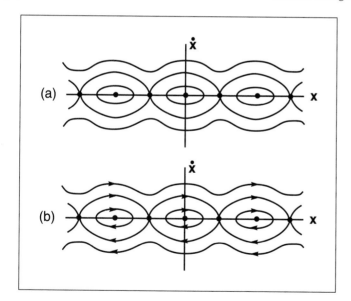

Figure 7.14 Solution curves of the undamped pendulum.
(a) Level curves of the energy function. (b) The phase plane of the pendulum. The solutions move along level curves; equilibria are denoted by dots. The variable x is an angle, so what happens at x also happens at $x + 2\pi$. As a results, (a) and (b) are periodic in x with period 2π.

Keeping in mind that a solution of (7.34) is constrained to one energy level, we can turn the sketch of level curves into a phase plane of (7.34) merely by putting arrows on the level curves to indicate the direction of motion, as in Figure 7.14(b). Notice that solutions above the x-axis move from left to right, and solutions below the x-axis move from right to left. Of particular interest are the points in E_0. These are precisely the equilibria whose stability could not be determined by Theorem 7.18: $(n\pi, 0)$, where n is even. In this case, the phase plane shows us that these are stable equilibria, since the trajectories of nearby points lie on closed level curves around the equilibria.

The pendulum equation (7.33) is a special case of the more general equation

$$\ddot{x} + \frac{\partial P}{\partial x} = 0 \tag{7.38}$$

governing motion in a potential field. This is another way of viewing Newton's second law of motion—acceleration is proportional to the force, which is the negative of the gradient of the potential field. The potential energy field of the pendulum equation with $k = 1$ is a series of potential wells whose minima are spaced at $2n\pi$ for all integers n, shown in Figure 7.15(a).

In the general case we multiply (7.38) by $\dot{x}$ and integrate both sides:

$$\ddot{x}\dot{x} + \frac{\partial P}{\partial x}\frac{dx}{dt} = 0$$

$$\frac{1}{2}\dot{x}^2 + P(x) = E_1 \tag{7.39}$$

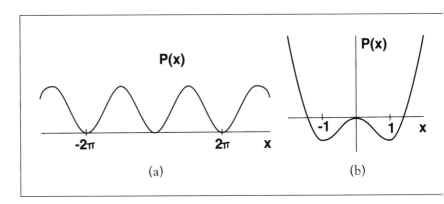

(a) (b)

Figure 7.15 Potential energy functions.
(a) The potential function for the pendulum is $P(x) = 1 - \cos x$. There are infinitely many wells. (b) The double-well potential $P(x) = x^4/4 - x^2/2$.

Color Plates

Plate 1.

The Ikeda attractor (5.6) has regions of varying density. Here one orbit is shown, with denser accumulations of points shown in purple and lighter accumulations in blue. The orbit plotted here spends most of its time in the inner purple region with intermittent excursions to the outside blue region. This phenomenon develops from a crisis, discussed in Chapter 10.

Plate 2.

Extreme and pervasive sensitivity to initial conditions as seen in the plane map (4.3). The second iterate of the map has six finite attractors, whose basins are shown in red, yellow, green, aqua, blue and violet; the black points diverge to infinity. The basins are riddled, in the sense that each disk in the colored region, no matter how small, contains points whose orbits move to different attractors.

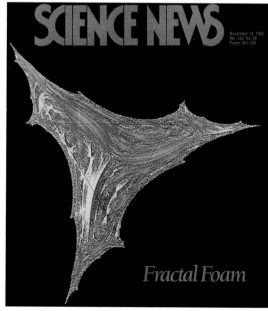

Plates 3—10.
Basins of the four
attractors for
the force damped
pendulum show
fractal structure
at successive
magnifications.
The equations are
given in (2.10)
with parameter
settings $c = 0.1$
and $\rho = 1.75$.
(**3**) shows the
entire phase space.
(**4—10**) show
blowups centered at
a particular point.

Plate 11.
The Mandelbrot set,
shown in black, is
the subset of the
complex plane
consisting of num-
bers c such that
the iteration $z^2 + c$,
starting with $z = 0$,
does not diverge
to infinity. See
Chapter 4 for
more details. Points
outside of the
Mandelbrot set are
colored according
to their divergence
rate. Dark blue is the
fastest divergence,
red denotes slowest
divergence.

Plate 12.
A magnification
of the middle right
edge of the
Mandelbrot set.

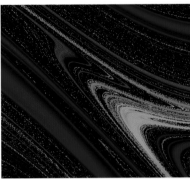

3

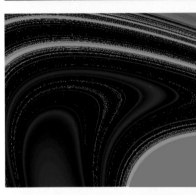

5

7

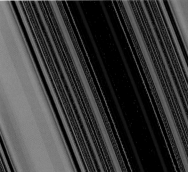

9

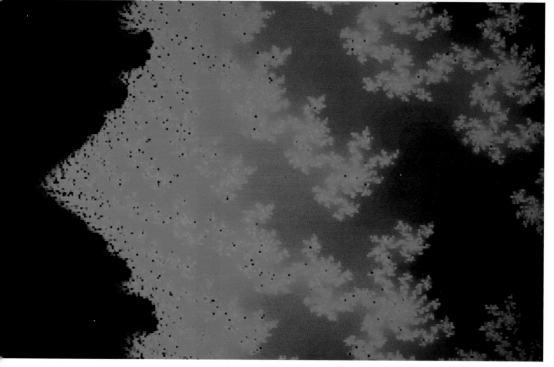

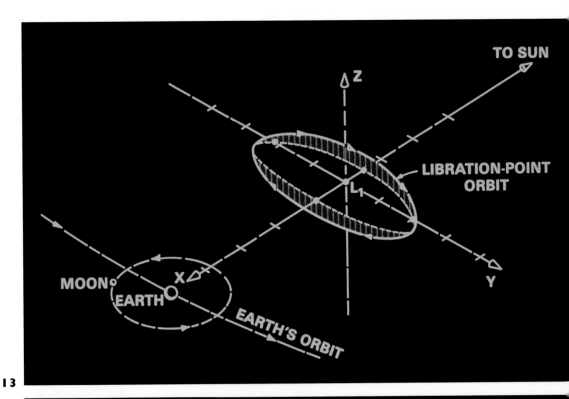

13

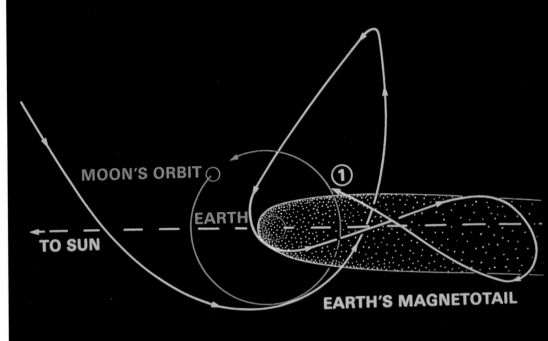

14

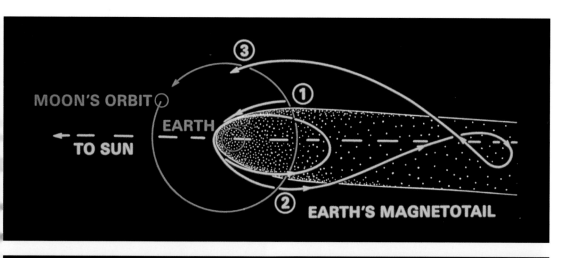

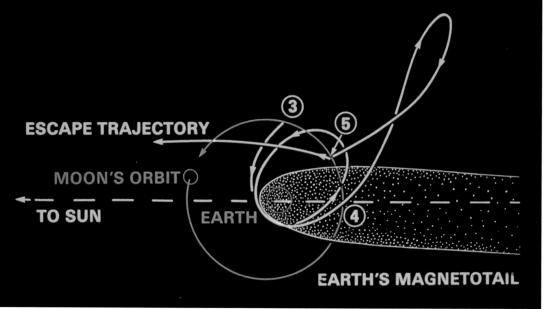

Plates 13—16.

NASA's use of sensitive dependence on initial conditions in orbital mechanics allowed it to visit the Giacobini-Zinner comet in 1985 with minimal expenditure of fuel. See Lab Visit 2 for a fuller description. (**13**) The ISEE-3 satellite, launched in 1978 to study the solar wind, was originally parked in a halo orbit around the Lagrange point L_1. (**14—16**) show the trajectory of ISEE-3, starting in 1982, carefully designed to intersect the tail of the comet. In all, 37 relatively small thruster burns, and five close passes by the moon were needed. Poincaré would have been pleased.

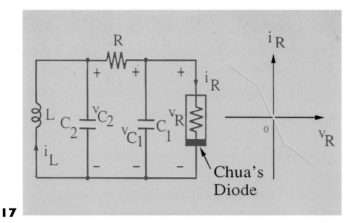

17

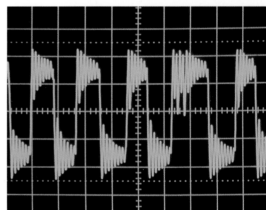

19

Plates 17—20.
The Chua attractor. (**17**) The circuit consists of an inductor L, capacitors C_1 and C_2, a linear resistor R and a nonlinear resistor (Chua's diode, in the diagram) with a piecewise linear voltage-current function. (**18**) The Chua attractor in computer simulation (see Chapter 9 for the equations). (**19**) An experimentally realized version of the attractor, made by plotting voltage drop at C_1 versus current at L. (**20**) A time series of the voltage at C_1. These pictures are reprinted from (Shil'nikov, 1994).

Plates 21, 22.
Photographs from an eccentric cylinder apparatus built at Northwestern University to study coherent structures in mixing. (**21**) A horseshoe in the time-T map. (**22**) The advection pattern resulting from further mixing. These pictures were provided by L. Bresler, T. Shinbrot and J. M. Ottino.

Plate 23.
The bifurcation diagram shows a period-doubling cascade from the output of a CO_2 laser at the Laboratoire de Spectroscopie Hertzienne in Lille, France. See Lab Visit 12 for a fuller description. This photograph of the oscilloscope screen was provided by P. Glorieux.

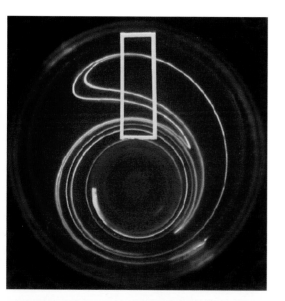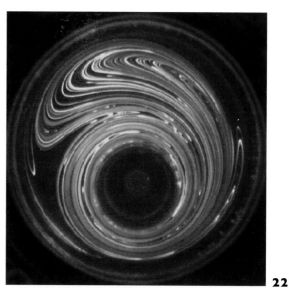

22

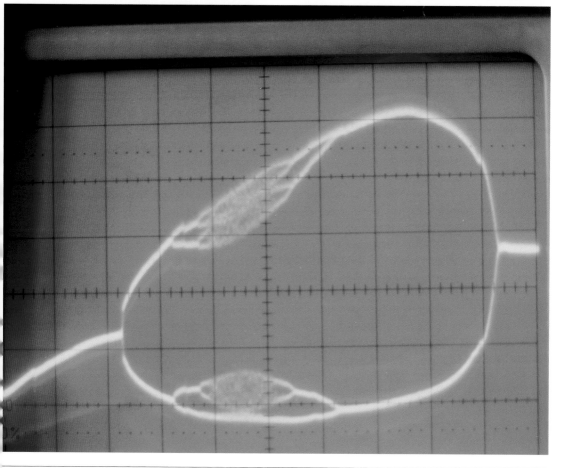

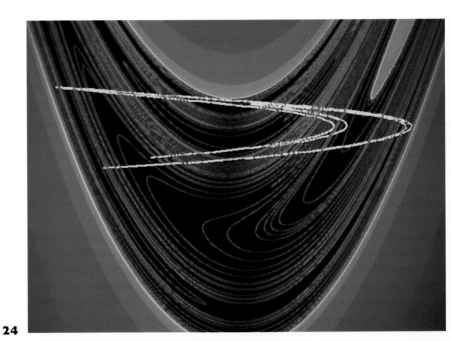

24

25

Plates 24, 25.

The stable manifold of a saddle fixed point of the Hènon map (2.8) is drawn in red. The Hènon attractor is plotted in yellow inside its basin (black), while the basin of infinity is blue. (**24**) shows the first part of the stable manifold; in (**25**) the manifold has been extended further.

where E_1 is a constant of integration. This leads to a simple technique for drawing phase plane solutions of motion in a potential field (7.38). The key is to follow the difference between total and potential energy as x varies; this difference is the kinetic energy $\dot{x}^2/2$. In Figure 7.16(a) we illustrate a typical single-well potential energy function. According to (7.39), for a fixed energy level E_1, the difference $E_1 - P(x)$ is proportional to the square of the velocity $\dot{x}$. As x increases from x_1 (where $P = E_1$) to x_2, the derivative $\dot{x}$ increases from 0 to some maximum value (at the minimum of P) and then decreases to 0 at x_2. The phase plane solution at energy level E_1 is depicted in Figure 7.16(b). Solutions to single-well potential problems in the absence of damping are periodic orbits.

Figure 7.15(b) shows the double-well potential $P(x) = x^4/4 - x^2/2$, which by substitution in (7.38) leads to the double-well Duffing equation

$$\ddot{x} - x + x^3 = 0. \tag{7.40}$$

Most of the solutions are periodic orbits. If the initial conditions $(x, \dot{x})$ are set so that the total energy $E_1 = P(x) + \dot{x}^2/2$ is less than zero, then the orbit is trapped in one of the two potential wells. If $E_1 > 0$, orbits will move periodically through both wells, reaching a maximum of $P(x) = E_1$ on the far sides of the wells.

As might be expected from our experiences with the pendulum equation throughout this book, the Duffing equation becomes even more interesting if

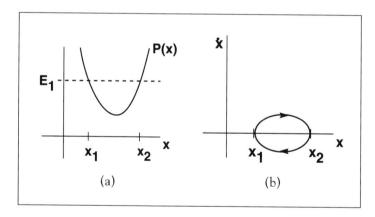

(a)   (b)

Figure 7.16 Drawing phase plane curves from the potential.
(a) Graph of the potential energy function $P(x)$. Each trajectory of the system is trapped in a potential energy well. The total energy $\dot{x}^2/2 + P(x)$ is constant for trajectories. As a trajectory with fixed total energy E_1 tries to climb out near x_1 or x_2, the kinetic energy $\dot{x}^2/2 = E_1 - P(x)$ goes to zero, as the energy E converts completely into potential energy. (b) A periodic orbit results: The system oscillates between positions x_1 and x_2.

damping and periodic forcing are added. With damping, (7.38) becomes

$$\ddot{x} + c\dot{x} + \frac{\partial P}{\partial x} = 0. \qquad (7.41)$$

Energy is no longer conserved, for the time derivative of total energy $E = \dot{x}^2/2 + P(x)$ is

$$\dot{E} = \dot{x}\ddot{x} + \frac{\partial P}{\partial x}\dot{x}$$

$$= \dot{x}\left(-c\dot{x} - \frac{\partial P}{\partial x}\right) + \frac{\partial P}{\partial x}\dot{x}$$

$$= -c\dot{x}^2. \qquad (7.42)$$

Total energy decreases along orbits. Typical orbits will move progressively slower and stop at the bottom of one of the two energy wells. See the Computer Experiment 7.3 and Section 9.5 for the results of periodically forcing the Duffing double well oscillator.

✎ **EXERCISE T7.11**

Not all orbits of (7.41) are attracted to the bottom of one of the wells. Describe as many as you can that end up elsewhere.

⇨ **COMPUTER EXPERIMENT 7.3**

Write a computer program to plot numerical solutions of the forced damped double-well Duffing oscillator $\ddot{x} + 0.1\dot{x} - x + x^3 = 2\sin t$ in the $(x, \dot{x})$-plane. In particular, locate and plot the attracting periodic orbit of period 2π and the two attracting periodic orbits of period 6π that lie in the region $-5 \le x, \dot{x} \le 5$.

7.6 LYAPUNOV FUNCTIONS

The basic idea of using energy-like functions to investigate the dynamics of solutions can be applied to equations more general than (7.38). The theory of Lyapunov functions, a generalization of potential energy functions, gives us a global approach toward determining asymptotic behavior of solutions. The main stability result of this chapter, Theorem 7.18, tells about local stability. In the neighborhood of an equilibrium, solution trajectories are attracted to the equilbrium if the eigenvalues of the linear part of the equation have negative real part. Lyapunov functions can tell us that initial values from a large region converge to an equilibrium. In addition, they can sometimes be used to determine stability of equilibria where the eigenvalues of $\mathbf{Df}$ have real part 0, as in Example 7.20.

Let $\mathbf{v}(t) = (x_1(t), \ldots, x_n(t))$ be a solution of the n-dimensional system (7.29) of differential equations. Suppose we pick a real-valued function of the state, $E(\mathbf{v})$, which we would like to consider to be the energy of the system when it is in the state $\mathbf{v}$. To measure the time rate of change of E along a solution trajectory, we need to take the derivative of E with respect to t. Using the chain rule and the differential equation, we find:

$$\dot{E}(x_1, \ldots, x_n) = \frac{\partial E}{\partial x_1} \frac{dx_1}{dt} + \cdots + \frac{\partial E}{\partial x_n} \frac{dx_n}{dt}$$

$$= \frac{\partial E}{\partial x_1} f_1(x_1, \ldots, x_n) + \cdots + \frac{\partial E}{\partial x_n} f_n(x_1, \ldots, x_n). \quad (7.43)$$

That is, the derivative of E with respect to time can be expressed in terms of the differential equation itself—the solutions do not explicitly appear in this formula.

The derivative $\dot{E}(\mathbf{v})$ measures the rate of change of E along a solution trajectory of (7.29) as it passes through the point $\mathbf{v}$. In the example of the pendulum, we found $\dot{E} = 0$ along trajectories. Using the total energy function to determine the stability of equilibria for a conservative system (one in which energy is conserved) is an example of the technique of Lyapunov functions.

Definition 7.22 Let $\bar{\mathbf{v}}$ be an equilibrium of (7.29). A function $E : \mathbb{R}^n \to \mathbb{R}$ is called a **Lyapunov function** for $\bar{\mathbf{v}}$ if for some neighborhood W of $\bar{\mathbf{v}}$, the following conditions are satisfied:

1. $E(\bar{\mathbf{v}}) = 0$, and $E(\mathbf{v}) > 0$ for all $\mathbf{v} \neq \bar{\mathbf{v}}$ in W, and
2. $\dot{E}(\mathbf{v}) \leq 0$ for all $\mathbf{v}$ in W.

If the stronger inequality

$$2'. \ \dot{E}(\mathbf{v}) < 0 \text{ for all } \mathbf{v} \neq \bar{\mathbf{v}} \text{ in } W$$

holds, then E is called a **strict** Lyapunov function.

Condition (1) says that $\bar{\mathbf{v}}$ is at the bottom of the well formed by the graph of the Lyapunov function E, as shown in Figures 7.17(a) and (b). Condition (2) says that solutions can't move up, but can only move down the side of the well

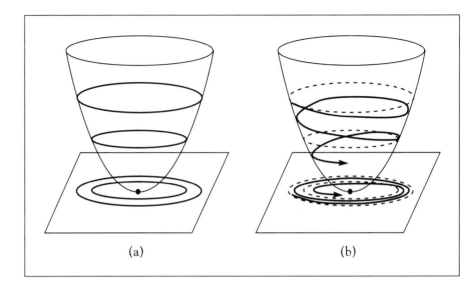

Figure 7.17   Behavior of solution trajectories with a Lyapunov function.
The bowl in the figure is the graph of E. In both parts, we plot $E(\mathbf{v}(t))$ versus the solution trajectory $\mathbf{v}(t)$, which lies in the horizontal plane. (a) An equilibrium is at the critical point of the graph of the Lyapunov function E. The equilibrium is (Lyapunov) stable, since any nearby solution cannot go uphill, and can move away only a bounded distance dictated by its original energy level. (b) For a strict Lyapunov function, energy of solutions must continually decrease toward zero, cutting through energy level sets. The equilibrium is asymptotically stable.

or stay level. Figure 7.17(b) shows a strict Lyapunov function, where energy must keep decreasing toward the equilibrium.

In order to verify the conditions for a Lyapunov function, it is helpful to use (7.43). For example, the one-dimensional equation $\dot{x} = ax$, where $a < 0$, has the Lyapunov function $E(x) = x^2$. It is clear that condition (1) is satisfied for the equilibrium $x = 0$. Moreover, (7.43) shows

$$\dot{E}(x) = \frac{\partial E}{\partial x}f(x) = (2x)(ax) = 2ax^2 < 0,$$

which verifies (2'). Therefore $E(x) = x^2$ is a strict Lyapunov function.

✎ EXERCISE T7.12

Show that for each even integer n, total energy $E(x, y) = (1/2)y^2 + 1 - \cos x$ is a Lyapunov function for the equilibria $(n\pi, 0)$ of (7.34).

The stability analysis illustrated above for the pendulum relies on Theorem 7.23, due to the Russian mathematician Alexander Mikhailovich Lyapunov.

Theorem 7.23 *Let $\bar{v}$ be an equilibrium of $\dot{v} = f(v)$. If there exists a Lyapunov function for $\bar{v}$, then $\bar{v}$ is stable. If there exists a strict Lyapunov function for $\bar{v}$, then $\bar{v}$ is asymptotically stable.*

We refer the reader to (Hirsch and Smale, 1974) for a proof of the theorem. Note that the energy function of Exercise T7.12 is not a strict Lyapunov function. It cannot be, since the equilibria are not attracting.

We can now return to the one-dimensional Example 7.19: $\dot{x} = -x^3$. Linear analysis is inconclusive in determining the stability of the equilibrium $x = 0$. The next exercise implies that $x = 0$ is asymptotically stable.

✎ **EXERCISE T7.13**

Show that $E(x) = x^2$ is a strict Lyapunov function for the equilibrium $x = 0$ in $\dot{x} = -x^3$.

✎ **EXERCISE T7.14**

Let

$$\dot{x} = -x^3 + xy$$

$$\dot{y} = -y^3 - x^2. \tag{7.44}$$

Prove that $(0,0)$ is an asymptotically stable equilibium of (7.44).

Figure 7.17(b) depicts bounded level curves of a strict Lyapunov function for an equilibrium $\bar{v}$. Since E is strictly decreasing (as a function of t) along solutions, the solutions (shown with arrows) cut through the level curves of E and converge to $\bar{v}$. The sets $W_c = \{v \in W : E(v) \leq c\}$ can help us understand the extent of the set of initial conditions whose trajectories converge to an asymptotically stable equilibrium.

Definition 7.24 Let $\bar{v}$ be an asymptotically stable equilibrium of $\dot{v} = f(v)$. Then the **basin of attraction** of $\bar{v}$, denoted $B(\bar{v})$, is the set of initial conditions v_0 such that $\lim_{t \to \infty} F(t, v_0) = \bar{v}$.

Note that any set W on which V is a strict Lyapunov function for $\bar{\mathbf{v}}$ (as in Definition 7.22) will be a subset of the basin $B(\bar{\mathbf{v}})$.

✎ EXERCISE T7.15

Show that the basin of attraction of the equilibrium $(0, 0)$ for the system (7.44) is $\mathbb{R}^2$.

A somewhat weaker notion of containment than a basin is that of a "trapping region"—a set in $\mathbb{R}^n$ where solutions, once they enter, cannot leave as time increases.

Definition 7.25 A set $U \subset \mathbb{R}^n$ is called a **forward invariant set** for (7.29) if for each $\mathbf{v}_0 \in U$, the forward orbit $\{\mathbf{F}(t, \mathbf{v}_0) : t \geq 0\}$ is contained in U. A forward invariant set that is bounded is called a **trapping region**. We also require that a trapping region be an n-dimensional set.

✎ EXERCISE T7.16

Let E be a Lyapunov function on W, let c be a positive real number, and let

$$W_c = \{\mathbf{v} \in W : E(\mathbf{v}) \leq c\}.$$

(a) Show that, for each c, W_c is a forward invariant set. (b) Show that if in addition $E(\mathbf{v}) \to \infty$ as $|\mathbf{v}| \to \infty$, then W_c is a trapping region.

Returning to the pendulum example, we reconsider the system under the effects of a damping force, such as air resistance. Once the pendulum is set into motion, it will lose energy and eventually come to rest. We assume that the damping force is proportional to, but opposite in direction from, the velocity of the pendulum, so that the motion of the pendulum is governed by the equation

$$\ddot{x} + b\dot{x} + \sin x = 0, \tag{7.45}$$

for a constant $b > 0$.

From our observations, we would like to be able to conclude that what were stable equilibria for the undamped pendulum are now asymptotically stable and attract nearby initial conditions. The fact these equilibria are asymptotically stable follows from evaluating the Jacobian (check this); however, we are not able to conclude from the local analysis that all trajectories converge to an equilibrium. Understanding the technique of Lyapunov functions for conservative systems, such as the undamped pendulum, enabled us to draw the phase plane for this system and view it globally. In that case we were able to understand the motions for all initial conditions. Unfortunately, total energy is not a strict Lyapunov function for any equilibria of the damped system. Instead, we compute $\dot{E}(\mathbf{v}) = -by^2$, so that the inequality $\dot{E}(\mathbf{v}) < 0$ fails to be strict for points arbitrarily near the equilibrium.

Notice that trajectories (other than equilibrium points) do not stay on the x-axis, the set on which $\dot{E} = 0$. Therefore we might expect trajectories to behave as if E were a strict Lyapunov function. The following corollary, often called "LaSalle's Corollary" to the Lyapunov Theorem 7.23, not only provides another means of deducing asymptotic stability for equilibria of the damped system, but also gives information as to the extent of the basin of attraction for each asymptotically stable equilibrium. We postpone the proof of the theorem until Chapter 8 when we study limit sets of trajectories.

Corollary 7.26 (Barbashin-LaSalle) Let E be a Lyapunov function for $\bar{\mathbf{v}}$ on the neighborhood W, as in Definition 7.22. Let $Q = \{\mathbf{v} \in W : \dot{E}(\mathbf{v}) = 0\}$. Assume that W is forward invariant. If the only forward-invariant set contained completely in Q is $\bar{\mathbf{v}}$, then $\bar{\mathbf{v}}$ is asymptotically stable. Furthermore, W is contained in the basin of $\bar{\mathbf{v}}$; that is, for each $\mathbf{v}_0 \in W$, $\lim_{t\to\infty}(\mathbf{F}(t, \mathbf{v}_0)) = \bar{\mathbf{v}}$.

✎ **EXERCISE T7.17**

Let $\ddot{x} + b\dot{x} + \sin x = 0$, for a constant $b > 0$. (a) Convert the differential equation to a first-order system. (b) Use Corollary 7.26 to show that the equilibria $(n\pi, 0)$, for even integers n are asymptotically stable. (c) Sketch the phase plane for the associated first-order system.

7.7 LOTKA-VOLTERRA MODELS

A family of models called the Lotka-Volterra equations are often used to simulate interactions between two or more populations. Interactions are of two types. Competition refers to the possibility that an increase in one population is bad for the other populations: an example would be competition for food or habitat. On the other hand, sometimes an increase in one population is good for the other. Owls are happy when the mouse population increases. We will consider two cases of Lotka-Volterra equations, called competing species models and predator-prey models.

We begin with two competing species. Because of the finiteness of resources, the reproduction rate per individual is adversely affected by high levels of its own species and the other species with which it is in competition. Denoting the two populations by x and y, the reproduction rate per individual is

$$\frac{\dot{x}}{x} = a(1 - x) - by, \tag{7.46}$$

where the carrying capacity of population x is chosen to be 1 (say, by adjusting our units). A similar equation holds for the second population y, so that we have the **competing species** system of ordinary differential equations

$$\dot{x} = ax(1 - x) - bxy$$
$$\dot{y} = cy(1 - y) - dxy \tag{7.47}$$

where a, b, c, and d are positive constants. The first equation says the population of species x grows according to a logistic law in the absence of species y (i.e., when $y = 0$). In addition, the rate of growth of x is negatively proportional to xy, representing competition between members of x and members of y. The larger the population y, the smaller the growth rate of x. The second equation similarly describes the rate of growth for population y.

The **method of nullclines** is a technique for determining the global behavior of solutions of competing species models. This method provides an effective means of finding trapping regions for some differential equations. In a competition model, if a species population x is above a certain level, the fact of limited resources will cause x to decrease. The nullcline, a line or curve where $\dot{x} = 0$, marks the boundary between increase and decrease in **x**. The same characteristic is true of the second species y, and it has its own curve where $\dot{y} = 0$. The next two examples show that the relative orientation of the x and y nullclines determines which of the species survives.

EXAMPLE 7.27

(Species extinction) Set the parameters of (7.47) to be $a = 1, b = 2, c = 1$, and $d = 3$. To construct a phase plane for (7.47), we will determine four regions in the first quadrant: sets for which (I) $\dot{x} > 0$ and $\dot{y} > 0$; (II) $\dot{x} > 0$ and $\dot{y} < 0$; (III) $\dot{x} < 0$ and $\dot{y} > 0$; and (IV) $\dot{x} < 0$ and $\dot{y} < 0$. Other assumptions on a, b, c, and d lead to different outcomes, but can be analyzed similarly.

In Figure 7.18(a) we show the line along which $\dot{x} = 0$, dividing the plane into two regions: points where $\dot{x} > 0$ and points where $\dot{x} < 0$. Analogously, Figure 7.18(b) shows regions where $\dot{y} > 0$ and $\dot{y} < 0$, respectively. Combining the information from these two figures, we indicate regions (I)–(IV) (as described above) in Figure 7.19(a). Along the **nullclines** (lines on which either $\dot{x} = 0$ or

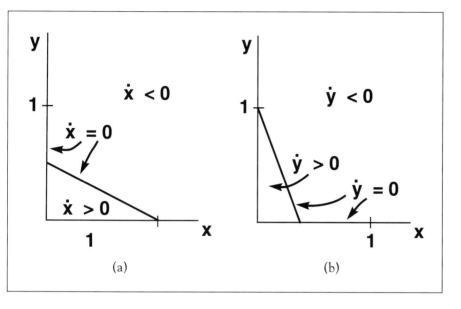

Figure 7.18 Method of nullclines for competing species.
The straight line in (a) shows where $\dot{x} = 0$, and in (b) it shows where $\dot{y} = 0$ for $a = 1, b = 2, c = 1$, and $d = 3$.

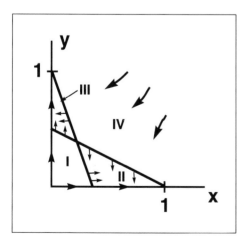

Figure 7.19 Competing species: Nullclines.
The vectors show the direction that trajectories move. The nullclines are the lines along which either $\dot{x} = 0$ or $\dot{y} = 0$. In this figure, the x-axis, the y-axis, and the two crossed lines are nullclines. Triangular regions II and III are trapping regions.

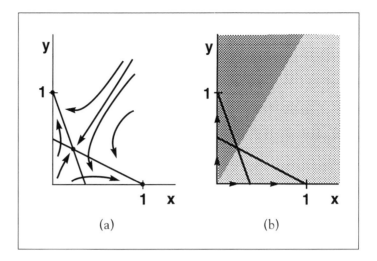

Figure 7.20 Competing species, extinction.
(a) The phase plane shows attracting equilibria at $(1, 0)$ and $(0, 1)$, and a third, unstable equilibrium at which the species coexist. (b) The basin of $(0, 1)$ is shaded, while the basin of $(1, 0)$ is the unshaded region. One or the other species will die out.

$\dot{y} = 0$), arrows indicate the direction of the flow: left/right where $\dot{y} = 0$ or up/down where $\dot{x} = 0$. Notice that points where the two different types of nullclines cross are equilibria. There are four of these points. The equilibrium $(0, 0)$ is a repellor; $(1/5, 2/5)$ is a saddle; and $(1, 0)$ and $(0, 1)$ are attractors.

The entire phase plane is sketched in Figure 7.20(a). Almost all orbits starting in regions (I) and (IV) move into regions (II) and (III). (The only exceptions are the orbits that come in tangent to the stable eigenspace of the saddle $(1/5, 2/5)$. These one-dimensional curves form the "stable manifold" of the saddle and are discussed more fully in Chapter 10.) Once orbits enter regions (II) and (III), they never leave. Within these trapping regions, orbits follow the direction indicated by the derivative toward one or the other asymptotically stable equilibrium. The basins of attraction of the two possibilities are shown in Figure 7.20(b). The stable manifold of the saddle forms the boundary between the basin shaded gray and the unshaded basin. We conclude that for almost every choice of initial populations, one or the other species eventually dies out.

EXAMPLE 7.28

(Coexistence) Set the parameters to be $a = 3$, $b = 2$, $c = 4$, and $d = 3$. The nullclines are shown in Figure 7.21(a). In this case there is a steady state at $(x, y) = (2/3, 1/2)$ which is attracting. The basin of this steady state includes

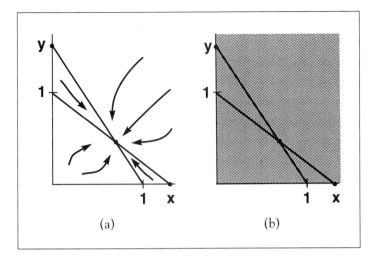

(a) (b)

Figure 7.21 Competing species, coexistence.
(a) The phase plane shows an attracting equilibrium in which both species survive. The x-nullcline $y = \frac{3}{2} - \frac{3}{2}x$ has smaller x-intercept than the y-nullcline $y = 1 - \frac{3}{4}x$. According to Exercise T7.18, the equilibrium $(\frac{2}{3}, \frac{1}{2})$ is asymptotically stable. (b) All initial conditions with $x > 0$ and $y > 0$ are in the basin of this equilibrium.

the entire first quadrant, as shown in Figure 7.21(b). Every set of nonzero starting populations moves toward this equilibrium of coexisting populations.

✎ **EXERCISE T7.18**

Consider the general competing species equation (7.47) with positive parameters a, b, c, d. (a) Show that there is an equilibrium with both populations positive if and only if either (i) both a/b and c/d are greater than one, or (ii) both a/b and c/d are less than one. (b) Show that a positive equilibrium in (a) is asymptotically stable if and only if the x-intercept of the x-nullcline is less than the x-intercept of the y-nullcline.

EXAMPLE 7.29

(Predator-Prey) We examine a different interaction between species in this example in which one population is prey to the other. A simple model of this interaction is given by the following equations:

$$\dot{x} = ax - bxy$$
$$\dot{y} = -cy + dxy \tag{7.48}$$

where a, b, c, and d are positive constants.

✎ **EXERCISE T7.19**

Explain the contribution of each term, positive or negative, to the predator-prey model (7.48).

System (7.48) has two equilibria, $(0, 0)$ and $(\frac{c}{d}, \frac{a}{b})$. There are also nullclines; namely, $\dot{x} = 0$ when $x = 0$ or when $y = \frac{a}{b}$, and $\dot{y} = 0$ when $y = 0$ or when $x = \frac{c}{d}$. Figure 7.22(a) shows these nullclines together with an indication of the flow directions in the phase plane.

Unlike previous examples, there are no trapping regions. Solutions appear to cycle about the nontrivial equilibrium $(\frac{c}{d}, \frac{a}{b})$. Do they spiral in, spiral out, or are they periodic? First, check the eigenvalues of the Jacobian at $(\frac{c}{d}, \frac{a}{b})$.

✎ **EXERCISE T7.20**

Find the Jacobian **Df** for (7.48). Verify that the eigenvalues of $\mathbf{Df}(\frac{c}{d}, \frac{a}{b})$ are pure imaginary.

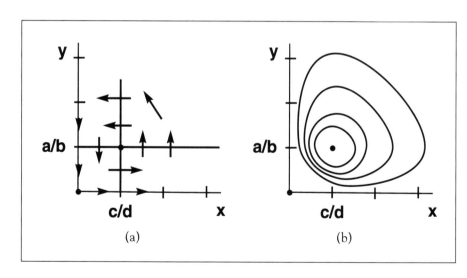

Figure 7.22 Predator-prey vector field and phase plane.
(a) The vector field shows equilibria at $(0, 0)$ and $(\frac{c}{d}, \frac{a}{b})$. Nullclines are the x-axis, the y-axis, and the lines $x = \frac{c}{d}$ and $y = \frac{a}{b}$. There are no trapping regions. (b) The curves shown are level sets of the Lyapunov function E. Since $\dot{E} = 0$, solutions starting on a level set must stay on that set. The solutions travel periodically around the level sets in the counterclockwise direction.

Since the system is nonlinear and the eigenvalues are pure imaginary, we can conclude nothing about the stability of $(\frac{c}{d}, \frac{a}{b})$. Fortunately, we have a Lyapunov function.

✎ EXERCISE T7.21

Let

$$E(x, y) = dx - c \ln x + by - a \ln y + K,$$

where $a, b, c,$ and d are the parameters in (7.48) and K is a constant. Verify that $\dot{E} = 0$ along solutions and that E is a Lyapunov function for the equilibrium $(\frac{c}{d}, \frac{a}{b})$.

We can conclude that $(\frac{c}{d}, \frac{a}{b})$ is stable. Solutions of (7.48) lie on level curves of E. Since $(\frac{c}{d}, \frac{a}{b})$ is a relative minimum, these level curves are closed curves encircling the equilibrium. See Figure 7.22(b). Therefore, solutions of this predator-prey system are periodic for initial conditions near $(\frac{c}{d}, \frac{a}{b})$. In fact, every initial condition with x and y both positive lies on a periodic orbit.

☞ CHALLENGE 7

A Limit Cycle in the Van der Pol System

THE LIMITING BEHAVIOR we have seen in this chapter has consisted largely of equilibrium states. However, it is common for solutions of nonlinear equations to approach periodic behavior, converging to periodic orbits, or cycles. The Van der Pol equation

$$\ddot{x} + (x^2 - 1)\dot{x} + x = 0 \tag{7.49}$$

is a model of a nonlinear electrical circuit that has a limit cycle.

Defining $y = \dot{x}$, the second-order equation is transformed to

$$\dot{x} = y$$

$$\dot{y} = -x + (1 - x^2)y. \tag{7.50}$$

The origin $(0, 0)$ is the only equilibrium of (7.50), and it is unstable. In this Challenge, you will show that all other trajectories of the system approach a single attracting periodic orbit that encircles the origin. This type of limiting behavior for orbits is a phenomenon of nonlinear equations. Although linear systems may have periodic orbits, they do not attract initial values from outside the periodic orbit.

We begin by introducing a change of coordinates. Let $z = y + F(x)$, where

$$F(x) = \frac{x^3}{3} - x.$$

Step 1   Show that the correspondence $(x, y) \rightarrow (x, z)$ is one-to-one (and therefore a change of coordinates), and that the system (7.50) is transformed to the following system:

$$\dot{x} = z - \left(\frac{x^3}{3} - x\right)$$

$$\dot{z} = -x. \tag{7.51}$$

Step 2   Draw a phase plane for the system (7.51), indicating the approximate direction of the flow. (Hint: Begin with Figure 7.23.) Argue that starting from a point on the positive z-axis, denoted z^+, a solution $\mathbf{v}(t) = (x(t), z(t))$ will go into region I until it intersects the branch F^+ (the graph of $F(x)$ for positive x),

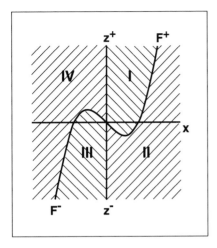

Figure 7.23   Phase plane structure for the Van der Pol system.
The plane separates into four regions according to the sign of $\dot{x}$ and $\dot{z}$. The direction of the flow (which you provide, see Step 2) implies a circulation of solution trajectories around the origin.

crosses F^+, moves into region II, and then intersects the negative z-axis, denoted z^-. Then the solution moves through region III, through region IV, and back to the half-line z^+. Conclude that the solution repeatedly cycles through regions I, II, III, and IV, in that order.

Define the map T on z^+, as follows: For a point $(0, a^+)$ on z^+, draw the trajectory $\mathbf{X}$ through $(0, a^+)$, and let $T(a^+)$ be the first intersection (after a^+) of $\mathbf{X}$ with z^+; that is, the intersection that occurs after $\mathbf{X}$ has cycled once through regions I–IV. The definition of T is illustrated in Figure 7.24.

Step 3   Show that the map T is one-to-one, and that a number a^+ is fixed under T if and only if the solution trajectory through $(0, a^+)$ is a periodic solution of (7.51). Explain why an orbit of (7.51) is periodic if and only if the corresponding orbit of (7.50) is periodic.

The map T is another example of the Poincaré map, first introduced in Chapter 2. In Steps 4–10 we show that T has an attracting fixed point.

Step 4   Let $a^+ > 0$. Show that because of symmetry, if the solution through $(0, a^+)$ intersects z^- at $(0, -a^+)$, then $T(a^+) = a^+$.

Notice that (7.49) bears some resemblance to the linear oscillator $\ddot{x} + x = 0$. The coefficient of the middle term $\dot{x}$ goes from negative to positive with increasing

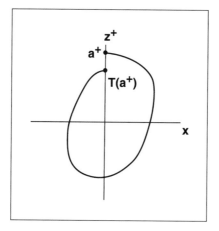

Figure 7.24 The first-return map T.
The point $T(a^+)$ is defined to be the intersection of the solution through a^+ with the positive z-axis.

x, resulting in a variable "damping" effect. We define an energy function, as follows:

$$E(x, z) = (1/2)(x^2 + z^2).$$

Starting with the point $(0, a^+)$, we want to measure the change in energy over a segment $\mathbf{X_0}$ of $\mathbf{X}$ from $(0, a^+)$ to the first crossing of $\mathbf{X}$ with z^-, denoted by a^-. We define this change as follows: Let

$$h(a^+) = E(0, a^-) - E(0, a^+) = (1/2)[(a^-)^2 - (a^+)^2].$$

According to Step 4, the problem reduces to showing that there exists a real number c such that $h(c) = 0$.

Step 5 Show that

$$\dot{E} = \left(\frac{x^3}{3} - x\right)\dot{z},$$

and so

$$h(a^+) = \int_{\mathbf{X_0}} \left(\frac{x^3}{3} - x\right)\dot{z}\, dt.$$

Let r be the x-coordinate of the point where the solution $\mathbf{X}$ intersects F^+, and let $(q, 0)$ be the point on the positive x-axis where F^+ intersects the axis. For this example, $q = \sqrt{3}$. (See Figure 7.25.)

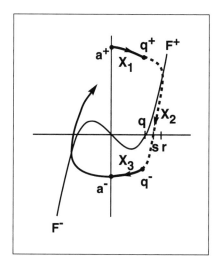

Figure 7.25 A schematic view of a Van der Pol trajectory.
The initial value is $(0, a^+)$; the intersection with the negative z-axis is a^-. The trajectory between is the union $\mathbf{X}_0$ of three pieces denoted $\mathbf{X}_1$, $\mathbf{X}_2$, and $\mathbf{X}_3$.

Step 6 Show that if $r < q$, then

$$\int_{\mathbf{X}_0} \left(\frac{x^3}{3} - x \right) \dot{z}\, dt > 0.$$

Refer to Figure 7.25. Let $\mathbf{q}^+$ (respectively, $\mathbf{q}^-$) be the first point (respectively, second point) where $\mathbf{X}$ intersects the line $x = q$. For $r > q$, we write the path $\mathbf{X}_0$ as $\mathbf{X}_1 \cup \mathbf{X}_2 \cup \mathbf{X}_3$, where $\mathbf{X}_1$ goes from $(0, a^+)$ to $\mathbf{q}^+$, $\mathbf{X}_2$ goes from $\mathbf{q}^+$ to $\mathbf{q}^-$, and $\mathbf{X}_3$ goes from $\mathbf{q}^-$ to $(0, a^-)$.

Step 7 Show:

$$\int_{\mathbf{X}_1} dE > 0, \int_{\mathbf{X}_2} dE < 0, \text{ and } \int_{\mathbf{X}_3} dE > 0.$$

Step 8 Show that $\int_{\mathbf{X}_1} \dot{E}\, dt$ and $\int_{\mathbf{X}_3} \dot{E}\, dt$ are decreasing functions of a^+.
[Hint: For example,

$$\int_{\mathbf{X}_1} \dot{E}\, dt = \int_0^{\sqrt{3}} \left(\frac{x^3}{3} - x \right) \frac{dz}{dx}\, dx = \int_0^{\sqrt{3}} \left(\frac{x^3}{3} - x \right) \frac{-x}{z - (\frac{x^3}{3} - x)}\, dx.]$$

Step 9 Show that $\int_{X_2} \dot{E}\, dt$ monotonically decreases to $-\infty$ as $a^+ \to \infty$. [Hint: The integral

$$\int_{X_2} \dot{E}\, dt = -\int_{X_2} \left(\frac{x^3}{3} - x \right) x\, dt.$$

By uniqueness, the solutions must move farther to the right, and the terms $(\frac{x^3}{3} - x)$ and x in the integral must increase, as a^+ is increased.]

Step 10 Conclude that there exists a unique positive number c such that $h(c) = 0$, and that c is a fixed point of T.

Step 11 Show that c is an attracting fixed point of T. [Hint: Show that uniqueness of solutions implies that if $a^+ < c$, then $-a^- < c$. Show that the change in E from a^+ to $T(a^+)$ is greater than 0 for $a^+ < c$, and less than 0 for $a^+ > c$.]

Postscript. An attracting periodic orbit, or limit cycle, is the first distinctly non-linear phenomenon we have seen for differential equations. Although Figure 7.9 shows periodic orbits circling a Lyapunov stable equilibrium, the cycles are "neutrally stable": They do not attract nearby initial conditions. The periodic orbits of the undamped pendulum in Figure 7.14 and the predator-prey system in Figure 7.22 are also not attracting. In Challenge 7 we have explored the Van der Pol equation in detail because it has a rigorously-provable attracting periodic orbit.

Attracting periodic orbits are interesting subjects of study because they exemplify the notion that certain periodic oscillations in nature can be "locked in", so as to be fairly impervious to drift. If a small amount of noise were to push the current state away from a particular periodic orbit in the neutrally stable examples of the previous paragraph, there is no mechanism to return it to the original cycle. An attracting cycle, on the other hand, has such a mechanism built in. This is consistent with overwhelming evidence that periodic cycles, like your heartbeat, can oscillate more or less unperturbed, even in systems exposed to external disturbances. Chapter 8 is dedicated to the study of periodic cycles. Lab Visit 8 examines experimental evidence for the usefulness of the limit cycle paradigm in the context of neurons.

EXERCISES

7.1. The differential equation governing a linear oscillator, such as a spring moving according to Hooke's law, is $\ddot{x} + kx = 0$.

    (a) Convert to a system of first-order equations.

    (b) Sketch the phase plane.

7.2. Consider the second-order equation $\ddot{x} + 3\dot{x} - 4x = 0$.

    (a) Convert to a system of first-order equations.

    (b) Sketch the phase plane.

7.3. Consider the system

$$\dot{x} = 2x - y$$
$$\dot{y} = x^2 + 4y.$$

Find equilibria and classify them as asymptotically stable, stable, or unstable.

7.4. Consider the equation $\dot{x} = a(x - b)(x - c)$, where $b < c$, with initial condition $x(0) = x_0$.

    (a) Sketch the slope field and phase portrait. For each initial condition x_0, use this qualitative information to determine the long-term behavior of the solutions. Treat the cases $a < 0$, $a = 0$, and $a > 0$ separately.

    (b) Solve the equation and initial values problem by separation of variables and partial fractions. Verify that the behavior of this solution is consistent with the qualitative information in (a).

    (c) Show that if $a > 0$ and $x_0 > c$, or if $a < 0$ and $x_0 < b$, the solutions blow up in finite time. Find the largest interval $[0, T)$ on which the solution exists. Does this example contradict the existence theorem?

    (d) What happens if $b = c$?

    (e) A chemical reaction $Y + Z \rightarrow X$ proceeds at a rate proportional to the product of the concentration of the reactants Y and Z. Each time the reaction occurs, one molecule of species X appears and one each of Y and Z disappears. Therefore the reaction is modeled by the differential equation $\dot{x} = a(b - x)(c - x)$, $x(0) = 0$, where $x(t)$ is the relative molar concentration of species X at time t, and b and c are the relative molar concentrations of Y and Z, respectively, at time $t = 0$. The constant a is called the **rate constant** of the reaction. Which parts of the above analysis (which combinations of a and x_0) correspond to this physical problem, and which do not? Describe the eventual outcome of the reaction in these cases.

7.5. Sketch phase portraits for the following linear systems.

(a)
$$\dot{x} = 3x - y$$
$$\dot{y} = 2x + 4y$$

(b)
$$\dot{x} = -2x + 3y$$
$$\dot{y} = 7x - 6y$$

(c)
$$\dot{x} = 3x - y$$
$$\dot{y} = x + 3y$$
$$\dot{z} = -2z$$

7.6. Show by examples that there are equilibria that are attracting but not stable, and equilibria that are stable but not attracting.

7.7. Verify that if the independent variable t does not appear in the right side of (7.29), then solutions do not cross in the phase portrait.

7.8. Consider the second-order equation $\ddot{x} + x + x^3 = 0$.

(a) Find a Lyapunov function for the equilibrium $x = \dot{x} = 0$. Verify your answer.

(b) Convert the equation to a first-order system and sketch the phase plane.

7.9. Let $\ddot{x} + \dot{x} + x + x^3 = 0$. Show that the equilibrium $(0, 0)$ is globally asymptotically stable.

7.10. Each graph in Figure 7.26 represents the graph of a potential function $P(x)$. In each case, sketch the phase plane for the system $\ddot{x} + f(x) = 0$, where $f(x) = P'(x)$. Identify the equilibria and discuss their stability.

7.11. Let $\ddot{x} + b\dot{x} + f(x) = 0$, where b is a positive constant and $f(x) = P'(x)$ where $P(x)$ is shown in Figure 7.26. Write the equation as a first-order system and sketch the phase plane.

7.12. Figure 7.14(b) shows the phase portrait of the undamped pendulum. Graph the six qualitatively different solutions $x(t)$ as functions of t.

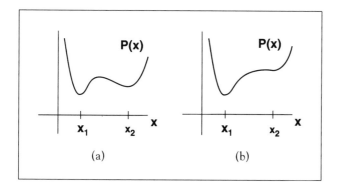

Figure 7.26 Potential functions for Exercise 7.10.
In part (b), $P'(x_2) = 0$.

7.13. Use the method of nullclines to determine the global behavior of solutions of the following competing species models. Describe sets of initial conditions that evolve to distinct final states.

(a)

$$\dot{x} = 3x(1 - x) - xy$$

$$\dot{y} = 5y(1 - y) - 2xy$$

(b)

$$\dot{x} = 2x(1 - x) - xy$$

$$\dot{y} = y(1 - y) - 3xy$$

7.14. Consider the system

$$\dot{x} = -z + \left(\frac{x^3}{3} - x\right)$$

$$\dot{z} = x.$$

(a) By comparing with (7.51), explain why this system can be called the backwards-time Van der Pol system.

(b) Show that $V(x, z) = x^2 + z^2$ is a Lyapunov function.

(c) Find the largest possible open disk contained in the basin of attraction of the origin.

(d) What can you conclude from this about the solutions of (7.51)?

7.15. Find two different solutions of the differential equation

$$\dot{x} = \begin{cases} 0 & \text{if } x < 0 \\ \sqrt[3]{x} & \text{if } x \geq 0 \end{cases}$$

with initial condition $x(0) = 0$. Explain why this does not contradict Theorem 7.14.

7.16. Sketch the phase plane in $(x, \dot{x})$ for the system $\ddot{x} = x - ax^3$ for various values of a. Find all stable and unstable equilibrium points.

Fly vs. Fly

DURING THE PAST ten years a competition has developed between two mosquito species in North America. The tiger mosquito *Aedes albopictus*, which occupies much of Asia, was unknown in the United States before being observed in Houston, Texas in 1985, apparently as the result of imports of used automobile tires.

The invasion of the tiger mosquito has caused alarm in the United States public health community because it is a vector for dengue fever. Dengue is a viral disease known as "breakbone fever" in Africa, a reflection of the level of pain that accompanies it. Even in its mildest form it causes intense headaches and extreme muscle spasms. The severe form, known as dengue hemorrhagic fever, causes internal bleeding, coma and shock, killing over 10% of those afflicted. Due in part to reduced mosquito eradication programs, Latin America experienced a sixty-fold increase in cases of dengue hemorrhagic fever in 1989–1994, compared with the previous five-year period.

A key scientific question is to address the potential success of the new mosquito's invasion into North American mosquito habitats. Here we describe an experiment, which was designed to gauge the possibility that the new mosquito species will be able to coexist with or displace currently-existing mosquito species that are less susceptible to carrying viruses, and less aggressive biters. An example of a native species, *Aedes triseriatus*, was used in the study.

Two common mosquito habitats were simulated, treeholes and automobile tires. The critical time for development is the larval stage; the species that can compete best for resources at this stage will have a competitive survival advantage. In Experiment A, one-day-old larvae of both species were put in plastic cups containing leaf litter and 100 ml of stagnant water taken from holes in maple trees, which represents a typical habitat for mosquito larvae. Experiment B was the same except that "treehole fluid" was replaced with stagnant water from discarded tires.

The parameters in the competing species model (7.47) were estimated as the result of these experiments. The nullclines derived from the estimated parameters

Livdahl, T. P., Willey, M. S. 1991. Propects for an invasion: Competition between *Aedes albopictus* and native *Aedes triseriatus*. Science 253:189–91.

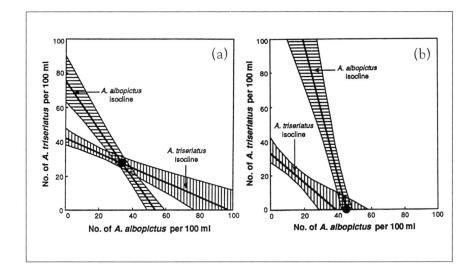

Figure 7.27   Nullclines of two mosquito species A. triseriatus and A. albopictus.

Hatched regions depict 95% confidence levels for the nullclines for Experiment A (treehole habitat) and Experiment B (tire habitat). Circles denote the equilibria of the competing species equations. The relative slopes of the nullclines for Experiment A indicate that the equilibrium is stable. Experiment B points toward extinction of A. *triseriatus*.

are graphed in Figure 7.27, along with confidence limits based on the estimates. The true nullclines lie within the shaded regions with 95% certainty. The units in the axes of the figure differ from our assumptions for (7.47). If we rescale units so that the x-nullcline has an x-intercept of 1 and the y-nullcline has an y-intercept of 1, then we can interpret the nullclines exactly as in (7.47).

In any case, Exercise T7.18 shows how to interpret the experimentally-measured nullclines shown in the Figure. In Experiment A (treehole habitat), there is an asymptotically stable state of coexisting nonzero populations. In Experiment B (tire habitat), there is no equilibrium of coexisting populations, and all initial populations are attracted to the horizontal axis, which represents extinction of A. *triseriatus*.

The significance of the predictions that follow from the experiment has increased since 1991. Now ten years since it appeared in Houston, the tiger mosquito has been identified in half of the United States, and as far north as Chicago. It is now the dominant mosquito in many urban areas, including New Orleans and several Florida cities.

Corroboration of the laboratory result has come from ecological studies. Coexistence of the two species within treehole communities is suggested by a

1988 study done in a Louisiana forest, finding A. *triseriatus* and A. *albopictus* as the two dominant species, at 81% and 17% of larvae collected, respectively.

Since the completion of this project, the authors of the study have begun to search for the mechanism behind this pronounced difference in the habitats. For example, it is not yet known whether there is a biochemical difference between treehole water and tire water that can account for the differing results. In any case, the experiment suggests that the tiger mosquito has a distinct competitive advantage in the man-made environment of automobile tires.

CHAPTER EIGHT

Periodic Orbits and Limit Sets

THE BEGINNING of this book was devoted to an understanding of the asymptotic behavior of orbits of maps. Here we begin a similar study for solution orbits of differential equations. Chapter 7 contains examples of solutions that converge to equilibria and solutions that converge to periodic orbits called limit cycles. We will find that the dimension and shape of the phase space put serious constraints on the possible forms that asymptotic behavior can take.

For autonomous differential equations on the real line, we will see that solutions that are bounded must converge to an equilibrium. For autonomous differential equations in the plane, a new limiting behavior is possible—solutions that are bounded may instead converge to closed curves, called periodic orbits or cycles. However, nothing more radical can happen for solutions of autonomous differential equations in the plane. Solutions cannot be chaotic. The topological rule about plane geometry that enforces this fact is the Jordan Curve Theorem. Winding solutions can wind only in a single direction, as shown in Figure 8.7. This is the subject of the Poincaré-Bendixson Theorem, which is one of the main topics of this chapter. In three-dimensional space, there is no such rule. The one extra dimension makes a difference. We will investigate several chaotic three-dimensional equations in Chapter 9.

Typical limiting behavior for planar systems can be seen in the equation

$$\dot{r} = r(1 - r)$$

$$\dot{\theta} = 8, \tag{8.1}$$

where r and θ are polar coordinates. There is an equilibrium at the origin $r = 0$ and a periodic orbit circling at $r = 1$.

✎ **EXERCISE T8.1**

Show that all nonequilibrium solutions of (8.1) have form $(r, \theta) = (ce^t/(ce^t - 1), 8t + d)$ for some constants c, d.

The solution of (8.1) with initial condition $(r, \theta) = (2, 0)$ is shown in Figure 8.1. It spirals in forever toward the periodic solution $r = 1$, also shown.

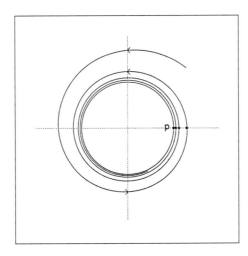

Figure 8.1   The definition of an ω-limit set.
The point **p** is in the ω-limit set of the spiraling trajectory because there are points $F(t_1, v_0)$, $F(t_2, v_0)$, $F(t_3, v_0) \ldots$ of the trajectory, indicated by dots, that converge to **p**. The same argument can be made for any point in the entire limiting circle of the spiral solution, so the circle is the ω-limit set.

For this reason, we will say that the periodic solution $r = 1$ is the limit set of the spiraling solution. We will usually use the terminology ω-limit set, to emphasize that we are interested in final, or asymptotic, behavior. In Section 8.1 we define limit set and give examples in one and two dimensions.

8.1 LIMIT SETS FOR PLANAR DIFFERENTIAL EQUATIONS

We begin with an autonomous differential equation

$$\dot{\mathbf{v}} = \mathbf{f}(\mathbf{v}), \tag{8.2}$$

where $\mathbf{f}$ is a map of $\mathbb{R}^n$, which is differentiable or at least Lipschitz, to guarantee uniqueness. Throughout this chapter all differential equations will be autonomous; the function $\mathbf{f}$ of (8.2) does not depend explicitly on t. For $v \in \mathbb{R}$ that means f is a scalar (one-dimensional) map. When $\mathbf{v}$ is a vector in $\mathbb{R}^2$, $\mathbf{f}$ is a vector-valued function of two scalar variables, and the phase portrait of (8.2) is a phase plane.

Recall the definition of flow in Chapter 7. For a point $\mathbf{v}_0$ in $\mathbb{R}^n$, we write $\mathbf{F}(t, \mathbf{v}_0)$ for the unique solution of (8.2) satisfying $\mathbf{F}(0, \mathbf{v}_0) = \mathbf{v}_0$.

Definition 8.1 A point z in $\mathbb{R}^n$ is in the ω-**limit set** $\omega(\mathbf{v}_0)$ of the solution curve $\mathbf{F}(t, \mathbf{v}_0)$ if there is a sequence of points increasingly far out along the orbit (that is, $t \to +\infty$) which converges to z. Specifically, z is in $\omega(\mathbf{v}_0)$ if there exists an unbounded increasing sequence $\{t_n\}$ of real numbers with $\lim_{n\to\infty} \mathbf{F}(t_n, \mathbf{v}_0) = z$. A point z is in the α-**limit set** $\alpha(\mathbf{v}_0)$ if there exists an unbounded decreasing sequence $\{t_n\}$ of real numbers $(t_n \to -\infty)$ with $\lim_{n\to\infty} \mathbf{F}(t_n, \mathbf{v}_0) = z$.

The concept of ω-limit set (or forward limit set) first appeared for maps in Chapter 6. For maps, the unbounded, increasing sequence $\{t_n\}$ of real numbers in Definition 8.1 is replaced by an (unbounded) increasing sequence $\{m_n\}$ of positive integers (iterates of the map).

If $\mathbf{v}_0$ is an equilibrium, then $\omega(\mathbf{v}_0) = \alpha(\mathbf{v}_0) = \{\mathbf{v}_0\}$. Furthermore, for any $\mathbf{v}_0$ the α-limit set of the equation $\dot{\mathbf{v}} = \mathbf{f}(\mathbf{v})$ is the ω-limit set of $\dot{\mathbf{v}} = -\mathbf{f}(\mathbf{v})$. See Exercise 8.13.

EXAMPLE 8.2

Recall the one-dimensional example

$$\dot{x} = x(a - x)$$

from Chapter 7. Assume that $a > 0$. There are two equilibria in $\mathbb{R}$, $x = 0$ and $x = a$. All trajectories with $x_0 > 0$ converge to the stable equilibrium $x = a$. According to the above definition, $\omega(x_0) = \{a\}$ for $x_0 > 0$. Since 0 is an equilibrium, $\omega(0) = \{0\}$. If $x_0 < 0$, the trajectory diverges to $-\infty$, and therefore $\omega(x_0)$ is the empty set.

With this example in mind, it is easy to describe the asymptotic behavior (the ω-limit set) for any bounded orbit of an autonomous differential equation on the real line.

Theorem 8.3 *All solutions of the scalar differential equation $\dot{x} = f(x)$ are either monotonic increasing or monotonic decreasing as a function of t. For $x_0 \in \mathbb{R}$, if the orbit $\mathbf{F}(t, x_0)$, $t \geq 0$, is bounded, then $\omega(x_0)$ consists solely of an equilibrium.*

Figure 8.2 shows a *nonmonotonic* scalar function $x(t)$, together with an explanation (in the caption) as to why it cannot be the solution to a scalar equation $\dot{x} = f(x)$.

Proof of Theorem 8.3: Assume that for a fixed initial condition x_0, there is some t_* at which the solution $\mathbf{F}(t, x_0)$ has a local maximum or local minimum x_*, as a function of t. Then $\dot{x}(t_*) = f(x_*) = 0$. Hence x_* is an equilibrium, and there is a solution identically equal to x_*. By uniqueness of solutions, $\mathbf{F}(t, x_0)$ must be the equilibrium solution; $\mathbf{F}(t, x_0) \equiv x_*$, for all t. This means that no solution can "turn around". All solutions are monotonic.

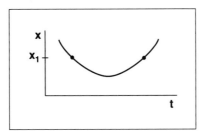

Figure 8.2 Why nonmonotonic functions cannot be solutions of a scalar autonomous equation.
The function $x(t)$ depicted here cannot be the solution of a scalar equation $\dot{x} = f(x)$. Notice that the solution passes through the value x_1 twice, once with positive slope and once with negative slope. The value of $f(x_1)$ is either positive or negative (or zero), however, and it must equal the slope at both points. The point x_1 can be chosen so that $f(x_1)$ is not 0.

On an interval $[x_0, x_1]$ not containing an equilibrium (so that f is not zero), we can rewrite the equation as

$$\frac{dt}{dx} = \frac{1}{f(x)}.$$

Integrating both sides from x_0 to x_1 yields

$$t(x_1) - t(x_0) = \int_{x_0}^{x_1} \frac{dx}{f(x)} < \infty.$$

We conclude that a trajectory traverses any closed, bounded interval without equilibria in finite time.

If $f(x_0)$ is greater than 0, the orbit $F(t, x_0)$ must increase as t increases, as long as it is below the first equilibrium. It cannot stop at any point below the equilibrium, because it traverses an interval centered at such a point. It cannot reach the equilibrium, by uniqueness of solutions. Therefore it converges monotonically to the equilibrium as $t \to \infty$. The same argument can be turned around to handle the case of $f(x_0) < 0$. □

Theorem 8.3 describes the ω-limit sets for scalar autonomous differential equations and shows they are rather simple. Either the orbit diverges to infinity, or if bounded, converges to a single equilibrium. For differential equations of dimension two and higher, there is another asymptotic behavior, called a periodic orbit. Like equilibria, periodic orbits can attract solutions, and therefore be ω-limit sets.

Definition 8.4 If there exists a $T > 0$ such that $F(t + T, v_0) = F(t, v_0)$ for all t, and if v_0 is not an equilibrium, then the solution $F(t, v_0)$ is called a **periodic orbit**, or **cycle**. The smallest such number T is called the **period** of the orbit.

A periodic orbit traces out a simple closed curve in $\mathbb{R}^n$. The term **closed curve** refers to a path that begins and ends at the same point. Examples include circles, triangles, and figure-eights. A **simple closed curve** is one that does not cross itself. Hence, a figure-eight is not a simple closed curve. By uniqueness of solutions, the closed curve of a periodic orbit must be simple. We have already seen examples of cycles. In Example 7.8, the solution $F(t, v_0)$ is a circle whose radius is $|v_0|$. Every initial condition lies on a circular periodic orbit of period 2π, except for the origin, which is an equilibrium.

The pendulum equation of Section 7.5 also has cycles when there is no friction. These solutions correspond physically to full back-and-forth periodic swings of the pendulum. Figure 7.14 shows orbits encircling equilibria that are simple closed curves traversed by the system.

EXAMPLE 8.5

Here is a general version of (8.1). Let r and θ be polar coordinates in the plane. For $a, b > 0$, consider

$$\dot{r} = r(a - r)$$

$$\dot{\theta} = b. \tag{8.3}$$

There is an equilibrium at the origin. Therefore $\omega(0) = \{0\}$. See Figure 8.3. For every other initial condition $(r_0, \theta_0) \neq 0$, the trajectory moves counterclockwise around the origin and the ω-limit set is the circle $r = a$. Figure 8.3(a) shows the phase plane for this system.

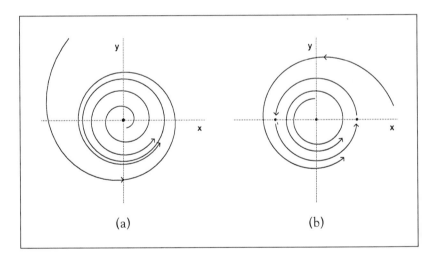

(a)   (b)

Figure 8.3   Examples of ω-limit sets for planar flows.
(a) The phase plane for Example 8.5 shows the circle $r = a$ as an attracting periodic orbit of system (8.3). The origin is an unstable equilibrium. The ω-limit set of every trajectory except the equilibrium is the periodic orbit. (b) The phase plane for Example 8.6 looks very similar to the phase plane in (a), except that in this example there are no periodic orbits. There are three equilibria: the origin and the points $(a, 0)$ and (a, π). Every other point on the circle $r = a$ is on a solution called a connecting arc, whose α- and ω-limit sets are the equilibria. The ω-limit set of each nonequilibrium solution not on the circle is the circle $r = a$.

✎ **EXERCISE T8.2**

Solve the differential equation (8.3), and verify the statements made about ω-limit sets in Example 8.5.

⇨ **COMPUTER EXPERIMENT 8.1**

Write a computer program to plot numerical solutions of the Van der Pol equation $\ddot{x} + (x^2 - 1)\dot{x} + x = 0$. Plot the attracting periodic orbit that was proved to exist in Challenge 7. Find the ω-limit sets for all planar orbits.

EXAMPLE 8.6

Example 8.5 can be modified so that it does not have any periodic orbits. We can destroy the circular orbit at $r = a$ by making an adjustment in the equation for $\dot{\theta}$. For example, replace b in the equation by a function $g(r, \theta)$ that is always positive except for the two points $(r, \theta) = (a, 0)$ and (a, π), where it is zero. If we use $g(r, \theta) = \sin^2 \theta + (r - a)^2$, the equation is

$$\dot{r} = r(a - r)$$
$$\dot{\theta} = \sin^2 \theta + (r - a)^2.$$

This equation is much harder to solve explicitly than (8.3), but we can derive lots of qualitative information without an explicit formula for the solution. For any initial condition not on the circle $r = a$, the limit set has not changed from the previous example. Except for the equilibrium trajectory at the origin, these trajectories limit on the entire circle. The novel aspect of this example is that trajectories on the circle itself do not have the entire circle as the limit set. Figure 8.3(b) shows the phase plane for this system. On the circle $r = a$ there are two equilibria, $(a, 0)$ and (a, π). The trajectory through any other point on the circle has as its α- and ω-limit sets these equilibria. Nonequilibrium solutions such as these, whose limit sets contain only equilibria, are called **connecting arcs**.

EXAMPLE 8.7

The pendulum solutions in Figure 7.14 show two types of ω-limit sets. One type is an equilibrium, which is its own ω-limit set and which is also the limit set of the connecting arcs. The connecting arcs are solutions that approach the equilibria $((2k + 1)\pi, 0)$, $(k = \pm1, \pm2, \ldots)$, as $t \to \pm\infty$. The other type of ω-limit set is a cycle, which is again its own limit set. The pendulum also has unbounded orbits (which have no limit sets).

We again raise the question: What types of sets in $\mathbb{R}^2$ can be ω-limit sets of a bounded solution of an autonomous differential equation? We have seen examples of points (equilibria) and closed curves (periodic orbits), and sets containing equilibria and connecting arcs, as in Example 8.6. Can flows on $\mathbb{R}^2$ have chaotic limit sets? The answer to this question is "no". We shall see in the remainder of this chapter that the types of ω-limit sets are severely restricted by intrinsic topological properties of the plane.

↻   **C O M P U T E R   E X P E R I M E N T   8 . 2**

Write a computer program to compute numerical solutions of the double-well Duffing oscillator $\ddot{x} + 0.1\dot{x} - x + x^3 = 0$. There are two stable equilibria corresponding to the bottoms of the wells. Plot the set of initial values in the $(x, \dot{x})$-plane whose ω-limit set is the equilibrium point $(1, 0)$. How would you describe the set of points whose ω-limit set is the point $(0, 0)$?

The celebrated Poincaré-Bendixson Theorem, which we state in this section and prove in Section 8.3, gives a classification of planar ω-limit sets. An ω-limit set of a two-dimensional autonomous equation must be one of the following: (1) a set of equilibria; (2) a periodic orbit; or (3) a set containing only equilibria and connecting arcs. Figure 8.3(b) shows a circle with two connecting arcs on it. The entire circle is a limit set of the trajectory spiraling in towards it. This example has ω-limit sets of types (1) and (3).

We usually deal with equations whose equilibria are **isolated**, meaning that there are only finitely many in any bounded set. In particular, this implies that each equilibrium has a surrounding neighborhood with no other equilibria. For such equations, a type (1) limit set contains only one equilibrium point. None of these three types of ω-limit sets can be a chaotic set. In each case, as an orbit converges to one of these sets, its behavior is completely predictable. In particular, such a set may contain a dense orbit only if the limit set is either a periodic orbit or an equilibrium point. Thus we can conclude from the following theorem that for autonomous differential equations, chaos can occur only in dimension three and higher. As we have seen, there can be chaos in one-dimensional maps, and for invertible maps it can occur in dimensions two and higher.

Theorem 8.8   (Poincaré-Bendixson Theorem.) *Let f be a smooth vector field of the plane, for which the equilibria of* $\dot{\mathbf{v}} = \mathbf{f}(\mathbf{v})$ *are isolated. If the forward orbit* $\mathbf{F}(t, \mathbf{v}_0), t \geq 0$, *is bounded, then either*

1. $\omega(\mathbf{v}_0)$ *is an equilibrium, or*
2. $\omega(\mathbf{v}_0)$ *is a periodic orbit, or*
3. *For each* $\mathbf{u}$ *in* $\omega(\mathbf{v}_0)$, *the limit sets* $\alpha(\mathbf{u})$ *and* $\omega(\mathbf{u})$ *are equilibria.*

The hypothesis that the equilibria are isolated is included to simplify the statement of the theorem. If this assumption is omitted, then we have to include the possibility that either $\omega(\mathbf{v}_0)$ or $\omega(\mathbf{u})$ is a connected set of equilibria.

The three possibilities allowed by Theorem 8.8 are illustrated in Figure 8.4. In (a), the ω-limit sets of the solution shown is an equilbrium; in (b) both solutions have the circular periodic solution as ω-limit set. In (c), the ω-limit set of the outermost orbit is the equilibrium $\mathbf{P}$ together with the two connecting arcs that begin and end at $\mathbf{P}$. As required by Theorem 8.8, any point $\mathbf{u}$ in $\omega(\mathbf{v}_0)$ has the property that $\omega(\mathbf{u}) = \mathbf{P}$.

In the next section we discuss properties of limit sets, not just for planar flows, but for autonomous equations in any dimension. These properties are then used in the proof of the Poincaré-Bendixson Theorem, which is given in Section 8.3.

8.2 PROPERTIES OF ω-LIMIT SETS

Now that we have seen some common examples of ω-limit sets, we turn to a more theoretical investigation and establish five important properties of all ω-limit sets. The statements and proofs of these properties involve the concept of "limit point", a concept we have previously seen, although not explicitly, in all our discussions of limit sets. Specifically, a point $\mathbf{v}$ in $\mathbb{R}^n$ is called a **limit point** of a set A if every neighborhood $N_\epsilon(\mathbf{v})$ contains points of A distinct from $\mathbf{v}$. This means that there is a sequence of points in A that converge to $\mathbf{v}$. A limit point $\mathbf{v}$ of A may be in A or it may not be. A set A that contains all its limit points is called a **closed set**. Thus, for example, if a and b are real numbers, then the intervals $[a, b]$, $[0, \infty)$, and $(-\infty, \infty)$ are closed, while the intervals $[a, b)$, (a, b), and $(0, \infty)$ are not. In the plane, the unit disk (with its boundary circle) is a closed set, while the interior of the disk, all points (x, y) such that $x^2 + y^2 < 1$, is not

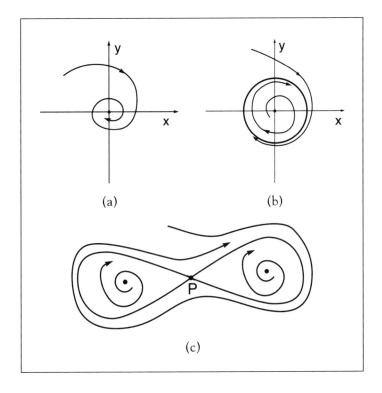

Figure 8.4 Planar limit sets.
The three pictures illustrate the three cases of the Poincaré-Bendixson Theorem. (a) The limit set is one point, the origin. (b) The limit set of each spiraling trajectory is a circle, which is a periodic orbit. (c) The limit set of the outermost trajectory is a figure eight. This limit set must have an equilibrium point P at the vertex of the "eight". It consists of two connecting arcs plus the equilibrium. Trajectories on the connecting arcs tend to P as $t \to \infty$ and as $t \to -\infty$.

closed. (Take away any point on the boundary circle $x^2 + y^2 = 1$ and the set will not be closed.)

An important fact about limit points is that any set that is both infinite and bounded will have limit points. This fact is called the Bolzano-Weierstrass Theorem, and it can be found in any standard advanced calculus text—for example, (Fitzpatrick, 1996).

Now we can proceed with the properties of ω-limit sets. Let $\dot{\mathbf{v}} = \mathbf{f}(\mathbf{v})$, $\mathbf{v} \in \mathbb{R}^n$, be an autonomous differential equation, and let $\omega(\mathbf{v}_0)$ be the ω-limit set of an orbit $\mathbf{F}(t, \mathbf{v}_0)$.

- **Existence Property.** If the orbit $\mathbf{F}(t, \mathbf{v}_0)$ is bounded for all $t \geq 0$, then $\omega(\mathbf{v}_0)$ is non-empty.

Proof: The sequence $\{F(n, v_0) : n = 1, 2, 3, \ldots\}$ either is a bounded infinite set (which must have a limit point) or repeats the same point infinitely many times. In either case, a point is in $\omega(v_0)$. □

- **Property of Closure.** $\omega(v_0)$ is closed; that is, $\omega(v_0)$ contains all of its limit points.

✎ EXERCISE T8.3

Prove the Property of Closure.

In order to prove the next property, we will have to reparametrize orbits. Since the solution of an autonomous equation goes through a point independent of the time that it reaches the point, we could solve an equation up to time t in different ways: either directly find $F(t, v_0)$, or write t as $t_1 + t_2$ and solve by first finding $F(t_1, v_0)$, then using $v_1 = F(t_1, v_0)$ as our new initial point, find $F(t_2, v_1)$. In other words, we can stop and start the process, reparametrizing along the way, and still end up with the same solution. In mathematical terms, $F(t_1 + t_2, v_0) = F(t_2, F(t_1, v_0))$. We refer to this equivalence as the **composition property** (also called the **semi-group property**) of autonomous equations.

- **Property of Invariance.** $\omega(v_0)$ is invariant under the flow; that is, if u is in $\omega(v_0)$, then the entire orbit $F(t, u)$ is in $\omega(v_0)$.

✎ EXERCISE T8.4

Prove the Property of Invariance.

The next property involves the topological concept called connectedness. Loosely speaking, a set is "connected" if it is geometrically of one piece. For a closed set (one that contains all its limit points) that is also bounded, we have the following characterization: a closed, bounded set S is called **connected** if it is not possible to write $S = A \cup B$, where the sets A and B are a positive distance apart. More precisely, there are no sets A and B such that $S = A \cup B$ and there is a distance $d > 0$ with each point of A separated from each point of B by a distance of at least d.

- **Property of Connectedness.** If $\{F(t, v_0)\}$ is a bounded set, then $\omega(v_0)$ is connected.

Proof: Suppose $\omega(\mathbf{v}_0) = A \cup B$, where sets A and B are d apart. There are infinitely many disjoint segments of solution curves moving from within $d/4$ of A to within $d/4$ of B. Since A and B are separated by d, there must be a point on each such segment that is $d/2$ from A. Such points are at least $d/2$ from B. These points form a bounded infinite set that has a limit point lying at least $d/2$ from $\omega(\mathbf{v}_0)$. Therefore, d must be 0. $\qquad\square$

• **Property of Transitivity.** If $z \in \omega(\mathbf{u})$ and $\mathbf{u} \in \omega(\mathbf{v}_0)$, then $z \in \omega(\mathbf{v}_0)$.

✎ **EXERCISE T8.5**

Prove the Property of Transitivity.

The same five properties hold for α-limit sets since, as mentioned previously, every α-limit set of the equation $\dot{\mathbf{v}} = \mathbf{f}(\mathbf{v})$ is an ω-limit set of $\dot{\mathbf{v}} = -\mathbf{f}(\mathbf{v})$. (Exercise 8.13.) We summarize the properties of ω-limit sets in a table for easy reference.

Properties of ω-limit sets
1. **Existence:** The ω-limit set of a bounded orbit is non-empty.
2. **Closure:** An ω-limit set is closed.
3. **Invariance:** If $\mathbf{y}$ is in $\omega(\mathbf{v}_0)$, then the entire orbit $\mathbf{F}(t, \mathbf{y})$ is in $\omega(\mathbf{v}_0)$.
4. **Connectedness:** The ω-limit set of a bounded orbit is connected.
5. **Transitivity:** If z is in $\omega(\mathbf{y})$ and $\mathbf{y}$ is in $\omega(\mathbf{v}_0)$, then z is in $\omega(\mathbf{v}_0)$.

We return now to the subject of Lyapunov functions (introduced in Chapter 7) to determine the behavior of these functions on limit sets. Recall that for a Lyapunov function $E : \mathbb{R}^n \to \mathbb{R}$, the value of E along an orbit decreases with time: $\frac{dE}{dt}(\mathbf{F}(t, \mathbf{v})) \leq 0$. The following lemma says that the value of E is constant on limit sets. For any positive real number c, we let

$$W_c = \{\mathbf{v} \in \mathbb{R}^n : E(\mathbf{v}) \leq c\}.$$

In Exercise T7.16 it was shown that, for each c, W_c is a forward invariant set.

Lemma 8.9 Let $\mathbf{v}_0 \in W_c$, for some $c > 0$. Then there is a number $d, 0 \leq d \leq c$, such that $E(\mathbf{v}) = d$, for every $\mathbf{v} \in \omega(\mathbf{v}_0)$.

Since E is constant on $\omega(\mathbf{v}_0)$ and $\omega(\mathbf{v}_0)$ is an invariant set, we have the following corollary.

Corollary 8.10 $\dot{E}(\mathbf{v}) = 0$ for each $\mathbf{v}$ in $\omega(\mathbf{v}_0)$.

✎ **EXERCISE T8.6**

Prove Lemma 8.9.

Lyapunov functions provide a method of showing that an equilibrium is asymptotically stable and measuring the extent of the basin of attraction for such an equilibrium. We use Lemma 8.9 to prove LaSalle's Corollary (Corollary 7.26).

Let E be a Lyapunov function for $\bar{\mathbf{v}}$, let W be a neighborhood of $\bar{\mathbf{v}}$ as in the hypothesis of Corollary 7.26. Let $Q = \{\mathbf{v} \in W : \dot{E}(\mathbf{v}) = 0\}$. Assume that $\bar{\mathbf{v}}$ is an isolated equilibrium and that W is forward invariant. We prove that if the only forward invariant set contained completely in Q is $\bar{\mathbf{v}}$, then $\bar{\mathbf{v}}$ is asymptotically stable. Furthermore, W is contained in the basin of $\bar{\mathbf{v}}$; for each $\mathbf{v}_0 \in W$, $\lim_{t \to \infty} \mathbf{F}(t, \mathbf{v}_0) = \bar{\mathbf{v}}$.

Proof of Corollary 7.26 (LaSalle's Corollary): Since E is constant on $\omega(\mathbf{v}_0)$, the limit set $\omega(\mathbf{v}_0)$ is contained in Q. Also, $\omega(\mathbf{v}_0)$ is forward invariant. Therefore, $\omega(\mathbf{v}_0) = \bar{\mathbf{v}}$. □

8.3 PROOF OF THE POINCARÉ-BENDIXSON THEOREM

The proof of the Poincaré-Bendixson Theorem is illuminating. We will break down the ideas into a series of lemmas. In the first lemma below, we develop a picture of the flow in a small neighborhood of a nonequilibrium point $\mathbf{u}$. In such a neighborhood, solution curves run roughly parallel to the orbit through $\mathbf{u}$.

For a nonequilibrium point $\mathbf{u}$, let L be a short line segment containing $\mathbf{u}$ but no equilibrium points, which is both perpendicular to the vector $\mathbf{f}(\mathbf{u})$ and short enough so that for each $\mathbf{v}$ in L, the vector $\mathbf{f}(\mathbf{v})$ is not tangent to L. We call such a segment a **transversal** at $\mathbf{u}$.

In Figure 8.5, L is a transversal at $\mathbf{u}$, but L^* is not a transversal at $\mathbf{u}$. Notice, in particular, that all solution curves that intersect the transversal cross from the same side as the orbit through $\mathbf{u}$. If instead some solutions crossed from left to

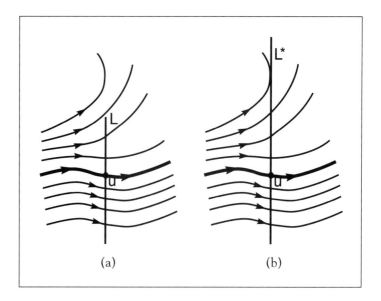

Figure 8.5 Illustration of a transversal line segment at a non-equilibrium point.
(a) A transversal line segment L through a non-equilibrium point $\mathbf{u}$ is shown.
(b) The segment L^* is not a transversal because it is tangent to the vector field at a point on the top solution curve shown.

right and others from right to left, then $\frac{dx}{dt} > 0$ at some points on L, and $\frac{dx}{dt} < 0$ at other points on L. Then there must be at least one point on L where $\frac{dx}{dt} = 0$, implying the existence of a tangency or equilibrium.

It is convenient to look now at neighborhoods around $\mathbf{u}$ that are diamond-shaped, rather than round. We define a δ-**diamond** about $\mathbf{u}$, denoted $D_\delta(\mathbf{u})$, to be the set of points in a square centered at $\mathbf{u}$ whose diagonals are of length 2δ and lie, respectively, on the transversal L and on a line segment perpendicular to L (that is, it is parallel to the vector $f(\mathbf{u})$). A δ-diamond is illustrated in Figure 8.6.

The following lemma states that the solution through each initial point in a sufficiently small δ-diamond about $\mathbf{u}$ must cross a transversal at $\mathbf{u}$ in either forward or backward time before leaving the diamond. To simplify notation, let $v(t)$ denote the orbit $\mathbf{F}(t, \mathbf{v}_0)$. In the proof of the following lemma, we let f_1 and f_2 be the coordinate functions of the vector function $\mathbf{f}$ of (8.2); $\mathbf{f}(\mathbf{v}) = (f_1(\mathbf{v}), f_2(\mathbf{v}))$. Sets of the form $S = \{\mathbf{F}(t, \mathbf{v}_0) : t_1 \leq t \leq t_2\}$ (for some $\mathbf{v}_0 \in \mathbb{R}^2$ and real numbers t_1 and t_2) are called **trajectory segments** (or **orbit segments**).

Lemma 8.11   (The Diamond Construction.) Let L be a transversal at $\mathbf{u}$. For $\delta > 0$ sufficiently small there exists a δ-diamond $D_\delta(\mathbf{u})$ about $\mathbf{u}$ such that if

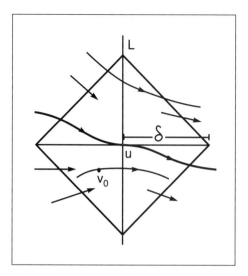

Figure 8.6 Illustration of a δ-diamond.
A δ-diamond is a square neighborhood of a nonequilibrium point **u**. Its diagonals
are of length 2δ, and which lie on the transversal L and on the line segment in
the direction of $\mathbf{f}(\mathbf{u})$. The Diamond Construction guarantees that there is a δ-
diamond such that for each $\mathbf{v}_0$ in the interior of the diamond there is an orbit
segment completely contained in the diamond that passes through $\mathbf{v}_0$ with positive
horizontal speed and intersects L.

$\mathbf{v}_0$ is in the interior of $D_\delta(\mathbf{u})$, then there is a trajectory segment that contains $\mathbf{v}_0$,
is completely contained in $D_\delta(\mathbf{u})$, and intersects L.

Proof: We choose coordinates (x, y) so that $\mathbf{u} = (0, 0)$ and $\mathbf{f}(\mathbf{u}) = (a, 0)$,
for some $a > 0$. Then L is on the y-axis. Choose $\delta > 0$ sufficiently small so that
for z in $D_\delta(\mathbf{u})$ the following conditions hold:

1. The slope of $\mathbf{f}(z)$ (that is, $f_2(z)/f_1(z)$) is strictly between -1 and 1. This
   restriction is possible since the slope is 0 at $\mathbf{u}$.
2. $f_1(z) > a/2$.

Notice, in particular, that the slope is between -1 and 1 on the edges of the
$D_\delta(\mathbf{u})$. A solution can only exit the diamond in forward time through the right-
hand side of the diamond. Similarly, it can only exit the diamond in backward
time through the left-hand side. Now

$$x(t) - x(0) = \int_0^t f_1(x(s)) \, ds > \frac{a}{2}t.$$

Since the maximum value of $x(t) - x(0)$ is 2δ, the solution through an initial
condition $\mathbf{v}_0$ in the diamond must exit the right side of the diamond within time

$t < 2\frac{2\delta}{a}$. Let t_2 be the smallest t value with $t > 0$ for which the orbit through $\mathbf{v}_0$ intersects the right side of $D_\delta(\mathbf{u})$. Similarly, the backward solution (for t decreasing from 0) must exit the left side within time t, $-2\frac{2\delta}{a} < t < 0$. Let t_1 be the largest value of t with $t < 0$ for which $\mathbf{v}(t)$ intersects the left side of $D_\delta(\mathbf{u})$. Then the trajectory segment $S = \{\mathbf{v}(t) : t_1 \le t \le t_2\}$ is contained in $D_\delta(\mathbf{u})$ and intersects L. □

The proof of the Poincaré-Bendixson Theorem depends on the next four lemmas. These results are strictly for planar flows and, unlike the previous lemma, have no natural analogue in higher dimensions. First we show that if one trajectory repeatedly crosses a transversal L, then the crossings must occur in order along L for increasing t. We will make use of the Jordan Curve Theorem from plane topology (see, e.g., (Guillemin and Pollack, 1974)). Recall that a simple closed curve is a path which begins and ends at the same point and does not cross itself. This theorem says that a simple closed curve divides the plane into two parts: a bounded region (the "inside") and an unbounded region (the "outside"). In order for a path to get from a point inside the curve to a point outside the curve, it must cross the curve.

Lemma 8.12 (The "In the Bag" or Monotonicity Lemma.) Given $t_1 < t_2 < t_3$ for which $\mathbf{v}(t_1)$, $\mathbf{v}(t_2)$, and $\mathbf{v}(t_3)$ are three distinct points on L. Then $\mathbf{v}(t_2)$ is between $\mathbf{v}(t_1)$ and $\mathbf{v}(t_3)$.

Proof: Let L be a transversal at $\mathbf{u}$. Since the orbit $\mathbf{v}(t)$ crosses L only finitely many times between t_1 and t_3 (see Exercise 8.18), it suffices to prove the conclusion in the case that $t_1 < t_2 < t_3$ are consecutive times for which $\mathbf{v}(t)$ crosses L. Let C' be the orbit segment of $\mathbf{v}(t)$ between $\mathbf{v}(t_1)$ and $\mathbf{v}(t_2)$, $C' = \{\mathbf{v}(t), t_1 \le t \le t_2\}$, and let L' be the segment of L connecting these points. See Figure 8.7. Let C be the closed curve $C' \cup L'$. Note that C is simple by the uniqueness of solutions.

From the Jordan Curve Theorem, there are two cases. In case 1, the vector field on L' points into the bounded region, and in case 2, it points into the unbounded region. Figure 8.7 shows case 1. We will assume case 1 holds and leave to the reader any adjustments that are needed for case 2. Now, the orbit $\mathbf{v}(t)$ is trapped inside C for $t > t_2$, (beyond $\mathbf{v}(t_2)$). It cannot cross C' by uniqueness of solutions. Also, any solution moving from inside C to outside C at a point on L' would have to be going in an opposite direction to the flow arrows, which is impossible. Therefore, the next and all subsequent crossings of L by $\mathbf{v}(t)$ occur inside C. This fact implies that the crossings $\mathbf{v}(t_1), \mathbf{v}(t_2), \mathbf{v}(t_3)$, etc., occur in order on L. □

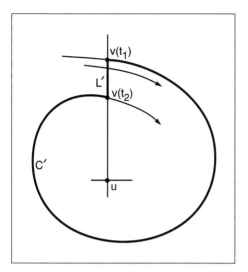

Figure 8.7   The "In the Bag" Lemma.
A solution is trapped inside the Jordan curve and cannot get out. A simple closed
curve is made up of an orbit segment between successive crossings $v(t_1)$ and $v(t_2)$
of a transversal L, together with the piece L' of L between these points. Once the
orbit enters this "bag" (at $v(t_2)$), it cannot leave. Therefore, any later crossings of L
by the orbit must occur below $v(t_2)$.

Now let $\mathbf{u}$ be a nonequilibrium point in $\omega(v_0)$, and let $D_\delta(\mathbf{u})$ be a sufficiently
small δ-diamond about $\mathbf{u}$. We will show that $\omega(v_0)$ intersects $D_\delta(\mathbf{u})$ in a trajectory
segment $S_{\mathbf{u}}$ through $\mathbf{u}$. The Invariance Property of ω-limit sets provides part of this
picture—namely, that each point of $\mathbf{u}(t)$ is in $\omega(v_0)$. The Diamond Construction
Lemma tells us that the solution through $\mathbf{u}$ must extend across the diamond. In
the following lemma, we prove that the the the only points of $\omega(v_0)$ in $D_\delta(\mathbf{u})$ are
in $S_{\mathbf{u}}$.

Lemma 8.13   (The "Locally a Curve" Lemma.) Assume there is a
nonequilibrium point $\mathbf{u}$ in $\omega(v_0)$. Then there is a δ-diamond $D_\delta(\mathbf{u})$ such that
$\omega(v_0) \cap D_\delta(\mathbf{u})$ is a trajectory segment.

Proof: Let L be a transversal at $\mathbf{u}$, let δ be as in the proof of the Diamond
Construction Lemma, and let $S_{\mathbf{u}}$ be the trajectory segment containing $\mathbf{u}$ which
extends across $D_\delta(\mathbf{u})$, as guaranteed by the Diamond Construction. Now for an
arbitrary point z in $\omega(v_0) \cap D_\delta(\mathbf{u})$, there exists a trajectory segment S_z such that
S_z contains z, is contained in $D_\delta(\mathbf{u})$, and intersects L at a point z_L. Let $t_1, t_2, \ldots$
be the times $(t_1 < t_2 < \cdots)$ for which $v(t) \in L \cap D_\delta(\mathbf{u})$. Then z_L and $\mathbf{u}$ are both

limit points of the set $\{v(t_i)\}$, $i = 1, 2, \ldots$. Since the points $v(t_i)$ are monotonic on L and have both z_L and $\mathbf{u}$ as limit points, it must be that $z_L = \mathbf{u}$. By uniqueness of solutions (see Theorem 7.14 of Chapter 7), z is in $S_{\mathbf{u}}$. □

✎ **EXERCISE T8.7**

Show that if $\mathbf{v_0} \in \omega(\mathbf{v_0})$, then $\omega(\mathbf{v_0})$ is either an equilibrium or a periodic orbit. Specifically, let $\mathbf{v} = \mathbf{v}(0)$ be a nonequilibrium point, let L be a transversal at $\mathbf{v}$, and let $\{t_n\}$ be an increasing, unbounded sequence of real numbers. Suppose that $\mathbf{v}(t_n)$ is in L, for each n, and $\mathbf{v}(t_n) \to \mathbf{v}$. Then $\mathbf{v}(t_n) = \mathbf{v}$, for each n, and $\mathbf{v}(t)$ is a periodic orbit.

Corollary 8.14 If $\mathbf{u} \in \omega(\mathbf{v_0})$, $z \in \omega(\mathbf{u})$, and z is not an equilibrium, then $\mathbf{u}$ is on a periodic orbit.

Proof: In this case, $z \in \omega(\mathbf{v_0})$ by the Transitivity Property. Also, the entire orbit through $\mathbf{u}$ is in $\omega(\mathbf{v_0})$ by the Invariance Property. By Lemma 8.13 ("Locally a curve"), there is a $\delta > 0$ such that the only points of $\omega(\mathbf{v_0})$ in $D_\delta(z)$ are on a trajectory segment S_z through z. But since z is in $\omega(\mathbf{u})$, there are points on the orbit of $\mathbf{u}$ that are arbitrarily close to z. In particular, there are points of $\mathbf{u}(t)$ within $D_\delta(z)$. Since $\omega(\mathbf{v_0})$ intersects $D_\delta(z)$ only in S_z, the orbit through $\mathbf{u}$ and the orbit through z must be the same orbit. Thus $\mathbf{u}$ is in $\omega(\mathbf{u})$, and, by Exercise T8.7, $\mathbf{u}$ is a periodic orbit. □

Lemma 8.15 (The "Has One, Is One" Lemma.) If $\omega(\mathbf{v_0})$ contains a periodic orbit, then $\omega(\mathbf{v_0})$ is a periodic orbit.

Proof: Assume there is a periodic orbit G in $\omega(\mathbf{v_0})$. Let z be a point in $\omega(\mathbf{v_0})$. Assume further that z is not on G. Let $d \geq 0$ be the (minimum) distance from z to G. Since $\omega(\mathbf{v_0})$ is connected, there must be another point z_1 of $\omega(\mathbf{v_0})$ which is not on G, but which is within distance $d/2$ of G. Analogously, for each $n > 1$, let z_n be a point of $\omega(\mathbf{v_0})$ which is not on G, but which is within $d/2^n$ of G. The sequence $\{z_n\}$ necessarily contains infinitely many distinct points, which must have an accumulation point, since the sequence is bounded. Call this point $\mathbf{u}$. Then $\mathbf{u}$ is on G, but $\omega(\mathbf{v_0})$ is not locally a curve in any diamond neighborhood of $\mathbf{u}$, contradicting Lemma 8.13. Therefore, z is on G. □

Now we have all the pieces in place to prove the Poincaré-Bendixson Theorem.

Proof of Theorem 8.8 (Poincaré-Bendixson Theorem):   The ω-limit set $\omega(\mathbf{v}_0)$ is not empty by the Existence Property. We begin by assuming that $\omega(\mathbf{v}_0)$ consists entirely of equilibria. Since $\omega(\mathbf{v}_0)$ is connected (the Connectedness Property) and since one of our hypotheses is that equilibria are isolated, $\omega(\mathbf{v}_0)$ consists of exactly one equilibrium, and case 1 of Theorem 8.8 holds. If case 1 does not hold, there exists a non-equilibrium point $\mathbf{u}$ in $\omega(\mathbf{v}_0)$. Case 3 holds if $\omega(\mathbf{u})$ is an equilibrium.

The basic idea of the proof is to show that if neither cases 1 nor 3 holds, then case 2 must hold, and the proof is complete. Suppose, therefore, that cases 1 and 3 do not hold. Then there is a nonequilibrium point $\mathbf{u}$ in $\omega(\mathbf{v}_0)$ for which there is a nonequilibrium point $\mathbf{z}$ in $\omega(\mathbf{u})$. By Corollary 8.14, the orbit of $\mathbf{u}$ is a periodic orbit. By Lemma 8.15 ("Has one, is one"), $\omega(\mathbf{v}_0)$ is a periodic orbit.   □

The Poincaré-Bendixson Theorem does not give a complete characterization of limit sets. In particular, some sets that are not ruled out by the theorem still cannot be limit sets of planar flows. We illustrate this fact in the following example.

EXAMPLE 8.16

Each of the following figures shows a set consisting of equilibria and connecting arcs. Figure 8.8 (a) has one equilibrium and three connecting arcs. The ω-limit set of any point in the set shown is the equilibrium. An orbit not shown must be contained either completely inside one of the loops formed by the connecting arcs or completely outside the set of loops. In any case, the entire figure cannot be an ω-limit set. Figure 8.8 (b) has two equilibria and one connecting arc. The ω-limit set of any point in the set shown is one of the two equilibria.

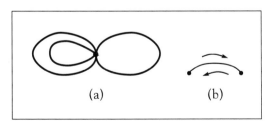

(a)   (b)

Figure 8.8   Sets that cannot be limit sets of planar flows.
(a) No single orbit can limit on the set of connecting arcs and equilibrium shown.
(b) This set consists of two equilibria and a connecting arc. If a trajectory has this figure as its limit set, then the flow must be going in opposite directions on either side of the connecting arc, an impossibility unless all points on the arc are equilibria.

If an orbit not shown limits on this set, it must wind around and around the connecting arc, with arrows indicating the direction of the flow going in opposite directions on either side of the arc. This type of trajectory is only possible if every point on the arc is an equilibrium, a situation ruled out by the hypothesis that equilibria are isolated.

The structure and dynamics of ω-limit sets in higher dimensions include many more possibilities than in $\mathbb{R}^2$. What are the simplest examples of bounded ω-limit sets that do not contain equilibria and are not periodic orbits? We describe a differential equation defined on a bounded subset of $\mathbb{R}^3$, none of whose trajectories limit on a periodic orbit or an equilibrium.

EXAMPLE 8.17

Begin with the unit square in the plane. Identify the left and right edges by gluing them together, and likewise identify the top and bottom edges, as shown in Figure 8.9(a). The result is a torus, a two-dimensional surface shaped like the surface of a doughnut, as shown in Figure 8.9 (b).

Also shown in the figure is a picture of a vector field on the square. Each vector has slope q, where q is an irrational number. When the corresponding edges

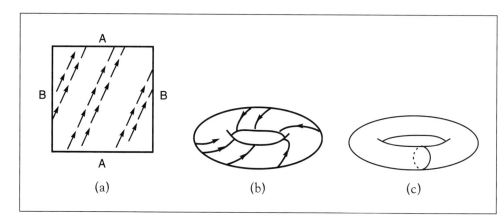

(a) (b) (c)

Figure 8.9 A dense orbit on the torus.
(a) The vector field has the same irrational slope at each point in the unit square.
(b) The square is made into a torus by gluing together the top and bottom (marked A) to form a cylinder and then gluing together the ends (marked B). Each orbit winds densely around the torus. For each point $\mathbf{u}$ of the torus, $\mathbf{u}$ belongs to $\omega(\mathbf{u})$ and $\alpha(\mathbf{u})$. (c) There is no analogue of the Poincaré-Bendixson Theorem on the torus because there is no Jordan Curve Theorem on the torus. A simple closed curve that does not divide the torus into two parts is shown.

of the square are matched to form the torus, the vectors on either side of the glued edges match up, so that the vector field still is continuous, considered as a vector field on the torus. A solution of the differential equation described by this vector field wraps around the torus and cannot intersect itself. Such a solution is drawn in Figure 8.9(b).

Given an initial location (x_0, y_0) on the torus, we describe intersections of the orbit $F(t, (x_0, y_0))$ with the vertical line segment $I = \{(x_0, y) : 0 \le y \le 1\}$. First notice that successive intersections are given by $(x_0, (y_0 + nq) \mod 1)$, $n = 1, 2, 3, \ldots$. This set of points is dense in the interval I. Hence each point in I is in the ω-limit set $\omega(x_0, y_0))$. Since x_0 and y_0 are arbitrary numbers, the entire torus is in the ω-limit set.

The torus is a type of ω-limit set that can occur in $\mathbb{R}^n$ when $n \ge 3$. We see that the conclusions of the Poincaré-Bendixson Theorem do not hold on the torus, even though it is a two-dimensional manifold. Since there is no analogue of the Jordan Curve Theorem on the torus (see Figure 8.9(c)), we can't expect the arguments we made in the planar case to apply here.

☞ **CHALLENGE 8**

Two Incommensurate Frequencies Form a Torus

THE POINCARÉ-BENDIXSON THEOREM says that possible limit sets for trajectories of an autonomous system in the plane are limited. They can be equilibria, connecting arcs asymptotic to equilibria, or periodic orbits, which are topological circles. Once we move out of the plane, however, new phenomena emerge. For example, for autonomous equations in three dimensions, it is possible for a single trajectory to wind indefinitely around a two-dimensional torus, filling up the torus while never returning to itself. Although this type of complicated behavior is possible on a two-dimensional torus, it is strictly prohibited in a two-dimensional plane.

Step 1 Begin with the equation

$$\ddot{x} + 4x = 0 \tag{8.4}$$

for scalar x. Show that all solutions can be written in the form $x(t) = a \sin(2t + b)$, where a and b can be computed from initial conditions.

Step 2 Write (8.4) as a first order autonomous system in $\mathbb{R}^2$. In this context, the Poincaré-Bendixson Theorem applies. Show that the geometric form of the solution curve $(x(t), \dot{x}(t))$ in $\mathbb{R}^2$ is an ellipse, a periodic orbit. The ellipse is traversed every π radians, therefore the angular frequency of the solution is $2\pi/\text{period} = 2$ radians per unit time.

Step 3 Now introduce a second frequency by forcing the system with natural frequency 2 at a frequency of 1. For example, consider the equation

$$\ddot{x} + 4x = 3 \sin t \tag{8.5}$$

$$x(0) = 0$$

$$\dot{x}(0) = 3.$$

Find the solution of this initial value problem.

Step 4 Consider the equation as a first-order system in $\mathbb{R}^2$. It is nonautonomous, since the vector field depends on t, so Poincaré-Bendixson does not apply. Sketch the solution trajectory $(x(t), \dot{x}(t))$ in $\mathbb{R}^2$. Is the fact that the trajectory crosses itself consistent with what we have learned? Prove that $x(t) = \sin t + \sin 2t$ cannot be the solution of any equation of type

$$\ddot{x} + a\dot{x} + bx = f(x) \tag{8.6}$$

$$x(0) = x_0$$

$$\dot{x}(0) = x_1.$$

Step 5   More generally, for any real c, show that $x(t) = \sin t + \sin ct$ is the solution of

$$\ddot{x} + c^2 x = (c^2 - 1)\sin t \tag{8.7}$$

$$x(0) = 0$$

$$\dot{x}(0) = c + 1.$$

Step 6   In order to construct a phase space for this process, we need to move to $\mathbb{R}^3$, where the system of equations can be made autonomous. Consider the following autonomous system in $\mathbb{R}^3$:

$$\dot{x} = y \tag{8.8}$$

$$\dot{y} = -c^2 x + (c^2 - 1)\sin\theta$$

$$\dot{\theta} = 1.$$

with initial conditions $x(0) = 0$, $y(0) = c + 1$, $\theta(0) = 0$. Check that the solution of system (8.8) is

$$x(t) = \sin t + \sin ct \tag{8.9}$$

$$y(t) = \cos t + c\cos ct$$

$$\theta(t) = t.$$

Step 7   In moving to $\mathbb{R}^3$, we have fictionalized the original problem in one way. The angle θ should only have meaning modulo 2π. In other words, the true phase space for our original system is the infinite slab $\mathbb{R}^2 \times [0, 2\pi]$, with $\theta = 0$ and $\theta = 2\pi$ identified together. We want to examine the shape of the solution trajectory

$$x(t) = \sin t + \sin ct \tag{8.10}$$

$$y(t) = \cos t + c\cos ct$$

$$\theta(t) = t \text{ modulo } 2\pi$$

of equation (8.7), which we have reconceived as equation (8.8). There are two cases. The first case is when 1 and c are **commensurate**, meaning that their ratio is a rational number. Show that in this case, the trajectory is a periodic orbit. If c is a ratio of the two integers p/q in reduced terms, what is the period?

Step 8   To see the torus, suppose that c is an irrational number. Fix an angle $\theta = \theta_0$. Show that the trajectory at times $t = \theta_0 + 2\pi k$, for each integer k, lies on an ellipse in the $\theta = \theta_0$ plane centered at $(\sin \theta_0, \cos \theta_0)$. The ellipses fit together for $0 \leq \theta \leq 2\pi$ into a two-dimensional torus.

Postscript.   A trajectory is dense in the torus in the irrational case. A trajectory with two or more incommensurate natural frequencies is called quasiperiodic. The Poincaré map defined on a cross-sectional circle is given by rotation through an angle incommensurate with its circumference. This map was introduced in Chapter 3 as a quasiperiodic circle map.

EXERCISES

8.1. Explain why $x(t) = \sin t$ cannot be the solution of a one-dimensional equation $\dot{x} = f(x)$.

8.2. Find the ω-limit sets of the orbits of

$$\dot{r} = r(r-1)(3-r)$$

$$\dot{\theta} = 1$$

with initial conditions $(r, \theta) = (0, 0), (1/2, 0), (1, 0), (2, 0)$.

8.3. Let A be a 2×2 (real) matrix.

(a) Describe the ω-limit sets for orbits of $\dot{v} = Av$.

(b) What condition(s) on a matrix A guarantee(s) that the equilibria of $\dot{v} = Av$ are isolated?

(c) Repeat parts (a) and (b) for a 3×3 matrix A.

8.4. Assume a differential equation in the plane has a single equilibrium v_0, which is attracting. Describe the possible limit sets of a trajectory $v(t)$ that is bounded for $t \geq 0$, but does not converge to the single point v_0 as $t \to \infty$.

8.5. Find the α-limit sets and ω-limit sets for the flows depicted in Figure 8.4.

8.6. Assuming equilibria are isolated, which of the following letters can be ω-limit sets of planar flows? Explain. (Compare with Example 8.16.)

A B C D P Q Y

8.7. Repeat Exercise 8.6, without the assumption that equilibria are isolated.

8.8. Find the ω-limit sets for the phase planes of $x'' + P'(x) = 0$ where the potentials $P(x)$ are as given in Figure 7.26.

8.9. Let $x'' + bx' + g(x) = 0$, where b is a positive constant.

(a) Show that this system has no non-equilibrium periodic solutions.

(b) Find the ω-limit sets for $x'' + bx' + P'(x) = 0$, where the potentials $P(x)$ are as given in Figure 7.26.

8.10. The version of the Poincaré-Bendixson Theorem stated here assumes that equilibria are isolated. State a more general theorem in which this hypothesis is omitted. Describe how the proof must be adapted for this more general case. In particular, if a limit set of a point contains a periodic orbit, must the periodic orbit be the entire limit set?

8.11. Use Continuous Dependence on Initial Conditions (Theorem 7.16) to give an alternate proof of Lemma 8.15.

8.12. Consider a vector field in the plane.

(a) Assume that p is a point in the plane, and that $\alpha(p)$ and $\omega(p)$ are the same periodic orbit. Prove that p lies on this periodic orbit.

(b) For an initial condition p, define the **width** of the forward (respectively, backward) orbit of p to be the maximum distance between two points on the forward (respectively, backward) orbit (this could be infinite). For example, a point is an equilibrium if and only if its forward orbit has zero width. Prove that the existence of a periodic orbit of width w implies the existence of another (forward or backward) orbit of width less than w.

(c) Show that any vector field in the plane that has a periodic orbit must also have an equilibrium.

8.13. (a) Show that the α-limit set of the equation $\mathbf{v}' = \mathbf{f}(\mathbf{v})$ is the ω-limit set of $\mathbf{v}' = -\mathbf{f}(\mathbf{v})$.

(b) State and prove a theorem analogous to the Poincaré-Bendixson Theorem for α-limit sets.

8.14. Let $\mathbf{v}(t)$ be a trajectory for which $|\mathbf{v}(t)| \to \infty$ as $t \to \infty$. Show that $\omega(\mathbf{v}_0)$ is the empty set.

8.15. Sketch a vector field in the plane for which the ω-limit set of a single trajectory is two (unbounded) parallel lines.

8.16. Consider the two-dimensional equation $\dot{\mathbf{v}} = \mathbf{f}(\mathbf{v})$, and let L be a Lipschitz constant for $\mathbf{f}$; that is, $|\mathbf{f}(\mathbf{v}) - \mathbf{f}(\mathbf{y})| \leq L|\mathbf{x} - \mathbf{y}|$. Let $x(t)$ be a periodic orbit with period T. Show that $T \geq \frac{2\pi}{L}$. This theorem is also true in higher dimensions, but the proof is considerably more difficult. See (Busenberg, Fisher and Martelli, 1989).

8.17. State an analogue of the Diamond Construction Lemma for flows in $\mathbb{R}^n$ with $n \geq 3$.

8.18. Let L be a transversal to a nonequilibrium point of a planar vector field. Show that if an orbit crosses L at times t_1 and t_2, then it crosses L at most a finite number of times between t_1 and t_2.

Steady States and Periodicity in a Squid Neuron

STEADY STATES and periodic orbits are the two simplest forms of behavior for discrete and continuous dynamical systems. For two-dimensional systems of differential equations, all bounded solutions must converge either to periodic orbits or to sets containing equilibrium points and perhaps connecting arcs, according to the Poincaré-Bendixson Theorem. In this Lab Visit, we will see evidence of coexisting attractors in a biological system, one a steady state and the other a periodic orbit.

Periodic orbits are ubiquitous in biological systems, and interruption of periodicity is often a symptom of malfunction or disease. In physiology alone, a significant proportion of medical problems exhibit abnormal dynamical behavior, including tremor and Parkinsonism, respiratory and cardiac disorders, sleep apnea, epileptic seizures, migraine, manic-depressive illness, endocrinological irregularities, and hiccups. (For the last, see (Whitelaw et al., 1995).)

In particular, the nervous systems of animals thread together a vast hierarchy of oscillatory processes. Cyclic behavior goes on at many scales, and neurophysiologists search for mechanisms that form the control point of these processes. The eventual goal is to find the critical parameters that affect the temporal pattern of a given process, and to determine how to recover from an abnormal setting.

The experiment reported here demonstrated the possibility of intervening in a biological system to move a trajectory from one of the coexisting attractors to the other. The system is a neuronal circuit that is well-studied in the scientific literature, the giant axon from a squid. The axon is the part of the neuron that is responsible for transmitting information in the form of pulses to other neurons. The squid giant axon is often chosen because of its size, and the fact that it can be made to continue firing in a fairly natural manner after it is dissected from the squid.

The axon was bathed in an artificial seawater solution with abnormally low calcium concentration, held at approximately 22° C. Periodic firing was

Guttman, R., Lewis, S., Rinzel, J., 1980. Control of repetitive firing in squid axon membrane as a model for a neuroneoscillator. J. Physiology **305**:377–395.

initiated in the axon by applying a fixed current over a 30 msec interval (1 msec, or millisecond, is 0.001 seconds). Within this interval, a very brief (0.15 msec) pulse of current was added on top of the fixed bias current. This small shock was often able to move the system from periodic to steady behavior.

Figure 8.10 shows four different experimental runs. In the upper half of part (a), the upper trace shows the step of applied bias current of 30 msec duration. The lower trace shows the electrical output of the axon, which has a repetitive spiking behavior. This periodic motion is assumed to be the result of the states of the axon moving along a periodic trajectory in its phase space.

The lower half of Figure 8.10(a) shows the 30 msec step current with a dot above it, representing the 0.15 msec added pulse. The pulse, applied after two oscillations of the system, annihilates the repetitive spiking. The experimental interpretation is that the perturbation provided by the pulse is sufficient to move the system trajectory from a periodic orbit to the basin of a nearby stable equilibrium, where it subsequently stays.

Figure 8.10(b) shows two more experimental runs, and the effect of the pulse strength on the spike annihilation. The upper pair of traces show the step current with added pulse, which stops the periodic spiking, but leaves behind a low-amplitude transient that eventually damps out completely. The interpretation here is that the perturbation moved the trajectory less directly to the equilibrium

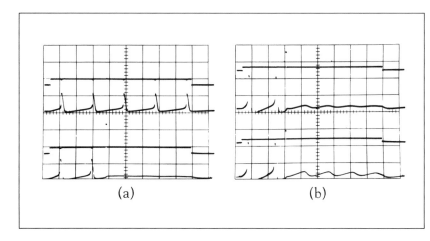

(a)   (b)

Figure 8.10   Oscilloscope traces of the firing of a squid axon.
Four experimental runs are shown, each represented by a pair of curves. (a) The upper pair of traces represents the fixed applied current and the spiking axon output. In the lower half, the repetitive spiking is interrupted by a large applied pulse (shown as a single dot) on top of the fixed current. (b) Annihilation of spiking for two different pulse magnitudes. The degree of damping of the oscillations depends on the size of the pulse.

in this case, and that we are seeing the spiraling in to the stable equilibrium. The lower pair of traces represent the effect of a low pulse strength, which moves the trajectory into the steady state basin but even farther from the steady state, so that more time is needed to spiral in to fixed behavior.

The authors of this study compared their results with a computer simulation of the Hodgkin-Huxley equations, which are widely considered to be a reasonably accurate model of neuronal circuit dynamics. Although they did not attempt to directly match the experimental conditions of the squid axon to the parameters of the computer simulation, they exhibit computer trajectories in order to schematically represent possible interpretations using ideas from nonlinear dynamics.

Figure 8.11 is an illustration of a computer simulation that may explain the experimental results, at least in a qualitative sense. The curves shown are trajectories of the voltage level in the simulated neuron plotted against the time derivative of the voltage on the vertical axis. Part (a) shows two periodic orbits together with an equilibrium denoted with a plus sign. (None of the three intersect but they are very close in the picture.) The outside orbit is a stable, attracting periodic orbit, and the inner orbit repels nearby trajectories. The + equilibrium lies

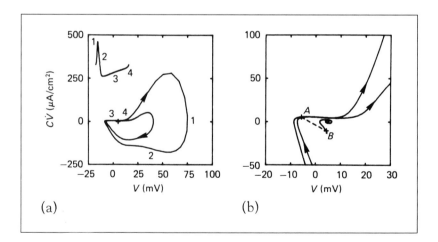

(a)   (b)

Figure 8.11 Diagram of solutions of the Hodgkin-Huxley nerve conduction equations.

This diagram serves as a suggested qualitative explanation of the experimental results of Figure 8.10. (a) Two periodic orbits and an equilibrium denoted by a plus sign. The numbers 1-4 along the outside orbit correspond to parts of the axon spike as shown in the schematic inset in the upper left corner. (b) Magnification of part (a), showing the vicinity of the equilibrium. The brief pulse moves the trajectory from A to B, which lies in the basin of the stable equilibrium.

inside the inner orbit. This is the picture we might expect following a subcritical Hopf bifurcation, discussed in Chapter 11.

A magnification of this picture is shown in Figure 8.11(b). The pulse moves the trajectory from point A on the stable periodic orbit to point B, after which the system dynamics causes the trajectory to spiral in to the equilbrium.

Almost two decades after this set of experiments, it is widely suspected that coexisting attractors are one of the nature's principal means of efficient regulation and communication. Once the periodic (spiking) behavior of such a cell is established, it is stable against small noise perturbations. The cell can be used as a timekeeper or pacemaker for a finite duration, while prepared to turn off instantly when signalled to stop. Although this research involved neuronal cells, the basic principle is probably more widespread. (Jalife and Antzelevitch, 1979) is a contemporaneous article on coexisting periodic and equilibrium attractors in cardiac tissue.

CHAPTER NINE

Chaos in Differential Equations

9.1 THE LORENZ ATTRACTOR

In the late 1950s, a meteorologist at MIT named Edward Lorenz acquired a Royal-McBee LGP-30 computer. It was the size of a refrigerator carton and contained 16KB of internal memory in a thicket of vacuum tubes. It could calculate at the rate of 60 multiplications per second. For the time, it was a staggering cache of computational power to be assigned to a single scientist.

Lorenz set the new computer to work calculating approximate solutions of a system of 12 differential equations that model a miniature atmosphere. The statistical weather forecasting community at the time was developing sophisticated linear methods for prediction. Lorenz had come to the conclusion that there had to be a fundamental factor, as yet unacknowledged, limiting the success of linear models in weather prediction. He was trying to demonstrate this point by finding solutions to his miniature atmosphere that were not periodic nor asymptotically periodic; in short, trajectories that would confound linear prediction techniques.

The atmosphere model included many parameters that he did not know how to set. The use of a computer allowed him to explore parameter space in a way that would have been impossible otherwise. He tinkered, for example, with parameters that affected the way the atmosphere was heated from below by the (sun-warmed) oceans and continents. The LGP-30 was soon producing longer and longer trajectories that seemed to be aperiodic. Moreover, they shared many qualitative features with real weather, such as long persistent trends interrupted by rapid changes. The computer printed out the trajectories on rolls of paper at its top printing speed of 6 lines of numbers per minute.

The flash of insight came from an unexpected direction. To speed up the output, Lorenz altered the program to print only three significant digits of the approximate solution trajectories, although the calculation was being done internally using several more digits of accuracy. After seeing a particularly interesting computer run, he decided to repeat the calculation to view it in more detail. He typed in the starting values from the printed output, restarted the calculation, and went down the hall for a cup of coffee. When he returned, he found that the restarted trajectory had gone somewhere else entirely—from initial conditions that were unchanged in the first three significant digits. Originally suspecting a vacuum tube failure, he was surprised to find that the discrepancy occurred *gradually*: First in the least significant decimal place, and then eventually in the next, and so on. Moreover, there was order in the discrepancy: The difference between the original and restarted trajectories approximately doubled in size every four simulated days. Lorenz concluded that he was seeing sensitive dependence on initial conditions. His search for aperiodicity had led to sensitive dependence.

Realizing the wide scope of the discovery, Lorenz tried to reduce the complexity of the 12-equation model, to verify that the effect was not simply an idiosyncrasy of one particular model. Due to the Poincaré-Bendixson Theorem, a chaotic solution could not be found in a model with fewer than 3 differential equations, but if the effect were general, it should be present in smaller, simpler systems than the 12-equation model. He would not succeed in the reduction of the miniature atmosphere model until 20 years later.

In the meantime, on a 1961 visit to Barry Saltzman of the Travelers Insurance Company Weather Center in Hartford, Connecticut, Lorenz was shown a 7-equation model of convective motion in a fluid heated from below and cooled from above. Saltzman's seven equations were themselves the reduction from a set of partial differential equations describing Rayleigh-Bénard convection, which

THE COMPUTER AGE

Lorenz used his computer to explore parameter space. Eventually, he was able to find the settings in the meteorological model that corresponded to a chaotic attractor. This was one of the early successes of computer simulation.

In fact, the motivating interest of John Mauchly, a physics professor in Pennsylvania who together with John Eckert built the first general-purpose electronic digital computer in 1946, was the solution of complex meteorological equations. The Electronic Numerical Integrator and Computer (ENIAC), weighing 30 tons and containing 18,000 vacuum tubes, first operated on St. Valentine's Day, 1946. It is said that ENIAC, in its 10 years of operation, did more floating-point calculations than had been done by the entire human race before 1946.

A stunning public demonstration of the use of computers occurred in 1969, when three astronauts traveled to the moon and back. Advances in computer hardware and software provided the telemetry and communications necessary for the flight. Progress in miniaturization allowed computers to be aboard the lunar lander to assist in the descent to the moon.

Miniaturization problems were solved by the invention of the integrated circuit, and by the 1980s cheap computation was generally available in the form of personal computers. In the 1990s, the embedding of computing chips in automobiles, toasters, and credit cards is routine. The modern car has more on-board computing power than the lunar lander of 1969.

study how heat rises through a fluid like air or water. Assume that there are two parallel horizontal plates with a fluid between them. Gravity is in the downward direction, and the temperature of the lower plate, T_l, is maintained at a higher value than temperature of the upper plate T_u, as shown in Figure 9.1.

A possible equilibrium of this system is one in which the fluid is at rest and heat is transported upward via thermal conduction. Lord Rayleigh studied the linear stability of this equilibrium and found that if the difference in temperature $T_l - T_u$ exceeds a critical value, then the equilibrium becomes unstable,

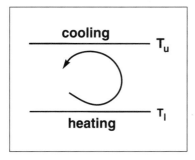

Figure 9.1   Rayleigh-Bénard convection.
The way in which heat rises in a fluid from the lower warm plate to the higher cool plate depends on the temperature difference $T_u - T_l$ of the plates. If the difference is small, heat is transferred by conduction. For a larger difference, the fluid itself moves, in convection rolls.

and convection rolls appear. Further instability occurs as $T_l - T_u$ is increased. Considering variations in only two dimensions, (Saltzman, 1962) derived a set of nonlinear ordinary differential equations by expanding the solution functions in a Fourier series, substituting the series into the original governing set of partial differential equations, and truncating the infinite sum to a finite number of terms. Lorenz then set all but three Fourier coefficients equal to zero, and obtained a system of three ordinary differential equations:

$$\dot{x} = -\sigma x + \sigma y$$

$$\dot{y} = -xz + rx - y \tag{9.1}$$

$$\dot{z} = xy - bz.$$

In this highly idealized model of a fluid, the warm fluid below rises and the cool fluid above sinks, setting up a clockwise or counterclockwise current. The Prandtl number σ, the Rayleigh (or Reynolds) number r, and b are parameters of the system. The variable x is proportional to the circulatory fluid flow velocity. If $x > 0$, the fluid circulates clockwise while $x < 0$ means counterclockwise flow. The width of the flow rolls in Figure 9.1 is proportional to the parameter b. The variable y is proportional to the temperature difference between ascending and descending fluid elements, and z is proportional to the distortion of the vertical temperature profile from its equilibrium (which is linear with height).

For $\sigma = 10$, $b = 8/3$, Lorenz found numerically that the system behaves "chaotically" whenever the Rayleigh number r exceeds a critical value $r \approx 24.74$; that is, all solutions appear to be sensitive to initial conditions, and almost all of them are apparently neither periodic solutions nor convergent to periodic solutions or equilibria.

✎ **EXERCISE T9.1**

Show that equations (9.1) exhibit the following symmetry: The equations are unchanged when (x, y, z) is replaced by $(-x, -y, z)$. As a result, if $(x(t), y(t), z(t))$ is a solution, then so is $(-x(t), -y(t), z(t))$.

✎ **EXERCISE T9.2**

Find all equilibrium points of the Lorenz equations (9.1).

For the rest of the discussion, we will assume that $\sigma = 10, b = 8/3$, and r is greater than 0. The equilibrium $(0, 0, 0)$ exists for all r, and for $r < 1$, it is a stable attractor. The origin corresponds to the fluid at rest with a linear temperature profile, hot at the bottom and cool on top. Two new equilibria exist for $r \geq 1, C_+ = (\sqrt{b(r-1)}, \sqrt{b(r-1)}, r-1)$ and $C_- = (-\sqrt{b(r-1)}, -\sqrt{b(r-1)}, r-1)$, representing steady convective circulation (clockwise or counterclockwise flow). This pair of equilibria branch off from the origin at $r = 1$ and move away as r is increased. For $r \geq 1$, the origin is unstable. The two equilibria representing convective rolls, C_+ and C_-, are stable at their birth at $r = 1$ and remain stable for $r < r_u = 24\frac{14}{19} \approx 24.74$. For $r > r_u$, all three equilibria are unstable.

For r in a range of values greater than $r_* \approx 24.06$, the chaotic attractor shown in Figure 9.2 is observed numerically. This figure depicts the orbit of a single initial condition, with several different rotated views. We say that there is a numerically observed attractor since the choice of almost any initial condition in a neighborhood of the depicted set results in a similar figure (after disregarding an initial segment of the orbit). This observation also indicates that a generating orbit is dense.

Lyapunov numbers and exponents can be assigned to orbits of autonomous differential equations in a straightforward way: They are the Lyapunov numbers and exponents of the time-1 map of the flow. See Section 9.6 for a discussion. Calculations indicate that a typical orbit of the Lorenz attractor has one positive Lyapunov exponent, calculated to be approximately 0.905 ± 0.01.

⇨ **COMPUTER EXPERIMENT 9.1**

An important feature of the Lorenz attractor is its robustness—it persists in the basic form of Figure 9.2 over a significant range of parameters. Fix $r = 28$ and plot the set of (σ, b) in the plane for which there exists an apparently chaotic attractor.

As Figure 9.2 illustrates, the attractor occupies a very thin subset of $\mathbb{R}^3$. We know, of course, that it cannot be constrained to a two-dimensional subset due to

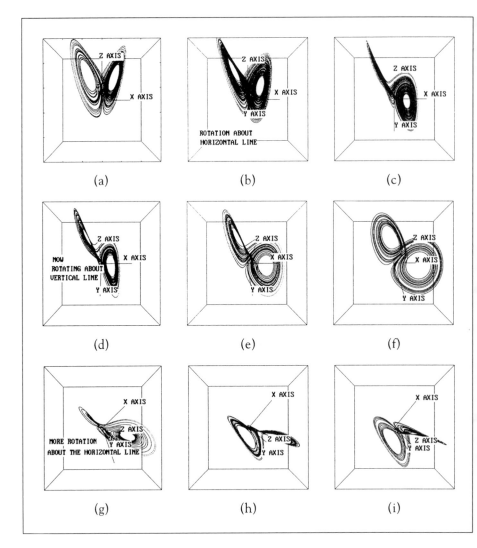

Figure 9.2 Several rotated views of the Lorenz attractor with $r = 28$.
In frames (a)–(c), the attractor is tipped up (rotated about the horizontal x-axis) until the left lobe is edge-on. In frames (d)–(f), the attractor is rotated to the left, around a vertical line. In frames (g)–(i), more rotation about a horizontal line.

r	Attractor
$[-\infty, 1.00]$	$(0, 0, 0)$ is an attracting equilibrium
$[1.00, 13.93]$	C_+ and C_- are attracting equilibria; the origin is unstable
$[13.93, 24.06]$	Transient chaos: There are chaotic orbits but no chaotic attractors
$[24.06, 24.74]$	A chaotic attractor coexists with attracting equilibria C_+ and C_-
$[24.74, ?]$	Chaos: Chaotic attractor exists but C_+ and C_- are no longer attracting

Table 9.1 Attractors for the Lorenz system (9.1).
For $\sigma = 10, b = 8/3$, a wide range of phenomena occur as r is varied.

the Poincaré-Bendixson Theorem. A trajectory will appear to spiral out around one equilibria, C_1 or C_2, until its distance from that equilibrium exceeds some critical distance. Thereafter, it spirals about the other equilibrium with increasing amplitude oscillations until the critical distance is again exceeded.

Other phenomena are observed in different parameter ranges. For r values between approximately 24.06 and 24.74, a stable chaotic attractor and stable attracting equilibria coexist. The fate of the system depends upon whether initial conditions lie in the basin of attraction of the chaotic attractor or one of the equilibria. For some much larger values of r, ($r > 50$), Lorenz has found stable periodic orbits, and for such values no chaotic attractor is observed.

The observations of the phenomena of the Lorenz attractor were first described in detail in the original article by (Lorenz, 1963) and later expanded by (Kaplan and Yorke, 1979). C. Sparrow wrote an entire book about these observations and theoretical aspects of the Lorenz system (Sparrow, 1982). The Rayleigh–Bénard convection geometry that originally motivated the Lorenz equations does not follow the route to chaos just described because of the three-dimensional nature of that geometry. The fluid moves in more complicated patterns involving swirling behavior on many length scales. This is ignored in the simplification of the convection to two spatial dimensions, but plays a crucial role in the transition to time-dependent motion. In the meantime it has been suggested that the Lorenz equations might better describe convective flow in a closed circular tube. The tube forces the fluid to be constrained to move in large circular patterns, so the Lorenz equations are more appropriate. Derivation of the equations from this model is given in (Yorke, Yorke, and Mallet-Paret, 1979).

9.2 STABILITY IN THE LARGE, INSTABILITY IN THE SMALL

In this section we discuss three typical properties of chaotic attractors and show how they are illustrated by the Lorenz attractor. More precisely, we will discuss attractors that are dissipative (volume-decreasing), locally unstable (orbits do not settle down to stationary, periodic, or quasiperiodic motion) and stable at large scale (in the sense that they have a trapping region).

First is the property of volume contraction, or dissipation. We have seen this basic principle before: With area-contracting maps of the plane, it is common to have an expanding local direction as well as a perpendicular, contracting local direction so that area ends up being decreased by the map. If the product of the contraction rate and expansion rate is less than one, then the attractor will have zero area.

The attractors we study in this chapter have the same property, although in each case the phase space is three-dimensional, so we talk about volume decreasing. The Lorenz equations decrease volume with a factor of $e^{-\sigma-1-b} \approx$ 0.00000116 each time unit. We will show where this expression comes from in Section 9.6.

The second important property is local instability. The typical trajectory of the Lorenz attractor shown in Figure 9.2 is evidently a chaotic trajectory, meaning that it has a positive Lyapunov exponent and is not asymptotically periodic. Neither of these statements can be rigorously proved at the present time. In place of a rigorous proof we will show evidence that we hope will be persuasive if not convincing.

Best estimates for the Lyapunov exponents of the Lorenz trajectory shown in Figure 9.2 are $\lambda_1 = 0.905 \pm 0.005$, $\lambda_2 = 0.0$, and $\lambda_3 = -14.57 \pm 0.01$. We will discuss how these numbers were computed in Section 9.6. The sum of the three Lyapunov exponents must be $-(\sigma + 1 + b) = -13\frac{2}{3}$. The Lyapunov numbers are found by exponentiating the Lyapunov exponents, and are $L_1 \approx 2.47, L_2 = 1$, and $L_3 \approx 0.00000047$. Therefore the distance between a pair of points that start out close together on the Lorenz attractor increases by the factor of ≈ 2.47 per time unit (often called a "second"). One time unit is roughly the time required for a typical loop of a Lorenz trajectory. Volume decreases by a factor of 0.00000116 per time unit. This explains the extreme flatness of the Lorenz attractor. It is almost a two-dimensional surface. The Lyapunov dimension is $2 + 0.905/14.57 \approx 2.062$.

To argue that the attractor is not periodic, Lorenz decided to examine the behavior of successive maxima of the z-coordinate of the trajectory. This is the vertical direction in Figure 9.2(a). He made an interesting discovery: If one plots the next vertical maximum z_{n+1} as a function f of the current z_n, a very simple diagram emerges. It is shown in Figure 9.3. The result is almost a curve; there is a tiny thickness to the graph. For practical purposes, we can view Figure 9.3 as a one-dimensional map.

Lorenz's idea was to reduce the question of whether the Lorenz attractor is periodic to the same question about the one-dimensional map. The map f has similarities to a tent map; the absolute value of the slope is greater than one at all points. Any periodic orbit must be unstable, since the derivative of f^k at any point of the periodic orbit must be a product of numbers whose magnitudes are greater than one. The shape of the z-maximum return map makes it impossible for an attracting periodic orbit to exist.

Viewing the attractor by its z-maxima gives us a way to draw a bifurcation diagram of the Lorenz system as we change a parameter. Fixing $\sigma = 10$ and $b = 8/3$, we look at the attractors of the Lorenz system for a range of r values by plotting the attractors of the z-maxima return map. As r varies, the shape of the map in Figure 9.3 changes. For some r there are periodic attractors. Figure 9.4 shows the results for $25 \leq r \leq 325$. The left side of the graph corresponds to the traditional Lorenz attractor at $r = 28$. Windows corresponding to periodic attractors are clearly visible. At $r = 400$ there is a single loop periodic orbit

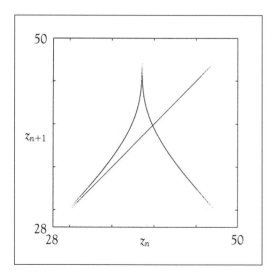

Figure 9.3 Successive maxima of z-coordinate of Lorenz attractor.
Each plotted dot on the tent-like map is a pair (z_n, z_{n+1}) of maximum z-coordinates of loops of the trajectory, one following the other. The nearly one-dimensional nature of the map arises from the very strong volume contraction.

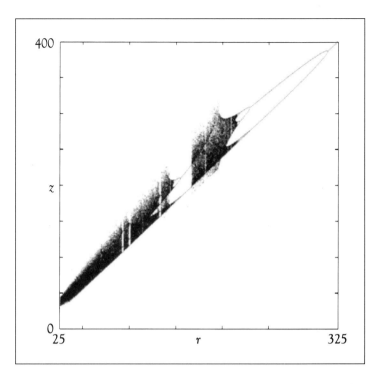

Figure 9.4   Bifurcation diagram of the Lorenz tent map.
The asymptotic behavior of the tent map of Figure 9.3 is plotted as a function of the bifurcation parameter r. The points plotted above each r correspond to the z-maxima of the orbit, so that 1 point means a period-T orbit, 2 points correspond to a period-2T orbit, and so on.

that attracts all initial conditions. As r is decreased from 400, there is a period-doubling bifurcation that results in a double-loop attractor, one with two different z-maxima before returning to repeat the orbit. As r is decreased further, the double loop again period-doubles to a loop with four different z-maxima, and so on. This is a period-doubling cascade, which we will discuss in detail in Chapter 12.

Finally, we discuss global stability, which means simply that orbits do not diverge to infinity, but stay trapped in some finite ball around the origin. Lorenz showed that all solutions to the equations (9.1) were attracted into a ball. He did this by finding a function $E(x, y, z)$ that is decreasing along any trajectory that meets the boundary of the bounded region defined by $E \leq C$, for some constant C. This means it can never leave the region, since E must increase to get out.

Lemma 9.1   (Existence of a trapping region.) Let $\mathbf{u}(t) = (x(t), y(t), z(t))$, and let $E(\mathbf{u})$ be a smooth real-valued function with the property that $E(\mathbf{u}) \to \infty$

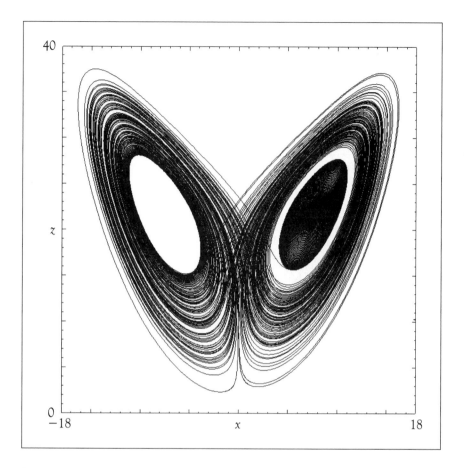

Figure 9.5 Transient chaos in the Lorenz equations.
A trajectory of the Lorenz system has been plotted using $b = 8/3$ and $\sigma = 10$, the same values Lorenz used, but here $r = 23$. When $r < r_1 \approx 24.06$ there is no longer a chaotic attractor, but if $r > 13.926\ldots$ there are chaotic trajectories, producing "transient chaos", where the trajectory behaves like a chaotic trajectory and then suddenly leaves the chaotic region. In this case, after a long transient time it finally approaches one of the asymptotically stable equilibrium points. The part of this picture that looks like the Lorenz attractor is actually the early part of the trajectory; the black spot shows the trajectory spiraling down to the equilibrium point. For $r = 23$, the mean lifetime is about 300 (which means that the variable $z(t)$ oscillates slightly over 300 times on the average before the decay sets in. For $r = 23.5$, the mean number of oscillations is over 2400. As r is chosen closer to the critical value r_1, the number of oscillations increases so that it becomes increasingly difficult to distinguish between a chaotic attractor and transient chaos. Even at $r = 23.0$, there are chaotic trajectories that oscillate chaotically forever, never spiraling in to the equilibrium.

as $||\mathbf{u}|| \to \infty$. Assume that there are constants a_i, b_i, c, for $i = 1, 2, 3$, where $a_1, a_2, a_3 > 0$ such that for all x, y, z,

$$\dot{E}(x, y, z) \leq -a_1 x^2 - a_2 y^2 - a_3 z^2 + b_1 x + b_2 y + b_3 z + c. \qquad (9.2)$$

Then there is a $B > 0$ such that every trajectory $\mathbf{u}(t)$ satisfies $|\mathbf{u}(t)| \leq B$ for all sufficiently large time t.

✎ **EXERCISE T9.3**

Provide a proof of the Trapping Region Lemma 9.1.

We will try to apply the Lemma to the Lorenz equations. A typical first guess for a function E that will satisfy (9.2) is $E(x, y, z) = \frac{1}{2}(x^2 + y^2 + z^2)$. Then $\dot{E} = x\dot{x} + y\dot{y} + z\dot{z}$. Using the Lorenz system to provide $\dot{x}$, $\dot{y}$, $\dot{z}$ yields

$$x\dot{x} = \sigma xy - \sigma x^2$$

$$y\dot{y} = rxy - y^2 - xyz$$

$$z\dot{z} = -bz^2 + xyz. \qquad (9.3)$$

The sum of these terms includes $(\sigma + r)xy$, which is not allowed in (9.2). However, $\dot{z}$ does have such a term. If we change to $E(x, y, z) = \frac{1}{2}(x^2 + y^2 + (z - \sigma - r)^2)$, we can replace the bottom line above by

$$(z - \sigma - r)\dot{z} = -(\sigma + r)xy - bz^2 + xyz + b(\sigma + r)z.$$

Then $\dot{E} = -\sigma x^2 - y^2 - bz^2 + b(\sigma + r)z$, which has the form required by Lemma 9.1. We conclude that all trajectories of the Lorenz system enter and stay in some bounded ball in three-dimensional space.

9.3 THE RÖSSLER ATTRACTOR

The Lorenz attractor has been studied in detail because it is a treasure trove of interesting phenomena. It was the first widely known chaotic attractor from a set of differential equations. The equations are simple, yet many different types of dynamical behavior can be seen for different parameter ranges. Subsequently, many other chaotic systems of differential equations have been identified. In this section and the next, we discuss two other systems that produce chaotic attractors.

There is a symmetry in the Lorenz attractor about the z-axis, as seen in Figure 9.2, and as explained by Exercise T9.1. While this symmetry undoubtedly contributes to the beauty of the Lorenz attractor, it should be noted that symmetry is not a necessity. The German scientist O. Rössler found a way to create a chaotic attractor with an even simpler set of nonlinear differential equations.

The Rössler equations (Rössler, 1976) are

$$\dot{x} = -y - z$$
$$\dot{y} = x + ay$$
$$\dot{z} = b + (x - c)z. \tag{9.4}$$

For the choice of parameters $a = 0.1$, $b = 0.1$, and $c = 14$, there is an apparent chaotic attractor, shown in Figure 9.6. The Lyapunov exponents for this attractor have been measured by computational simulation to be approximately $0.072, 0$ and -13.79. The corresponding Lyapunov dimension is 2.005. Rössler primarily considered a slightly different set of parameters, $a = 0.2$, $b = 0.2$, and $c = 5.7$, but the properties are not much different for these values.

We can understand much of the behavior of the Rössler equations since all but one of the terms are linear. We begin by looking at the dynamics in the xy-plane only. Setting $z = 0$ yields

$$\dot{x} = -y$$
$$\dot{y} = x + ay. \tag{9.5}$$

The origin is an equilibrium. To find its stability, we calculate the eigenvalues of the Jacobian matrix at $\bar{v} = (0, 0)$

$$\mathbf{Df}(0, 0) = \begin{pmatrix} 0 & -1 \\ 1 & a \end{pmatrix}$$

to be $(a \pm \sqrt{a^2 - 4})/2$. For $a > 0$, there is at least one eigenvalue with positive real part, so the origin is unstable, for the dynamics in the xy-plane. For $0 < a < 2$, the eigenvalues are complex, implying a spiraling out from the origin along the xy-plane.

Now let's make the assumption that $0 < a < 2$ and $b, c > 0$, and turn the z-direction back on. Assume for the moment that $z \approx 0$, so that we are near the xy-plane. The orbit will spiral out from the origin and stay near the xy-plane as long as x is smaller than c, since the third equation in (9.4) has a negative coefficient for z. When x tries to pass c, the z-variable is suddenly driven to large positive values. This has the effect of stopping the increase of x because of the negative z-term in the first equation of (9.4). The back and forth damping

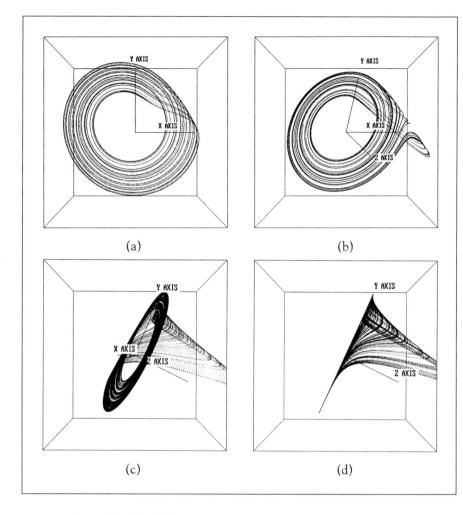

Figure 9.6   The Rössler attractor.
Parameters are set at $a = 0.1$, $b = 0.1$, and $c = 14$. Four different views are shown. The dynamics consists of a spiraling out from the inside along the xy-plane followed by a large excursion in the z-direction, followed by re-insertion to the vicinity of the xy-plane. Part (d) shows a side view. The Lyapunov dimension is 2.005—indeed it looks like a surface.

influences between the x- and z-variable keep the orbit bounded. This motion is shown in the four rotated views of the Rössler attractor in Figure 9.6.

In Figure 9.7, we fix $a = b = 0.1$, and vary the parameter c. For each c we plot the attractor. A variety of different types of attractors can be seen, beginning with a single loop periodic orbit, abruptly doubling its period when it turns into a double loop at a bifurcation value of c, and then period-doubling twice more.

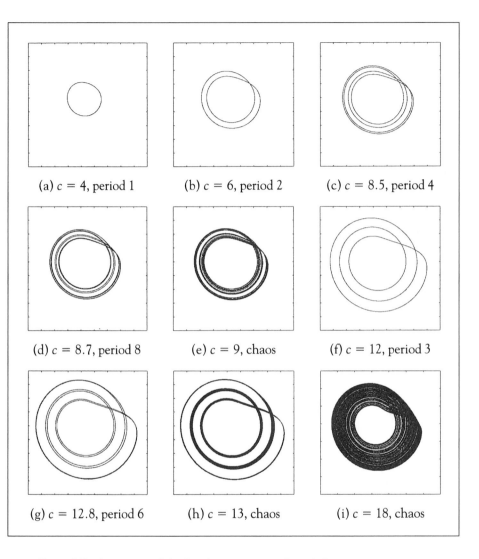

Figure 9.7 Attractors of the Rössler system as c is varied.
Fixed parameters are $a = b = 0.1$. (a) $c = 4$, periodic orbit. (b) $c = 6$, period-doubled orbit. (c) $c = 8.5$, period four. (d) $c = 8.7$, period 8. (e) $c = 9$, thin chaotic attractor. (f) $c = 12$, period three. (g) $c = 12.8$, period six. (h) $c = 13$, chaotic attractor. (i) $c = 18$, filled-out chaotic attractor

The period-doublings continue infinitely. This is another example of the period-doubling route to chaos, which we have encountered previously in Figures 1.6 and 1.8 of Chapter 1 and Figure 2.16 of Chapter 2. We will study the period-doubling bifurcation along with others in Chapter 11, and study period-doubling cascades (infinite sequences of doublings, of which this is an example) in Chapter 12.

Figure 9.8 shows a bifurcation diagram for the Rössler system. Again a and b are fixed at 0.1. For each c-value on the horizontal axis, we have plotted the local maxima of the x-variable of the attracting solution. A single loop (period one) orbit will have a single local maximum, and double loop (period two) will typically have two separate local maxima, etc. We use period one and period two in a schematic sense; the period-one orbit will have some fixed period T (not necessarily $T = 1$), and the so-called period-two orbit will have period approximately double that of the period-one orbit. A chaotic attractor will ordinarily have infinitely many local maxima due to its fractal structure.

Starting at the left edge of Figure 9.8, we can see the existence of an attracting period-one orbit which for larger c period-doubles into a period-two orbit, which eventually period-doubles as well as part of a period-doubling cascade. The result is a chaotic attractor, shown in Figure 9.7(e), followed by a

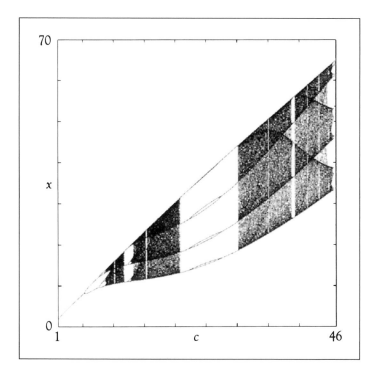

Figure 9.8 Bifurcation diagram for the Rössler equations.
The parameters $a = b = 0.1$ are fixed. The horizontal axis is the bifurcation parameter c. Each vertical slice shows a plot of the local maxima of the x-variable of an attractor for a fixed value of the parameter c. A single point implies there is a periodic orbit; two points mean a periodic orbit with "two loops", the result of a period doubling, and so on. Near $c = 46$ the attractor disappears abruptly.

period-three window, corresponding to Figure 9.7(f). The period-three orbit then period-doubles as part of another period-doubling cascade. In the center of the figure a period doubling of a period-four attractor is followed by a period-halving bifurcation.

9.4 CHUA'S CIRCUIT

A rather simple electronic circuit became popular for the study of chaos during the 1980's (Matsumoto et al., 1985). It allows almost all of the dynamical behavior seen in computer simulations to be implemented in an electronics lab and viewed on an oscilloscope. As designed and popularized by L. Chua, an electronic engineering professor at the University of California at Berkeley, and the Japanese scientist T. Matsumoto, it is an RLC circuit with four linear elements (two capacitors, one resistor, and one inductor) and a nonlinear diode, which can be modeled by a system of three differential equations. The equations for Chua's circuit are

$$\dot{x} = c_1(y - x - g(x))$$
$$\dot{y} = c_2(x - y + z)$$
$$\dot{z} = -c_3 y, \tag{9.6}$$

where $g(x) = m_1 x + \frac{m_0 - m_1}{2}(|x + 1| - |x - 1|)$.

Another way to write $g(x)$, which is perhaps more informative, is

$$g(x) = \begin{cases} m_1 x + m_1 - m_0 & \text{if } x \leq -1 \\ m_0 x & \text{if } -1 \leq x \leq 1 \\ m_1 x + m_0 - m_1 & \text{if } 1 \leq x \end{cases} \tag{9.7}$$

The function $g(x)$, whose three linear sections represent the three different voltage-current regimes of the diode, is sketched in Figure 9.9. This piecewise linear function is the only nonlinearity in the circuit and in the simulation equations. We will always use slope parameters satisfying $m_0 < -1 < m_1$, as drawn in the figure.

Typical orbits for the Chua circuit equations are plotted in Figure 9.10. All parameters except one are fixed and c_3 is varied. Two periodic orbits are created simultaneously in a Hopf bifurcation, which we will study in Chapter 11. They begin a period-doubling cascade, as shown in Figure 9.10(b)-(c) and reach chaos in Figure 9.10(d). The chaotic attractors fill out and approach one another as c_3 is varied, eventually merging in a crisis, one of the topics of Chapter 10.

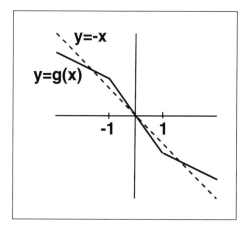

Figure 9.9   The piecewise linear $g(x)$ for the Chua circuit.
Equilibria correspond to intersections of the graph with the dotted line $y = -x$.

Color Plate 17 shows a circuit diagram of Chua's circuit. Color Plate 18 shows the computer-generated attractor for parameter settings $c_1 = 15.6, c_2 = 1$, $c_3 = 25.58, m_0 = -8/7, m_1 = -5/7$. Color Plate 19 shows a projection of the experimental circuit attractor in the voltage-current plane, and (d) shows an oscilloscope trace of the voltage time series.

9.5 FORCED OSCILLATORS

One way to produce chaos in a system of differential equations is to apply periodic forcing to a nonlinear oscillator. We saw this first for the pendulum equation in Chapter 2. Adding damping and periodic forcing to the pendulum equation

$$\ddot{\theta} + \sin \theta = 0$$

produces

$$\ddot{x} + c\dot{\theta} + \sin \theta = \rho \cos t,$$

which has apparently chaotic behavior for many parameter settings.

A second interesting example is the double-well Duffing equation from Chapter 7:

$$\ddot{x} - x + x^3 = 0. \tag{9.8}$$

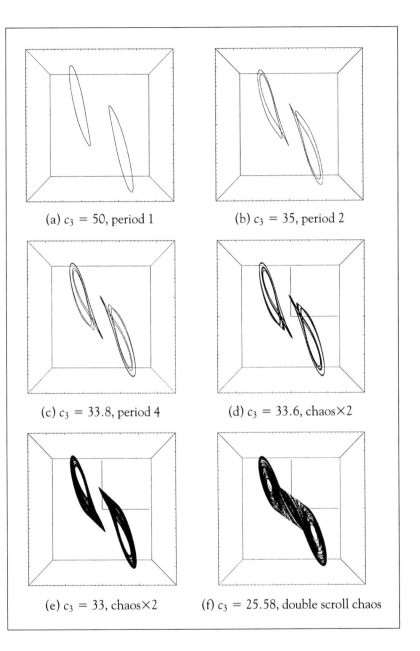

(a) $c_3 = 50$, period 1

(b) $c_3 = 35$, period 2

(c) $c_3 = 33.8$, period 4

(d) $c_3 = 33.6$, chaos×2

(e) $c_3 = 33$, chaos×2

(f) $c_3 = 25.58$, double scroll chaos

Figure 9.10 Chua circuit attracting sets.
Fixed parameters are $c_1 = 15.6$, $c_2 = 1$, $m_0 = -8/7$, $m_1 = -5/7$. The attracting set changes as parameter c_3 changes. (a) $c_3 = 50$, two periodic orbits. (b) $c_3 = 35$, the orbits have "period-doubled". (c) $c_3 = 33.8$, another doubling of the period. (d) $c_3 = 33.6$, a pair of chaotic attracting orbits. (e) $c_3 = 33$, the chaotic attractors fatten and move toward one another. (f) $c_3 = 25.58$, a "double scroll" chaotic attractor. This attractor is shown in color in Figure COLFIG6(b).

Equation (9.8) describes motion under a potential energy field which has minima at $x = -1$ and $x = 1$. We can think of a ball rolling up and down the potential wells of Figure 7.15(b). To model a ball that slowly loses energy because of friction, we add a damping term:

$$\ddot{x} + c\dot{x} - x + x^3 = 0. \tag{9.9}$$

Most orbits of (9.9) end up in one or the other of the wells, no matter how much energy the system begins with.

Both (9.8) and (9.9) are autonomous two-dimensional first-order systems in the variables x and $\dot{x}$, and as such their solution behavior falls under the enforcement of the Poincaré-Bendixson Theorem. For (9.8), most solutions are periodic, either confined to one well or rolling through both, depending on the (constant) energy of the system. They are periodic orbits, which falls under case 2 of Theorem 8.8. If the system has total energy equal to 0, the peak between the two wells in Figure 7.15(b), solutions will end up approaching the peak infinitesimally slowly (reaching the peak in "infinite time"). The ω-limit set of these orbits is the unstable equilibrium at the peak; this falls under case 1 of Theorem 8.8. For (9.9), a solution can never reach a previous $(x, \dot{x})$ position because it would have the same energy; but the energy is decreasing, as shown in Chapter 7. Therefore periodic orbits are impossible, and each orbit converges to one of the three equilibrium points, either the peak or one of the two well bottoms.

The forced damped double-well Duffing equation

$$\ddot{x} + c\dot{x} - x + x^3 = \rho \sin t \tag{9.10}$$

is capable of sustained chaotic motion. As a system in the two variables x and $\dot{x}$, it is nonautonomous; the derivative of $y = \dot{x}$ involves time. The usual trick is to declare time as a third variable, yielding the autonomous three-dimensional system

$$\dot{x} = y$$
$$\dot{y} = -cy + x - x^3 + \rho \sin t$$
$$\dot{t} = 1 \tag{9.11}$$

Solutions of (9.10) are not bound by the Poincaré-Bendixson Theorem.

The physical interpretation of the forced damped double-well Duffing equation is fairly clear. The system, because of damping, is trying to settle in one of the energy minima at the the bottom of the wells; due to the sinusoidal buffetting by the forcing function, it cannot settle down. The bronze ribbon of Lab Visit 5 is a mechanical device that follows this behavior in a qualitative way. In the absence of forcing, the magnet on the end of the ribbon will be trapped by one or the other of the permanent magnets as it loses energy due to friction. When an alternating magnetic field is set up by turning on the coil, the ribbon oscillates aperiodically for the duration of the experiment. The plot of $(x, \dot{x})$ in Lab Visit 5 is the time-2π map of the experiment. It looks qualitatively similar to the time-2π map of (9.10), which is shown in Figure 9.11. Here we have set the forcing amplitude $\rho = 3$ and investigated two different settings for the damping parameter.

⇨   C O M P U T E R   E X P E R I M E N T   9 . 2

The attractors plotted in Figure 9.11 are sensitive to moderate-sized changes in the parameters. Change the damping parameter c to 0.01 or 0.05 and explore the attracting periodic behavior that results. Show parametric plots of the periodic orbits in $(x, \dot{x})$ space as well as plots of the attractor of the time-2π maps.

⇨   C O M P U T E R   E X P E R I M E N T   9 . 3

Plot orbits of the forced Van der Pol equation $\ddot{x} + c(x^2 - 1)\dot{x} + x^3 = \rho \sin t$. Set parameters $c = 0.1$, $\rho = 5$ and use initial value $(x, \dot{x}) = (0, 0)$ to plot an attracting periodic orbit. What is its period? Repeat for $\rho = 7$; what is the new period? Next find out what lies in between at $\rho = 6$. Plot the time-2π map.

9.6 LYAPUNOV EXPONENTS IN FLOWS

In this section, we extend the definition of Lyapunov exponents for maps, introduced in Chapter 3, to the case of flows. A chaotic orbit can then be defined to be a bounded aperiodic orbit that has at least one positive Lyapunov exponent.

First recall the concept of Lyapunov exponents for maps. The local behavior of the dynamics varies among the many directions in state space. Nearby initial conditions may be moving apart along one axis, and moving together along another. For a given point, we imagined a sphere of initial conditions of infinitesimal radius evolving into an ellipse as the map is iterated. The average growth rate (per iteration) of the longest orthogonal axis of the ellipse was defined to be the first

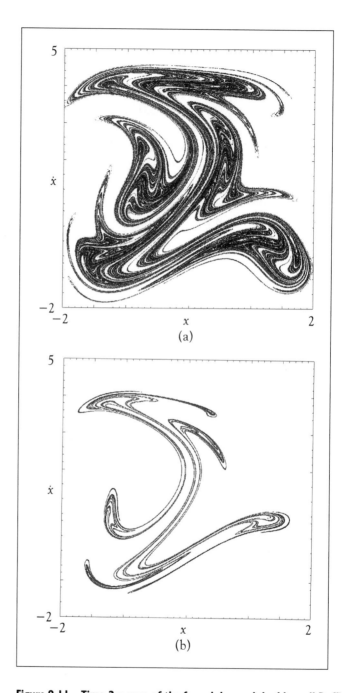

Figure 9.11   Time-2π map of the forced damped double-well Duffing equation.
(a) The variables $(x, \dot{x})$ of (9.10) with $c = 0.02$, $\rho = 3$ are plotted each 2π time units.
One million points are shown. (b) Same as (a), but $c = 0.1$. Compare with Figure
5.24, which was measured from experiment with a qualitatively similar system.

Lyapunov number of the orbit, and its natural logarithm was called the Lyapunov exponent. A positive Lyapunov exponent signifies growth along that direction, and therefore exponential divergence of nearby trajectories. The existence of a local expanding direction along an orbit is the hallmark of a chaotic orbit.

For flows, the concept is the same, once we replace the discrete iterations of the map with the continuous flow of a differential equation. Recall the definition of the time-T map of a flow $\mathbf{F}_T(\mathbf{v})$. The flow $\mathbf{F}_T(\mathbf{v})$ is defined to be the point at which the orbit with initial condition $\mathbf{v}$ arrives after T time units.

Let

$$\dot{\mathbf{v}} = \mathbf{f}(\mathbf{v}) \tag{9.12}$$

be a system of n autonomous differential equations in $\mathbf{v} = (v_1, \ldots, v_n)$. We define the Lyapunov exponent of a flow as the Lyapunov exponent of its time-T map for $T = 1$.

Definition 9.2 The **Lyapunov numbers** (respectively, **exponents**) of the flow $\mathbf{F}_T(\mathbf{v})$ are defined to be the Lyapunov numbers (respectively, exponents) of the associated time-1 map.

It is straightforward to define the time-T map $\mathbf{F}_T$, and to define the Lyapunov numbers and exponents of a flow, by simply falling back on our previous definitions in the map case. We begin with a tiny sphere of initial conditions around some $\mathbf{v}_0$, and imagine the evolution of the sphere as the initial conditions follow the flow of the differential equation. The only problem arises if you want to actually determine the Lyapunov numbers and exponents. To do so, you will need to know

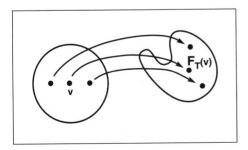

Figure 9.12 The time-T map F_T of a flow.
A ball of initial conditions are followed from time $t = 0$ to time $t = T$. The image of the point $\mathbf{v}$ under F_T is the position of the solution at time T of the initial value problem with $\mathbf{v}_0 = \mathbf{v}$.

$DF_1(v)$, the derivative of the the time-1 map $F_1(v)$ with respect to the initial value v.

If we fix T for the moment, $DF_T(v)$ is a linear map on $\mathbb{R}^n$, represented by an $n \times n$ matrix. Intuitively, the vector $DF_T(v)(w)$ is the small variation in the solution of (9.12) at time T caused by a small change in the initial value at $t = 0$ from v to $v + w$.

Although there is no explicit formula for the matrix $DF_T(v)$, we can find a differential equation involving it that can be solved in parallel with (9.12). Since $\{F_t(v) : t \text{ in } \mathbb{R}\}$ is the solution of (9.12) with initial value v, we have by definition

$$\frac{d}{dt}F_t(v) = f(F_t(v)). \tag{9.13}$$

This equation has two variables, time t and the initial value v in $\mathbb{R}^n$. Differentiating with respect to v, the chain rule yields

$$\frac{d}{dt}DF_t(v) = Df(F_t(v)) \cdot DF_t(v), \tag{9.14}$$

which is known as the **variational equation** of the differential equation. The name comes from the fact that if we could solve the equation for $DF_t(v)$, we would know the derivative matrix of F_t, and therefore know how F_t acts under small variations in the initial value v.

To simplify the looks of the variational equation, define

$$J_t = DF_t(v)$$

to be the Jacobian of the time-t map evaluated at initial value v, and

$$A(t) = Df(F_t(v))$$

to be the matrix of partial derivatives of the right-hand side f of the differential equation (9.12) evaluated along the solution. Note that $A(t)$ can be computed explicitly from knowledge of the original differential equation. Then we can rewrite the variational equation (9.14) as

$$\dot{J}_t = A(t)J_t. \tag{9.15}$$

In writing (9.14) this way, we have fixed v, the initial value of the orbit under consideration, and so have not explicitly written it into (9.15). In order to uniquely define J_t from (9.15), we need to add an initial condition, which is $J_0 = I$, the identity matrix. This follows from the fact that the flow satisfies $F_0(v) = v$ by definition. The variational equation (9.15) is a linear differential equation, even when the original differential equation has nonlinear terms. Unlike the original equation (9.12), it is not autonomous, since $A(t)$ is time-dependent in general.

EXAMPLE 9.3

In order to calculate the output of the time-T map $\mathbf{F}_T(\mathbf{v})$ for the Lorenz equations, it suffices to solve the equations with initial condition $\mathbf{v}_0 = (x_0, y_0, z_0)$ and to follow the trajectory to time T; then

$$\mathbf{F}_T(x_0, y_0, z_0) = (x(T), y(T), z(T)).$$

Next we show how to calculate the 3×3 Jacobian matrix J_T of the three-dimensional map $\mathbf{F}_T$. Differentiating the right-hand-side of the Lorenz equations (9.1) yields

$$A(t) = \begin{pmatrix} -\sigma & \sigma & 0 \\ r - z(t) & -1 & -x(t) \\ y(t) & x(t) & -b \end{pmatrix}. \tag{9.16}$$

Each column of J_T can be calculated individually from the variational equation (9.15). For example, the first column of J_T is the solution of the differential equation

$$\begin{pmatrix} \dot{J}_{11}(t) \\ \dot{J}_{21}(t) \\ \dot{J}_{31}(t) \end{pmatrix} = A(t) \begin{pmatrix} J_{11}(t) \\ J_{21}(t) \\ J_{31}(t) \end{pmatrix} \tag{9.17}$$

at time T, and the other two columns satisfy a similar equation. Notice that the current $A(t)$ needs to be available, which involves the current $(x(t), y(t), z(t))$. So the variational equation must be solved simultaneously with a solution of the original differential equation (9.12).

An important fact about the Jacobian matrix $J_T = \mathbf{DF}_T(\mathbf{v})$ of the time-T map evaluated at $\mathbf{v}$ is that it maps small variations tangent to the orbit at time 0 to small variations tangent to the orbit at time T. More precisely,

$$\mathbf{DF}_T(\mathbf{v}) \cdot \mathbf{f}(\mathbf{v}) = \mathbf{f}(\mathbf{F}_T(\mathbf{v})). \tag{9.18}$$

As usual, $\mathbf{v}$ denotes the initial value of the flow at time 0, and $\mathbf{F}_T(\mathbf{v})$ the value at time T. Since $\mathbf{f}$ is the right-hand-side of the differential equation, it defines the direction of the orbit at each time.

The derivation of this fact follows from applying the variational equation (9.14) to the vector $f(v)$:

$$\frac{d}{dt}DF_t(v)f(v) = Df(F_t(v)) \cdot DF_t(v)f(v). \tag{9.19}$$

Setting $w = DF_t(v)f(v)$, we see that $w(t)$ satisfies the initial value problem

$$\dot{w} = Df(F_t(v))w$$

$$w(0) = f(v). \tag{9.20}$$

On the other hand, differentiating the original equation (9.13) with respect to t yields

$$\frac{d^2}{dt^2}F_t(v) = Df(F_t(v))\frac{d}{dt}F_t(v)$$

$$\frac{d}{dt}f(F_t(v)) = Df(F_t(v))f(F_t(v)), \tag{9.21}$$

and $f(F_t(v))$ equals $f(v)$ at time 0. Since w and $f(F_t(v))$ satisfy the same initial value problem, they are equal.

The important consequence of (9.18) is that **a bounded orbit of an autonomous flow (9.12) either has one Lyapunov exponent equal to zero, or else it has an equilibrium in its ω-limit set**. If the latter does not occur, then $0 < b < |f(F_t(v))| < B$ for all t, for some positive bounds b and B. If $r(n)$ is the expansion in the direction of $f(v)$ after n time units (n steps of the time-1 map), then

$$0 \le \lim_{n\to\infty} \frac{1}{n}\ln b \le \lim_{n\to\infty} \frac{1}{n}\ln r(n) \le \lim_{n\to\infty} \frac{1}{n}\ln(B) \le 0.$$

Therefore the Lyapunov exponent in the direction tangent to the orbit is zero.

The change in volume due to the flow can be found with the help of a famous formula due to Liouville. Define $\Delta(t) = \det J_t$ where $\dot{J}_t = A(t)J_t$ as in (9.15). Liouville's Formula (Hartman, 1964) says that $\Delta(t)$ satisfies the differential equation and initial condition

$$\Delta_t' = \text{Tr}(A(t))\Delta_t$$

$$\Delta_0 = \det J_0 = 1, \tag{9.22}$$

where Tr denotes the trace of the matrix $A(t)$. It follows directly from (9.22) that

$$\det J_t = \exp\left(\int_0^t \text{Tr}\, A(t)\, dt\right). \tag{9.23}$$

Definition 9.4 A system of differential equations is **dissipative** if its time-T map decreases volume for all $T > 0$.

Note that $\text{Tr}(A(t)) < 0$ for all t implies that the system is dissipative. For the Lorenz equations, $\text{Tr}\, A(t) = -(\sigma + 1 + b)$, so that

$$\det J_t = e^{-(\sigma+1+b)t}.$$

The volume decrease is constant for all $\mathbf{v}$. In one time unit, a ball of initial conditions decreases in volume by a factor of $e^{-(\sigma+1+b)}$. For the standard Lorenz parameters $\sigma = 10$ and $b = 8/3$, this factor is 0.00000116 per second, so it is a dissipative system.

EXAMPLE 9.5

(Forced damped pendulum.) Liouville's formula can also be used to find the area contraction rate for the time-2π map of the forced damped pendulum, a dissipative system introduced in Chapter 2. First, write the equation $\ddot{x} + c\dot{x} + \sin x = b \cos t$ in the form of a first-order system:

$$\dot{x} = y$$
$$\dot{y} = -cy - \sin x + b\cos t$$
$$\dot{t} = 1. \tag{9.24}$$

Then

$$A(t) = \begin{pmatrix} 0 & 1 & 0 \\ -\cos x & -c & -b\sin t \\ 0 & 0 & 0 \end{pmatrix}. \tag{9.25}$$

Equation (9.23) shows us that the area contraction rate per iteration of the time-2π map is

$$\exp\left(\int_0^{2\pi} \text{Tr}\, A(t)\, dt\right) = \exp\left(\int_0^{2\pi} -c\, dt\right) = e^{-2\pi c}. \tag{9.26}$$

With the definition of Lyapunov exponent in hand, we can go on and define chaotic orbit in a straightforward way.

Definition 9.6 Let $F_t(\mathbf{v}_0)$ be a solution of $\dot{\mathbf{v}} = f(\mathbf{v})$, where $\mathbf{v}_0 \in \mathbb{R}^n$. We say the orbit $F_t(\mathbf{v}_0)$ is **chaotic** if the following conditions hold:

1. $\mathbf{F}_t(\mathbf{v}_0)$, $t \geq 0$, is bounded;
2. $\mathbf{F}_t(\mathbf{v}_0)$ has at least one positive Lyapunov exponent; and
3. $\omega(\mathbf{v}_0)$ is not periodic and does not consist solely of equilibrium points, or solely of equilibrium points and connecting arcs (as in the conclusion of the Poincaré–Bendixson Theorem).

In order to be precise, one might also rule out higher-dimensional ω-limit sets on which the map exhibits some sort of patterned behavior, such as the example of the irrational flow on the torus in Chapter 8. If the torus itself is repelling in $\mathbb{R}^3$, then a well-behaved dense orbit on the torus can have a positive Lyapunov exponent, but we would not want to consider it a chaotic orbit.

Synchronization of Chaotic Orbits

A SURPRISING FACT about chaotic attractors is their susceptibility to synchronization. This refers to the tendency of two or more systems which are coupled together to undergo closely related motions, *even when the motions are chaotic.*

There are many types of synchronization, depending on whether the motions are identical or just related in some patterned way. Synchronization can be local, meaning that the synchronized state is stable, and that once synchronized, small perturbations will not desynchronize the systems; or global, meaning that no matter where the systems are started in relation to one another, they will synchronize. There are also different ways to couple the systems. Coupling can be one-way, in which outputs from one system affect the second system but not vice versa, or two-way, in which each affects the other.

In Challenge 9, you will first establish a theorem that explains local synchronization for two-way coupled identical nonlinear systems. It states that if the coupling is strong enough, then identical systems which are started close enough together will stay close forever. Secondly, there is an example of global synchronization for one-way coupling, in which two identical Lorenz systems, started with arbitrary different initial conditions, will synchronize exactly: their (x, y, z) states are eventually (asymptotically) equal as a function of time. Both of these behaviors are different from the behavior of identical *uncoupled* (that is, independent) chaotic systems. If the latter are started with approximately equal but nonidentical initial conditions, we know that sensitive dependence will cause the two systems to eventually move far apart.

Here is synchronization in its simplest form. Consider the two-way coupled system of autonomous differential equations

$$\dot{x} = ax + c(y - x)$$

$$\dot{y} = ay + c(x - y). \tag{9.27}$$

Assume that $a > 0$. We consider the original identical systems to be $\dot{x} = ax$ and $\dot{y} = ay$. The coupling coefficient c measures how much of x to replace with y in the x-equation, and the reverse in the y-equation. First notice that if the coupling is turned off, there is no synchronization. The solutions for $c = 0$ are $x(t) = x_0 e^{at}$ and $y(t) = y_0 e^{at}$, and the difference between them is $|x(t) - y(t)| = |x_0 - y_0|e^{at}$, which increases as a function of time because of our assumption $a > 0$. In this case the synchronized state $x(t) = y(t)$ is unstable: if $x_0 = y_0$, any small difference caused by perturbing the systems will grow exponentially.

As we turn on the coupling, at first we see little difference from the uncoupled case. Figure 9.13(a) shows plots of $x(t)$ and $y(t)$ for $a = 0.1$ and coupling parameter $c = 0.03$. The difference between the two trajectories, started from two different initial values, again grows exponentially. Figure 9.13(b) is the result of stronger coupling $c = 0.07$. The trajectories move towards one another and stay together as t increases. This is an example of global synchronization. Your first assignment is to find the mechanism that explains the difference between the two cases.

Step 1 Write (9.27) as the linear system

$$\begin{pmatrix} \dot{x} \\ \dot{y} \end{pmatrix} = A \begin{pmatrix} x \\ y \end{pmatrix} \quad \text{where } A = \begin{pmatrix} a - c & c \\ c & a - c \end{pmatrix}. \tag{9.28}$$

Find the eigenvalues and eigenvectors of A. Define $\mathbf{u} = S^{-1} \begin{pmatrix} x \\ y \end{pmatrix}$, where S is the matrix whose columns are the eigenvectors of A. Write down and solve the corresponding differential equation for $\mathbf{u}$.

Step 2 Using Step 1, find the solution of system (9.27). Show that $|x(t) - y(t)| = |x_0 - y_0|e^{(a-2c)t}$.

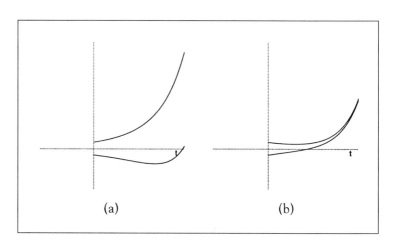

(a)   (b)

Figure 9.13 Synchronization of scalar trajectories.
(a) Two solutions of the coupled system (9.27) with $a = 0.1$ and weak coupling $c = 0.03$. Initial values are 5 and -4.5. The distance between solutions increases with time. (b) Same as (a), but with stronger coupling $c = 0.07$. The solutions approach synchronization.

Step 2 explains the difference between the weak coupling in Figure 9.13(a) and stronger coupling in Figure 9.13(b). The coupling parameter must be greater than $a/2$ to cause the two solutions to synchronize.

Now that we see how to synchronize coupled scalar equations, let's consider coupled linear systems. Let A be an $n \times n$ matrix. Define the linear system

$$\dot{\mathbf{v}} = A\mathbf{v} \tag{9.29}$$

and the coupled pair of linear systems

$$\dot{\mathbf{v}}_1 = A\mathbf{v}_1 + c(\mathbf{v}_2 - \mathbf{v}_1)$$

$$\dot{\mathbf{v}}_2 = A\mathbf{v}_2 + c(\mathbf{v}_1 - \mathbf{v}_2)$$

which can be rewritten as

$$\begin{pmatrix} \dot{\mathbf{v}}_1 \\ \dot{\mathbf{v}}_2 \end{pmatrix} = \begin{pmatrix} A - cI & cI \\ cI & A - cI \end{pmatrix} \begin{pmatrix} \mathbf{v}_1 \\ \mathbf{v}_2 \end{pmatrix}. \tag{9.30}$$

Step 3 Show that if $\mathbf{v}$ is an eigenvector of A, then both $(\mathbf{v}_1, \mathbf{v}_2)^T = (\mathbf{v}, \mathbf{v})^T$ and $(\mathbf{v}, -\mathbf{v})^T$ are eigenvectors of the matrix in (9.30). Denote the eigenvalues of A by $\lambda_1, \ldots, \lambda_n$. Show that the eigenvalues of the matrix in (9.30) are $\lambda_1, \ldots, \lambda_n, \lambda_1 - 2c, \ldots, \lambda_n - 2c$.

Step 4 Define the difference vector $\mathbf{u} = \mathbf{v}_1 - \mathbf{v}_2$. Show that $\mathbf{u}$ satisfies the differential equation

$$\dot{\mathbf{u}} = (A - 2cI)\mathbf{u}. \tag{9.31}$$

Step 5 Show that if all eigenvalues of A have real part less than $2c$, then the origin of (9.31) is globally asymptotically stable (see Chapter 7). Conclude that for sufficiently large coupling, the solutions $\mathbf{v}_1(t)$ and $\mathbf{v}_2(t)$ of (9.29) undergo global synchronization, or in other words, that for any initial values $\mathbf{v}_1(0)$ and $\mathbf{v}_2(0)$,

$$\lim_{t \to \infty} |\mathbf{v}_1(t) - \mathbf{v}_2(t)| = 0.$$

When the equilibrium $\mathbf{u} = \mathbf{v}_1 - \mathbf{v}_2 = 0$ is globally asymptotically stable, we say that the system undergoes global synchronization. If instead $\mathbf{u} = 0$ is only stable, we use the term local synchronization.

Now we move on to nonlinear differential equations. Consider a general autonomous system

$$\dot{\mathbf{v}} = \mathbf{f}(\mathbf{v}) \tag{9.32}$$

and the coupled pair of systems

$$\dot{\mathbf{v}}_1 = \mathbf{f}(\mathbf{v}_1) + c(\mathbf{v}_2 - \mathbf{v}_1)$$

$$\dot{\mathbf{v}}_2 = \mathbf{f}(\mathbf{v}_2) + c(\mathbf{v}_1 - \mathbf{v}_2) \tag{9.33}$$

The variational equations for a synchronized trajectory of (9.33) are

$$\begin{pmatrix} \dot{\mathbf{v}}_1 \\ \dot{\mathbf{v}}_2 \end{pmatrix} = \begin{pmatrix} A(t) - cI & cI \\ cI & A(t) - cI \end{pmatrix} \begin{pmatrix} \mathbf{v}_1 \\ \mathbf{v}_2 \end{pmatrix}, \tag{9.34}$$

where $A(t) = D_v\mathbf{f}(\mathbf{v}_1(t))$ is the matrix of partial derivatives of $\mathbf{f}$ evaluated along the synchronized trajectory $\mathbf{v}_1(t) = \mathbf{v}_2(t)$. To determine whether this trajectory is stable, we will investigate $\mathbf{u} = \mathbf{v}_1 - \mathbf{v}_2$. Let $J(t)$ denote the Jacobian matrix of the time-t map of the flow of $\mathbf{v}$ of (9.32), and $K(t)$ denote the Jacobian matrix of the time-t map of the flow of $\mathbf{u}$.

Step 6 Show that $J'(t) = A(t)J(t)$ and $K'(t) = (A(t) - 2cI)K(t)$. Using initial conditions $J(0) = K(0) = I$, show that $K(t) = J(t)e^{-2ct}$.

Step 6 shows the relationship between the matrix $J(1)$, the Jacobian derivative of the time-1 map for the original system, and the matrix $K(1)$, the derivative of the time-1 map of the difference vector $\mathbf{u}$.

Step 7 Prove:

Theorem 9.7 (**Synchronization Theorem.**) *Let λ_1 be the largest Lyapunov exponent of the system (9.32). Assume two-way coupling as in (9.34). If $c > \lambda_1/2$, then the coupled system satisfies local synchronization. That is, the synchronized state $\mathbf{u}(t) = \mathbf{v}_1(t) - \mathbf{v}_2(t) = 0$ is a stable equilibrium.*

Figure 9.14 shows an application of the Synchronization Theorem. For the Chua circuit which generates the double scroll attractor of Figure 9.10(f), the x-coordinate oscillates between negative and positive values. The largest Lyapunov exponent of the chaotic orbit is approximately $\lambda_1 = 0.48$.

Figure 9.14(a) shows the x-coordinates plotted from a coupled pair of Chua circuit systems as in (9.32) with $c = 0.15$. The initial values used were $(0, 0.3, 0)$ for $\mathbf{v}_1$ and $(-0.1, 0.3, 0)$ for $\mathbf{v}_2$. Although the trajectories begin very close together, they soon diverge and stay apart. Part (c) of the Figure shows a scatter plot of the simultaneous x-coordinates of $\mathbf{v}_1$ and $\mathbf{v}_2$. It starts out by lying along the identity diagonal, but eventually roams around the square. The lack of synchronization is expected because of the weak coupling strength. In part (b) of the Figure, the coupling parameter is $c = 0.3 > \lambda_1/2$, and synchronization is observed.

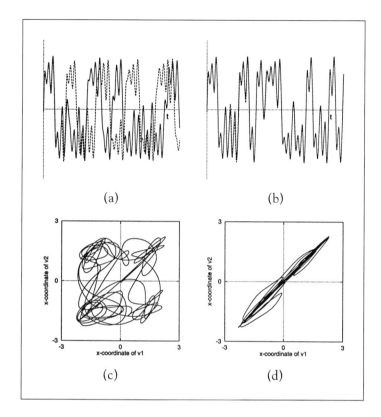

Figure 9.14 Synchronization of the Chua attractor.
(a) Time traces of the x-coordinates of v_1 (solid) and v_2 (dashed) for coupling strength $c = 0.15$. (b) Same as (a), but for $c = 0.30$. (c) A simultaneous plot of one curve from (a) versus the other shows a lack of synchronization. (d) Same as (c), but using the two curves from (b). The plot lines up along the diagonal since the trajectories are synchronized.

One-way coupling can also lead to synchronization. Next we write a pair of Lorenz systems. The first is to be considered the sender:

$$\dot{x}_1 = -\sigma x_1 + \sigma y_1$$
$$\dot{y}_1 = -x_1 z_1 + r x_1 - y_1 \qquad (9.35)$$
$$\dot{z}_1 = x_1 y_1 - b z_1,$$

and the second the receiver:

$$\dot{x}_2 = -\sigma x_1 + \sigma y_2$$
$$\dot{y}_2 = -x_1 z_2 + r x_1 - y_2 \qquad (9.36)$$
$$\dot{z}_2 = x_1 y_2 - b z_2,$$

Notice that the receiver contains a signal x_1 from the sender, but that the sender is autonomous. Using the Lyapunov function ideas from Chapter 7, you will show how to guarantee that the difference vector $\mathbf{u} = (u_y, u_z) = (y_1 - y_2, z_1 - z_2)$ tends to 0 for any set of initial conditions.

Step 8 Assume that $b > 0$. Show that the function $E(u_y, u_z) = u_y^2 + u_z^2$ is a strict Lyapunov function for the synchronized state $\mathbf{u} = 0$. Conclude that the receiver trajectory (x_2, y_2, z_2) globally synchronizes to the sender trajectory (x_1, y_1, z_1).

Postscript. Since the discovery of one-way coupling results like Step 8 by (Pecora and Carroll, 1990), there has been engineering interest in the use of the synchronization of chaos for the purpose of communications. The Rössler and Chua attractors have also been shown to admit global synchronization by one-way coupling.

Various schemes have been suggested in which the signal transmitted by the sender can be used as a message carrier. For example, very small intermittent perturbations (perhaps a secret message m coded in binary numbers) could be added to the signal x_1 and sent to the receiver. Since the signal plus message is mostly chaos, reading the message would presumably be difficult. However, if the receiver could synchronize with the sender, its x_2 would be equal to x_1 with the perturbations m greatly reduced (since they are small compared to x_1, the receiver acts as a type of filtering process). Then the receiver could simply subtract its reconstructed x_2 from the received $x_1 + m$ to recover m, the coded message. For more details on this and similar approaches, see the series of articles in Chapter 15, Synchronism and Communication, of (Ott, Sauer, and Yorke, 1994).

EXERCISES

9.1. Find out what happens to trajectories of the Lorenz equations (9.1) whose initial conditions lie on the z-axis.

9.2. Find all equilibrium points of the Chua circuit system (9.6). Show that there are three if $m_0 < -1 < m_1$, and infinitely many if $m_0 = -1$.

9.3. Find the area-contraction rate per iteration of the time-2π map of the forced damped double-well Duffing equation (9.10). For what values of damping parameter c is it dissipative?

9.4. Find a formula for the area-contraction rate per iteration of the time-1 map of the linear system $\dot{v} = Av$, where A is an $n \times n$ matrix.

9.5. Damped motion in a potential field is modeled by $\ddot{x} + c\dot{x} = \frac{\partial p}{\partial x} = 0$, as in (7.41). Use (9.23) to prove that the sum of the two Lyapunov exponents of any orbit in the $(x, \dot{x})$-plane is $-c$.

9.6. Find the Lyapunov exponents of any periodic orbit of undamped motion in a potential field, modeled by $\ddot{x} + \frac{\partial p}{\partial x} = 0$.

9.7. How are the Lyapunov exponents of an orbit of the differential equation $\dot{v} = f(v)$ related to the Lyapunov exponents of the corresponding orbit of $\dot{v} = cf(v)$, for a constant c?

9.8. Show that if an orbit of (9.12) converges to a periodic orbit, then they share the same Lyapunov exponents, if they both exist.

9.9. Consider the unforced, undamped pendulum equation $\ddot{x} + \sin x = 0$.
(a) Write as a first-order system in x and $y = \dot{x}$, and find the variational equation along the equilibrium solution $(x, y) = (0, 0)$.
(b) Find the Jacobian J_1 of the time-1 map at $(0,0)$, and find the Lyapunov exponents of the equilibrium solution $(0,0)$.
(c) Repeat (a) for the equilibrium solution $(x, y) = (\pi, 0)$.
(d) Repeat (b) for the equilibrium solution $(x, y) = (\pi, 0)$.
(e) Find the Lyapunov exponents of each bounded orbit of the pendulum.

☞ LAB VISIT 9

Lasers in Synchronization

THE EQUATIONS governing the output of a laser are nonlinear. There is a large amount of interest in the field of nonlinear optics, in which researchers study laser dynamics and related problems. The natural operating state of a laser consists of very fast periodic oscillations. During the last 20 years, it has become relatively straightforward to design a laser that operates in a chaotic state.

In his lab at the Georgia Institute of Technology, R. Roy and coworkers have studied many aspects of the nonlinear dynamics of Nd:YAG lasers. (The acronym stands for neodymium-doped yttrium aluminum garnet.) The intensity of the laser fluctuates, making a complete oscillation in several microseconds (1 microsecond = 10^{-6} seconds). Depending on parameter settings, the pattern of oscillations can be either periodic or chaotic. Systems of nonlinear differential equations exist that do a very precise job of modeling the instantaneous electric and magnetic field and population inversion of the laser. (See (Haken, 1983) for example.)

In this synchronization experiment, two Nd:YAG lasers exhibiting chaotic intensity fluctuations are placed side by side on a lab table. The two laser beams are coupled through overlap of their electromagnetic fields, which fluctuate with time. The width of each beam is much less than a millimeter. The closer the two beams, the stronger the mutual couplings of their respective differential equations. In units relevant to the experiment, the mutual coupling strength is $\approx 10^{-2}$ when the beam separation is 0.6 mm and $\approx 10^{-12}$ when the separation is 1.5 mm.

Figure 9.15 shows a diagram of the pair of pumped lasers. The two beams are driven by a single argon laser, shown at left. Beam-splitters and mirrors divide the argon laser beam into two beams, each of which lases in the Nd:YAG crystal. The two beams lase far enough apart that the population inversions of the two lasers do not overlap—the coupling occurs only through the overlap of the electromagnetic fields of the beams.

The beam separation can be changed using the beam combiner, marked BC in Figure 9.15. The two separate laser beams are marked with a single arrow and

Roy, R., and Thornburg, K.S. 1994. Experimental synchronization of chaotic lasers. Physical Review Letters **72**:2009–2012.

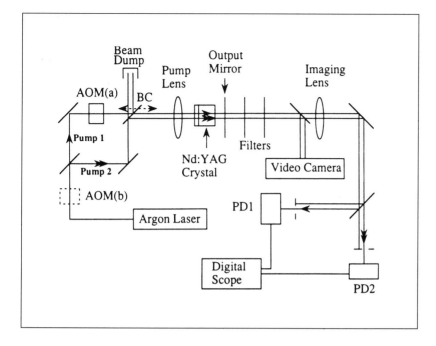

Figure 9.15   Diagram of two spatially-coupled lasers.
The two beams, denoted by a single arrow and double arrows, are driven periodically
by the same argon laser at left. The separation distance is controlled by BC, the
beam combiner. Lasing of the two beams takes place in the crystal, and output
intensity is measured at the photodiodes PD1 and PD2.

double arrows, respectively. The laser output intensity is measured by the video
camera and the photodiodes PD1 and PD2, which feed an oscilloscope for data
recording.

The chaos in the laser beam is caused by making the argon laser output
oscillate periodically. This is achieved by putting the argon laser beam through the
acousto-optic modulator, denoted AOM. When the AOM is placed in position
(a), only one beam is chaotic; in position (b), both beams are chaotic. For large
separations of the beams within the crystal, say $d = 1.5$ mm, the two lasers operate
independently and are unsynchronized.

For smaller separations ($d < 1$ mm), the effects of synchronization begin
to appear. Figure 9.16(a) shows the intensity as a function of time for the two
beams, for $d = 1$ mm. Although the two beams are not synchronized, the weak
coupling with laser 1 causes the previously quiet laser 2 to fluctuate in an erratic
fashion. Figure 9.16(c) plots simultaneous intensities versus one another. The
wide scattering of this plot shows that no synchronization is occurring. This

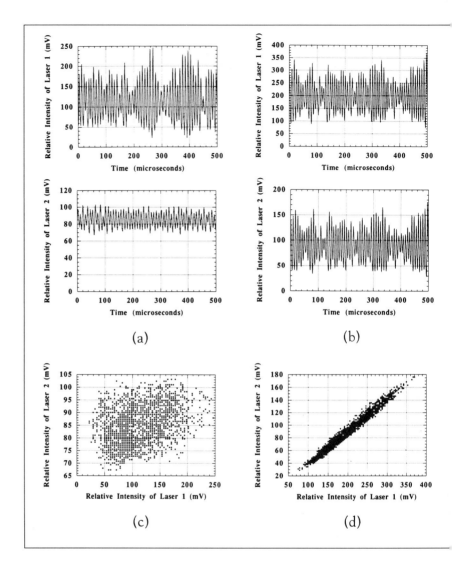

Figure 9.16 Synchronization of intensities for two coupled lasers.
(a) The beams are separated by $d = 1$ mm. Laser 1 is chaotic, as shown by the time trace of its intensity. The coupling has caused Laser 2, which is nominally at equilibrium, to oscillate erratically. The coupling is not strong enough to cause synchronization. (b) The beam separation is reduced to $d = 0.75$ mm, causing significant interaction of the two electromagnetic fields. Laser 1 is chaotic as before but now the mutual coupling has caused Laser 2 to oscillate in synchronization with Laser 1. (c) At each time, the intensities of lasers 1 and 2 in (a) are plotted as an xy-point. The lack of pattern shows that there is no synchronization between the two beams at the weak coupling strength. (d) The xy-plot from (b) lies along the line $y = x$, verifying synchronicity.

figure can be compared with the similar Figure 9.14, which shows analogous behavior with the Chua circuit.

In Figure 9.16(b), the beam separation has been decreased to $d = 0.75$ mm. The intensities are now in synchronization, as shown by the match of the time traces as well as the fact that the scatter plot is restricted to the diagonal. In both the weak and strong coupling cases in Figure 9.16, the AOM was in position (a), although similar behavior is found for position (b).

Besides possible uses of synchronization in communication applications, as suggested in Challenge 9, there are many other engineering possibilities. For advances in electronics that rely on the design of devices that harness large numbers of extremely small oscillators, split-second timing is critical. Synchronization of chaos may lead to simple ways to make subunits operate in lockstep.

CHAPTER TEN

Stable Manifolds and Crises

WE INTRODUCED the subject of stable and unstable manifolds for saddles of planar maps in Chapter 2. There we emphasized that Poincaré used properties of these sets to predict when systems would contain complicated dynamics. He showed that if the stable and unstable manifolds crossed, there was behavior that we now call chaos. For a saddle fixed point in the plane, these "manifolds" are curves that can be highly convoluted. In general, we cannot hope to describe the manifolds with simple formulas, and we need to investigate properties that do not depend on this knowledge. Recall that for an invertible map of the plane and a fixed point saddle **p**, the stable manifold of **p** is the set of initial points whose forward orbits (under iteration by the map) converge to **p**, and the unstable manifold of **p** is the set whose backward orbits (under iteration by the inverse of the map) converge to **p**.

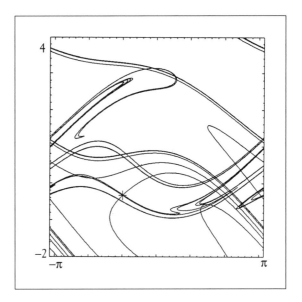

Figure 10.1 Stable and unstable manifolds for a fixed point saddle the forced, damped pendulum.

A cross marks a saddle fixed point of the time-2π map of the forced, damped pendulum with equation of motion $\ddot{x} + .2\dot{x} + \sin x = 2.5 \cos t$. The stable manifold emanates from the saddle in the direction of an eigenvector $V_s \approx (1, 0.88)$, and the unstable manifold emanates from the saddle in the direction of an eigenvector $V_u \approx (1, -0.59)$. A finite segment of each of these manifolds was computed. Larger segments would show more complex patterns.

Figure 10.1 shows numerically calculated stable and unstable manifolds of a saddle fixed point of the time-2π map of a forced damped pendulum, whose motion satisfies the differential equation $\ddot{x} + .2\dot{x} + \sin x = 2.5 \cos t$. The saddle fixed point $\mathbf{p} \approx (-.99, -.33)$ is marked with a cross. The eigenvalues of the Jacobian evaluated at $\mathbf{p}$ are $s \approx -0.13$ and $u \approx -2.26$. The stable manifold $S(\mathbf{p})$ emanates from $\mathbf{p}$ in the direction of an eigenvector $V^s \approx (1, 0.88)$ associated with s, and the unstable manifold $\mathcal{U}(\mathbf{p})$ emanates from $\mathbf{p}$ in the direction of an eigenvector $V^u \approx (1, -0.59)$ associated with u. Although these manifolds are far too complicated to be described by a simple formula, they still retain the form of a one-dimensional curve we observed of all the examples in Chapter 2. Don't forget that points on the left side of the figure (at $x = -\pi$) match up with points on the right side (at $x = \pi$), so that the curves, as far as they are calculated, have no endpoints. It has been conjectured that the stable manifold comes arbitrarily close to every point in the cylinder $[-\pi, \pi] \times \mathbb{R}$. Of course, here we have plotted only a finite segment of that manifold.

The stable and unstable manifolds shown here look deceptively like trajectories of a differential equation, except for the striking difference that these curves cross each other. (The stable manifold does not cross itself, and the unstable manifold does not cross itself.) We stress that there is no contradition here: although distinct solutions of an autonomous differential equation in the plane cannot cross, the stable and unstable manifolds of a saddle fixed point of a plane map are made up of infinitely many distinct, discrete orbits. Points in the intersection of stable and unstable manifolds are points whose forward orbits converge to the saddle (since they are in the stable manifold) and whose backward orbits converge to the saddle (since they are in the unstable manifold). As we shall see in Section 10.2, when stable and unstable manifolds cross, chaos follows.

We begin this chapter with an important theorem which guarantees that the stable and unstable manifolds of a planar saddle are one-dimensional curves.

10.1 THE STABLE MANIFOLD THEOREM

For a linear map of the plane, the stable and unstable manifolds of a saddle are lines in the direction of the eigenvectors. For nonlinear maps, as we have seen, the manifolds can be curved and highly tangled. Just as with nonlinear sinks and sources, however, more can be said about the structure of stable and unstable manifolds for a nonlinear saddle by looking at the derivative, the Jacobian matrix evaluated at the fixed point. If, for example, 0 is a fixed-point saddle of a map f, then the stable and unstable manifolds of 0 under f are approximated in a neighborhood of 0 by the stable and unstable manifolds of 0 under $L(x) = Ax$, where $A = Df(0)$. The relationship between the stable manifold of 0 under f and of the stable manifold under $Df(0)$ is given by the Stable Manifold Theorem, the main result of this chapter.

Suppose, for example, we look at the map

$$f(x, y) = (.5x + g(x, y), 3y + h(x, y)),$$

where all terms of the functions g and h are of order two or greater in x and y; functions like $x^2 + y^2$ or $xy + y^3$. Then the eigenvalues of $Df(0)$ are .5 and 3, and 0 is a fixed point saddle. We will see that the initial piece of the stable manifold of 0, called the local stable manifold, emanates from 0 as the graph of a function $y = \phi(x)$. In addition, the x-axis is tangent to S at 0, that is $\phi'(0) = 0$. See Figure 10.2(a), which shows local stable and unstable manifolds. Globally, (that is, beyond this initial piece), the stable manifold S may wind around and not be expressible as a function of x. It will, however, retain its one-dimensionality: S is a smooth curve with no endpoints, corners, or crossing points. See Figure

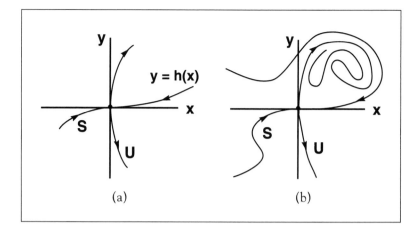

Figure 10.2 Stable and unstable manifolds for a saddle in the plane.
(a) The local stable and unstable manifolds emanate from 0. (b) Globally, the stable and unstable manifolds are one-dimensional manifolds.

10.2(b). We saw in Chapter 2 that such set is called a one-manifold. In the case of a saddle in the plane, the stable manifold is the image of a one-to-one differentiable function $\mathbf{r}$, where $\mathbf{r} : \mathbb{R} \rightarrow \mathbb{R}^2$. The unstable manifold $\mathcal{U}(0)$ is also a one-manifold that emanates from the origin in the direction of the y-axis. Figure 10.3 illustrates one-dimensional stable and unstable manifolds emanating from the origin as described.

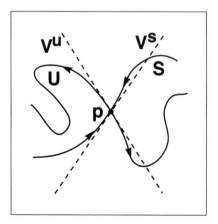

Figure 10.3 Illustration of the Stable Manifold Theorem.
The eigenvector V^s is tangent to the stable manifold S at $\mathbf{p}$, and the eigenvector V^u is tangent to the unstable manifold $\mathcal{U}$. The manifolds are curves that can wind through a region infinitely many times. Here we show a finite segment of these manifolds.

We state the theorem for a fixed-point saddle in the plane and discuss the higher-dimensional version in Sec. 2.5. The theorem says that the stable and unstable manifolds are one-manifolds, in the topological sense, and that they emanate from the fixed point in the same direction as the corresponding eigenvectors of the Jacobian. Figure 10.3 illustrates the theorem: The stable manifold S emanates from the saddle $\mathbf{p}$ in the direction of an eigenvector V^s, while the unstable manifold $\mathcal{U}$ emanates in the direction of an eigenvector V^u. A corresponding version of the theorem holds for periodic points, in which case each point in the periodic orbit has a stable and an unstable manifold. We assume that maps are one-to-one with continuous inverses. Such maps are called **homeomorphisms**. Smooth homeomorphisms (in which both the map and its inverse are smooth) are called **diffeomorphisms.**

Theorem 10.1 (**Stable Manifold Theorem.**) *Let* $\mathbf{f}$ *be a diffeomorphism of* $\mathbb{R}^2$. *Assume that* $\mathbf{f}$ *has a fixed-point saddle* $\mathbf{p}$ *such that* $D\mathbf{f}(\mathbf{p})$ *has one eigenvalue* s *with* $|s| < 1$ *and one eigenvalue* u *with* $|u| > 1$. *Let* V^s *be an eigenvector corresponding to* s, *and let* V^u *be an eigenvector corresponding to* u.

Then both the stable manifold S *of* $\mathbf{p}$ *and the unstable manifold* $\mathcal{U}$ *of* $\mathbf{p}$ *are one-dimensional manifolds (curves) that contain* $\mathbf{p}$. *Furthermore, the vector* V^s *is tangent to* S *at* $\mathbf{p}$, *and* V^u *is tangent to* $\mathcal{U}$ *at* $\mathbf{p}$.

Section 10.4 is devoted to a proof of the Stable Manifold Theorem. We end this section with examples illustrating the theorem.

EXAMPLE 10.2

Let $\mathbf{f}(x, y) = ((4/\pi)\arctan x, y/2)$. This map has two fixed-point attractors, $(-1, 0)$ and $(1, 0)$, and a fixed-point saddle $(0, 0)$. The stable manifold of $(0, 0)$ is the y-axis. See Figure 10.4. The unstable manifold of $(0, 0)$ is the set $\{(x, y) : -1 < x < 1 \text{ and } y = 0\}$. The orbits of all points in the left half-plane are attracted to $(-1, 0)$, and those of points in the right half-plane are attracted to $(1, 0)$; i.e., these sets form the basins of attraction of the two attractors. (See Chapter 3 for a more complete treatment of basins of attraction.) The stable manifold of the saddle forms the boundary between the two basins. Focusing here on the local behavior around the saddle fixed-point $(0, 0)$, we calculate the eigenvalues of $D\mathbf{f}(0, 0)$ as $s = 1/2$ and $u = 4/\pi$, with corresponding eigenvectors $V^s = (0, 1)$ and $V^u = (1, 0)$, respectively. Thus the unstable and stable manifolds emanate from $(0, 0)$ in the directions of these vectors.

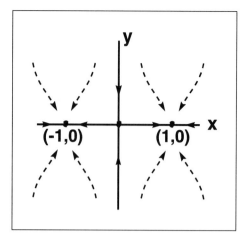

Figure 10.4   Action of the orbits for the map $f(x, y) = ((4/\pi)\arctan x, y/2)$.
The points $(-1, 0)$ and $(1, 0)$ are fixed-point sinks, while the origin is a saddle.
The stable manifold of $(0, 0)$ is the y-axis. The unstable manifold is the set $\{(x, 0) : -1 < x < 1\}$.

EXAMPLE 10.3

Let $f(r, \theta) = (r^2, \theta - \sin \theta)$, where (r, θ) are polar coordinates in the plane. There are three fixed points: the origin and $(r, \theta) = (1, 0)$ and $(1, \pi)$. See Figure 10.5. The origin is a sink, attracting all points in the interior of the unit disk since $r \to r^2$. There are two ways to compute the eigenvalues away from the origin. One is to work in polar coordinates. Then

$$Df(r, \theta) = \begin{pmatrix} 2r & 0 \\ 0 & 1 - \cos \theta \end{pmatrix},$$

for $r > 0$. The eigenvalues of each of the two fixed points with $r > 0$ are easily read from this diagonal matrix. The other way is to compute the eigenvalues in rectangular coordinates. Since eigenvalues are independent of coordinates, the results are the same. Checking the stability of the other two points in rectangular coordinates allows us to review the chain rule for two variables.

The conversion between xy-coordinates and polar coordinates is $x = r \cos \theta, y = r \sin \theta$. The map f in terms of xy-coordinates is given by

$$F(x, y) = \begin{pmatrix} F_1 \\ F_2 \end{pmatrix} = \begin{pmatrix} f_1 \cos f_2 \\ f_1 \sin f_2 \end{pmatrix},$$

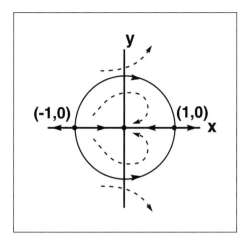

Figure 10.5 Action of orbits for the map $f(r, \theta) = (r^2, \theta - \sin \theta)$.
Here (r, θ) are polar coordinates in the plane. In rectangular (x, y) coordinates, the
fixed point $(0, 0)$ is a sink; $(-1, 0)$ is a source; and $(1, 0)$ is a saddle. The stable
manifold of $(1, 0)$ is the unit circle minus the fixed point $(-1, 0)$. The unstable
manifold of $(1, 0)$ is the positive x-axis.

where $f_1(r, \theta) = r^2, f_2(r, \theta) = \theta - \sin \theta$. The Jacobian of $\mathbf{F}$ with respect to rectangular coordinates is given by the chain rule:

$$\frac{\partial(F_1, F_2)}{\partial(x, y)} = \frac{\partial(F_1, F_2)}{\partial(f_1, f_2)} \frac{\partial(f_1, f_2)}{\partial(r, \theta)} \frac{\partial(r, \theta)}{\partial(x, y)}$$

$$= \begin{pmatrix} \cos f_2 & -f_1 \sin f_2 \\ \sin f_2 & f_1 \cos f_2 \end{pmatrix} \begin{pmatrix} 2r & 0 \\ 0 & 1 - \cos \theta \end{pmatrix} \begin{pmatrix} \cos \theta & -r \sin \theta \\ \sin \theta & r \cos \theta \end{pmatrix}^{-1}$$

where we use the fact that the matrices of partial derivatives satisfy

$$\frac{\partial(r, \theta)}{\partial(x, y)} = \left(\frac{\partial(x, y)}{\partial(r, \theta)} \right)^{-1}.$$

Now we can evaluate the rectangular coordinate Jacobian at the fixed points
without actually converting the map to rectangular coordinates, which would be
quite a bit more complicated. At $(r, \theta) = (1, 0)$, or equivalently $(x, y) = (1, 0)$,
we have

$$\frac{\partial(F_1, F_2)}{\partial(x, y)}(1, 0) = \begin{pmatrix} 1 & 0 \\ 0 & 1 \end{pmatrix} \begin{pmatrix} 2 & 0 \\ 0 & 0 \end{pmatrix} \begin{pmatrix} 1 & 0 \\ 0 & 1 \end{pmatrix}^{-1} = \begin{pmatrix} 2 & 0 \\ 0 & 0 \end{pmatrix}.$$

Thus $\mathbf{D_x f}(1, 0)$ has eigenvalues $s = 0$ and $u = 2$, with corresponding eigenvectors $V^s = (0, 1)$ and $V^u = (1, 0)$. The stable manifold of this fixed-point saddle is given by the formula $r = 1$, $-\pi < \theta < \pi$. The unstable manifold is given by $r > 0$, $\theta = 0$.

✎ **EXERCISE T10.1**

Repeat the computations of Example 10.3 for the other fixed point $(r, \theta) = (1, \pi)$, or $(x, y) = (-1, 0)$.

EXAMPLE 10.4

The phase plane of the double-well Duffing equation

$$\ddot{x} - x + x^3 = 0 \qquad (10.1)$$

is shown in Figure 10.6. This equation was introduced in Chapter 7. Here we investigate the stable and unstable manifolds under a time-T map. The equilibrium $(0, 0)$ of (10.1) is a saddle with eigenvectors $V^u = (1, 1)$ and $V^s = (1, -1)$. These vectors are tangent to a connecting arc as it emanates from the equilibrium and then returns to it. Under the time-T map $\mathbf{F}_T$, all forward and backward iterates

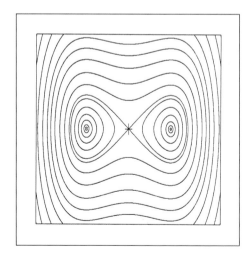

Figure 10.6   Phase plane of the undamped Duffing equation.
The phase plane of the two-well Duffing equation $\ddot{x} - x + x^3 = 0$ is shown. The equilibrium 0 (marked with a cross) is a fixed point saddle of the time-T map. The origin, together with the connecting arcs, form both the stable and the unstable manifolds of 0 under the time-T map.

of initial conditions on a connecting arc remain on the arc. The origin is a fixed point saddle of $\mathbf{F}_T$, and the origin, together with the two connecting arcs, are both the stable manifold and the unstable manifold of 0 under $\mathbf{F}_T$.

Although the previous examples illustrated the Stable Manifold Theorem, we didn't really use the theorem, since the stable and unstable manifolds could be explicitly determined. (Recall that solution curves of the Duffing phase plane are level curves of the potential $P(x) = x^4/4 - x^2/2$.) We end this section with a Hénon map, an example in which the stable and unstable manifolds must be approximated numerically. We outline the method used in all the numerically calculated manifolds pictured in this book. The approximation begins by moving in the direction of an eigenvector, as the theorem indicates.

EXAMPLE 10.5

Let $f(x, y) = (2.12 - x^2 - .3y, x)$, one of the Hénon family of maps. Figure 10.7 shows stable and unstable manifolds of a fixed-point saddle $\mathbf{p} \approx (.94, .94)$. The eigenvalues of $\mathbf{Df}(\mathbf{p})$ are $s \approx -0.18$ and $u \approx -1.71$. The corresponding eigenvectors are $V^s \approx (1, -5.71)$ and $V^u \approx (1, -.58)$. We describe a practical method for approximating $\mathcal{U}(\mathbf{p})$; $S(\mathbf{p})$ can be approximated using the same algorithm and $\mathbf{f}^{-1}$. First, find (as we have done) an eigenvector V^u associated with the eigenvalue u. Choose a point $\mathbf{a}$ on the line through V^u so that $\mathbf{a}$ and $\mathbf{b} = f(\mathbf{a})$ are within 10^{-6} of $\mathbf{p}$. (If u happens to be negative, which is the case above, replace $\mathbf{f}$ with $\mathbf{f}^2$ here and throughout this discussion.)

If we assume that the unstable manifold is given locally as a quadratic function of points on V^u, then since $|\mathbf{a} - \mathbf{p}| < 10^{-6}$, the distance of $\mathbf{b} = f(\mathbf{a})$ from $\mathcal{U}(\mathbf{p})$ is on the order of 10^{-12}. See Figure 10.8. There might be extreme cases where the distances 10^{-6} and 10^{-12} are too large, but such cases are very rare in practice. Then apply $\mathbf{f}$ to the line segment $\mathbf{ab} = J$. This involves choosing a grid of points $\mathbf{a} = \mathbf{a}_0, \mathbf{a}_1, \ldots, \mathbf{a}_n = \mathbf{b}$ along the segment J. Let $\mathbf{b}_1 = f(\mathbf{a}_1)$. The rule used here is that the distance $|\mathbf{b}_1 - \mathbf{b}|$ should be less than 10^{-3}. Otherwise, move $\mathbf{a}_1$ closer to $\mathbf{a}$. Repeat this procedure when choosing each grid point. (Continue with $\mathbf{b}_2$ so that $|\mathbf{b}_1 - \mathbf{b}_2| \leq 10^{-3}$, and so on.)

Using this method, calculate $\mathbf{f}, \mathbf{f}^2, \ldots, \mathbf{f}^n$ of segment J. (Plot $\mathbf{f}(J)$ after it is computed, then ignore the computation of $\mathbf{f}(J)$ when computing $\mathbf{f}^2(J)$, etc.) At some points, the computed value of $\mathbf{f}^n(\mathbf{q})$ may be far from the actual value, due to sensitive dependence. To avoid this problem, make sure that $\mathbf{Df}^n(\mathbf{q})$ is not too large. Plot only points $\mathbf{q}$ where each of the four entries in $\mathbf{Df}^n(\mathbf{q})$ is less than 10^8. This will ensure that the error in $\mathbf{f}^n(\mathbf{q})$, the distance from $\mathbf{q}$ to $\mathcal{U}(\mathbf{p})$, is on the order of at most 10^{-4}, assuming that $\mathbf{f}$ can be computed with an error of much less

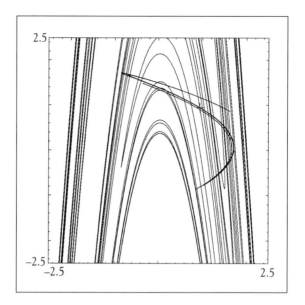

Figure 10.7 Stable and unstable manifolds for a fixed point saddle of the Hénon map $f(x, y) = (2.12 - x^2 - .3y, x)$.
The fixed point is marked with a cross. The unstable manifold is S-shaped; the stable manifold is primarily vertical.

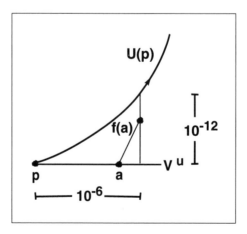

Figure 10.8 Calculating an unstable manifold.
Select a point **a** along an eigenvector V^u so that **a** and $f(a)$ are within 10^{-6} of the saddle **p**. Assuming that the unstable manifold is given locally as a quadratic function of points on V^u, then the distance of $f(a)$ from $\mathcal{U}(p)$ is on the order of at most 10^{-12}. Figure 10.7 and Color Plates 24–25 were created using the method described in this section.

than 10^{-12}. Usually 10^{-4} is smaller than the diameter of a dot. Ignoring the "10^8 rule", however, will probably give excellent longer plots. Also, the use of 10^{-6} and 10^{-12} are very conservative; in practice, one can usually use $|a - b| \approx 0.1$. This method is most successful when the unstable manifold is bounded; refinements are necessary otherwise. See (You, Kostelich, and Yorke, 1991).

10.2 HOMOCLINIC AND HETEROCLINIC POINTS

At first glance, the picture of crossing stable and unstable manifolds shown in Figure 10.9 looks bizarre, if not impossible. Perhaps drawing the manifolds as curves makes us incorrectly think of them as solutions of differential equations which, by uniqueness of solutions through a given point (Theorem 7.14 of Chapter 7), can never cross. For diffeomorphisms, however, stable and unstable manifolds are each composed of infinitely many orbits, some of which can belong to both invariant manifolds. Suppose, for example, that x is a point belonging to both the stable and unstable manifolds of a fixed point p. According to the definitions, both forward and backward iterates of x converge to p.

Definition 10.6 Let f be an invertible map of $\mathbb{R}^n$, and let p be a fixed point saddle. A point that is in both the stable and the unstable manifold of p and that is distinct from p is called a **homoclinic point**. If x is a homoclinic point,

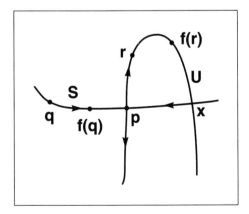

Figure 10.9 Crossing stable and unstable manifolds.
The stable manifold S and unstable manifold $\mathcal{U}$ of a saddle fixed point or periodic point p cross at a homoclinic point x. If q is a point on S, then $f(q)$ is also on S; if r is a point on $\mathcal{U}$, then $f(r)$ is also on $\mathcal{U}$. The "homoclinic" point x is on both S and $\mathcal{U}$.

then $\mathbf{f}^n(\mathbf{x}) \to \mathbf{p}$ and $\mathbf{f}^{-n}(\mathbf{x}) \to \mathbf{p}$, as $n \to \infty$. The orbit of a homoclinic point is called a **homoclinic orbit.** A point in the stable manifold of a fixed point $\mathbf{p}$ and in the unstable manifold of a different fixed point $\mathbf{q}$ is called a **heteroclinic point.** The orbit of a heteroclinic point is called a **heteroclinic orbit.**

✎ EXERCISE T10.2

Show that homoclinic points map to homoclinic points under $\mathbf{f}$ and $\mathbf{f}^{-1}$.

The existence of homoclinic orbits has complex and profound consequences. We explore a few of these consequences in the remainder of this section. A point where stable and unstable manifolds cross maps to another such point; its pre-image is also a crossing point. We can only begin to indicate the intricate web of these crossings, as in Figure 10.10. Certainly Poincaré appreciated the complicated dynamics implied here. S. Smale's celebrated work on the horseshoe map in the 1960s (Smale, 1967) greatly simplified the understanding of these dynamics.

In Chapter 5 we described the prototype or ideal horseshoe map. In that discussion we alluded to the fact that in order for a map of the plane, such as the Hénon map, to have exactly the dynamics of the ideal horseshoe, there must be uniform stretching and contraction at points in the invariant set. Smale showed that the presence of homoclinic points implies the existence of a **hyperbolic**

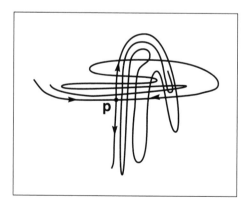

Figure 10.10 Tangle of stable and unstable manifolds implied by homoclinic points.
If the stable and unstable manifolds of a fixed-point saddle or periodic point $\mathbf{p}$ cross in one homoclinic point, then they cross infinitely many times: each forward and backward iterate of a homoclinic point is a homoclinic point.

horseshoe, a horseshoe that is identical (after a change in coordinates) to the one in Chapter 5. In particular, the set of points that remain in the horseshoe under both forward and backward iteration of the appropriate map is a Cantor set, an uncountable set with no connected subsets (except individual points). This invariant Cantor set will be in one-to-one correspondence with the set of bi-infinite binary sequences, as in Chapter 5.

The construction of a horseshoe near a homoclinic orbit is indicated in Figure 10.11. Start with a box R containing a fixed point $\mathbf{p}$ of an invertible map $\mathbf{f}$. Under iterates of $\mathbf{f}$, R stretches out along the unstable manifold of $\mathbf{p}$; under iterates of $\mathbf{f}^{-1}$, it stretches out along the stable manifold. In particular, there are numbers l and k such that $\mathbf{f}^{-l}(R)$ extends along the stable manifold to include a homoclinic point $\mathbf{x}$, and $\mathbf{f}^k(R)$ extends along the unstable manifold to include $\mathbf{x}$, as shown in Figure 10.11. Thus $\mathbf{f}^{k+l}$ is a horseshoe map with domain $\mathbf{f}^{-l}(R)$ and its image $\mathbf{f}^k(R)$. Geometrically, this construction is clear. What is not so easy to see is just how small R and how large k and l must be in order to have the uniform contraction and expansion necessary for a hyperbolic horseshoe.

We summarize Smale's result in Theorem 10.7. A stable manifold and an unstable manifold are said to **cross transversally** if the two manifolds intersect with a positive angle between them; if the angle between lines tangent to the two manifolds at the point of crossing is nonzero. If the curves are tangent at their crossing, then the angle between them is zero and the curves do not cross transversally. Figure 10.12 illustrates transversal and nontransversal intersections.

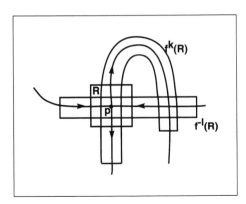

Figure 10.11   Construction of a horseshoe near a homoclinic point.
The stable and unstable manifolds of a saddle $\mathbf{p}$ intersect in a homoclinic point $\mathbf{x}$. A rectangle R is centered at $\mathbf{p}$. Then for some positive integers k and l, k forward iterates of R and l backward iterates of R intersect at $\mathbf{x}$, so that $\mathbf{f}^{k+l}$ forms a horseshoe.

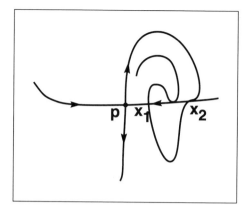

Figure 10.12 **Transversal and nontransversal crossings of stable and unstable manifolds.**

The crossing at x_1 is a transversal crossing, while the crossing at x_2 is not, since the stable and unstable manifolds have the same tangent line at x_2.

Theorem 10.7 *Let* **f** *be a diffeomorphism of the plane, and let* **p** *be a fixed point saddle. If the stable and unstable manifolds of* **p** *cross transversally, then there is a hyperbolic horseshoe for some iterate of* **f**.

Many of the computer representations of stable and unstable manifolds in this book indicate homoclinic points. Figure 10.1 shows apparent transversal crossings of the manifolds in the time-2π map of the forced, damped pendulum. They are also apparent in the time-2π map of the forced, damped Duffing system, as described in the following example.

EXAMPLE 10.8

We return to the Duffing equation of Example 10.4—this time adding damping, and then external forcing, to the system. Figure 10.13 shows the stable and unstable manifolds of a saddle fixed point under the time-2π map of the damped Duffing equation

$$\ddot{x} + 0.1\dot{x} - x + x^3 = g(t), \tag{10.2}$$

first with no external forcing ($g(t)$ set equal to 0) in (a), and then with $g(t) = 0.3 \sin t$ in (b). In the presence of damping (represented by the nonzero $\dot{x}$ term), the connecting arcs of the undamped system (Figure 10.6) have split into distinct stable and unstable manifolds. With $g(t) = 0$, each branch of the unstable manifold of 0 spirals into one of two fixed point sinks.

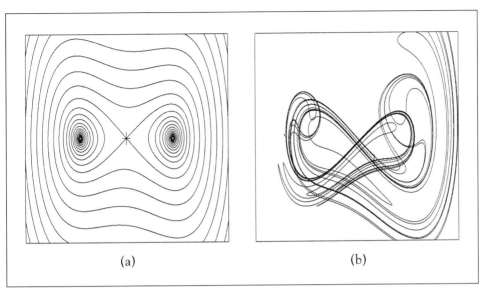

(a)   (b)

Figure 10.13 Stable and unstable manifolds for the time-2π map of the damped and periodically forced Duffing equation.
(a) Motion is governed by the autonomous equation $\ddot{x} + 0.1\dot{x} - x + x^3 = 0$. Stable and unstable manifolds of the fixed point 0 (marked with a cross) under the time-2π map are shown. Each branch of the unstable manifold converges to one of two fixed point sinks. The stable manifold of 0 forms the boundary between the basins of attraction of these sinks. In (b) stable and unstable manifolds cross, as an external force is added to the system, now governed by the equation $\ddot{x} + 0.1\dot{x} - x + x^3 = 0.3\sin t$. The fixed point saddle (marked with a cross) has moved from the origin.

When a periodic force $g(t) = 0.3\sin t$ is applied to the system, orbits that previously converged to the sinks may no longer converge. Figure 10.13(b) shows the stable and unstable manifolds of a saddle fixed point (moved slightly from the origin shown in (a)), which appear to cross transversally at **transverse homoclinic points**. By Theorem 10.7, these points imply the system has chaotic orbits. A numerically observed chaotic attractor for this system is shown in Chapter 9.

10.3 CRISES

Much of the theory of dynamical systems describes changes that can occur in the dynamics as a parameter is varied past critical values. How are the significant features of a dynamical system affected when the system undergoes a slight perturbation—as, for example, when the parameters in a map are varied? A hyperbolic fixed point may move slightly, but it will persist as long as it remains

413

hyperbolic. In this section, we turn our attention to chaotic attractors and show how small variations in a map can result in sudden drastic changes in an attractor. These changes are called **crises** and can include the sudden appearance or disappearance of the attractor or a discontinuous change in its size or shape.

Figure 10.14 shows numerically observed chaotic attractors for the one-parameter Ikeda family

$$f_a(z) = 0.84 + 0.9z \exp\left[i\left(0.4 - \frac{a}{1 + |z|^2}\right)\right], \quad z \in C,$$

at $a = 7.1$ in (a) and $a = 7.3$ in (b). The key to writing this map in (x, y)-coordinates (where $z = x + iy$) is to use the fact that $e^{ir} = \cos r + i \sin r$, for any real number r.

✎ **EXERCISE T10.3**

Find the inverse of $f_a(z)$.

The shape of the attractor varies only slightly from that shown in (a) for a in the range $7.1 < a \le a_c \approx 7.24$. For $a > a_c$, however, the attractor in this

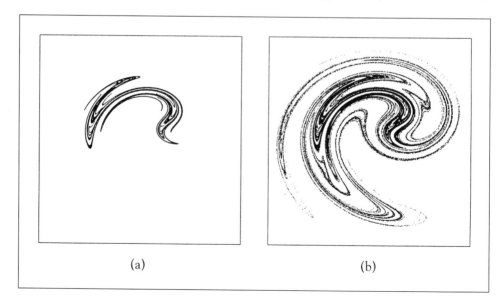

(a)   (b)

Figure 10.14 Chaotic attractor of the Ikeda map.
The shape of the attractor observed for a range of parameters up to and including the crisis value $a = a_c$ is shown in (a). For all a strictly greater than and near a_c, the attractor is significantly larger, as shown in (b). Notice that the attractor in (b) has a dark central part that is similar to the attractor in (a).

region of the plane is suddenly much larger. Figure 10.14(b) shows the extent of the larger attractor for $a = 7.3$.

⇨ COMPUTER EXPERIMENT 10.1

Create at least six additional plots of the chaotic attractor of the Ikeda map for six values of a near a_c. Plot 10^5 points of a trajectory on the attractor for each. Then describe what you can see concerning how the smaller attractor in Figure 10.14(a) changes discontinuously into the larger one in (b). Warning: there is another attractor—a fixed point.

The structure of the smaller attractor is still apparent in Figure 10.14(b). It appears darker, since orbits spend a larger percentage of iterates in this region. Surprisingly, the attractor does not continuously increase in size as a passes a_c. The closer $a > a_c$ is to the crisis parameter value a_c, however, the longer orbits typically stay on the smaller structure before bursting free. The time the trajectory stays in the new region is largely independent of a for $a - a_c$ small.

What happens at a crisis value, such as a_c in this example? We observe a saddle periodic orbit $\mathbf{p}$ that exists for all a near a_c: for $a < a_c$, this periodic orbit is not in the attractor θ_a, but as a approaches a_c (from below), the distance between them goes to 0, and at $a = a_c$, the attractor and the periodic orbit collide. The periodic orbit is in no way affected by this collision. To understand why the attractor in our example suddenly increases in size for $a > a_c$, we want to examine carefully what happens when the attractor crosses a stable manifold—in this case, the stable manifold of the orbit $\mathbf{p}$. This step is provided by Theorem 10.9. Recall that one curve crosses another transversally if the angle between them at the point of intersection is nonzero.

Theorem 10.9 (The Lambda or Inclination Lemma.) *Let $\mathbf{f}$ be a diffeomorphism of the plane, and let $\mathbf{p}$ be a hyperbolic fixed-point saddle of $\mathbf{f}$. Suppose that a curve L crosses the stable manifold of $\mathbf{p}$ transversally. Then each point in the unstable manifold of $\mathbf{p}$ is a limit point of $\bigcup_{n>0} \mathbf{f}^n(L)$.*

The proof of Theorem 10.9 can be found, for example, in (Palis and de Melo, 1982). When L is a segment of the unstable manifold itself, for example when there is a transverse crossing of the stable and unstable manifolds of $\mathbf{p}$, then Theorem 10.9 says that each segment of $\mathcal{U}(\mathbf{p})$ has other segments of $\mathcal{U}(\mathbf{p})$

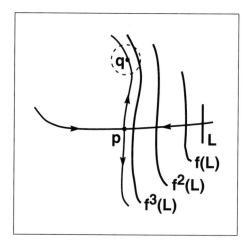

Figure 10.15 Illustration of the Lambda Lemma.
A curve L crosses the stable manifold of **p** transversally. Forward iterates of L limit on the entire unstable manifold of **p**. Specifically, the Lambda Lemma says that for each point **q** on the unstable manifold of **p** and for each ϵ-neighborhood $N_\epsilon(\mathbf{q})$, there are points of $\mathbf{f}^n(L)$ in $N_\epsilon(\mathbf{q})$, if n is sufficiently large.

limiting on it (see Figure 10.15). Specifically, the Lambda Lemma says that for each point **q** on the unstable manifold of **p** and for each ϵ-neighborhood $N_\epsilon(\mathbf{q})$, there are points of $\mathbf{f}^n(L)$ in $N_\epsilon(\mathbf{q})$, if n is sufficiently large. Of course, in this case, each $\mathbf{f}^n(L)$ is a different segment of $\mathcal{U}(\mathbf{p})$. Recall that each point in a Cantor set is a limit point of (other points in) the set. Hence, typical cross sections of the unstable manifold will have a Cantor set structure. An analogous result holds for curves that cross the unstable manifold.

Using Theorem 10.9, we can interpret the crisis in the Ikeda example above. Figure 10.16 shows that as a parameter a nears a crisis value a_c, the attractor θ_a approaches the stable manifold of a period-five point $\mathbf{p_5}$. At a_c, the outer edge of θ_{a_c} is tangent to this stable manifold; for every $a > a_c$, θ_a has crossed the stable manifold. See Figure 10.16. Once there is a crossing, the λ-Lemma tells us that forward iterates of portions of θ_a limit on the entire unstable manifold of the periodic point $\mathbf{p_5}$. Thus for each $a > a_c$, θ_a contains both branches of $\mathcal{U}(\mathbf{p_5})$. We see a jump in the size of the attractor as θ_a fills out to $\mathcal{U}(\mathbf{p_5})$. It is important that $\mathcal{U}(\mathbf{p_5})$ is contained in the basin of attraction of θ_a for a near a_c. In this case we say that there is an **interior crisis** at $a = a_c$. Specifically, at an interior crisis, the attractor jumps discontinuously in size. As a passes a_c, the attractor collides with a saddle fixed point or periodic point **p** and suddenly incorporates the outward branch of the unstable manifold of **p**.

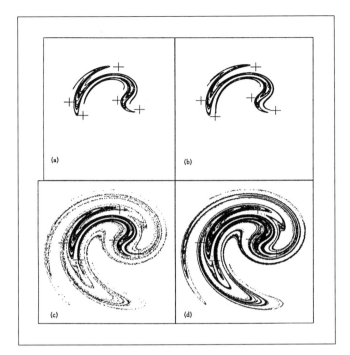

Figure 10.16 Crisis of the Ikeda attractor.
The numerically observed chaotic attractor of the Ikeda map is plotted for parameter values (a) $a = 7.1$ (b) $a = 7.2$ (c) $a = 7.25$ (d) $a = 7.3$. One million points are plotted in each part. The crisis parameter value occurs between (b) and (c). The five crosses in each picture show the location of a period-five saddle with which the attractor collides at the crisis. Another version of (c) appears in Color Plate 1.

EXERCISE T10.4

Let **p** be a fixed point or periodic point. Show that if a point $\mathbf{v} \in S(\mathbf{p})$ is a limit point of an attractor θ, then **p** is also a limit point of θ.

Sometimes one branch of an unstable manifold is not in the basin of the attractor θ_a for $a < a_c$, but then crosses the basin boundary for $a > a_c$ and is drawn into θ_a. Such is the case in the crises shown in Figure 10.17, in which the pieces of a two-piece chaotic attractor are observed to merge. The result is an interior crisis in which each piece (under the second iterate of the map) suddenly jumps in size. Prior to the crisis, each branch of the unstable manifold of the boundary saddle goes to a distinct piece of the attractor. After the crisis, both branches of the unstable manifold are contained in the resulting one-piece attractor. Exercise 10.3 refers to Figures 10.17 and 10.18.

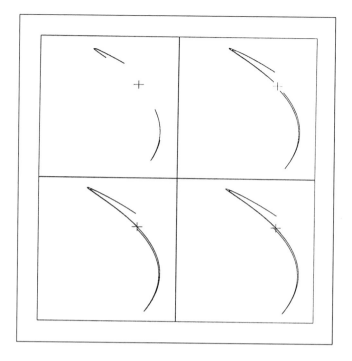

Figure 10.17 Crisis of a two-piece Hénon attractor.
Numerically observed chaotic attractors of the Hénon map are shown for parameter values $a = 2.00, 2.01, 2.02, 2.03$. Each figure is created by plotting a single trajectory. A fixed point saddle is also plotted. The two pieces of the attractor join when the pieces simultaneously collide with the saddle.

✎ **EXERCISE T10.5**

Figure 10.17 shows what is numerically observed to be a chaotic attractor for the Hénon map $f_a(x, y) = (a - x^2 - 0.3y, x)$, at parameter values $a = 2.00, 2.01, 2.02$, and 2.03. The attractor has two pieces for $a < a_c \approx 2.018$. Then the two pieces merge into one at $a = a_c$. Although the attractor appears to change continuously, there is a crisis at a_c at which point the two pieces of the attractor collide with a fixed point $\mathbf{p} \approx (.91, .91)$. Figure 10.18 shows the attractor together with $\mathbf{p}$ and $S(\mathbf{p})$. Describe the evolution of the attractor as it passes through the crisis.

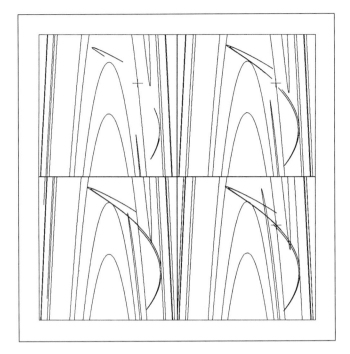

Figure 10.18 A crisis in which a two-piece Hénon attractor hits the stable manifold of a saddle.
The stable manifold of a saddle is shown together with the chaotic attractor for the same parameters as in Figure 10.17.

⇨ COMPUTER EXPERIMENT 10.2

Recreate Figure 10.17 so that each picture includes the plot of only the even-numbered iterates of some initial point, $\mathbf{p}, \mathbf{f}^2(\mathbf{p}), \mathbf{f}^4(\mathbf{p}), \ldots$.

⇨ COMPUTER EXPERIMENT 10.3

Create additional pictures of the two-piece Hénon attractor in Figure 10.17. Then make a graph of the distance between the two pieces as a function of a near the crisis value. Use this graph to give an estimate of the crisis value.

Other outcomes are possible. If, in particular, **p** is on the boundary of the basin of θ_a for $a \le a_c$ (in this case one branch of $\mathcal{U}(\mathbf{p})$ is not in the basin of θ_a), we say there is a **boundary crisis** at $a = a_c$. In this case there are points on $\mathcal{U}(\mathbf{p})$ that go to another attractor (perhaps infinity). Then for a slightly greater than a_c, the attractor (and its basin) no longer exist. However, for a only slightly larger than a_c, typical orbits spend many iterates on the "ghost" of θ_a before escaping to the other attractor. Figure 10.19 illustrates this phenomenon in the orbit of one map f_r for $r = 1.01$, in the Ikeda family

$$f_r(z) = r + 0.9z \exp\left[i\left(0.4 - \frac{6.0}{1 + |z|^2} \right) \right], \quad z \in C,$$

right after a boundary crisis at $r = r_c \approx 1.00$. For $r < r_c$, the stable manifold of a fixed point **p** (not shown) forms the boundary between the basins of a fixed-point sink (shown in the upper center of the figure) and another attractor (which is observed to be chaotic for many r values less than r_c). In this parameter range, one branch of $\mathcal{U}(\mathbf{p})$ goes to the chaotic attractor, and the other branch goes to the fixed-point attractor. For $r > r_c$, the chaotic attractor no longer exists. The orbit shown in Figure 10.19 spends many iterates on what was the structure of the chaotic attractor. Then it crosses the stable manifold $S(\mathbf{p})$ and converges to

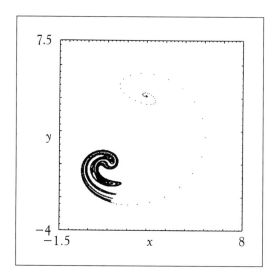

Figure 10.19 Transient chaos.
One orbit of the Ikeda map $f_r(z) = r + 0.9z e^{i(0.4 - \frac{6.0}{1+|z|^2})}$ is plotted. The parameter value is $r = 1.003$, immediately following a crisis at which the chaotic attractor disappears. The orbit spends many iterates on the "ghost" of what was the chaotic attractor before escaping and converging to a fixed-point attractor. The closer the parameter to the crisis value, the longer the orbit appears to be chaotic.

he fixed-point attractor. This behavior is called **transient chaos**. The closer the parameter to the crisis value, the longer the orbit appears to be chaotic. See Figure 9.5 for an example of transient chaos in the Lorenz attractor.

As a final example, Figure 10.20 illustrates crises in the Chua circuit system 9.6). The crises are described as a parameter a is *decreased*. In (a) there are two

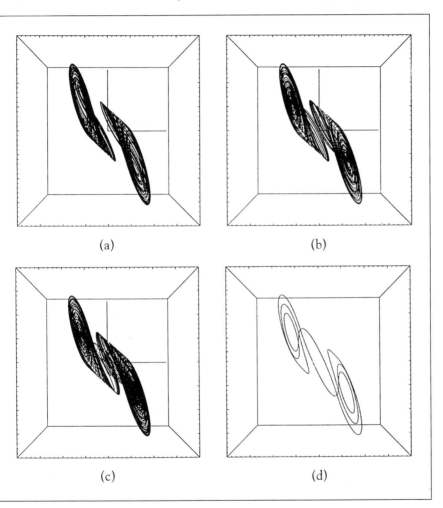

(a)

(b)

(c)

(d)

Figure 10.20 Interior crisis and boundary crisis in the Chua circuit.
Fixed parameters are $c_1 = 15.6$, $c_2 = 1$, $m_0 = -8/7$, $m_1 = -5/7$. The attracting set changes as parameter c_3 changes. (a) $c_3 = 32$, two coexisting chaotic attractors with separate basins. (b) $c_3 = 31.5$, the attractors move toward one another, but do not yet touch (although their projections to the plane overlap). (c) $c_3 = 31$, the attractors have merged into a single attractor. (d) $c_3 = 30$, a boundary crisis has caused a periodic orbit to attract the basin of the previous chaotic double-scroll attractor. As c_3 is decreased further, the double-scroll attractor reappears.

distinct chaotic attractors, which approach each other in (b) (overlapping only in the two-dimensional projection), and then merge in (c). Due to symmetries in the system, this crisis is similar to the interior crisis of the Hénon family shown in Figure 10.17. A boundary crisis follows, where the chaotic (double scroll) attractor moves into the basin of a periodic attractor, as shown in (d).

One-dimensional families can also undergo crises as a parameter is varied. See, for example, Figure 6.3(a), the bifurcation diagram of the logistic family. (Several other bifurcation diagrams appear in Chapters 11 and 12.) Chaotic attractors are observed to jump in size or suddenly appear or disappear. For one-dimensional maps, basin boundary points are repelling fixed points or periodic points or pre-images of these points. Interior and boundary crises occur as an attractor collides with a repelling fixed point or periodic orbit. For example, in the logistic family $g_a(x) = ax(1 - x)$, $1 \le a \le 4$, the points $x = 0$ and $x = 1$ form the boundary between points with bounded orbits and those in the basin of infinity. At $a = 4$ the chaotic attractor has grown to fill the unit interval and contains these boundary points. For $a > 4$, there are no (finite) attractors. Thus there is a boundary crisis at $a = 4$.

10.4 PROOF OF THE STABLE MANIFOLD THEOREM

A fixed-point saddle of a planar diffeomorphism has a stable manifold and an unstable manifold, which are smooth one-dimensional curves (that is, one-manifolds). In this section we prove that the stable manifold is a one-dimensional curve, although we do not discuss its smoothness. The proof of smoothness is rather technical and is not within the scope of this book. We direct the interested reader to (Devaney, 1986). The existence of the unstable manifold as a one-dimensional curve follows by applying the result to the inverse of the map.

Recall that a point $\mathbf{v} = (x_1, x_2)$ is in the stable manifold of a fixed point $\mathbf{p} = (p_1, p_2)$ if $\mathbf{f}^n(\mathbf{v}) \to \mathbf{p}$ as $n \to \infty$. As Figure 10.21 illustrates, given a neighborhood B of $\mathbf{p}$, the stable manifold may enter and leave B many times. In fact, when there is a homoclinic point, the stable manifold $S(\mathbf{p})$ intersects B in infinitely many pieces. We focus on the one piece of the intersection which contains $\mathbf{p}$, and call this (connected) piece the **local stable manifold** of $\mathbf{p}$ in B. We will show in the proof of the Stable Manifold Theorem that there exists a neighborhood B such

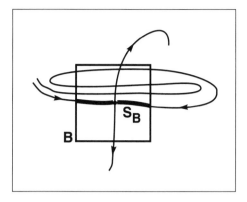

Figure 10.21   The local stable manifold.
For a sufficiently small neighborhood B of $\mathbf{p}$, the local stable manifold S_B is the set of points $\mathbf{v}$ in the stable manifold of $\mathbf{p}$ such that $\mathbf{f}^n(\mathbf{v})$ is in B, for all $n = 0, 1, 2, \ldots$.

that the local stable manifold of $\mathbf{p}$ in B can be characterized as the set

$$S_B = \{\mathbf{v} : \mathbf{f}^n(\mathbf{v}) \in B, \text{ for all } n = 0, 1, 2, \ldots, \text{ and } \mathbf{f}^n(\mathbf{v}) \to \mathbf{p}\},$$

that is, the set of points whose entire forward orbits remain in B and tend to P. For the purposes of this proof, we call the set S_B the **local stable set** of $\mathbf{p}$ in B, and show that it has the properties desired of the local stable manifold.

To analyze the behavior near a saddle, we choose coordinates so that $(0, 0)$ is the fixed point, and $(1, 0)$ and $(0, 1)$ are eigenvectors of $\mathbf{Df}(\mathbf{0})$. The eigenvector $(1, 0)$ has eigenvalue s, with $|s| < 1$, and $(0, 1)$ has eigenvalue u, with $|u| > 1$. We prove the following restatement of Theorem 10.1. Recall that a map $\mathbf{f}$ is called "twice differentiable" if all the first- and second-order partial derivatives of $\mathbf{f}$ exist and are continuous.

Theorem 10.10   *Let $\mathbf{f}$ be a diffeomorphism of the plane that is twice differentiable. In the coordinate system given above, there exists a number $\delta > 0$ and a square neighborhood $B = [-\delta, \delta] \times [-\delta, \delta]$ of $(0, 0)$ in which the local stable set S_B of $(0, 0)$ is given by the graph of a continuous function; that is, there is a continuous function ϕ on $[-\delta, \delta]$ such that*

$$S_B = \{(x, \phi(x)) : x \in [-\delta, \delta]\}.$$

Furthermore, $\phi'(0) = 0$.

Theorem 10.10 says that the curve S_B extends through a neighborhood of $(0, 0)$. In order to show that the entire stable manifold is a curve (with no self-intersections and no endpoints), see Exercise 10.6. The proof of Theorem 10.10 is given in seven lemmas. In the first, we choose a square neighborhood of the origin on which we can estimate how much the function differs from its derivative.

From our hypothesis, $\mathbf{Df}(0) = \begin{pmatrix} s & 0 \\ 0 & u \end{pmatrix}$. Call this matrix $\mathbf{A}$. Since

$$\mathbf{f}(\mathbf{p}) - \mathbf{f}(0) \approx \mathbf{Df}(0)(\mathbf{p}) \quad \text{and} \quad \mathbf{f}(0) = 0,$$

we can say that $\mathbf{f}(\mathbf{p})$ is approximated by $\mathbf{Ap}$. We now choose a square centered at the origin to be small enough that $\mathbf{Ap}$ is a good approximation to $\mathbf{f}(\mathbf{p})$ and $\mathbf{A}(\mathbf{q} - \mathbf{p})$ is a good approximation to $\mathbf{f}(\mathbf{q}) - \mathbf{f}(\mathbf{p})$.

Lemma 10.11   For any $\epsilon > 0$, there is a $\delta > 0$ such that

$$|\mathbf{f}(\mathbf{q}) - \mathbf{f}(\mathbf{p}) - \mathbf{A}(\mathbf{q} - \mathbf{p})| \le \epsilon|\mathbf{q} - \mathbf{p}|,$$

for all $\mathbf{p}, \mathbf{q}$ in the box $B = [-\delta, \delta] \times [-\delta, \delta]$.

This result should seem quite reasonable to the reader. The proof, however, is quite technical. We recommend postponing a careful reading of the proof until the structure of the proof is clear. (In fact, we recommend skipping all proofs on first reading of this chapter.)

Proof: We can estimate how $\mathbf{f}(\mathbf{q}) - \mathbf{f}(\mathbf{p})$ compares with $\mathbf{q} - \mathbf{p}$, provided $\mathbf{p}$ and $\mathbf{q}$ are close to the origin. From the two-dimensional Taylor's Theorem, we have

$$|\mathbf{f}(\mathbf{q}) - \mathbf{f}(\mathbf{p}) - \mathbf{Df}(\mathbf{p})(\mathbf{q} - \mathbf{p})| \le C_1|\mathbf{q} - \mathbf{p}|^2,$$

for some positive constant C_1. Since $\mathbf{f}$ is twice differentiable, the matrix $\mathbf{Df}(\mathbf{p})$, which is the derivative of $\mathbf{f}$ at $\mathbf{p}$, and the derivative $\mathbf{Df}(0)$ differ by a matrix whose norm (and the entries in the matrix) differ by a quantity at most proportional to $|\mathbf{p} - 0| = |\phi|$. That is,

$$|\mathbf{Df}(\mathbf{p})(\mathbf{q} - \mathbf{p}) - \begin{pmatrix} s & 0 \\ 0 & u \end{pmatrix}(\mathbf{q} - \mathbf{p})|$$

$$\le \left|\mathbf{Df}(\mathbf{p}) - \begin{pmatrix} s & 0 \\ 0 & u \end{pmatrix}\right| |\mathbf{q} - \mathbf{p}|$$

$$\le C_2|\mathbf{p}| |\mathbf{p} - \mathbf{q}|,$$

for some positive constant C_2. Hence

$$\left| f(q) - f(p) - \begin{pmatrix} s & 0 \\ 0 & u \end{pmatrix} (q - p) \right|$$

$$\leq C_2 |p| |p - q| + C_1 |p - q|^2$$

$$= |p - q|(C_2 |p| + C_1 |p - q|). \tag{10.3}$$

Let B be the square $[-\delta, \delta] \times [-\delta, \delta]$. Then for p and q in the square B, the right-hand side multiplier of $|p - q|$ in Equation (10.3) is no bigger than $2\sqrt{2}C_1\delta + \sqrt{2}C_2\delta$. We choose δ small enough that this quantity is less than ϵ.
□

We now have a bound on how much $f(p) - f(q)$ can differ from the linear approximation $A(p - q)$, for p and q in B. For the remainder of the proof assume that p and q are points in B. We use subscripts to denote the x- and y- coordinates of vectors; $(q - p)_x$ and $(q - p)_y$ mean, respectively, the x- and y- coordinates of $q - p$.

We say p and q are *horizontally aligned* if

$$|(q - p)_x| \geq |(q - p)_y|$$

and *vertically aligned* if

$$|(q - p)_x| \leq |(q - p)_y| .$$

In Figure 10.22 the points p and q are horizontally aligned and points r and s are vertically aligned. We will assume throughout the remainder of this section that

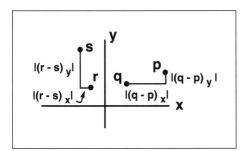

Figure 10.22   Horizontal and vertical alignment.
The points s and r are vertically aligned since the x-component of $r - s$ is smaller than the y-component; the points p and q are horizontally aligned since the y-component of $p - q$ is smaller than the x-component.

the eigenvalues u and s are positive. The proofs of the remaining lemmas can be adapted easily for the general case.

In Lemma 10.12, we show that there is a neighborhood B of the origin such that if two points are vertically aligned and in B, then their images are also vertically aligned.

Lemma 10.12 Let $\hat{u} = \frac{u+1}{2} > 1$ and $\hat{s} = \frac{s+1}{2} < 1$. There exists a square neighborhood B of $(0, 0)$, $B = [-\delta, \delta] \times [-\delta, \delta]$, such that the following properties hold for all $\mathbf{p}, \mathbf{q} \in B$:

1. If $\mathbf{p}$ and $\mathbf{q}$ are vertically aligned, then $\left|(\mathbf{f}(\mathbf{q}) - \mathbf{f}(\mathbf{p}))_y\right| \geq \hat{u}\left|(\mathbf{q} - \mathbf{p})_y\right|$.
2. If $\mathbf{p}$ and $\mathbf{q}$ are horizontally aligned, then $\left|(\mathbf{f}(\mathbf{q}) - \mathbf{f}(\mathbf{p}))_x\right| \leq \hat{s}\left|(\mathbf{q} - \mathbf{p})_x\right|$.
3. If $\mathbf{p}$ and $\mathbf{q}$ are vertically aligned, then so are $\mathbf{f}(\mathbf{p})$ and $\mathbf{f}(\mathbf{q})$.

Before studying the proof, the reader should first check that (1), (2), and (3) hold in the special case $\mathbf{f}(\mathbf{p}) = \mathbf{A}\mathbf{p}$, and that they also hold if $\mathbf{f}(\mathbf{p}) = \mathbf{A}_1\mathbf{p}$, for a matrix $\mathbf{A}_1$ sufficiently close to $\mathbf{A}$.

Proof: Let $\epsilon < \min\{\frac{u-1}{2}, \frac{1-s}{2}\}$, and let δ and B be given (for this ϵ, as in Lemma 10.11). Notice with this choice of $\epsilon > 0$ that $u - \epsilon\sqrt{2} > \hat{u}$ and $s + \epsilon\sqrt{2} < 1$.

We begin with the proof of (1). Lemma 10.11 implies that each coordinate of $\mathbf{f}(\mathbf{q}) - \mathbf{f}(\mathbf{p})$ differs from $(s(\mathbf{q} - \mathbf{p})_x, u(\mathbf{q} - \mathbf{p})_y)$ by at most $\epsilon\|\mathbf{q} - \mathbf{p}\|$. We assume $\mathbf{p} \neq \mathbf{q}$. Since $\mathbf{p}$ and $\mathbf{q}$ are vertically aligned, it differs by at most $\epsilon\sqrt{2}\left|(\mathbf{q} - \mathbf{p})_y\right|$. Therefore,

$$\left|u(\mathbf{q} - \mathbf{p})_y\right| - \left|(\mathbf{f}(\mathbf{q}) - \mathbf{f}(\mathbf{p}))_y\right| \leq \left|u(\mathbf{q} - \mathbf{p})_y - (\mathbf{f}(\mathbf{q}) - \mathbf{f}(\mathbf{p}))_y\right|$$
$$\leq \epsilon\sqrt{2}\left|(\mathbf{q} - \mathbf{p})_y\right|,$$

which implies

$$\left|(\mathbf{f}(\mathbf{q}) - \mathbf{f}(\mathbf{p}))_y\right| \geq u\left|(\mathbf{q} - \mathbf{p})_y\right| - \epsilon\sqrt{2}\left|(\mathbf{q} - \mathbf{p})_y\right|$$
$$= (u - \epsilon\sqrt{2})\left|(\mathbf{q} - \mathbf{p})_y\right|$$
$$> \hat{u}\left|(\mathbf{q} - \mathbf{p})_y\right|.$$

The second inequality follows from the choice of ϵ. The proof of (2) is similar.

To prove (3), we continue to assume that $\mathbf{p}$ and $\mathbf{q}$ are vertically aligned and refer to (1). Then

$$|(\mathbf{f}(\mathbf{q}) - \mathbf{f}(\mathbf{p}))_x| - s|(\mathbf{q} - \mathbf{p})_x| \le |(\mathbf{f}(\mathbf{q}) - \mathbf{f}(\mathbf{p}))_x - s(\mathbf{q} - \mathbf{p})_x|$$
$$\le \epsilon\sqrt{2}\,|(\mathbf{q} - \mathbf{p})_y|,$$

which implies

$$|(\mathbf{f}(\mathbf{q}) - \mathbf{f}(\mathbf{p}))_x| < s|(\mathbf{q} - \mathbf{p})_x| + \epsilon\sqrt{2}|(\mathbf{q} - \mathbf{p})_y|$$
$$\le s|(\mathbf{q} - \mathbf{p})_y| + \epsilon\sqrt{2}|(\mathbf{q} - \mathbf{p})_y|$$
$$\le (s + \epsilon\sqrt{2})|(\mathbf{q} - \mathbf{p})_y|$$
$$< |(\mathbf{q} - \mathbf{p})_y|$$
$$< \hat{u}|(\mathbf{q} - \mathbf{p})_y|$$
$$\le |(\mathbf{f}(\mathbf{q}) - \mathbf{f}(\mathbf{p}))_y|.$$

Therefore, $\mathbf{f}(\mathbf{q})$ and $\mathbf{f}(\mathbf{p})$ are vertically aligned. □

In Lemmas 10.13–10.16, let B be defined as in Lemma 10.12.

Lemma 10.13 If $\mathbf{f}^n(\mathbf{p})$ is in B for all $n = 0, 1, 2, \ldots$, then $\mathbf{p}$ is in the local stable set S_B.

Proof: We need to show that $\mathbf{f}^n(\mathbf{p}) \to 0$, as $n \to \infty$. If $\mathbf{p} = 0$, we are done. Thus assume $\mathbf{p} \ne 0$. First, we argue that $\mathbf{f}^n(\mathbf{p})$ and 0 are horizontally aligned for all $n > 0$. Suppose otherwise: Assume there exists a k such that $\mathbf{f}^k(\mathbf{p})$ and 0 are vertically aligned. Since $\mathbf{p} \ne 0$, $\mathbf{f}^k(\mathbf{p})_y \ne 0$, and by Lemma 10.12 (1), $|\mathbf{f}^m(\mathbf{p})_y| > \hat{u}^{m-k}|\mathbf{f}^k(\mathbf{p})_y|$, for all $m > k$. Hence $|\mathbf{f}^m(\mathbf{p})_y| \to \infty$ as $m \to \infty$. But $\mathbf{f}^n(\mathbf{p})$ is in B, for all n, a contradiction. Thus $\mathbf{f}^n(\mathbf{p})$ and 0 are horizontally aligned, for all n.

Now, by Lemma 10.12(2), $|\mathbf{f}^n(\mathbf{p})_x| \to 0$. Since $\mathbf{f}^n(\mathbf{p})$ and 0 are horizontally aligned, $|\mathbf{f}^n(\mathbf{p})_x| \ge |\mathbf{f}^n(\mathbf{p})_y| \ge 0$, for each n. Therefore, $\mathbf{f}^n(\mathbf{p}) \to 0$. □.

Lemma 10.14 Every pair of points in S_B is horizontally aligned.

Proof: Let $\mathbf{p}$ and $\mathbf{q}$ be in S_B. Suppose instead that $\mathbf{p}$ and $\mathbf{q}$ are vertically aligned. Then, by Lemma 10.12 (1) and (3), the y-coordinates of forward iterates of $\mathbf{p}$ and $\mathbf{q}$ separate until at least one orbit leaves B, contradicting the definition of S_B. □

Lemma 10.15 For each x in $[-\delta, \delta]$, there is exactly one value $y = \phi(x)$ such that $(x, \phi(x))$ is in S_B.

Proof: Let $L_x = \{(x, y) : -\delta < y < \delta\} \subset B$ be the vertical line segment through x, as shown in Figure 10.23. Let T_x be the set of points in L_x whose orbits eventually leave B through the top of B, (whose y-coordinates become greater than δ), and let B_x be the set of points in L_x whose orbits eventually leave B through the bottom of B, (whose y-coordinates become smaller than $-\delta$). Notice that T_x and B_x are both nonempty. Then if $\mathbf{p}$ is in T_x, there is an $\epsilon > 0$ such that points in $N_\epsilon(\mathbf{p}) \cap L_x$ are also in T_x. Therefore, T_x is the union of a finite number of open intervals. So is B_x. Since T_x and B_x are disjoint, there must be a point $\mathbf{q}$ in L_x such that $\mathbf{q}$ is neither in T_x nor in B_x. (For example, we could take $\mathbf{q} = \inf T_x$.)

Now all points whose orbits leave B must leave through the top or bottom. Therefore, $\mathbf{f}^n(\mathbf{q}) \in B$, for all $n > 0$. By Lemma 10.13, $\mathbf{q}$ is in S_B. From Lemma 10.14, there is only one point on L_x whose orbit stays in B. Let $\phi(x) = q_y$. □

Lemma 10.16 The function ϕ is continuous.

Proof: Let a be a number in $[-\delta, \delta]$. Suppose $\lim_{x \to a} \phi(x) \neq \phi(a)$. Then there exists $\epsilon > 0$ and a sequence $x_n \to a$ such that $|\phi(x_n) - \phi(a)| > \epsilon$, for each n. Let x_k be an element of the sequence such that $|x_k - a| < \epsilon$. Then $(x_k, \phi(x_k))$ and $(a, \phi(a))$ are not horizontally aligned, contradicting Lemma 10.14. □

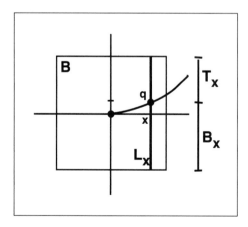

Figure 10.23 Illustration for the Proof of Lemma 10.15.
Points on the vertical line segment L_x through x are either in T_x, if the orbit of the point leaves the box B through the top, or in B_x, if it leaves through the bottom. The point $\mathbf{q}$ stays in B for all forward iterates.

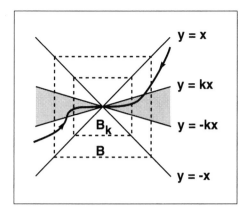

Figure 10.24   Cones containing the stable manifold.
The proof of Lemma 10.17 shows that for any $k > 0$ the local stable set enters the cone determined by the lines $y = \pm kx$.

Lemmas 10.13–10.16 show (among other properties) that S_B is contained within the cone determined by the lines $y = \pm x$. See Figure 10.24. If we decrease the magnitude of the slope of these lines, we can similarly show that there is a (perhaps smaller) square B_k such that S_{B_k} lies in the cone determined by $y = \pm kx$, $0 < k < 1$. Thus $\frac{\phi(x)}{x} \to 0$, as $x \to 0$. This is the idea behind the proof of Lemma 10.17.

Lemma 10.17 $\phi'(0) = 0$.

Proof: Given $k > 0$, choose $\delta > 0$ (the width of the square B_k) small enough so that $\epsilon < (u - 1)/2\sqrt{1 + \frac{1}{k^2}}$. (Recall that δ and ϵ are related as in Lemma 10.11.) Let $\mathbf{p} = (x, y)$ be a point in B_k but outside the $y = \pm kx$ cone. The vertical position of $\mathbf{f}(\mathbf{p})$ can be approximated by Lemma 10.11, as follows:

$$|\mathbf{f}(\mathbf{p})_y| \geq u|y| - \epsilon\sqrt{x^2 + y^2}$$

$$\geq u|y| - \epsilon\sqrt{\frac{y^2}{k^2} + y^2}$$

$$= |y|\left(u - \epsilon\sqrt{\frac{1}{k^2} + 1}\right)$$

$$> |y|\left(u - \frac{(u - 1)}{2}\right)$$

$$= |y|\left(\frac{u + 1}{2}\right) = |\mathbf{p}_y|\left(\frac{u + 1}{2}\right).$$

Therefore, points outside the $y = \pm kx$ cone have y-coordinates that increase by a factor of $\frac{u+1}{2} > 1$ per iteration, and so eventually move out of the δ-box B_k. By the definition of local stable set, S_{B_k} is in the $y = \pm kx$ cone.

The definition of derivative of $\phi(x)$ at 0 is $\lim_{x \to 0} \frac{\phi(x)}{x}$. Since this ratio is bounded between $-k$ and k for x near 0, for arbitrarily small $k > 0$, the limit exists and equals 0.

10.5 STABLE AND UNSTABLE MANIFOLDS FOR HIGHER DIMENSIONAL MAPS

The Stable Manifold Theorem which is proved in Section 10.4 is for saddles in the plane. Actually, any hyperbolic periodic point has a stable and an unstable manifold. Recall that a periodic point of period k is called hyperbolic if $\mathbf{Df}^k$ at the point has no eigenvalues with absolute value 1. The definitions of stable and unstable manifolds for these points are identical to Definition 2.18 for saddles in the plane.

Definition 10.18 Let $\mathbf{f}$ be a smooth one-to-one map on $\mathbb{R}^n$, and let $\mathbf{p}$ be a hyperbolic fixed point or hyperbolic periodic point for $\mathbf{f}$. The **stable manifold** of $\mathbf{p}$, denoted $S(\mathbf{p})$, is the set of points $\mathbf{x} \in \mathbb{R}^n$ such that $|\mathbf{f}^n(\mathbf{x}) - \mathbf{f}^n(\mathbf{p})| \to 0$ as $n \to \infty$. The **unstable manifold** of $\mathbf{p}$, denoted $\mathcal{U}(\mathbf{p})$, is the set of points $\mathbf{x}$ such that $|\mathbf{f}^{-n}(\mathbf{x}) - \mathbf{f}^{-n}(\mathbf{p})| \to 0$ as $n \to \infty$.

The other two types of hyperbolic fixed points for maps of the plane are sinks and sources. For a fixed point $\mathbf{p}$ in the plane, if both eigenvalues of $D\mathbf{f}(\mathbf{p})$ are of absolute value less than one, then $\mathbf{p}$ is a sink, and the stable manifold contains a two-dimensional ϵ-neighborhood of $\mathbf{p}$. In this case, the entire stable manifold is a "two-dimensional manifold". In general, a k-dimensional manifold in $\mathbb{R}^n$ (or a k-manifold) is the image of a smooth, one-to-one function of $\mathbb{R}^k$ into $\mathbb{R}^n$. Thus a two-manifold is a smooth surface.

The unstable manifold of a sink $\mathbf{p}$ in $\mathbb{R}^2$ is only the point $\mathbf{p}$ itself (a 0-dimensional manifold). In the case of a repeller in $\mathbb{R}^2$, in which $D\mathbf{f}(\mathbf{p})$ has two eigenvalues outside the unit circle, the stable manifold is only $\mathbf{p}$ itself, while the unstable manifold is a two-manifold (which in this case is a two-dimensional subset of the plane).

EXAMPLE 10.19

Let $f(x, y, z) = (-3x, .5y, -.2z)$. Notice that $f^k(0, y, z) = (0, (.5)^k y, (-.2)^k z)$. Therefore, $\lim_{k \to \infty} |f^k(0, y, z)| = 0$, and the stable manifold of the fixed point $(0, 0, 0)$ is seen to be the (y, z)-plane. The unstable manifold is the x-axis, since $f^{-k}(x, 0, 0) = ((-3)^{-k}(x), 0, 0)$.

There is a higher-dimensional version of Theorem 10.1. For a hyperbolic fixed point p of a higher-dimensional map, the stable manifold will have the same dimension as the subspace of $\mathbb{R}^n$ on which the derivative contracts; that is, the dimension is equal to the number of eigenvalues of $Df(p)$ that have absolute value strictly smaller than 1, counted with multiplicities. This subspace, the eigenspace corresponding to these eigenvalues, will be tangent to the stable manifold at p. Similarly, the unstable manifold has dimension equal to the number of eigenvalues of $Df(p)$ with absolute value strictly larger than 1, counted with multiplicities; the linear subspace on which $Df(p)$ expands is tangent to the unstable manifold at p.

EXAMPLE 10.20

Let $f(x, y, z) = (.5x, .5y, 2z - x^2 - y^2)$. The origin is the only fixed point of f, and its eigenvalues are 0.5, 0.5, and 2. The linear map $Df(0)$ is contracting on the (x, y)-plane and expanding on the z-axis. The (x, y)-plane, however, is not the stable manifold for 0 under the nonlinear map f, as Exercise 10.6 shows.

✎ EXERCISE T10.6

Let $f(x, y, z) = (.5x, .5y, 2z - x^2 - y^2)$.

    (a) Show that $\mathcal{U}(0)$ is the z-axis.

    (b) Show that $S(0)$ is the paraboloid $\{(x, y, z) : z = \frac{4}{7}(x^2 + y^2)\}$.

☞ CHALLENGE 10

The Lakes of Wada

CONSIDER THE TWO open half-planes $L = \{(x, y) : x < 0\}$ and $R = \{(x, y)$: $x > 0\}$. Each of the two regions has a set of boundary points consisting of the y-axis. Therefore, any boundary point of either of the two regions is a boundary point of **both** regions. Can the same thing be done with three regions? It means getting each boundary point of any one of the three regions to be a boundary point of both of the others as well. A little thought should persuade you that such sets will be a little out of the ordinary.

In 1910 the Dutch mathematician L.E.J. Brouwer gave a geometric construction of three regions such that every boundary point is a boundary point of all three regions. Independently, the Japanese mathematician Yoneyama (1917) gave a similar example. Yoneyama attributed the example to "Mr. Wada". This example is described in (Hocking and Young, 1961); they called the example the "Lakes of Wada".

The first three steps in the construction of this fractal are shown in Figure 10.25. Start with an island of diameter 1 mile. The first lake is the exterior, white region, and the other two are shaded gray and black. Now the excavation begins. From the exterior region, dig a canal so that every point of land (the island) lies

Figure 10.25 Lakes of Wada.

no more than 1/2 mile from the white canal. Second, dig a gray canal so that no land lies more than 1/3 mile from it. Third, dig a black canal so that no land lies more than 1/4 mile from black water. Next, go back and extend the first (white) canal so that no land lies more than 1/5 mile from it (we have not shown this step). Continue in this way. Of course, the three lakes must never intersect. Be sure to pay the bulldozer operator by the amount of dirt moved and not by the mileage traveled by the bulldozer. In the limit of this process, the area of the remaining land is 0. Each shore point of any of the lakes is also a shore point for both of the other lakes.

As originally presented, the Lakes of Wada have nothing to do with dynamical systems. Although we have seen basins of attraction with extremely complicated, even fractal, boundaries, it is hard to imagine that such a configuration of three basins could exist for any but the most contrived dynamical systems. In this challenge, we present special trapping regions, called basin cells, whose boundaries are formed by pieces of stable and unstable manifolds, to show that "Wada basins" appear to exist in simple dynamical processes. We begin with a few definitions of topological terms.

Let A be a subset of $\mathbb{R}^n$. (Later concepts will pertain only to subsets of the plane.) The **boundary** of A, denoted **bd**(A), is the set of points that have both points in A and points not in A arbitrarily close to them. Specifically, $\mathbf{x}$ is in **bd**(A) if and only if for every $\epsilon > 0$, the neighborhood $N_\epsilon(\mathbf{x})$ contains points in A and points not in A. Boundary points may be in A or not. For example, the sets $D_c = \{\mathbf{x} : |\mathbf{x}|^2 \leq 1\}$ (the closed unit disk) and $D_o = \{\mathbf{x} : |\mathbf{x}|^2 < 1\}$ (the open unit disk) both have the same boundary; the unit circle $C = \{\mathbf{x} : |\mathbf{x}|^2 = 1\}$. The **interior** of A, denoted **int**(A), is the set of points in A that have neighborhoods completely contained in A. Specifically, $\mathbf{x}$ is in **int**(A) if and only if there is an $\epsilon > 0$ such that N_ϵ is a subset of A. Notice that **int**(A) and **bd**(A) are disjoint sets.

✎ **EXERCISE T10.7**

Show: $\mathsf{int}(D_c) = D_o$.

A set A is called **open** if it is equal to its interior. An open set does not contain any boundary points. Assuming that A is an open set, we say a point $\mathbf{y}$ in **bd**(A) is **accessible from** A if there is a curve J such that $\mathbf{y}$ is an endpoint of J, and all of J except $\mathbf{y}$ is in A. For our example of the disk, all points in **bd**(D_o) are accessible from D_o. A somewhat more interesting example is shown in Figure 10.26. There the boundary of the open set U is shown to wind around two limit

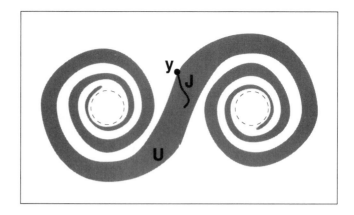

Figure 10.26 The open set U winds around the circles infinitely many times.
The point shown is accessible from U since it is the end of a curve contained in U.
Every point on the boundary of U is accessible except for those points on the limit
circles (the dashed circles).

circles. (The figure is intended to represent an infinite number of windings of the
boundary around these limit circles.) All points on the boundary, except the limit
circles, are accessible from U. The circles, while in $\mathbf{bd}(U)$, contain no points
accessible from U.

Now we introduce dynamics to the situation. Let $\mathbf{F}$ be a one-to-one smooth
map of the plane. Recall that a trapping region is a bounded set Q, with $\mathbf{int}(Q)$
not the empty set, with the property that $\mathbf{F}(Q)$ is a subset of Q and $\mathbf{F}(Q)$ is not
equal to Q. The **basin** of a trapping region Q, denoted $\mathbf{bas}(Q)$, is the set of all
points whose trajectories enter $\mathbf{int}(Q)$. Those trajectories then remain in $\mathbf{int}(Q)$.

Step 1   Show that if Q is a trapping region, a trajectory that enters Q must
stay in Q for all future iterates. [Hint: If a set A is a subset of B, the $\mathbf{F}(A)$ is a
subset of $\mathbf{F}(B)$.]

Step 2   Prove two topological properties of basins:
(1) Show that $\mathbf{bas}(Q)$ is an open set. [Hint: Given $\mathbf{x}$ in $\mathbf{bas}(Q)$, let j be a
positive integer such that $\mathbf{F}^j(\mathbf{x})$ is in $\mathbf{int}(Q)$. Use the continuity of $\mathbf{F}^j$ to obtain
the result.]
(2) A set A is called **path connected** if, given any two points $\mathbf{x}$ and $\mathbf{y}$ in A,
there is a path contained in A with endpoints $\mathbf{x}$ and $\mathbf{y}$. (Recall that a **path** is
a continuous function $s : [0, 1] \to \mathbb{R}^2$.) Show that if Q is path connected, then
$\mathbf{bas}(Q)$ is path connected.

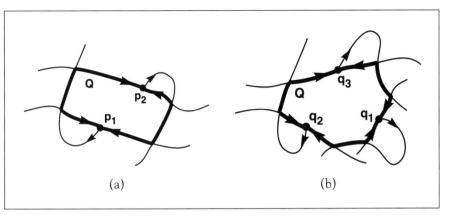

Figure 10.27 Basin cells.
In (a) a period-two orbit $\{p_1, p_2\}$ "generates" the cell, while in (b) a period-three orbit $\{q_1, q_2, q_3\}$ generates the cell. Notice that basin cells cannot exist if there are no homoclinic points.

The definition of basin cell is illustrated in Figure 10.27, where we see a four-sided cell with a period-two point on the boundary and a six-sided cell with a period-three point on its boundary. A cell is a region Q (homeomorphic to a disk) whose boundary is piecewise smooth and consists of a finite number of pieces of stable and unstable manifold of some orbit. Specifically, for a periodic orbit **p** of period k, a **cell** is a planar region bounded by $2k$ sides: included among these sides are pieces of both branches of the k stable manifolds as they emanate from the k points in the orbit together with k pieces of unstable manifolds, one from each of the k points in **p**. We denote by Q_s the union of the points in **p** together with points on the pieces of stable manifolds in the boundary of Q, and by Q_u, the union of points in pieces of unstable manifolds in the boundary that are not already points of Q_s. Then Q_s and Q_u are disjoint sets whose union is $\mathbf{bd}(Q)$.

In order to be able to verify whether a cell is a trapping region, the following is useful and almost obvious, but is not easy to show:

If $\mathbf{F}(\mathbf{bd}(Q))$ is in Q, then $\mathbf{F}(Q)$ is in Q.

Step 3 Show that a cell Q is a trapping region if and only if $\mathbf{F}(Q_u)$ is a subset of Q. Sketch an example of a cell that is not a trapping region.

A cell that is a trapping region is called a **basin cell**. From now on, we assume that Q is a basin cell and investigate **bas**(Q) and its boundary. The concept of basin cell was introduced and developed in (Nusse and Yorke, 1996).

Proposition 1 describes the fundamental structure of basin boundaries. When there are accessible periodic orbits, then all the accessible boundary points are on stable manifolds. Although we prove the theorem only for basins with basin cells, the result holds for a large class of planar diffeomorphisms. See, for example, (Alligood and Yorke, 1992).

Proposition 1. Let Q be a basin cell and let $\mathbf{x}$ be in $\mathbf{bd}(\mathbf{bas}(Q))$. If $\mathbf{x}$ is accessible from $\mathbf{bas}(Q)$, then $\mathbf{x}$ is on the stable manifold of $\mathbf{p}$.

The proof of Proposition 1 is given in Steps 4, 5, and 6.

Step 4 Let $\mathbf{x}$ be in $\mathbf{bd}(\mathbf{bas}(Q))$ and denote $\mathbf{F}^n(\mathbf{x})$ by $\mathbf{x_n}$. Show that $\mathbf{x}$ is accessible from $\mathbf{bas}(Q)$ if and only if each $\mathbf{x_n}, n = 1, 2, 3, \ldots$, is accessible from $\mathbf{bas}(Q)$. (The fact that $\mathbf{bd}(\mathbf{bas}(Q))$ is invariant under $\mathbf{F}$ should be included in the proof.)

We can assume $\mathbf{x}$ is not a point of the orbit $\mathbf{p}$, since such points are trivially in the stable manifold of $\mathbf{p}$. Let U be the union of the segments of unstable manifold that are on $\mathbf{bd}(Q)$ (the set Q_u) together with the pieces of unstable manifold that connect these segments to the orbit $\mathbf{p}$. (See Figure 10.28.) We assume further that $\mathbf{x}$ is not in U. (If it is, then for some positive integer n, the point $\mathbf{x_n}$ is not in U, and we can prove the result for $\mathbf{x_n}$.)

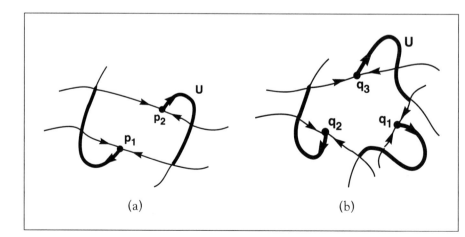

(a)   (b)

Figure 10.28   The sets U formed by pieces of unstable manifolds.
Segments of the unstable manifolds in the boundary of a cell, together with the pieces of the manifolds needed to connect them to the periodic points make up the set U.

Step 5 Since **x** is accessible from **bas**(Q), there is a curve J that is contained in **bas**(Q) except for its endpoint which is **x**. Show that J can be chosen so that it does not intersect U. Then prove that $\mathbf{F}^n(J)$ does not intersect U, for $n = 0, 1, 2, \ldots$.

Step 6 Let **y** denote a point in **bas**(Q) and in J. We can choose n so that $\mathbf{F}^n(\mathbf{y})$ is in the interior of Q. (Why can we do this?) Show then that the entire curve $\mathbf{F}^n(J)$ is in Q for this n. Conclude that $\mathbf{F}^n(\mathbf{x})$ is on the stable manifold of **p**.

Proposition 2. The accessible points are dense in the boundary of **bas**(Q).

Step 7 Prove Proposition 2. [Hint: On the straight line between a basin point and a boundary point, there is an accessible (boundary) point.]

Proposition 3. If the unstable manifold of the orbit **p** that generates the cell Q enters another basin B_2, then every point in the boundary of **bas**(Q) is also in the boundary of B_2.

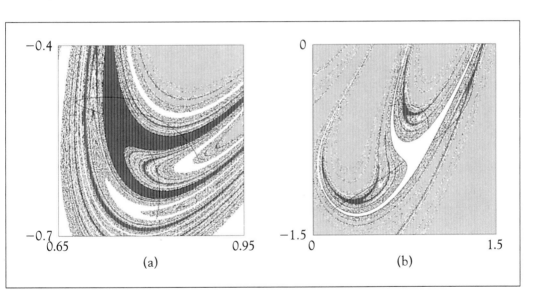

-0.4

0

-0.7

0.65 0.95

-1.5

0 1.5

(a)　　　　　　　　　　(b)

Figure 10.29　Basin cells for Hénon maps.
Parameters are set at $a = 0.71$ and $b = 0.9$. In (a) initial segments of unstable manifolds emanating from a period-three accessible orbit are shown. The orbit is accessible from the dark gray basin. The unstable manifolds intersect each of two other basins, shown in light gray and white. Therefore, the gray basin is a Wada basin. In (b) a similar configuration is shown for the basin indicated in white. The boundary of the white basin is also the boundary of the dark gray and the light gray basins.

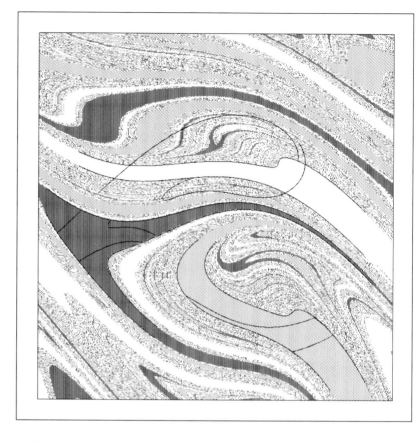

Figure 10.30 Three basin cells for the time-2π map of the forced, damped pendulum.

The white basin has one accessible periodic orbit of period two; the basin cell shown has four sides. Similarly, the light gray basin has a four-sided basin cell. The dark gray basin has one accessible periodic orbit of period three, producing a six-sided cell.

Step 8   Prove Proposition 3. [Hint: First show, by the Lambda Lemma, that the stable manifold of **p** is in the boundary of B_2; then use Proposition 2 to show that all of the boundary of **bas**(Q) is in the boundary of B_2.]

Finally, we say that a basin B is a **Wada basin** if there are two other basins B_2 and B_3 such that every point in the boundary of B is also in the boundary of B_2 and B_3. We assume here that B, B_2, and B_3 are pairwise disjoint. Wada basins for Hénon maps are shown in Figure 10.29. The natural occurrence of the Lakes of Wada phenomena in dynamical systems is described in (Kennedy and Yorke, 1991).

Corollary 4. If the unstable manifold of **p** in Proposition 3 enters two other basins B_2 and B_3 (where **bas**(Q), B_2, and B_3 are pairwise disjoint), then every boundary point of **bas**(Q) is also in the boundary of B_2 and B_3, and thus **bas**(Q) is a Wada basin.

Remark. In the basin picture for the pendulum in Chapter 2, the three basins shown are disjoint and each has a basin cell. Figure 10.30 depicts basin cells for each of the three basins. The orbits that generate each of these cells have unstable manifolds that enter all three basins. Hence, the boundary of each basin equals the boundary of all three basins.

EXERCISES

10.1. Prove that the stable manifold and unstable manifold is invariant under a diffeomorphism **f**.

10.2. Let **f** be a one-to-one linear map of $\mathbb{R}^2$. Describe the possible stable and unstable manifolds of **0**.

10.3. Figure 10.12 shows part of the stable and unstable manifolds of a saddle **p** of a planar diffeomorphism.

(a) Show that if **x** is a transverse homoclinic point, then **f**(**x**) and **f**$^{-1}$(**x**) are transverse homoclinic points.

(b) Extend the stable and unstable manifolds shown in the figure to include three forward and three backward iterates of each homoclinic point shown.

10.4. Let $f(x, y) = (x/2, 2y - 7x^2)$. Verify that $\{(x, 4x^2) : x \in \mathbb{R}\}$ is on the stable manifold of the fixed point $(0, 0)$. Why can you then conclude that $\{(x, 4x^2) : x \in \mathbb{R}\}$ is the stable manifold?

10.5. Let $f(x, y) = (x^2 - 5x + y, x^2)$. What information does the Stable Manifold Theorem give about the fixed points of **f**? In particular, plot the fixed points and indicate in the sketch the action of the map in a neighborhood of each fixed point.

10.6. Assume that **f** is an invertible map of the plane. Use the Stable Manifold Theorem to show that a stable manifold of a saddle has no endpoints and that it does not cross itself. (Of course, the same properties hold for an unstable manifold, as well.)

10.7. For the map $f(x, y) = (x + y(\text{mod}1), x + 2y(\text{mod}1))$, describe the unstable manifold of $(0, 0)$. Show that the stable and unstable manifolds meet at right angles.

10.8. State a Stable Manifold Theorem in the case that **f** is a diffeomorphism of $\mathbb{R}^n$.

10.9. Let **f** be a horseshoe map defined on the unit square which is linear for (x, y) in $[0, 1] \times ([0, 1/3] \cup [2/3, 1])$. To be specific, define

$$f(x, y) = \begin{cases} (x/3, 3y) & \text{if } 0 \leq y \leq 1/3 \\ (1 - x/3, 3 - 3y) & \text{if } 2/3 \leq x \leq 1. \end{cases}$$

For (x, y) in $[0, 1] \times (1/3, 2/3)$, define **f** in such a way that **f** is one-to-one and smooth. (The picture to keep in mind is the standard horseshoe image.) Notice that $(0, 0)$ is a fixed-point saddle of **f**. Calculate explicitly the stable manifold and unstable manifold of $(0, 0)$ (at least the parts that lie within the unit square—there will be many parts).

☞ LAB VISIT 10

The Leaky Faucet: Minor Irritation or Crisis?

ONE OF THE EARLIEST physical experiments to exhibit interesting dynamical behavior was the leaky faucet. Here is an experiment you *can* try at home. Some faucets work better than others; it helps to use a metal sink or in some other way exaggerate the noise of the falling drip. You should be able to just barely open the faucet and establish a pattern of drops evenly spaced in time. Now increase the flow rate slightly. The drops will speed up but remain evenly-spaced in time. If you can increase the flow rate finely enough, you should eventually reach a stage where the pattern changes qualitatively: the drops come in pairs, a short followed

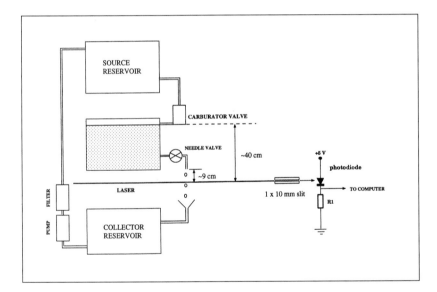

Figure 10.31 Diagram of the leaky faucet apparatus.
A carburetor valve is used to keep the main reservoir filled to a constant level, which holds the pressure at the needle valve constant. Drops are recorded when they break the laser beam which falls on the photodiode.

Martien, P., Pope, S. C., Scott, P. L., Shaw, R. S. 1985. The chaotic behavior of the leaky faucet. Physics Letters A **110**: 399–404.

Sartorelli, J. C., Goncalves, W. M., Pinto, R. D. 1994. Crisis and intermittence in a leaky-faucet experiment. Physical Review E **49**: 3963–3975.

by a long. Each short-long pair will take about as much time as two evenly-spaced drips before the transition. The change you have witnessed is a period-doubling bifurcation.

With laboratory equipment, a more quantitative study can be made, which uncovers a great variety of dynamical phenomena. Two groups, one from Santa Cruz, California, and one from Sao Paulo, Brazil, have done controlled experiments of the leaky faucet. The setups are similar. A reservoir of distilled water is kept at constant pressure at a needle valve, which can be controlled with great precision. Drips from the valve are detected by a laser beam trained on the drip stream and recorded by computer. Figure 10.31 shows the Sao Paulo apparatus; the Santa Cruz version is similar, although they used a carburetor valve from a Model A Ford automobile.

The period-doubling bifurcation we described above can be seen graphically in Figure 10.32, taken from the Santa Cruz experiment. The measured quantity in the experiments is the time interval T_n between drips. Each part of the figure

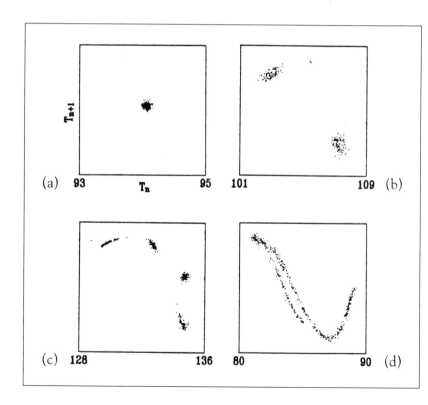

Figure 10.32 Scatter plots of successive interdrip intervals.
Pairs of form (T_n, T_{n+1}) are plotted. (a) Period-one. (b) Period-two. (c) Period-four. (d) Chaos.

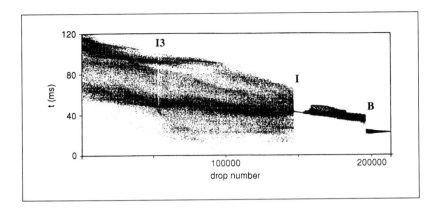

Figure 10.33   A bifurcation diagram for the leaky faucet.
Each vertical slice is 1024 dots corresponding to interdrip intervals. A period-three
window (I3), an interior crisis (I), and a boundary crisis (B) are identified.

shows a scatter plot of ordered pairs (T_n, T_{n+1}). A simple rhythmic drip pattern
corresponds to equal interdrip intervals $T_n = T_{n+1} = T_{n+2} = \ldots$, so that the
scatter plot of (T_n, T_{n+1}) is reduced to a single point. Figure 10.32(a) shows this
simple dynamics. Each interdrip interval is approximately .094 seconds.

The result of gradually increasing flow rate from the needle valve is the beginning of a period-doubling cascade of bifurcations. Figure 10.32(b) and (c) show
period two and four behavior, respectively. Figure 10.32(d) exhibits a more complicated relationship between successive intervals which presumably corresponds
to chaotic dynamics. This plot cannot exactly be compared to a one-dimensional
map $T_n \rightarrow T_{n+1}$, since there appear to be two distinct bands of points (T_n, T_{n+1})
over a range of T_n values.

Figure 10.33 shows a bifurcation diagram made by the Sao Paulo group. The
bifurcation parameter, the flow rate at the needle valve, was set at 208 different
values to make this plot. At each parameter value, a total of 1024 drips were
recorded, and the interdrip time intervals plotted. The resulting diagram shows a
period-three window, denoted I3, and two other interesting bifurcations denoted
by I and **B** in the figure.

Event I identifies an interior crisis. As this parameter value is passed, the
dynamical behavior changes from erratic to a period-one rhythm. The average
interdrip time interval does not change appreciably with event I, but is more or
less constant across the break. A scatter plot of (T_n, T_{n+1}) changes significantly
as the transition is crossed, as shown in Figure 10.34.

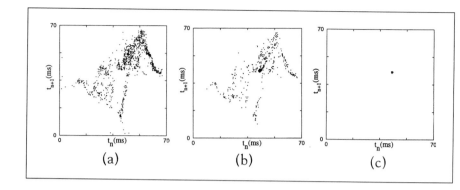

Figure 10.34 Scatter plots of successive time-interval pairs near the interior crisis.

(a) Complicated dynamics before the crisis parameter value I is followed by simpler dynamics in (b) and (c) as the flow rate is slowly increased.

Event **B** identifies a boundary crisis. Now the interdrip time interval changes abruptly, as a chaotic regime shifts into a periodic cycle of period five. The five states are close together; the points on the bifurcation diagram to the right of value **B** lack sufficient clarity to show the five separate states. Figure 10.35 shows the transition to a periodic orbit more clearly. Note the abrupt change of average interdrip time interval from 38.6 msec to 26.0 msec. This destruction of a basin of one attractor, resulting in movement to an alternative attractor perhaps far away in state space, is characteristic of a boundary crisis.

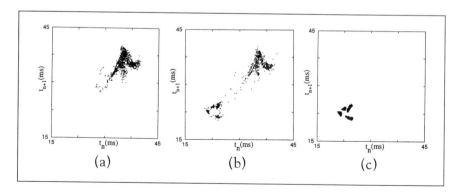

Figure 10.35 Scatter plots near a boundary crisis.

(a) Complicated dynamics before the transition yields to periodic behavior as the basin boundary of the original attractor is destroyed in (b) and a periodic attractor results in (c).

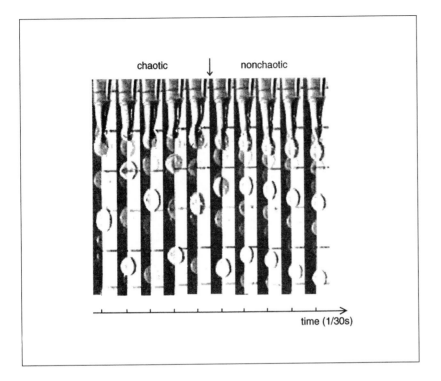

Figure 10.36 Transition from chaotic to periodic pattern.
A motion picture recorded at 30 frames per second shows the change from chaotic
(first five frames) to period-five (last five frames) behavior.

CHAPTER ELEVEN

Bifurcations

Of ALL POSSIBLE motions in dynamical systems, a fixed point or equilibrium is the simplest. If the fixed point is stable, the system is likely to persist in that state, even in the face of minor disturbances.

But systems change. Figure 1.6 of Chapter 1 shows that varying the parameter in the logistic family from $a = 2.9$ to $a = 3.1$ results in the preferred system behavior changing from a fixed population to an oscillation between high and low populations. This change is called a bifurcation. We saw this type of bifurcation again in the experiment of Lab Visit 1. Further system parameter changes are shown to result in even more extreme changes in behavior, including higher periodicities, quasiperiodicity and chaos. The forced, damped pendulum settles into periodic motions when the amplitude of the forcing function is small. When the amplitude is increased to moderate levels, chaotic motions develop. At still higher amplitudes, periodicity returns.

Just as it is helpful to classify the different types of motion in a system, it is helpful to categorize the ways that the motion can change as the system is modified. In fact, as a system parameter varies, these changes cannot occur capriciously, but only in a limited number of prescribed ways. This limited set of bifurcations is universal in the sense that it is the same for a large variety of systems. For this reason, bifurcation theory is a useful and widely studied subfield of dynamical systems. With the discovery of chaotic dynamics, the theory has become even more important, as researchers try to find mechanisms by which systems change from simple to highly complicated behavior. In Chapter 11 we will describe the most common bifurcations.

11.1 SADDLE-NODE AND PERIOD-DOUBLING BIFURCATIONS

In Chapter 1 we introduced the one-parameter family of logistic maps $g_a(x) = ax(1 - x)$ and investigated its dynamical properties at various fixed values of the parameter a. We found that at certain parameter values, including $0 < a < 2$, g_a has a fixed-point attractor; for larger a values, g_a can have periodic or chaotic attractors; finally, for $a > 4$, the logistic map g_a has infinitely many periodic points, but no attracting sets, and almost all orbits diverge to $-\infty$. If fixed points or periodic points exist at a certain parameter value and not at another, what has happened to the system at parameter values in between to cause the birth or death of these orbits? We call a parameter value at which the number or stability of fixed points or periodic points changes a **bifurcation value** of the parameter, and the orbit itself a **bifurcation orbit**.

Definition 11.1 A **one-parameter family** of maps on $\mathbb{R}^n$ is a set of maps $f_a(v)$, one for each value of a parameter a belonging to an interval I of real numbers. We refer to $\mathbb{R}^n$ as the **state space** and to I as the **parameter space**, and say f depends on a **scalar parameter** $a \in I$.

An alternate notation for $f_a(v)$ is $f(a, v)$, which we use when we want to emphasize the dependence of the family on the parameter. However, there is a difference between the parameter a and the state variable v. In order to calculate an orbit of a map f_a in such a family, the parameter a is fixed and successive iterates are calculated in state space for that fixed value of a. We concentrate in particular on maps with state space $\mathbb{R}^1$ and $\mathbb{R}^2$. For families of one-dimensional maps, we write $f_a{}'$ for the derivative of f_a with respect to x. For families of two-dimensional maps, we write $\mathbf{Df}_a$ for the Jacobian matrix of partial derivatives of $\mathbf{f}_a$ taken with respect to the state space variables (usually x and y).

Two types of bifurcations are most basic. In the first, called a saddle-node bifurcation, fixed points are born. The second is called a period-doubling bifurcation, where a fixed point loses its stability and simultaneously, a new orbit of doubled period is created. Both of these basic bifurcations occur in the one-dimensional quadratic family.

EXAMPLE 11.2

Let $f_a(x) = f(a, x) = a - x^2$, where a is a scalar (one-dimensional) parameter. When $a < -1/4$, there are no fixed points. At $a = -1/4$, the graph of f_a is tangent to the line $y = x$. There is exactly one fixed point at $x = -1/2$. For each a strictly larger than $-1/4$, the graph of f_a crosses the line $y = x$ in two points, giving rise to two fixed points of f_a. Figure 11.1 illustrates the three possibilities. The point $(a, x) = (-1/4, -1/2)$ is a bifurcation point for the map f since the number of fixed points, as a function of a, changes at $a = -1/4$. When $a = 1/2$, the map f has a repelling fixed point at $x = (-1 - \sqrt{3})/2$ and an attracting fixed

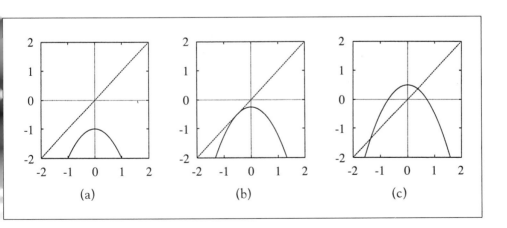

Figure 11.1 A saddle-node bifurcation.
The graph of the quadratic family $f(a, x) = a - x^2$ before, at, and following a saddle-node bifurcation is shown. (a) At $a = -1$, the graph does not intersect the dotted line $y = x$. (b) At $a = -0.25$, the graph and the line $y = x$ intersect in one point, the point of tangency; for $a > -0.25$, they intersect in two points. (c) At $a = 1/2$, f has a repelling fixed point and an attracting fixed point.

point at $x = (-1 + \sqrt{3})/2$. The case in which a pair of fixed points appear in a region where there were none, as a parameter is varied, is called a **saddle-node** bifurcation.

At a saddle-node bifurcation, two fixed points of f_a, one stable and one unstable, are born as the parameter a is increased. Of course, this is a matter of perspective. Alternatively, we could say that the two fixed points disappear as a is decreased. The term "saddle-node" derives from this type of bifurcation in a two-dimensional state space, where the unstable orbit is a saddle and the stable orbit is an attractor or "node". For maps in which the state space is one-dimensional, this bifurcation is also sometimes called a **tangent bifurcation**. This name comes from the fact that the derivative f_a' at a fixed point saddle-node bifurcation orbit must be $+1$, and thus the graph must be tangent to the line $y = x$, as we discuss later.

Figure 11.1 shows plots of the state space graphed against itself before, during, and after the bifurcation value. A different type of plot compresses the state space information so that variation as a function of the parameter a can be viewed. This type of graph, exemplified by Figure 11.2, is called a **bifurcation diagram**. The vertical axis corresponds to the state space. Only significant points in the state space are plotted.

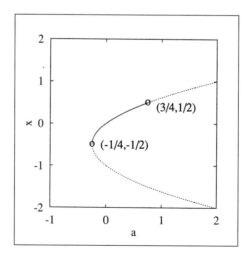

Figure 11.2 Bifurcation diagram of a saddle-node bifurcation.
Fixed points of the quadratic family $f(a, x) = a - x^2$ are shown. Attracting fixed points (sinks) are denoted by the solid curve, and repelling fixed points (sources) are on the dashed curves. The circle at $a = -1/4$ denotes the location of a saddle-node bifurcation. The other circle denotes where a period-doubling bifurcation occurs.

Figure 11.2 depicts the fixed points of f_a as a function of a near the saddle node. Attracting fixed points are shown on the solid curve and repelling fixed points are on the dashed curves. For example, when $a = 0$, the function f has two fixed points, one an attractor and one a repeller.

There are more bifurcations hiding in Figure 11.2. Notice that the point $(\frac{3}{4}, \frac{1}{2})$ represents a passage of the curve of fixed points from solid to dashed indicating a change in stability. How do the fixed points lose their stability? Recall that if the derivative $f_a{}'$ evaluated at a fixed point satisfies $|f_a{}'| < 1$, then the fixed point is an attractor, and if $|f_a{}'| > 1$, then it is a repeller. Clearly, the way fixed points change stability is to have $f_a{}'(x)$ move across the border ± 1 as the parameter a varies. As a increases from $-1/4$ to $3/4$, the derivative of f_a at the attracting fixed point decreases from $+1$ to -1. For $a > 3/4$, the derivative of f_a at this fixed point has crossed the -1 border, and the fixed point becomes a repeller. In the parameter range $[\frac{3}{4}, \infty)$, the map f_a has no fixed-point attractors.

However, f_a does have a period-two attractor for a in $(\frac{3}{4}, \frac{5}{4})$. At $a = 3/4$, the two points in this orbit split off from the fixed point $x = \frac{1}{2}$. For a slightly larger than $\frac{1}{2}$, one point in the orbit is greater than $\frac{1}{2}$ and one is less than $\frac{1}{2}$. This bifurcation is called a **period-doubling** bifurcation (or, sometimes, a **flip** bifurcation). In Figure 11.3 the graph of f_a is shown for a values before, at, and following the period-doubling bifurcation.

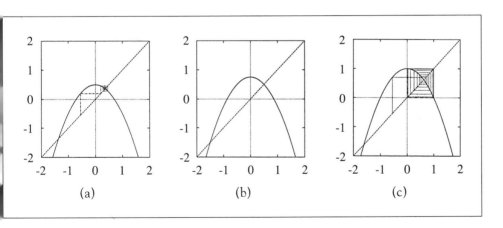

Figure 11.3 The quadratic map at a period-doubling bifurcation.
Graphs of the quadratic family $f_a(x) = a - x^2$ before, during and after a period-doubling bifurcation. (a) $a = 0.5$. Orbits flip back and forth around the sink as they approach. (b) $a = 0.75$. At the bifurcation point, $x = 1/2$ still attracts orbits, although the derivative of f_a is -1. (c) $a = 1$. Beyond the bifurcation, orbits are attracted by the period-two sink.

Since the derivative at the (positive) fixed point is negative for parameter values near $\frac{3}{4}$, iterates of points near the fixed point flip back and forth around the fixed point. We can observe this flipping in the cobweb plot (Figures 11.3(a) and (c)). The fixed point, which is unstable after the bifurcation, is called a **flip repeller** since its derivative is smaller than -1. A fixed point with derivative greater than $+1$ is called a **regular repeller**. (Compare these definitions with those of "flip saddle" and "regular saddle" in Chapter 2.) It is difficult to see precisely where the period-two points are located after the bifurcation without a careful cobweb plot. However, if we graph the second iterate f_a^2, as in Figure 11.4, the period-two orbit of f_a appears as two fixed points. We can also see from slopes on the graph of f_a^2 that the bifurcating fixed points are attractors.

✎ **EXERCISE T11.1**

Notice in Figure 11.4(b) that the period-doubling bifurcation point is an inflection point for f_a^2. Does $(f_a^2)''$ have to equal 0 at a period-doubling bifurcation point?

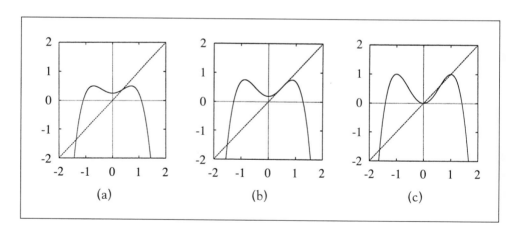

Figure 11.4 The second iterate of the quadratic map at a period-doubling bifurcation.

(a) The second iterate f_a^2 of the quadratic map $f_a(x) = a - x^2$ is graphed at $a = 0.5$, before the bifurcation. The fixed point above zero is an attractor. (b) The function f_a^2 is graphed at $a = .75$, the bifurcation parameter. The line $y = x$ is tangent to the graph at the positive fixed point, which is an attractor (although not a hyperbolic one). (c) At $a = 1$, following the bifurcation, there are three nonnegative fixed points of f_a^2: a flip repeller fixed point of f_a surrounded by a period-two attractor of f_a.

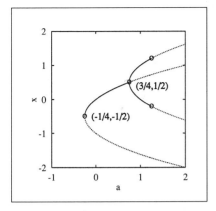

Figure 11.5 Diagram of a period-doubling bifurcation.
Fixed points and period-two orbits of the quadratic family $f(a, x) = a - x^2$ are shown. Attractors are on solid curves, while repellers are on dashed curves. The circles denote the location of the bifurcation orbits.

Figure 11.5 is a bifurcation diagram showing the fixed points and period-two points of f as a function of a and x near the period-doubling bifurcation point. Again, attracting orbits are shown with solid curves and repellers are indicated with dashed curves. At $a = 5/4$, the derivative of f_a^2 moves past the border at -1, causing another period-doubling bifurcation. For values of a greater than $5/4$ and less than the next period-doubling value, orbits are attracted to a period 4 sink.

11.2 BIFURCATION DIAGRAMS

Periodic orbits of periods greater than one can come into (or go out of) existence through saddle-node bifurcations, and they can undergo period-doubling bifurcations. Nothing new is involved—it is only a matter of applying the ideas we have already seen to a higher iterate f_a^k. For example, Figure 11.6 shows the graph of f_a^3 for three values of a before, at, and following a saddle-node bifurcation at $a = 1.75$. Note the simultaneous development of three new fixed points of f_a^3 at the bifurcation value. The map f_a is a degree-two polynomial and so can have at most 2 fixed points; they are already shown in Figure 11.6(a). Since the three new points are not fixed points, they must represent a new period-three orbit of f_a.

Figure 11.7 shows the bifurcation diagram of period-three points of f near $a = 1.75$. In the bifurcation diagram, two branches of fixed points of f_a^3 are shown to emanate from each of the three points in the bifurcation orbit. On one branch,

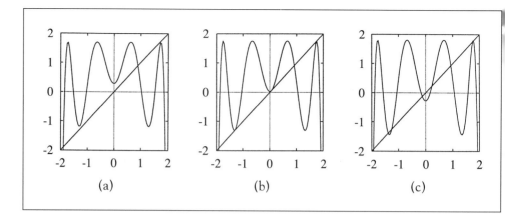

Figure 11.6 A period-three saddle-node bifurcation.
Graphs of the third iterate of the quadratic map $f_a(x) = a - x^2$ before, at, and following a period-three saddle-node bifurcation. (a) At $a = 1.7$, there are two solutions of $f_a^3(x) = x$. (b) The bifurcation point $a = 1.75$. (c) At $a = 1.8$, there are eight solutions of $f_a^3(x) = x$.

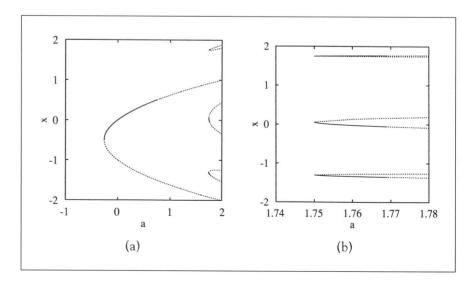

Figure 11.7 The bifurcation diagram of a period-three saddle-node bifurcation.
(a) Fixed points of f^3 are shown for the quadratic family $f(a, x) = a - x^2$; (b) a magnification of (a) near the saddle node shows only the period-three orbits.

the branch of stable periodic orbits, the derivative of f_a^3 decreases from $+1$ as a increases; on the other, the branch of unstable periodic orbits, the derivative increases from $+1$. In Figure 11.7(a), all fixed points of f_a^3 are shown for a in $[-1, 2]$ (including fixed points of f_a). Figure 11.7(b) shows only period-three points in a smaller parameter range containing the saddle-node bifurcation.

In Chapter 12 we investigate the phenomena of repeated period-doublings, called cascades, and explain why they are so prevalent. At a period-doubling bifurcation from a period-k orbit, two branches of period-$2k$ points emanate from a path of period-k points. When the branches split off, the period-k points change stability (going from attractor to repeller, or vice versa). This change is detected in the derivative of f^k which, when calculated along the path of fixed points, crosses -1.

For maps of dimension greater than 1 (maps whose state space is at least two-dimensional), a bifurcation that creates period-k points can only occur when an eigenvalue of the Jacobian $\mathbf{D}f^k$ passes through $+1$. Again, the letter $\mathbf{D}$ here means we are taking partial derivatives with respect to the coordinates x and y, not a. Figure 11.8(a) depicts a saddle-node bifurcation in a parametrized map of the plane. A period-doubling bifurcation is illustrated in Figure 11.8(b). As in the one-dimensional case, these two bifurcations are the ones most typically seen. The following example illustrates the saddle-node and period-doubling bifurcations for a Hénon family of two-dimensional maps.

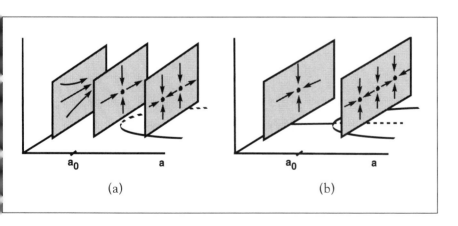

Figure 11.8 Bifurcations in a parametrized map of the plane.
(a) A fixed point appears at parameter $a = a_0$ in a saddle-node bifurcation. For $a > a_0$ there is an attracting fixed point and a saddle fixed point. The cross-sectional figures depict the action of the planar map at that parameter value. (b) An attracting fixed point loses stability at $a = a_0$ in a period-doubling bifurcation. For $a > a_0$ there is a saddle fixed point and a period-two attractor.

EXAMPLE 11.3

(Hénon map.) Let $h_a(x, y) = (a - x^2 + by, x)$, for fixed b, with $|b| < 1$ In principle, we might also consider b as a parameter, but our purpose here is to describe bifurcations of one-parameter families. See Example 11.14 for a discussion of the phenomena that can occur for fixed a and varying b. The map h has fixed points when the equations $a - x^2 + by = x$ and $y = x$ are satisfied when $a - x^2 + bx = x$. Fixed points occur at

$$\mathbf{p}_a = \left(\frac{-(1 - b) + \sqrt{(1 - b)^2 + 4a}}{2}, \frac{-(1 - b) + \sqrt{(1 - b)^2 + 4a}}{2} \right)$$

and

$$\mathbf{q}_a = \left(\frac{-(1 - b) - \sqrt{(1 - b)^2 + 4a}}{2}, \frac{-(1 - b) - \sqrt{(1 - b)^2 + 4a}}{2} \right),$$

when these expressions are defined. In particular, for $a < -\frac{1}{4}(1 - b)^2$, there are no fixed points; for $a > -\frac{1}{4}(1 - b)^2$, there are two fixed points; and at $a^+ = -\frac{1}{4}(1 - b)^2$ there is a saddle-node bifurcation. At a^+, the fixed point is $\mathbf{p}_{a^+} = \mathbf{q}_{a^+} = (-(1 - b)/2, -(1 - b)/2)$. Verify that the Jacobian $\mathbf{Dh}(\mathbf{p}_{a^+})$ has an eigenvalue of $+1$.

For a greater than (but near) a^+, the fixed point $\mathbf{p}_a$ is an attractor, and the fixed point $\mathbf{q}_a$ is a saddle. Then at $a^- = \frac{3}{4}(1 - b)^2$ there is a period-doubling bifurcation from the branch $\mathbf{p}_a$ of attractors. At a^-, the fixed point from which the bifurcation occurs is $\mathbf{p}_{a^-} = ((1 - b)/2, (1 - b)/2)$. Verify that the Jacobian $\mathbf{Dh}(\mathbf{p}_{a^-})$ has an eigenvalue of -1.

Figure 11.9(a) shows the bifurcation diagram (projected onto the x-axis) for $b = -0.3$ and a in the range $[1.9, 2]$. A magnified version of the range $[1.96, 1.97]$ is shown in Figure 11.9(b). Notice that the same type of cascade behavior observed in the one-dimensional quadratic map is hinted at in these figures.

EXAMPLE 11.4

(Logistic map.) Define $g_a(x) = ax(1 - x)$. As a increases from 0 to 4, the maximum of g (occuring at $x = 1/2$) increases from 0 to 1. As it does, the parabola stretches but still has a fixed point at $x = 0$. Since the equation

$$ax(1 - x) = x \tag{11.1}$$

has real root(s) for all a (one of them is $x = 0$), g_a has fixed points for all a.

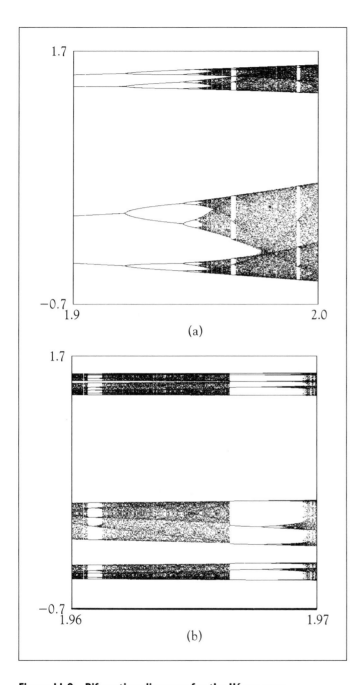

(a)

(b)

Figure 11.9 Bifurcation diagrams for the Hénon map.
The x-coordinate is plotted vertically, and the horizontal axis is the bifurcation
parameter a for the family $\mathbf{h}_a(x, y) = (a - x^2 - 0.3y, x)$. (a) A period-four orbit
at $a = 1.9$ period-doubles to chaos. At $a = 2$ the attractor is a two-piece chaotic
attractor. (b) A magnification of the period-twelve window in (a).

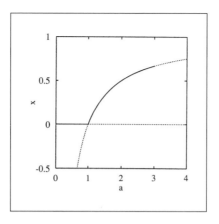

Figure 11.10 Transcritical bifurcation of fixed points in the logistic family.
The bifurcation shown here is atypical in that it is neither a saddle-node nor a period
doubling bifurcation. This type of bifurcation occurs since the graph is constrained
to pivot about the fixed point $(0, 0)$ as the parameter changes.

Equation (11.1) has exactly two roots, except at $a = 1$, when $x = 0$ is a
double root. The two paths of fixed points for $a < 1$ move toward one another
and cross at $a = 1$. At this point, the fixed point 0 loses stability and the other
fixed point becomes an attractor. Figure 11.10 shows the bifurcation diagram for
fixed points of this map. The term **transcritical** is sometimes used for this type
of bifurcation, when two paths meet and exchange stability. Elsewhere in the
bifurcation diagram for this family there are saddle-node bifurcations for periodic
orbits of periods greater than one, as well as period-doubling bifurcations. Both
can be seen in Figure 1.6 of Chapter 1.

✎ **EXERCISE T11.2**

Find a period-doubling bifurcation for the map $g_a(x) = ax(1 - x)$ of Example
11.4.

EXAMPLE 11.5

Let $f_a(x) = x^3 - ax$. Figure 11.11 shows the graph of f for values of a before
at, and following a bifurcation of fixed points at $a = -1$. The bifurcation diagram
of fixed points is shown in Figure 11.12. Notice that all maps in the family have at
least one fixed point. For each $a \leq -1$, there is one fixed point. For each $a > -1$
there are three fixed points.

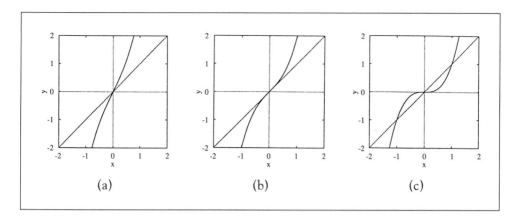

Figure 11.11   A pitchfork bifurcation.
The graphs of the cubic maps $f_a(x) = x^3 - ax$ before, during and after a pitchfork bifurcation. (a) $a = -2$ (b) $a = -1$ (c) $a = 0$.

We will see in Section 11.4 that this bifurcation, called a **pitchfork** bifurcation, is atypical, or "nongeneric", in the sense that it depends on special circumstances. In this case, the special circumstance is the symmetry the map has across the origin, since f_a is an odd function: $f_a(-x) = -f_a(x)$. This symmetry implies (1) that $x = 0$ is a fixed point (since $f_a(0) = -f_a(0)$) and (2) that if x is a fixed point, then so is $-x$.

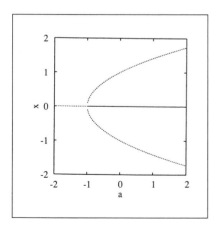

Figure 11.12   The bifurcation diagram for a pitchfork bifurcation.
Fixed points of the family of cubic maps $f_a(x) = x^3 - ax$ are shown. This type of bifurcation is not unusual for the special class of maps which are odd functions, for which $f_a(x) = -f_a(-x)$.

We investigate the subject of generic bifurcations at greater length in Section 11.4 and in Chapter 12. There we will see that the only "generic" bifurcations are saddle-nodes and period-doubling bifurcations. Both Examples 11.4 and 11.5 contain nongeneric bifurcations. We can verify that they are atypical by showing that the effect of adding a nonzero constant is to cause the map to revert to a generic bifurcation. For example, add a constant b to the function g_a to get $g_a(x) = ax(1 - x) + b$. It can be shown that for all b values but $b = 0$, the only bifurcations of fixed points are saddle-node and period-doubling bifurcations. Similarly, by adding a constant to the function f_a in Example 11.5, obtaining $f_a(x) = x^3 - ax + b$, we can say that fixed points come in through saddle nodes for all b except 0. These facts are summarized in Exercise 11.3.

✎ EXERCISE T11.3

(a) Let $g_a(x) = ax(1 - x) + b$. Show that when $b \neq 0$ is fixed and a is varied, the fixed points of g are born in saddle-node bifurcations.

(b) Let $f_a(x) = x^3 - ax + b$. Show that when $b \neq 0$, the fixed points of f are born in saddle-node bifurcations.

11.3 CONTINUABILITY

In the bifurcation diagrams shown in Section 11.2, there are many intervals of parameter values where the fixed or periodic points lie on a one-dimensional path above the parameter axis. By a "path", we mean a single-valued function of a. In this case, there is one fixed point (or periodic point) for each value of a in an interval. The paths are occasionally interrupted by bifurcation points.

For example, assume that f_a is a family of one-dimensional maps, and assume that x_a is a fixed point of f_a. In this section we will see that a path of fixed points x_a must continue as a is varied, without stopping or branching into two or more paths, as long as $f_a'(x_a)$ does not equal 1. In the next section, we will see the converse: that if $f_a'(x_a)$ crosses 1, then there must be extra fixed points in the vicinity.

Definition 11.6 Let $\mathbf{f}_a$ be a one-parameter family of maps on $\mathbb{R}^n$. We say a fixed point $\bar{\mathbf{v}}$ of $\mathbf{f}_{\bar{a}}$ is **locally continuable**, or simply **continuable**, if the fixed points of $\mathbf{f}_a$ for a near $\bar{a}$ lie on a continuous path (the graph of a continuous function). More precisely, the set of fixed points of $\mathbf{f}$ in some neighborhood

$(\bar{a} - d, \bar{a} + d) \times N_\epsilon(\bar{v})$ is the graph of a continuous function g from $(\bar{a} - d, \bar{a} + d)$ to $N_\epsilon(\bar{v})$:

$$\{(a, g(a)) : a \in (\bar{a} - d, \bar{a} + d)\}.$$

This definition is illustrated in Figure 11.13(b). A fixed point (a_c, v_c) on the path containing $(\bar{a}, \bar{v})$ is called the **continuation** of $(\bar{a}, \bar{v})$ at $a = a_c$.

We encountered examples of fixed points that are not locally continuable in Section 11.1. The saddle-node bifurcation point $(a, x) = (-1/4, -1/2)$ of the quadratic map $f(a, x) = f_a(x) = a - x^2$, shown in Figure 11.2, does not lie on a path of fixed points which extends to values of a smaller than $-1/4$, and thus does not satisfy Definition 11.6. The map f has a period-doubling bifurcation at $(\bar{a}, \bar{x}) = (3/4, 1/2)$. As we shall see from Theorem 11.7, this point is locally continuable as a fixed point of f because $f'_{3/4}(\frac{1}{2}) = -1$. But it is not locally continuable as a fixed point of f^2, since the set of fixed points of f^2 in any neighborhood of $(\bar{a}, \bar{x})$ does not form a one-dimensional path. Notice that the derivative of f_a at the saddle node $x = -1/2$ is $f'_{-1/4}(-1/2) = -2(-1/2) = +1$. The derivative of f^2 at the period-doubling bifurcation is $(f^2_{3/4})'(1/2) = f'_{3/4}(1/2)f'_{3/4}(1/2) = (-1)^2 = 1$.

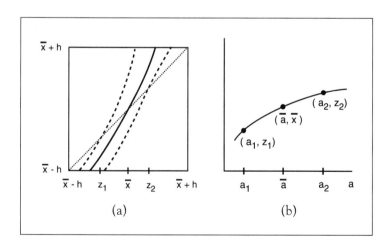

(a)

(b)

Figure 11.13   Graph of a scalar map f with a hyperbolic fixed point.
(a) An interval on which $f_a' > 1$ is shown. For small perturbations of $f_{\bar{a}}$ (small changes in a) the graph and derivatives of points near the fixed point remain close to those of the original map. The graph of $f_{\bar{a}}$ is indicated with a solid curve. The dashed curves are the graphs of maps f_{a_1} and f_{a_2}, where a_1 and a_2 are near $\bar{a}$, and $a_1 < \bar{a} < a_2$. The map f_{a_1} has one fixed point at z_1, and the map f_{a_2} has one fixed point at z_2. (b) The bifurcation diagram shows a continuous path of fixed points through $(\bar{a}, \bar{x})$.

When the derivative at a fixed point is not $+1$, the fixed point is continuable. This is the content of Theorem 11.7. Assume for example that $f_{\bar{a}}' > 1$. Then $f_{\bar{a}}$ is below the diagonal to the left of $\bar{x}$, above the diagonal to the right of $\bar{x}$, and the slope is strictly greater than one. If these three key facts are true for $f_{\bar{a}}$, and if f_a is continuous in the parameter a, then these three facts will also hold for all f_a where $\bar{a} - d < a < \bar{a} + d$, for some $d > 0$. The conclusion is that there is a single fixed point for f_a whenever a is near $\bar{a}$; these fixed points together form a continuous path in the parameter a.

We prove only part (a), the one-dimensional case, and illustrate part (b) in Example 11.8 following the proof. Theorem 11.7 is really a special case of the implicit function theorem. See, for example, (Fitzpatrick, 1996).

Theorem 11.7 (a) Let f be a smooth one-parameter family of maps defined on a subset of $\mathbb{R}$. If $\bar{x}$ is a fixed point of $f_{\bar{a}}$ and if the derivative $f_{\bar{a}}'(\bar{x})$ is not $+1$, then $(\bar{a}, \bar{x})$ is continuable.

(b) Let $\mathbf{f}$ be a smooth one-parameter family of maps defined on a subset of $\mathbb{R}^n$, $n > 1$. If $\bar{\mathbf{v}}$ is a fixed point of $\mathbf{f}_{\bar{a}}$ and if $+1$ is not an eigenvalue of $\mathbf{Df}_{\bar{a}}(\bar{\mathbf{v}})$, then $(\bar{a}, \bar{\mathbf{v}})$ is continuable.

Proof of (a): Assume that $f_{\bar{a}}'(\bar{x}) > 1$. The case $f_{\bar{a}}'(\bar{x}) < 1$ is analogous. The fixed point satisfies $f_{\bar{a}}(\bar{x}) = \bar{x}$. Because f is continuously differentiable in the two variables a and x, the fact that $f_{\bar{a}}'(\bar{x}) > 1$ implies that $f_a'(x) > 1$ for a and x in some square $[\bar{a} - h, \bar{a} + h] \times [\bar{x} - h, \bar{x} + h]$, where $h > 0$. Furthermore, because f is continuous in a and both

$$f_{\bar{a}}(\bar{x} - h) - (\bar{x} - h) < 0 \quad \text{and} \quad f_{\bar{a}}(\bar{x} + h) - (\bar{x} + h) > 0,$$

we know that

$$f_a(\bar{x} - h) - (\bar{x} - h) < 0 \quad \text{and} \quad f_a(\bar{x} + h) - (\bar{x} + h) > 0$$

for all a in a perhaps smaller interval of parameters $[a - d, a + d]$, where $d > 0$.

Now we are in the situation of Figure 11.13(a). For all a in a small interval, the function f_a has slope greater than one over a small interval of x, and straddles the diagonal. Thus for each a in the interval $[\bar{a} - d, \bar{a} + d]$ there is a unique fixed point p_a of f_a in the interval $[\bar{x} - h, \bar{x} + h]$.

Since the function f_a changes continuously with the parameter a, the fixed points p_a form a continuous path in a. To see this, let a be any point in the interval and let $a_i \to a$ be a sequence such that $\lim_{i \to \infty} p_{a_i} = p^*$. Then the continuity of f_a

implies that

$$f_a(p^*) = \lim_{i \to \infty} f_{a_i}(p_{a_i}) = \lim_{i \to \infty} p_{a_i} = p^*.$$

Therefore p^* is the unique fixed point p_a, and the path is continuous. □

In the following example we investigate the set of fixed points of the Hénon family of Example 11.3 via the Implicit Function Theorem. Not only is this theorem the mathematical basis of Theorem 11.7(b), but it also shows us that finding out whether an orbit has an eigenvalue of $+1$ reduces to checking the singularity of a 2×2 matrix.

EXAMPLE 11.8

Fix the parameter b and let $\mathbf{h} : \mathbb{R}^3 \to \mathbb{R}^2$ be given by

$$\mathbf{h}(a, x, y) = (a - x^2 + by, x). \tag{11.2}$$

Recall that the set of fixed points of $\mathbf{h}$ is the same as the set Z of zeroes of the related function $\mathbf{k}(a, x, y) = (x, y) - \mathbf{h}(a, x, y)$. As in the proof of Theorem 11.7, it turns out to be easier to deal with zeroes than with fixed points. Let $(\bar{a}, \bar{v}) = (\bar{a}, \bar{x}, \bar{y})$ be a zero of $\mathbf{k}$ (fixed point of $\mathbf{h}$). Now $+1$ is not an eigenvalue of $\mathbf{Dh}(\bar{a}, \bar{v})$ if and only if 0 is not an eigenvalue of $\mathbf{Dk}(\bar{a}, \bar{v})$, if and only if $\mathbf{Dk}(\bar{a}, \bar{v})$ is nonsingular. Then the Implicit Function Theorem says that there is a three-dimensional neighborhood V of $(\bar{a}, \bar{v})$ such that $Z \cap V$ is a (one-dimensional) path in a. In other words, $(\bar{a}, \bar{v})$ is a continuable fixed point of $\mathbf{k}$. This is just a restatement of Theorem 11.7(b).

In order to investigate Jacobian matrices containing various combinations of partial derivatives in this example, we write $\mathbf{D}_{(a,x,y)}$ for the 2×3 matrix of partial derivatives with respect to a, x, and y. The notations $\mathbf{D}_{(x,y)}$, $\mathbf{D}_{(a,x)}$, and $\mathbf{D}_{(a,y)}$ represent the Jacobian matrices of partial derivatives with respect to the two variables indicated. For $(\bar{a}, \bar{x}, \bar{y})$ in Z,

$$\mathbf{D}_{(a,x,y)}\mathbf{k}(\bar{a}, \bar{x}, \bar{y}) = \begin{pmatrix} 1 & -1 - 2\bar{x} & b \\ 0 & 1 & -1 \end{pmatrix}.$$

Since $\mathbf{D}_{(x,y)}\mathbf{k}(\bar{a}, \bar{x}, \bar{y})$ is singular for $1 + 2\bar{x} = b$, there is a path that can be parametrized by a through each point in Z except $(\bar{a}, \bar{x}, \bar{y}) = (-\frac{1}{4}(1 - b)^2, -(1 - b)/2, -(1 - b)/2)$, at which point there is a saddle node bifurcation. We can solve for the solution (x, y) for a greater than $-\frac{1}{4}(1 - b)^2$; namely,

$$(x, y) = \left(\frac{-(1 - b) + \sqrt{(1 - b)^2 + 4a}}{2}, \frac{-(1 - b) + \sqrt{(1 - b)^2 + 4a}}{2} \right)$$

$$(11.3)$$

Sometimes you may want to consider a coordinate other than a as the parameter. Since the 2×2 matrix $\mathbf{D}_{(a,x)}$ is nonsingular for all values of (a, x, y), the entire set of zeroes can be parametrized by y; namely,

$$(a, x) = (y^2 + (1 - b)y, y) \qquad (11.4)$$

Also since $\mathbf{D}_{(a,y)}$ is nonsingular everywhere, the entire set Z can be written as a function of x.

11.4 BIFURCATIONS OF ONE-DIMENSIONAL MAPS

We continue to investigate what can happen in a bifurcation diagram. We have just seen that paths of fixed points must continue unless the derivative passes through $+1$. In this section, we investigate the converse question for one-dimensional maps. If the derivative does pass through $+1$ along a path of fixed points, must there be a bifurcation? The answer is yes. Moreover, if the derivative passes through -1, there must be a period-doubling bifurcation, since the derivative of f_a^2 passes through $+1$.

To begin, we describe conditions on specific partial derivatives which imply the existence of a saddle-node bifurcation.

Theorem 11.9 (Saddle-node bifurcation.) *Let f_a be a smooth one-parameter family of one-dimensional maps. Assume that $\bar{x}$ is a fixed point of $f_{\bar{a}}$ such that $f_{\bar{a}}'(\bar{x}) = 1$. Assume further that*

$$A = \frac{\partial f}{\partial a}(\bar{a}, \bar{x}) \neq 0 \quad \text{and} \quad D = \frac{\partial^2 f}{\partial x^2}(\bar{a}, \bar{x}) \neq 0.$$

Then two curves of fixed points emanate from $(\bar{a}, \bar{x})$. The fixed points exist for $a > \bar{a}$ if $DA < 0$ and for $a < \bar{a}$ if $DA > 0$.

Proof: We will use Theorem 11.7, but with the roles of x and a reversed. Consider the map $g(a, x) = f(a, x) + a - x$. Note that (1) $g(a, x) = a$ if and only if $f(a, x) = x$, (2) $g(\bar{a}, \bar{x}) = \bar{a}$ and (3) $\partial g / \partial a(\bar{a}, \bar{x}) = A + 1 \neq 1$ by hypothesis. Theorem 11.7 implies that fixed points of $a \to g(a, x)$, and therefore the fixed

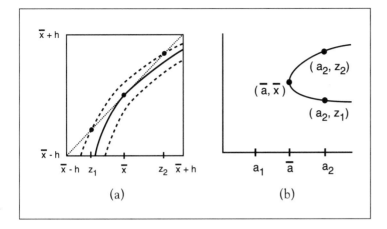

Figure 11.14 Graphs of a family of maps near a saddle-node bifurcation.
(a) The solid curve is the graph of $f_{\bar{a}}$, and the dashed curves show the graphs of
maps f_{a_1} and f_{a_2} for a_1 and a_2 near a, and $a_1 < \bar{a} < a_2$. The map f_{a_1} has no fixed
points, the map $f_{\bar{a}}$ has one fixed point (the bifurcation point), and the map f_{a_2} has
two fixed points, z_1 and z_2. (b) In the bifurcation diagram, two paths of fixed points
are created at $a = \bar{a}$.

points of $x \to f(a, x)$, can be continued in a path parametrized by x through the
point $(\bar{a}, \bar{x})$. Such a path is shown in Figure 11.14(b).

We will call the path $(p(x), x)$ in (a, x)-space. The fact that they are fixed
points of f_a is written as $f(p(x), x) = x$. By the chain rule, we calculate

$$\frac{\partial f}{\partial a}(\bar{a}, \bar{x})p'(\bar{x}) + \frac{\partial f}{\partial x}(\bar{a}, \bar{x}) = 1, \tag{11.5}$$

which implies that $p'(\bar{x}) = 0$. Therefore the curve $a = p(x)$ has a critical point at
$x = \bar{x}$, also shown in Figure 11.14(b).

Taking one more derivative of (11.5) with respect to x yields

$$\frac{\partial f}{\partial a}(\bar{a}, \bar{x})p''(\bar{x}) + f_a''(\bar{x}) = 0. \tag{11.6}$$

Therefore $DA < 0$ means that $p''(\bar{x}) > 0$, so that $p(x)$ is concave up at $x = \bar{x}$. The
path opens up to the right, as in Figure 11.14(b). If $DA > 0$, the path is reversed.

$\square$

✎ EXERCISE T11.4

Show that the saddle-node bifurcation of fixed points for $f(a, x) = a - x^2$
at $(\bar{a}, \bar{x}) = (-1/4, -1/2)$ opens to the right as in Figure 11.14(b).

Theorem 11.10   *Let f_a be a smooth one-parameter family of one-dimensional maps, and assume that there is a path of fixed points parametrized by a through $(\bar{a}, \bar{x})$. If $f_a'(x)$ evaluated along the path crosses $+1$ at $(\bar{a}, \bar{x})$, then every neighborhood of $(\bar{a}, \bar{x})$ in $\mathbb{R} \times \mathbb{R}$ contains a fixed point not on the path.*

Proof: For ease of exposition, we assume that the fixed points on the path have x-coordinate 0. Otherwise, replace (a, x) by (a, u), where $u = x - x(a)$. Notice then that $\frac{df}{du}(a, 0) = \frac{df}{dx}(a, x(a))$.

We argue the case in which $f_a'(0) > 1$ for $a < \bar{a}$, and $f_a'(0) < 1$ for $a > \bar{a}$. (The proof in the other case is analogous.) Then, as shown in Figure 11.15, there is an $h > 0$ such that

$$f_{\bar{a}-h}(h) - h > 0 \quad \text{and} \quad f_{\bar{a}+h}(h) - h < 0,$$

and also

$$f_{\bar{a}-h}(-h) - (-h) < 0 \quad \text{and} \quad f_{\bar{a}+h}(-h) - (-h) > 0.$$

Since $f_a(x)$ is continuous in a, there exist a_1 and a_2 in the interval $[\bar{a} - h, \bar{a} + h]$ satisfying $f_{a_1}(h) = h$ and $f_{a_2}(-h) = -h$. Since h can be chosen as small as desired

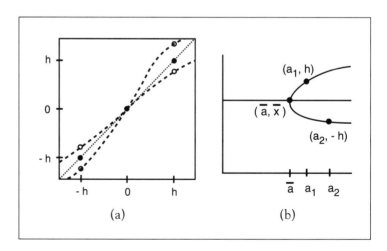

(a)   (b)

Figure 11.15   A family of maps where the derivative crosses 1.
(a) The dashed curves denote $f_{\bar{a}-h}$ and $f_{\bar{a}+h}$ in Theorem 11.10. The open circles straddle the diagonal, implying that f_a has fixed points other than 0 for parameter a between $\bar{a} - h$ and $\bar{a} + h$. (b) In the bifurcation diagram, a continuous path of fixed points bifurcates into paths of new fixed points. This is one of several possibilities when the derivative along a path of fixed points crosses 1.

but is nonzero, there are fixed points of f_a as near to $(\bar{a}, \bar{x})$ as needed. In fact, our argument shows that there are fixed points above and below the path. □

A *word of caution*: It is possible that the new fixed points in the conclusion of Theorem 11.10 exist for the single parameter value $a = \bar{a}$. For example, the family $f_a(x) = ax$ satisfies the hypotheses for the bifurcation value $\bar{a} = 1$. It is true that every neighborhood of $(\bar{a}, \bar{x}) = (1, 0)$ has fixed points other than 0, all of them occurring at $\bar{a} = 1$ because every point is a fixed point for $f_1(x)$.

The transcritical bifurcation diagram of Figure 11.10 is an illustration of Theorem 11.10. For example, pick either of the paths in Figure 11.10 and follow it toward the bifurcation point. The map is $f_a(x) = ax(1 - x)$ and so $f_a'(x) = a(1 - 2x)$. On the path of fixed points $x = 0$, the derivative $f_a'(0) = 1$ is reached at parameter value $a = 1$. In every neighborhood of this point there are points from the other path of fixed points. The pitchfork bifurcation of Figure 11.12 is another illustration of Theorem 11.10.

Theorem 11.11 *Let f be a smooth one-parameter family of maps. Assume that $\bar{x}$ is a fixed point of $f_{\bar{a}}$ such that $f_{\bar{a}}'(\bar{x}) = -1$. If $f_a'(x)$ evaluated along the path of fixed points through $(\bar{a}, \bar{x})$ crosses -1 at $(\bar{a}, \bar{x})$, then every neighborhood of $(\bar{a}, \bar{x})$ (in $\mathbb{R} \times \mathbb{R}$) contains a period-two orbit.*

Proof: Define $g_a = f_a^2$ and apply Theorem 11.10. The chain rule shows that the derivative of g_a passes through 1 at $a = \bar{a}$. Therefore g_a has fixed points other than the fixed point of f_a in any neighborhood of $(\bar{a}, \bar{x})$. Since $f_{\bar{a}}(\bar{x}) = -1$, the fixed point $\bar{x}$ is continuable, so the others must be period-two points of f_a. □

Table 11.1 summarizes information about the bifurcations we have described. If we follow a path of fixed points of a one-dimensional map, bifurcations can occur when the derivative becomes 1 or -1. Saddle-node and period-doubling

Bifurcation	$\dfrac{\partial f}{\partial x}(\bar{a}, \bar{x})$	$\dfrac{\partial f}{\partial a}(\bar{a}, \bar{x})$	$\dfrac{\partial^2 f}{\partial x^2}(\bar{a}, \bar{x})$
saddle node	1	$\neq 0$	$\neq 0$
pitchfork, transcritical	1	0	$\neq 0$
period-doubling	-1	$\neq 0$	$\neq 0$

Table 11.1 Identification of common bifurcations in one-dimensional maps. When the derivative at the fixed point is 1 or -1, the type of bifurcation depends on other partial derivatives.

bifurcations are the "generic", or commonly expected bifurcations, according to the table, because they don't depend on anything special happening at the bifurcation point. A pitchfork bifurcation can only happen if the bifurcation point $(\bar{a}, \bar{x})$ coincides with a point where $\partial f/\partial a(\bar{a}, \bar{x}) = 0$, a low probability event. However, certain special conditions that are forced on the family by symmetry and other reasons, can cause pitchfork bifurcations to occur, as in Example 11.5.

✎ **EXERCISE T11.5**

Let f_a be a family of one-dimensional maps, and assume that $(\bar{a}, \bar{x})$ is a fixed point that is either a pitchfork bifurcation point or a transcritical bifurcation point. Then the set Z of fixed points of f_a cannot be expressed as a (single-valued) function either of a or of x in any neighborhood of $(\bar{a}, \bar{x})$. We have seen that since Z is not locally continuable as a path in a, the derivative $f_a'(\bar{a}, \bar{x}) = \partial f/\partial x(\bar{a}, \bar{x})$ must be $+1$. Show that, in addition, since Z is not locally continuable as a path in x, the derivative $\partial f/\partial a(\bar{a}, \bar{x})$ is 0.

There is no end to the variety of behavior that can be seen in nongeneric bifurcations. As a mild indication, in Exercise 11.6, four fixed points are created at the bifurcation point.

✎ **EXERCISE T11.6**

Let $f_a(x) = x^4 - 3ax^2 + x + 2a^2$. Sketch the graph of f_a for $a < 0$ and $a > 0$ and conclude that $(\bar{a}, \bar{x}) = (0, 0)$ is a bifurcation point. Find the four fixed points of f_a for $a > 0$. Calculate the partial derivatives of Table 11.1 to determine in which ways this bifurcation fails to be generic.

11.5 BIFURCATIONS IN PLANE MAPS: AREA-CONTRACTING CASE

Many of the ideas we presented in Section 11.4 of this chapter carry over to the study of bifurcations in one-parameter families of two-dimensional maps. To illustrate this idea, we look at a simple extension of the logistic family to two-dimensions:

$$f_a(x, y) = (g_a(x), 0.2y),$$

where $g_a(x) = ax(1 - x)$. The Jacobian $\mathbf{D_v f}_a(x, y)$ has eigenvalues $e_1 = g_a'(x)$ and $e_2 = 0.2$. Again, the notation $\mathbf{D_v f}_a$, which we will sometimes simplify to $\mathbf{Df}_a$, denotes the matrix of partial derivatives with respect to the coordinates x and y (the coordinates of the $\mathbf{v}$ vector), not a.

The version of Theorem 11.7 for one-parameter families of two-dimensional maps says that a bifurcation of fixed points can only occur when $+1$ is an eigenvalue of $\mathbf{Df}_a$. For $\mathbf{f}_a$ this means that such bifurcations can only occur when g_a' is $+1$. Of course, because of the special form of the map, bifurcations of $\mathbf{f}_a$ occur at precisely those parameter values at which g_a has bifurcations. The hyperbolic fixed points of $\mathbf{f}_a$ are different than in the one-dimensional case: $\mathbf{f}_a$ has only attractors and saddles, whereas g_a has only attractors and sources. At a saddle-node bifurcation of $\mathbf{f}_a$ one branch of attractors and one branch of saddles emanate from the bifurcation orbit. At a period-doubling bifurcation, the eigenvalue e_1 goes through -1. A branch of fixed-point attractors changes to a branch of saddles as two branches of period-two attractors (one branch for each point of the orbit) bifurcate. These are the bifurcations pictured in Figure 11.8.

✎ **EXERCISE T11.7**

Let $\mathbf{h}_a(x, y) = (g_a(x), 3y)$. Classify the types of hyperbolic fixed points and periodic orbits of $\mathbf{h}_a$. Which hyperbolic orbits make up the branches emanating from saddle-node and period-doubling bifurcations?

Again, as in the one-dimensional case, we think of a periodic orbit as one entity. Recall (from Chapter 2) that for a period-k orbit the eigenvalues of $\mathbf{Df}^k = \mathbf{D}(\mathbf{f}^k)$ are the same when the Jacobian is evaluated at any of the k points in the orbit.

✎ **EXERCISE T11.8**

(a) Let $A = \begin{pmatrix} m_{11} & m_{12} \\ m_{21} & m_{22} \end{pmatrix}$. Show that the eigenvalues e_1 and e_2 are continuous functions of the variables $m_{11}, m_{12}, m_{21}, m_{22}$. (b) Let $\mathbf{f}$ be a smooth map of $\mathbb{R}^2$. Show that the eigenvalues of $\mathbf{Df}_a^k(\mathbf{v})$ depend continuously on $(a, \mathbf{v})$.

Describing saddle-node and period-doubling bifurcations for arbitrary families of two-dimensional maps is complicated by the fact that fixed and periodic points can have both contracting and expanding directions. Although the previous example is clearly a special case, the property of having one eigenvalue with absolute value smaller than 1 at all points (or, analogously, one eigenvalue with absolute value greater than 1 at all points) is one that holds for a large and important class of maps. For such maps, the types of bifurcations that can occur are limited to those seen in one-dimensional maps.

Definition 11.12 Let $\mathbf{f}$ be a smooth map of $\mathbb{R}^2$, and let $\mathbf{Df(v)}$ be the Jacobian matrix of partial derivatives of $\mathbf{f}$ with respect to the $\mathbf{v}$ coordinates evaluated at a point $\mathbf{v} \in \mathbb{R}^2$. We say $\mathbf{f}$ is **area-contracting** if $|\det \mathbf{Df(v)}| < 1$ for all $\mathbf{v} \in \mathbb{R}^2$. The map $\mathbf{f}$ is **area-preserving** if $|\det \mathbf{Df(v)}| = 1$ for all $\mathbf{v} \in \mathbb{R}^2$.

It is a fact from advanced calculus that the number $|\det \mathbf{Df(v)}|$ determines the change in area effected by the map $\mathbf{f}$ near the point $\mathbf{v}$. In fact, if S is a region in $\mathbb{R}^2$, then

$$\text{Area}(\mathbf{f}(S)) = \int\int_S |\det \mathbf{Df(v)}|\mathbf{dv}.$$

See (Fitzpatrick, 1995).

Area-contracting maps represent an important class of dynamical systems. Included in this class are time-T maps of dissipative systems of differential equations. An example is the time-2π map of the forced, damped pendulum with equation of motion $\ddot{x} + c\dot{x} + \sin x = b\cos t$. We saw in Example 9.5 that with damping $c > 0$, the Jacobian determinant is $e^{-2\pi c} < 1$ for each x and $\dot{x}$.

✎ **EXERCISE T11.9**

For which values of a and b is the Hénon map $\mathbf{f}(x, y) = (a - x^2 + by, x)$ area-contracting?

Let $\mathbf{p}$ be a periodic orbit of period k. We use the phrase "eigenvalues of $\mathbf{p}$" to refer to the eigenvalues of $\mathbf{D_v f^k(p)}$. For maps of the plane, $\mathbf{D_v f^k(p)}$ has eigenvalues e_1 and e_2. The area-contracting hypothesis implies $|e_1 e_2| < 1$. In particular, at least one of $\{e_1, e_2\}$ must be inside the unit circle in the complex plane. The unit circle is the set of complex numbers $u + iv$ with modulus $u^2 + v^2 = 1$. Thus the area-contracting hypothesis restricts the types of periodic orbits that can occur. There are only two types of hyperbolic orbits: attractors or sinks (both eigenvalues inside the unit circle) and saddles (one eigenvalue greater than $+1$ or smaller than −1). In Chapter 12, we need to distinguish between regular saddles (one eigenvalue greater than +1) and flip saddles (one eigenvalue less than −1). Notice that the unstable orbits are all saddles, regular or flip. There are no repellers for area-contracting maps.

Similarly, the types of nonhyperbolic orbits are restricted by the area-contracting hypothesis. There are only two types: orbits with one eigenvalue equal to +1, and orbits with one eigenvalue equal to −1. As in the one-dimensional case, the most frequently seen bifurcations are saddle-nodes and period-doublings. The difference in the two-dimensional area-contracting case is that what were paths of repellers in the one-dimensional case are replaced by paths of saddles, flip or regular, here. We discuss why the saddle-node and period-doubling bifurcations are those typically seen for families of one-dimensional maps and families of area-contracting maps of the plane in Chapter 12.

⇨ COMPUTER EXPERIMENT 11.1

Plot the period-6π attracting orbit of the driven double-well Duffing oscillator $\ddot{x} + 0.1\dot{x} - x + x^3 = \rho \sin t$ for $\rho = 5.9$. Next, increase ρ and locate the two period-doubling bifurcations that occur for $5.9 < \rho < 6$.

11.6 BIFURCATIONS IN PLANE MAPS: AREA-PRESERVING CASE

In previous sections we saw that bifurcations of fixed points or periodic points can occur only at parameter values for which the derivative or an eigenvalue of the orbit has absolute value one. For one-dimensional families, saddle nodes or period-doubling bifurcations can occur when the derivative is 1 or −1, respectively. For families of maps of two or more (real) variables, there can be eigenvalues of absolute value (magnitude) 1 which are not real numbers.

✎ **EXERCISE T11.10**

Let **f** be a map of $\mathbb{R}^n$, and let **p** be a fixed point of **f**.

(a) Show that if $a + bi$ is an eigenvalue of $\mathbf{Df(p)}$, then the complex conjugate $a - bi$ is also an eigenvalue.

(b) Show that a bifurcation of a period-k orbit from a fixed point can only occur at a bifurcation orbit that has an eigenvalue that is a kth root of unity.

In fact, a bifurcation of period-k orbits from a fixed point can only occur when a primitive kth root of unity is an eigenvalue. In families of one-dimensional maps this means $k = 1$ or $k = 2$. Bifurcations of higher-period orbits are similarly ruled out for families of area-contracting maps of the plane. If, however, we drop the area-contracting hypothesis, then bifurcations of higher-period orbits $(k > 2)$ can occur for families of two-dimensional maps. Unless there are special symmetries or restrictions on the Jacobian (as in the case of area-preserving maps), such bifurcations are rare compared to saddle nodes or period doublings. As we discuss in Chapter 12, the latter two bifurcations are "generic", in the sense that a large set of one-parameter families possess only these bifurcations.

To investigate how and when bifurcations of higher periods occur, we begin with a family of two-dimensional linear maps.

EXAMPLE 11.13

Let $f_a(x, y) = (\frac{a}{2}x - \frac{a\sqrt{3}}{2}y, \frac{a\sqrt{3}}{2}x + \frac{a}{2}y)$, for $a > 0$, and x, y in $\mathbb{R}$. When $a = 1$, the map rotates all nonzero vectors through an angle of $\pi/3$. The origin is fixed. When the eigenvalues of a fixed point all lie on the unit circle, it is called an **elliptic** fixed point. The origin is elliptic for $a = 1$. Although an elliptic fixed point of a linear map is not attracting, it is **stable**, meaning that for each neighborhood N_ε, there is a (perhaps smaller) neighborhood N_δ such that all orbits beginning in N_δ remain in N_ε. Orbits that start close stay close. For nonlinear maps, elliptic fixed points may or may not be stable.

It is worthwhile here to think of a "path" $e(a)$ of eigenvalues (parametrized by a); specifically, $e(a) = a(\frac{1}{2} + i\frac{\sqrt{3}}{2})$, which for $0 < a < 1$ begins inside the unit circle and crosses the unit circle when $a = 1$ at a sixth root of unity. Another path of eigenvalues consists of the complex conjugate of $e(a)$, for each $a > 0$. For $0 < a < 1$, the origin is a sink; for $a > 1$, the origin is a source; and at $a = 1$, the origin is an elliptic point around which all other points rotate on invariant circles through an angle of $\pi/3$. In the last case, all other points are period-six points.

For a one-parameter family f_a of nonlinear planar maps, we know that the eigenvalues of Df_a are the key to detecting changes of stability that might result in bifurcations. To locate bifurcations of period-k orbits along a path of fixed points, we look for points at which a path $e(a)$ of eigenvalues crosses the unit circle at a kth root of unity. If the path crosses the unit circle with nonzero speed (if the derivative de/da is nonzero), then a family of invariant closed curves will bifurcate from the path of fixed points. The family of closed curves may all occur at one parameter value, such as the invariant circles in the previous linear case, or they may occur in a path that emanates from the bifurcation parameter through either smaller or larger parameter values. We do not discuss the dynamics that occur on the invariant closed curves except to say that near the bifurcation point, on each closed curve, points rotate near the rate given by the Jacobian at the bifurcation orbit.

In Example 11.14, we investigate a bifurcation in the Hénon family which produces a family of invariant closed curves as we pass from area-contracting to area-expanding maps.

EXAMPLE 11.14

Let $h_B(x, y) = (-.75 - x^2 + by, x)$. Recall that $-b$ is the Jacobian determinant: h is area-contracting or area-expanding according to whether $|b|$ is less than or greater than 1. For $b = -1$, there is a fixed point at $(-1/2, -1/2)$ with eigenvalues $e = \frac{1}{2} \pm \frac{\sqrt{3}}{2}i$, (two sixth-roots of unity). Figure 11.16 shows a neighborhood of this fixed point (or its continuation) as b varies from $-.9$ to -1.1. The fixed point goes from sink (Figure 11.16(a)) to source (Figure 11.16(b)). In particular, Figure 11.16(c) shows invariant curves encircling the fixed point at $b = -1$. Notice that the crossing stable and unstable manifolds of a nearby fixed point saddle form the boundary of the "elliptic" behavior at the original fixed point. There are no invariant closed curves for $|b| \neq 1$. (See Exercise T11.11 below.)

✎ EXERCISE T11.11

Show that an invertible map of $\mathbb{R}^2$ which is either area-contracting or area-expanding can have no invariant closed curves (except points).

Families of area-preserving planar maps (see Defn. 11.12) can possess an extremely rich bifurcation structure. We still see periodic saddle orbits in these maps, but no sinks or sources. The stable orbits must be elliptic: Both eigenvalues are on the unit circle.

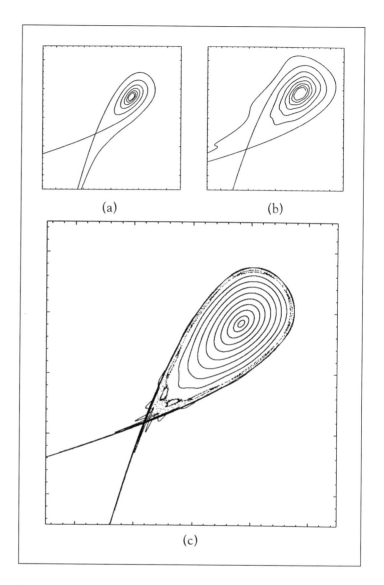

Figure 11.16 The continuation of two fixed points in the Hénon family.
Various orbits of the map $h_B(x, y) = (-.75 - x^2 + by, x)$ are shown for (x, y) in $[-2.5, 5.0] \times [-2.5, 5.0]$. (a) At $b = -0.9$, the map is area-contracting and there is an attracting fixed point. The stable and unstable manifolds of a nearby saddle fixed point are depicted. One branch of the unstable manifold spirals into the attractor. (b) At $b = -1.1$, the map is area-expanding and there is a repelling fixed point (the continuation of the elliptic point at this parameter). Now the stable manifold of the saddle is shown to spiral out of the repeller.(c) At the in-between value $B = -1.0$, the map is area-preserving and there is an elliptic fixed point (the continuation of the attractor at this parameter). The stable and unstable manifolds of the saddle now cross in infinitely many homoclinic points.

EXERCISE T11.12

Show that there are no hyperbolic periodic or fixed sinks or sources for area-preserving maps of the plane.

We investigate the Hénon family again—this time fixing b at -1 and varying a. Each map in the resulting family is an area-preserving map.

EXAMPLE 11.15

Let $h_a(x, y) = (a - x^2 - y, x)$. Verify that for each a, the plane map h_a is area-preserving. For $a < -1$ there are no fixed points. When $a = -1$, a fixed point appears at $(-1, -1)$. This orbit has a double eigenvalue of $+1$. For $a > -1$, there are two fixed points: one that is unstable (a saddle) and one that begins as a stable elliptic orbit for a near -1. We follow the orbits along the second, stable path. As a increases slightly beyond -1, eigenvalues of these orbits are constrained to the unit circle in the complex plane. Since for any fixed parameter the two eigenvalues must be complex conjugates, the two paths of eigenvalues move (with increasing a) around the unit circle. One moves along the top (imaginary part positive), and the other moves along the bottom (imaginary part negative) as shown in Figure 11.17. When $a = 3$, the two paths of eigenvalues join at -1, where there is a period-doubling bifurcation. What type of behavior do you expect near the elliptic fixed point as a increases from -1 to 3?

Our final example of this section is called the standard map. Like the Hénon family, it has been studied extensively as a prototype—in this case, as a model for understanding nonlinear phenomena in area-preserving maps. While Poincaré maps for dissipative systems of differential equations are area-contracting, Poincaré maps for conservative systems are area-preserving.

EXAMPLE 11.16

(Standard map.) Let

$$S_a(x, y) = (x + y(\mathrm{mod}\,2\pi), y + a\sin(x + y)(\mathrm{mod}\,2\pi)), \qquad (11.7)$$

for $x \in [-\pi, \pi], y \in [-\pi, \pi]$. Notice that the standard map S_a is periodic with period 2π in both variables. That is, $S_a(x + 2\pi, y) = S_a(x, y)$ and $S_a(x, y + 2\pi) = S_a(x, y)$. Therefore, like the cat map of Challenge 2, only the values of x and y modulo 2π are important, and we can consider the phase space to be a torus.

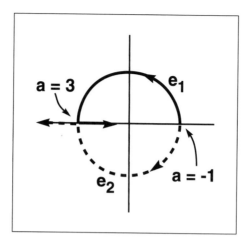

Figure 11.17 Complex eigenvalues for a family of elliptic fixed points.
Paths of eigenvalues in the complex plane for the continuation of an elliptic fixed point of the area-preserving Hénon family $h(a, x, y) = (a - x^2 - y, x)$ are depicted schematically. The elliptic point is born at $a = -1$, at which point both eigenvalues are $+1$. Then the paths split and move around the unit circle as complex conjugates. The paths join at a period-doubling bifurcation when $a = 3$. There the fixed point loses stability, becoming a saddle, as one path of eigenvalues moves outside the unit circle.

✐ **EXERCISE T11.13**

Let S_a denote the standard map. Show that $\det(DS_a(x, y)) = 1$. Therefore, S_a is a one-parameter family of area-preserving maps.

When the parameter a is zero, the torus is filled with invariant closed curves (one for each y value). In Figure 11.18, both axes are $[-\pi, \pi]$, so that one can imagine the right and left boundaries to be glued together, and the same for the top and bottom boundaries, to form the torus. The center of the square is the origin, which is a fixed point for the standard map.

The closed curves in Figure 11.18(a) are called KAM curves. KAM curves figure importantly in understanding stability within area-preserving maps. This stability analysis is the subject of KAM theory, a field whose major advances are due to the three mathematicians A. Kolmogorov, V. Arnold, and J. Moser. On each of these curves, all points move with the same rate of rotation, called the **rotation number**. The rotation number of a circle increases with y.

As the parameter a is increased from 0, other closed curves form that do not wrap all the way around the torus. These KAM curves appear as oval-shaped

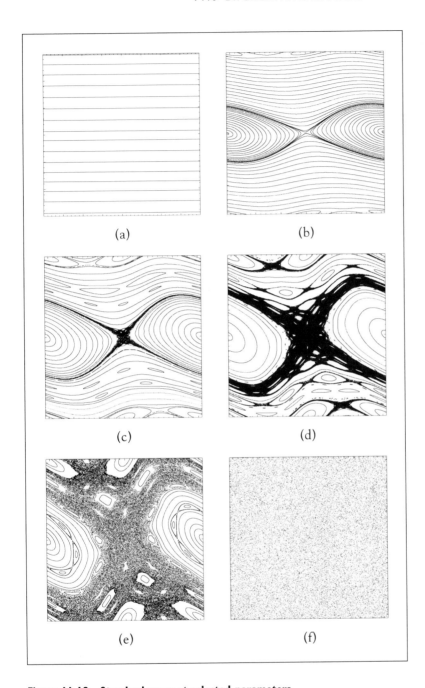

Figure 11.18 Standard maps at selected parameters.
Typical orbits of the standard family of maps $S_a(x, y) = (x + y \bmod 2\pi, y + a\sin(x + y) \bmod 2\pi)$ are shown for (x, y) in $[-\pi, \pi] \times [-\pi, \pi]$ and parameter values (a) $a = 0$ (b) $a = 0.3$ (c) $a = 0.6$ (d) $a = 0.9$. e) $a = 1.2$ and (f) $a = 7.0$. In all but (f), several orbits are plotted. In (f), one orbit is shown.

islands in Figure 11.18(b)–(e). Points on the closed curves are constrained to map to the same closed curve. Each island in Figure 11.18(b)–(e) was made by plotting a single orbit, which fills the curve densely. The KAM curves eventually give way to irregular orbits. The last KAM curve that wraps around the torus can be seen in the lower half of Figure 11.18(d). It disappears at $a \approx 0.97$.

✎ **EXERCISE T11.14**

Let $\mathbf{S}_a$ denote the standard map.
(a) Show that for $a > 0$, the origin $(0, 0)$ is a saddle fixed point.
(b) For what values of a is $(\pi, 0)$ an elliptic fixed point?

As a increases from zero, the invariant circles near $y = 0$ break up. We see the stable and unstable manifolds of the saddle $(0, 0)$ forming the boundary between the closed curves surrounding the elliptic point and the closed curves encircling the cylinder. A further increase in a shows regions of chaotic behavior developing around these stable and unstable manifolds. New elliptic orbits come into existence and disappear as chaotic regions expand to fill an increasingly larger space. Figure 11.18(f), at $a = 7.0$, shows no regions of stability. Only one orbit is plotted.

↻ **COMPUTER EXPERIMENT 11.2**

The standard map can be made area-contracting by altering one term. Define

$$S_{a,b}(x, y) = (x + y(\mathrm{mod}2\pi), by + a\sin(x + y)(\mathrm{mod}2\pi)). \qquad (11.8)$$

Investigate $S_{a,b}$ for $a = 4.0$ and $b = 0.98$. How many attracting periodic orbits can you find? (More than 100 exist for periods ≤ 5.)

11.7 BIFURCATIONS IN DIFFERENTIAL EQUATIONS

The main tool for analyzing the stability of periodic solutions of differential equations is the Poincaré map, which we first defined in Chapter 2. Let $\mathbf{v}$ be a point in $\mathbb{R}^n$, let $\mathbf{f}$ be a map defined on $\mathbb{R}^n$, and let γ be a periodic orbit of the autonomous differential equation

$$\dot{v} = f(v). \tag{11.9}$$

For a point v_0 on γ, the Poincaré map T is defined on an $n - 1$ dimensional disk D transverse to γ at v_0. Figure 11.19 shows a periodic orbit for a flow of $\mathbb{R}^3$ with a two-dimensional disk D. If v_1 is a point sufficiently close to v_0 on D, then the solution through v_1 will follow close to γ until it intersects D again (the first return to D) at a point v_2. This fact follows from continuous dependence on initial conditions, Theorem 7.16. Let $T(v_1) = v_2$. (We may have to restrict the domain of T to a smaller disk centered at v_0 in order to have the solution hit D upon first return.) For a periodic orbit of a planar differential equation, the cross-sectional disk is a line segment transverse to the periodic orbit.

Notice that v_0 is a fixed point or periodic point of T if and only if v_0 is on a periodic orbit of (11.9). Thus we can investigate the stability of periodic orbits through a linear analysis of T, by looking at the eigenvalues of $D_v T(v_0)$.

Definition 11.17 The eigenvalues of the $(n - 1) \times (n - 1)$ Jacobian matrix $D_v T(v_0)$ are called the (Floquet) **multipliers** of the periodic orbit γ.

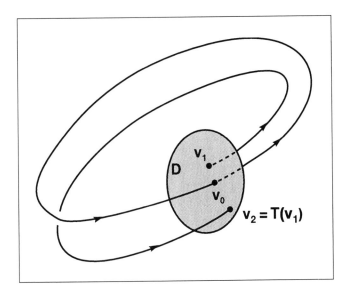

Figure 11.19 A Poincaré map.
The Poincaré map, called T here, is defined on a surface D which is transverse to the flow direction on a periodic orbit. The point v_0 on the periodic orbit maps to itself under the Poincaré map, while v_1 maps to v_2.

If all the multipliers of γ are inside (respectively, outside) the unit circle in the complex plane, then γ is called an **attracting** (respectively, **repelling**) periodic orbit. If γ has some multipliers inside and some multipliers outside the unit circle, then γ is called a **saddle** periodic orbit.

Finding multipliers of a periodic orbit is largely accomplished through computational methods, since cases in which a Poincaré map can be explicitly determined are rare. One such case follows.

EXAMPLE 11.18

Let

$$\dot{r} = br(1 - r) \tag{11.10}$$

$$\dot{\theta} = 1$$

where b is a positive constant, and (r, θ) are polar coordinates. Since $\dot{r} = 0$ when $r = 1$, the unit circle $r = 1$ is a periodic orbit. A segment L of the ray $\theta = \overline{\theta}$, (any fixed angle $\overline{\theta}$ will suffice), serves as the domain of a Poincaré map. Figure 11.20 illustrates one orbit $\{r_0, r_1, r_2, \ldots\}$ of the Poincaré map for the limit cycle

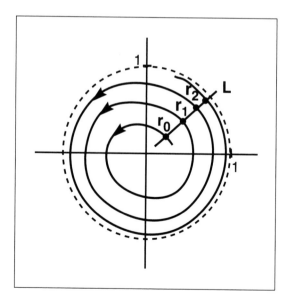

Figure 11.20 Poincaré map for a limit cycle of a planar system.
The system $\frac{dr}{dt} = 0.2r(1 - r)$, $\frac{d\theta}{dt} = 1$, has a limit cycle $r = 1$. The line segment L is approximately perpendicular to this orbit. Successive images $r_1, r_2, \ldots$, of initial point r_0 under the Poincaré map converge to $r = 1$.

$r = 1$ of the (11.10) with $b = 0.2$. Using the fact that a trajectory returns to the same angle after 2π time units, we separate variables in (11.10), and integrate a solution from one crossing, r_n, to the next, r_{n+1} of L:

$$\int_{r_n}^{r_{n+1}} \frac{dr}{r(1-r)} = \int_0^{2\pi} b\, dt = 2\pi b.$$

Evaluating the integral on the left (by partial fractions), we obtain

$$\frac{r_{n+1}(1-r_n)}{r_n(1-r_{n+1})} = e^{2\pi b}.$$

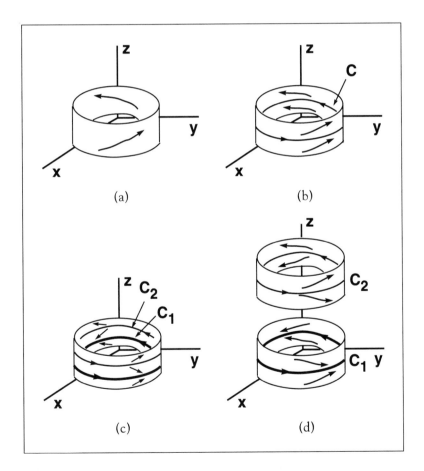

(a)

(b)

(c)

(d)

Figure 11.21 A saddle-node bifurcation for a three-dimensional flow.
(a) The system begins with no periodic orbits. (b) The parameter is increased, and a saddle-node periodic orbit C appears. (c) The saddle-node orbit splits into two periodic orbits C_1 and C_2, which then move apart, as shown in (d).

Then, solving for r_{n+1},

$$r_{n+1} = \frac{r_n e^{2\pi b}}{1 - r_n + r_n e^{2\pi b}}. \tag{11.11}$$

✎ **EXERCISE T11.15**

(a) Show that for $b > 0$, the sequence $\{r_n\}$, as defined by (11.11), converges to 1 as $n \to \infty$.

(b) Find the Floquet multiplier of the orbit $r = 1$.

(c) For which values of b is this orbit stable?

Through the Poincaré map and time-T map, we can exploit the bifurcation theory developed for maps to study the bifurcations of periodic orbits of parametrized differential equations. For $\mathbf{v} \in \mathbb{R}^n$ and a scalar parameter a, let

$$\dot{\mathbf{v}} = \mathbf{f}_a(\mathbf{v}) \tag{11.12}$$

denote a one-parameter family of differential equations. As before, a periodic orbit γ of (11.12) is classified as "continuable" or a "bifurcation" orbit, depending on whether $\mathbf{v}_0$ on γ is a locally continuable or bifurcation orbit for the one-parameter continuation of the Poincaré map $\mathbf{T}$ or time-T map $\mathbf{F}_T$.

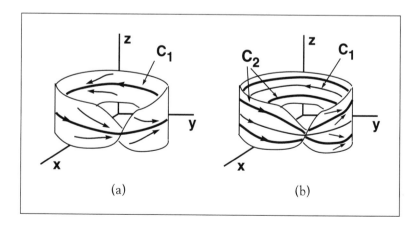

(a)   (b)

Figure 11.22   A period-doubling bifurcation for a three-dimensional flow.
The system begins with a periodic orbit C_1, which has one multiplier between 0 and -1, as shown in (a). As the parameter is increased, this multiplier crosses -1, and a second orbit C_2 of roughly twice the period as C_1 bifurcates. This orbit wraps twice around the Möbius strip shown in (b).

Figure 11.21 shows a saddle-node bifurcation of periodic orbits, and Figure 11.22 shows a period-doubling bifurcation. Recall that at a saddle-node bifurcation orbit, the linearized map has an eigenvalue of $+1$. In the case of a flow, a multiplier of the periodic orbit must be $+1$. For a period-doubling bifurcation, the orbit must have a multiplier of -1.

✎ **EXERCISE T11.16**

After a period-doubling bifurcation, are the periodic orbits linked or unlinked? (See, for example, the orbits C_1 and C_2, as shown in Figure 11.22(b).)

✎ **EXERCISE T11.17**

Show that a one-parameter family of autonomous plane differential equations cannot contain any period-doubling bifurcation orbits.

11.8 HOPF BIFURCATIONS

Saddle-node and period-doubling bifurcations are not the only ways in which periodic orbits are born in one-parameter families. There is one more "generic" bifurcation. In an Andronov-Hopf bifurcation (often shortened to Hopf bifurcation), a family of periodic orbits bifurcates from a path of equilibria.

EXAMPLE 11.19

(Hopf bifurcation.) Let

$$\dot{x} = -y + x(a - x^2 - y^2)$$
$$\dot{y} = x + y(a - x^2 - y^2) \tag{11.13}$$

for x, y, and a in $\mathbb{R}$. The bifurcation diagram for this system is illustrated in Figure 11.23. Verify that for $a > 0$, there are periodic solutions of the form

$$(x(t), y(t)) = (\sqrt{a}\cos t, \sqrt{a}\sin t).$$

In polar coordinates, equations (11.13) have the simple form

$$\dot{r} = r(a - r^2)$$
$$\dot{\theta} = 1. \tag{11.14}$$

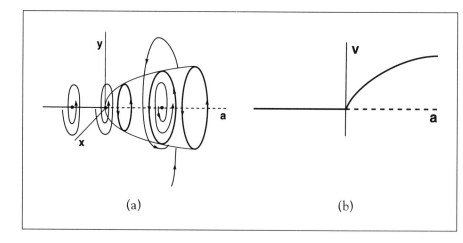

Figure 11.23   Supercritical Hopf bifurcation.
(a) The path $\{(a, 0, 0)\}$ of equilibria changes stability at $a = 0$. A stable equilibrium for $a < 0$ is replaced by a stable periodic orbit for $a > 0$. (b) Schematic path diagram of the bifurcation. Solid curves are stable orbits, dashed curves are unstable.

For each a, the origin $r = 0$ is an equilibrium of (11.14). When a is negative, $\dot{r}$ is negative, and all solutions decay to $\mathbf{0}$. For each $a > 0$, there is a nontrivial periodic solution $r = \sqrt{a}$. A path of periodic orbits bifurcates from $\mathbf{v}$ at $a = 0$. In this example, notice that there is exactly one periodic orbit for each planar flow when $a > 0$.

Common to bifurcations we have discussed, the family of periodic orbits bifurcates from a path of equilibria that changes stability at the bifurcation point. In this case, the equilibria go from attractors to repellers. We picture this planar nonlinear case in Figure 11.24. The variation in parameter brings about a small change in the vector field near the equilibrium, causing a change in its stability. Meanwhile, the vector field outside a small neighborhood N of the equilibrium remains virtually unchanged, with the vector field still pointing in towards the equilibrium. Orbits within N are trapped and, in the absence of any other equilibria, must converge to a periodic orbit.

Definition 11.20   Assume that $\bar{\mathbf{v}}$ is an equilibrium at $a = \bar{a}$ for the one-parameter family (11.12). We say that **a path of periodic orbits bifurcates from** $(\bar{a}, \bar{\mathbf{v}})$ if there is a continuous path of periodic orbits that converge to the equilibrium at $a = \bar{a}$.

In order to simplify the discussion here, we assume as in the previous example that $\mathbf{v} = \mathbf{0}$ is an equilibrium for all a (that is, $\mathbf{f}_a(\mathbf{0}) \equiv \mathbf{0}$ in (11.12)).

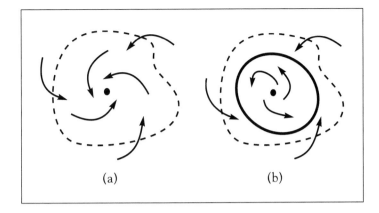

Figure 11.24   A periodic orbit bifurcates from an equilibrium in the plane.
As the equilibrium goes from attracting (a) to repelling (b), a periodic orbit appears.

Also, we assume that the bifurcation occurs at $\bar{a} = 0$. The eigenvalues of the matrix $\mathbf{D_v f}_a(\mathbf{0})$ vary continuously with a. We can refer to a path $c(a) + ib(a)$ of eigenvalues of the matrix. (In Example 11.19, one checks that $c(a) + ib(a) = a + i$.) A version of the following theorem was first proved by E. Hopf in 1942. For more details, see (Kocak and Hale, 1991).

Theorem 11.21   (Andronov-Hopf Bifurcation Theorem.) *Let* $\dot{\mathbf{v}} = \mathbf{f}_a(\mathbf{v})$ *be a family of systems of differential equations in* $\mathbb{R}^n$ *with equilibrium* $\bar{\mathbf{v}} = \mathbf{0}$ *for all* a. *Let* $c(a) \pm ib(a)$ *denote a complex conjugate pair of eigenvalues of the matrix* $\mathbf{Df}_a(\mathbf{0})$ *that crosses the imaginary axis at a nonzero rate at* $a = 0$; *that is,* $c(0) = 0$, $b = b(0) \neq 0$, *and* $c'(0) \neq 0$. *Further assume that no other eigenvalue of* $\mathbf{Df}_a(\mathbf{0})$ *is an integer multiple of* bi. *Then a path of periodic orbits of* (11.12) *bifurcates from* $(a, \mathbf{v}) = (0, \mathbf{0})$. *The periods of these orbits approach* $2\pi/b$ *as orbits approach* $(0, \mathbf{0})$.

✎ **EXERCISE T11.18**

Show that the system (11.13) satisfies the hypotheses of the Hopf Bifurcation Theorem, and describe the conclusion of the theorem for this system.

Remark 11.22   There is a weaker form of the Hopf Bifurcation Theorem that is useful for some purposes. The weaker version has a less stringent hypothesis, and delivers a weaker conclusion. Instead of assuming that the pair of eigenvalues crosses the imaginary axis with nonzero speed at $a = 0$, assume simply that it crosses the imaginary axis at $a = 0$. All other hypotheses remain unchanged.

(For example, the eigenvalues might move as $c(a) \pm ib(a) = a^3 \pm i$.) Then the conclusion is that every neighborhood of $(\bar{a}, \bar{v}) = (0, 0)$, no matter how small, contains a periodic orbit. They may not be parametrized by a continuous path in the parameter a. See (Alexander and Yorke, 1978).

Hopf bifurcations come in several types, depending on what happens to orbits near the equilibrium at the bifurcation point. Two types are important in experimental work, because they explain the important phenomenon of the creation of periodic behavior. At a **supercritical** bifurcation, the equilibrium at the bifurcation value of the parameter, which we might call the bifurcation orbit, is stable. A supercritical Hopf bifurcation is illustrated in Figure 11.23. When seen in experiment, a supercritical Hopf bifurcation is seen as a smooth transition. The formerly stable equilibrium starts to wobble in extremely small oscillations that grow into a family of stable periodic orbits as the parameter changes.

When the bifurcation orbit is unstable, the bifurcation is called **subcritical**. A subcritical Hopf bifurcation, illustrated in Figure 11.25, is seen experimentally as a sudden jump in behavior. As the parameter a is increased in the figure, the system follows the equilibrium $v = 0$ until reaching the bifurcation point $a = 0$. When this point is passed, no stable equilibrium exists, and the orbit is immediately attracted to the only remaining stable orbit, an oscillation of large amplitude. In this case, the bifurcating path of periodic orbits of the Hopf bifurcation exists for $a < 0$. The periodic orbits thrown off from the origin for negative a are unstable orbits. For $a < 0$, they provide a basin boundary between the attracting equilibrium and the attracting periodic orbit on the outside. When this boundary disappears at $a = 0$, immediate changes occur. Lab Visit 11 involves the interpretation of sudden changes in current oscillations from an electrochemistry experiment. In this case, the scientists conducting the experiment find that a subcritical Hopf bifurcation is the likely explanation.

An important nonlinear effect often seen in the presence of a subcritical Hopf bifurcation is hysteresis, which refers to the following somewhat paradoxical phenomenon. Assume a system is being observed, and shows a particular behavior at a given parameter setting. Next, the parameter is changed, and then returned to the original setting, whereupon the system displays a behavior completely different from the original behavior. This change in system behavior, which depends on the coexistence of two attractors at the same parameter value, is called **hysteresis**.

We have all seen this effect before. If you start with an egg at room temperature, raise its temperature to 100°C. for three minutes, and then return the egg to room temperature, the state of the egg will have changed. However, you

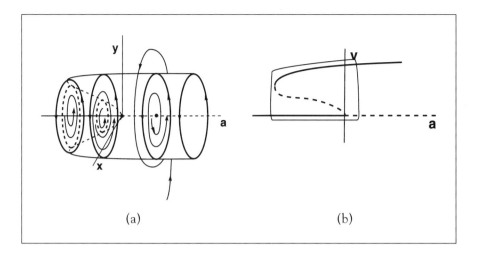

Figure 11.25   A subcritical Hopf bifurcation with hysteresis.
(a) There is a bifurcation at $a = 0$ from the path $r = 0$ of equilibria. At this point the equilibria go from stable to unstable, and a path of unstable periodic orbits bifurcates. The periodic orbits are unstable and extend back through negative parameter values, ending in saddle node at $a = -1$. An additional path of attracting periodic orbits emanates from the saddle node. (b) Schematic diagram of bifurcation paths. The rectangle shows a hysteretic path. The vertical segments correspond to sudden jumps.

may not be used to seeing this behavior in a reversible dynamical system such as a differential equation.

Figure 11.25 shows one way in which a system can undergo hysteresis. If the system is started at the (attracting) equilibrium at 0 for parameter $a < 0$, it will stay there. When the parameter is increased to $a > 0$, the equilibrium at 0 becomes unstable, and any small perturbation will cause the orbit to be attracted to the outside stable periodic orbit. When the parameter is decreased back to the starting value $a < 0$, the orbit stays on the periodic orbit, unless a large external perturbation moves it back inside the basin of the equilibrium. Hysteretic effects can be seen in the data collected in Lab Visit 11.

Figure 11.25 shows the stable and unstable periodic orbits being created together in a saddle-node bifurcation. If the parameter is decreased to extreme negative values, beyond the saddle-node point, then the system will of course return to equilibrium behavior.

EXAMPLE 11.23

A subcritical Hopf bifurcation is seen in the system

$$\dot{r} = ar + 2r^3 - r^5$$

$$\dot{\theta} = 1. \tag{11.15}$$

The bifurcation diagram of equilibria and periodic orbits of equations (11.15) is shown in Figure 11.25. There is a subcritical Hopf bifurcation at $a = 0$ from the path $r = 0$, at which point a path of unstable orbits bifurcates. This path contains unstable periodic orbits and extends through negative a values, ending in a saddle node bifurcation at $a = 1$. A path of stable orbits emanates from the saddle node. Thus for the interval $-1 < a < 0$ of parameter values there are two attractors: an attracting equilibrium and an attracting periodic orbit.

✎ EXERCISE T11.19

Find all periodic orbits in the system of equations (11.15), and check their stability. Verify that Figure 11.25 is the bifurcation diagram.

Example 11.19 was our first example of a one-parameter family of differential equations that satisfied the hypotheses of the Hopf Bifurcation Theorem. In the path of periodic orbits that bifurcates from the origin, there is exactly one periodic orbit for each parameter value when $a > 0$. However, the path of bifurcating orbits need not follow this pattern. In the following example, the entire path of orbits in $(a, \mathbf{v})$ space exists at only one parameter value, $a = 0$. The path must, however, emanate from $\mathbf{v} = 0$.

EXAMPLE 11.24

Let

$$\dot{x} = ax + y$$

$$\dot{y} = ay - x \tag{11.16}$$

for x, y, and a in $\mathbb{R}$. This one-parameter family of linear differential equations has a Hopf bifurcation from the path $\bar{\mathbf{v}} = \mathbf{0}$ of equilibria at $a = 0$. (Check the path of eigenvalues to show that they satisfy the hypotheses of Theorem 11.21.) In this case, however, instead of having one periodic orbit of the bifurcating family at each $a > 0$ (or, analogously, $a < 0$), the entire family of periodic orbits exists only at $a = 0$.

✎ EXERCISE T11.20

Find all periodic orbits of the one-parameter family (11.16).

EXAMPLE 11.25

(Lorenz Equations) As we discussed in Chapter 9, the system of equations below was first studied by Lorenz and observed to have chaotic attractors. For (x, y, z) in $\mathbb{R}^3$, let

$$\frac{dx}{dt} = \sigma(y - x) \tag{11.17}$$

$$\frac{dy}{dt} = rx - y - xz \tag{11.18}$$

$$\frac{dz}{dt} = -bz + xy,$$

for constant $\sigma > 0, r > 0$, and $b > 0$.

The origin is an equilibrium for all values of the parameters. For $r > 1$, the points $c^{\pm} = (\pm\sqrt{b(r-1)}, \pm\sqrt{b(r-1)}, r-1)$ are also equilibria. For $r < 1$, the origin is an attractor. For $r > 1$, the origin is unstable, while the points $c^{\pm}$ are attracting. The bifurcation at $r = 1$ is shown in Figure 11.26 for specific values of b and σ. There is also a subcritical Hopf bifurcation at $r = r_h = \frac{\sigma(\sigma+b+3)}{\sigma-b-1}$. This value is derived in Exercise 11.21.

✎ EXERCISE T11.21

(a) Show that the degree-three monic polynomial $\lambda^3 + a_2\lambda^2 + a_1\lambda + a_0$ has pure imaginary roots if and only if $a_1 a_2 = a_0$. Show that the roots are $\lambda = -a_2$ and $\lambda = \pm i\sqrt{a_1}$. (b) Evaluate the Jacobian of the Lorenz vector field at the equilibria $c^{\pm}$ and show that these equilibria have pure imaginary eigenvalues $\lambda = \pm i\sqrt{\frac{2\sigma(\sigma+1)}{\sigma-b-1}}$ at $r = \frac{\sigma(\sigma+b+3)}{\sigma-b-1}$.

Using the standard settings $\sigma = 10, b = 8/3$, the Hopf bifurcation point is $r_h = 24.74$. As the periodic orbits created at this subcritical Hopf bifurcation are followed to the left toward $r = 13.926$ in Figure 11.26, their periods continuously

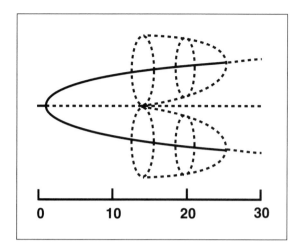

Figure 11.26 A Hopf bifurcation in the Lorenz equations.
A bifurcation diagram of the Lorenz equations for $\sigma = 10$, $b = 8/3$, and $0 \leq r \leq 30$, is shown. At $r = 1$, the origin goes from stable to unstable, as two attracting equilibria bifurcate. Paths of stable orbits are indicated by solid curves, while paths of unstable orbits are represented by dashed curves and circles. A subcritical Hopf bifurcation occurs at $r = 24.74$, at which point two families of unstable orbits bifurcate simultaneously as the two attracting equilibria lose their stability. Typical trajectories in computer simulations then move toward the chaotic attractor, first observed to occur at the crisis value $r = 24.06$.

increase and diverge to ∞. The behavior at $r = 13.926$ is quite complicated—this is the value for the onset of chaos in the Lorenz equations.

There are chaotic trajectories but no chaotic attractor for $13.926 < r < 24.06$. A chaotic attractor is observed to first occur near $r = 24.06$. See (Kaplan and Yorke, 1979) for specifics of this estimation. At exactly $r = 24.06$, a typical trajectory undergoes a long transient before being attracted by one of the sinks, as shown in Figure 9.5. Therefore, throughout the parameter range $24.06 < r < 24.74$, the Lorenz system possesses three attractors: one apparently chaotic and two which are not, the attracting equilibria.

☞ CHALLENGE 11

Hamiltonian Systems and the Lyapunov Center Theorem

WE HAVE SEEN that the stability of an equilibrium is often determined by the Jacobian of the vector field. Theorem 7.18 states that when the eigenvalues of the Jacobian all have negative real part, the equilibrium is asymptotically stable. When a single eigenvalue has positive real part, the equilibrium is unstable. What happens when the eigenvalues are pure imaginary (have real part equal to zero)? In general, an equilibrium is called a **center** if the Jacobian of the vector field has only pure imaginary eigenvalues.

We want to investigate the behavior of a system near a center. We can start to answer the question by looking at the linear case. For a two-dimensional linear system with eigenvalues $\pm bi$, the origin is an equilibrium, and all other solutions are periodic orbits of period $2\pi/b$. For a nonlinear system, it turns out that a center can be attracting, repelling, or neither. However, there is an important type of nonlinear system for which we can say that the behavior is stable near a center.

In Chapter 7, we saw examples of conservative systems such as the undamped pendulum. Trajectories of conservative systems are constrained to one energy level. In other words, energy remains constant along a solution. The set of equations that govern the motion of the pendulum is one example of a general class of differential equations called **Hamiltonian** systems. Specifically, for a smooth function $\mathbf{H} : \mathbb{R}^2 \to \mathbb{R}$, called the **Hamiltonian function** or **energy function**, Hamilton's equations are:

$$\dot{x} = \frac{\partial \mathbf{H}}{\partial y}$$

$$\dot{y} = -\frac{\partial \mathbf{H}}{\partial x}. \qquad (11.19)$$

Theorem 11.26 guarantees that in a Hamiltonian system, an equilibrium with pure imaginary eigenvalues will have a neighborhood containing periodic orbits. Your job in Challenge 11 is to construct a proof of this theorem in the two-dimensional case.

Theorem 11.26 (Lyapunov Center Theorem.) *Assume that 0 is a center equilibrium of the Hamiltonian system (11.19) and that $\pm bi$ are simple eigenvalues of* the Jacobian A of the vector field at 0. Assume further that no other eigenvalue of A is an integer multiple of bi. Then each neighborhood of the center contains periodic orbits whose periods approaches $2\pi/b$ as they approach the center.

The strategy is to show that the theorem is a consequence of the Hopf Bifurcation Theorem. To do this, you will build a one-parameter family of differential equations for which (11.19) is the member at $a = 0$. Then show that a Hopf bifurcation occurs at $a = 0$, but that no member of the family for $a \neq 0$ has periodic orbits. The only remaining possibility is that the periodic orbits emanating from the Hopf bifurcation point must all occur for $a = 0$, that is, they are periodic orbits surrounding the center.

Step 1 To warm up, verify that (11.19) implies that $\dot{H}(\mathbf{v}) = 0$, where $\mathbf{v} = (x, y)$ and

$$\dot{H}(\mathbf{v}) = \frac{\partial H}{\partial x}\frac{dx}{dt} + \frac{\partial H}{\partial y}\frac{dy}{dt}.$$

Therefore the Hamiltonian H is a constant of the motion.

These equations can be generalized to higher *even* dimensions in which case $\mathbf{H}$ is defined on $\mathbb{R}^{2n}$. Then for $(x_1, \ldots, x_n, y_1, \ldots, y_n) \in \mathbb{R}^{2n}$, the two equations of system 11.19 become the $2n$ equations $dx_i/dt = \partial H/\partial y_i$, $i = 1, \ldots, n$, and $dy_i/dt = -\partial H/\partial x_i$, $i = 1, \ldots, n$.

Now we assume the Hamiltonian satisfies the hypotheses of the two dimensional Lyapunov Center Theorem; that the origin 0 is an equilibrium and that $\pm bi$ are eigenvalues of the Jacobian at 0. Embed the Hamiltonian system in a one-parameter family of differential equations:

$$\dot{x} = a\frac{\partial H}{\partial x} + \frac{\partial H}{\partial y}$$

$$\dot{y} = a\frac{\partial H}{\partial y} - \frac{\partial H}{\partial x}, \tag{11.20}$$

where a is a scalar parameter. Notice that the original Hamiltonian system occurs at $a = 0$.

Step 2 Show that $\{(a, 0) : a \in \mathbb{R}\}$ is a path of equilibria.

Using the notation **gradH** for the gradient of $\mathbf{H}$, and letting $\mathbf{J} = \begin{pmatrix} 0 & 1 \\ -1 & 0 \end{pmatrix}$, we can rewrite the system (11.20) in the following matrix form:

$$\dot{\mathbf{v}} = (a\mathbf{I} + \mathbf{J})\mathbf{gradH}(\mathbf{v}).$$

Step 3 Show there are no periodic (nonequilibrium) orbits of (11.20) for $a \neq 0$. [Hint: Let $(x_a(t), y_a(t))$ be a solution of (11.20). Verify that $(dH/dt)(x_a(t), y_a(t)) = a|\text{grad}H|^2$.]

The goal now is to prove that (11.20) has a Hopf bifurcation at $a = 0$, in which case all bifurcating periodic orbits must occur at $a = 0$ (the original Hamiltonian). To that end, we need to show that a path of eigenvalues of the linearized system crosses the imaginary axis at ib when $a = 0$.

Let $f_a(v) = (a + J)\text{grad}H(v)$. Then $D_v f_a(0) = (aI + J)M$, where M is the 2×2 matrix of second partial derivatives of H (with respect to the phase variables) evaluated at $v = 0$.

Step 4 Show that M is a nonsingular symmetric matrix, and that its inverse M^{-1} is also.

Step 5 Show that

$$\dot{v} = JMv \tag{11.21}$$

is a Hamiltonian system with Hamiltonian function $E(v) = \frac{1}{2}v^T M v$.

Step 6 Let $c(a) + d(a)i$ be a path of eigenvalues of $D_v f_a(0) = (aI + J)M$. Conclude from Steps 3 and 4 that if $a \neq 0$ then $c(a) \neq 0$.

Step 7 Show that if λ is an eigenvalue of $(aI + J)M$, then $-\lambda$ is an eigenvalue of $(-aI + J)M$. [Hint: If $(aI + J)M - \lambda I$ is singular, so is $M^{-1}[(aI + J)M - \lambda I]^T M$.]

Step 8 Use Steps 6 and 7 to conclude that the eigenvalues of $D_v f_a(0)$ cross the imaginary axis at $\pm bi$ when $a = 0$.

Step 9 Use the weak version of the Hopf Bifurcation Theorem in Remark 11.22 to show that there are periodic orbits of period approaching $2\pi/b$ in every neighborhood of the origin.

EXERCISES

11.1. Find a saddle-node bifurcation and a period-doubling bifurcation for $f_a(x) = a \sin x$.

11.2. Find a period-doubling bifurcation for $g_a(x) = -x(a + x)$. For what parameter value a does the period-two orbit itself go through a period-doubling?

11.3. Find a period-doubling bifurcation for $f_a(x) = x^3 - ax$.

11.4. Let
$$T(a, x) = \begin{cases} ax & \text{if } x \le \frac{1}{2} \\ a(1 - x) & \text{if } x > \frac{1}{2} \end{cases}$$
Draw the tent map bifurcation diagram for (a, x) in $[0, 2] \times [0, 1]$.

11.5. Find all saddle-node and period-doubling bifurcations of fixed points for the following maps.

(a) $f_a(x) = a + \ln x$, for $x > 0$.

(b) $f_a(x) = a - \ln x$, for $x > 0$.

(c) $f_a(x) = a \ln x$, for $x > 0$.

(d) $f_a(x) = a + x - e^x$

11.6. Let $\mathbf{f}_a(x_1, x_2) = (ax_1 - x_2^2, x_1 + ax_2)$. Show that the family $\mathbf{f}_a$ has two paths of fixed points (paths in a). Find parameter values at which $+1$ is an eigenvalue of $D_v(\mathbf{f})$ evaluated along these paths. Explain what happens to the stability of the fixed points at these parameter values.

11.7. Find the saddle-node and period-doubling bifurcations of fixed points for the Hénon map $f_a(x, y) = (a - x^2 + 0.3y, x)$.

11.8. Give an example of a one-parameter family of scalar maps f and a fixed point $\bar{x}$ of $f_{\bar{a}}$ such that $f_{\bar{a}}'(\bar{x}) = -1$ and there is not a period-doubling bifurcation at $(\bar{a}, \bar{x})$.

11.9. (Period-doubling bifurcation.) In this exercise, a specific condition on derivatives is shown to guarantee that period-doubling bifurcations are isolated. Assume that $f_a(x) = f(a, x)$ is smooth (partial derivatives of order at least 2 exist and are continuous). Let $(\bar{a}, \bar{x})$ be a fixed point of f with $\frac{df}{dx}(\bar{a}, \bar{x}) = -1$. Since this value is not $+1$, there is a path of fixed points $(a, x(a))$ for a near $\bar{a}$ with $x(\bar{a}) = \bar{x}$.

(a) Show that if
$$\frac{d^2f}{dadx}(\bar{a}, \bar{x}) + \frac{1}{2}\frac{d^2f}{dx^2}(\bar{a}, \bar{x})\frac{df}{da}(\bar{a}, \bar{x}) \ne 0, \tag{11.22}$$

then $\dfrac{d^2f}{dadx}(a, x(a)) \ne 0$.

(b) Show that (11.22) implies there is a neighborhood of $(\bar{a}, \bar{x})$ in which there are no other fixed points (a^*, x^*) satisfying the period-doubling condition $\frac{df}{dx}(a^*, x^*) = -1$.

11.10. Unlike a saddle node orbit, a period-doubling bifurcation can be either supercritical or subcritical. In other words, the bifurcation orbit for a family of scalar maps can be either attracting or repelling. In this problem, two coefficients in the Taylor series expansion for the map at the bifurcation value are shown to determine the stability of the bifurcation orbit.

(a) Let $f(x) = x + cx^3$. Show that the fixed point $x = 0$ is attracting if $c < 0$ and repelling if $c > 0$.

(b) Assume that $x = 0$ is a fixed point of a scalar map g and that $g'(0) = -1$. (In this exercise there is no parameter: g is the map at the bifurcation parameter value.) Let

$$g(x) = -x + bx^2 + dx^3 + \text{higher order terms.}$$

Show that 0 is attracting if $b^2 + d > 0$ and repelling if $b^2 + d < 0$.

11.11. For a real number c, define the one-parameter family $f_a(x) = (x - a)(2x - 3a) + x + c$. For what values of c is there a bifurcation in this family? Describe the bifurcations and list the bifurcation points $(\bar{a}, \bar{x})$. Note that $c \neq 0$ corresponds to generic bifurcations while $c = 0$ corresponds to a special one.

11.12. Denote by $\dot{\mathbf{v}} = \mathbf{f}(\mathbf{v})$ the system which gives the supercritical Hopf bifurcation shown in Figure 11.23. Now draw the corresponding two diagrams for $\dot{\mathbf{v}} = -\mathbf{f}(\mathbf{v})$. Which type of Hopf bifurcation is this?

11.13. Consider the Chua circuit equations as defined in (9.6), with parameters set as in Figure 9.10. Use Exercise T11.21 to find the parameter c_3 for which the system undergoes a Hopf bifurcation.

11.14. Let

$$\dot{x} = r\left(\frac{c}{x} - d\right) + ax - bxy$$

$$\dot{y} = r\left(b - \frac{a}{y}\right) - cy + dxy \qquad (11.23)$$

where $a, b, c,$ and d are positive constants, and $r, x,$ and y are scalar variables, $x \neq 0, y \neq 0$. Notice that when $r = 0$, the system models a simple predator-prey interaction, as we described in Chapter 7.

(a) Show that $(x, y) = (\frac{c}{d}, \frac{a}{b})$ is an equilibrium for all r.

(b) Show there is a Hopf bifurcation at $(\frac{c}{d}, \frac{a}{b})$ when $r = 0$.

(c) Show that all bifurcating periodic orbits occur at $r = 0$.

495

☞ LAB VISIT 11

Iron + Sulfuric Acid ⟶ Hopf Bifurcation

THIS EXAMPLE of electric current oscillations in an electrochemical reaction exhibits a subcritical Hopf bifurcation and hysteresis. These dynamical ideas are fairly new compared to the long history of electrochemistry. The first electrochemical system showing periodic orbits dates from G. Fechner in 1828, who found an oscillating current caused by the competing forces of depositing and dissolving of silver on an iron electrode placed in a silver nitrate solution.

The experiment shown here was designed to explore oscillatory phenomena in the dissolving of iron in a sulfuric acid solution. A 99.99% iron rod with a diameter of 3 mm is lowered into 200 ml of the acid. When a potential difference is applied, the current of the electrochemical system is a measure of the overall reaction rate between the electrode surface and the electrolytic solution. The behavior of the current as a function of time shows considerable complication at certain parameter settings. The so-called electrodissolution problem is far from completely understood, due to the large number of coupled chemical reactions involved.

The electrical potential applied to the electrode is used as a bifurcation parameter in this experiment. Figure 11.27 shows a small sample of the interesting dynamical phenomena in this reaction. Parts (a)–(d) of the figure show the measured current (in milliamperes) as a function of time (in seconds), for four different settings of the potential. The potential increases from (a) to (d). Figure 11.27(a) shows a short transient followed by a constant current, which continues for the slightly higher potential in (b). In Figure 11.27(c), small irregular oscillations have developed, which seem aperiodic. Another small change in potential leads to clear periodic oscillations of larger amplitude in (d). Apparently, a Hopf bifurcation has occurred.

The time series in Figure 11.28 are from the same electrochemical experiment, except that in this case, the voltage parameter is decreased from part (a) to (d). Viewed together, the behaviors shown in Figures 11.27 and 11.28 can be

Sazou, D., Karantonis, A., and Pagitsas, M. 1993. Generalized Hopf, saddle-node infinite period bifurcations and excitability during the electrodissolution and passivation of iron in a sulfuric acid solution. Int. J. Bifurcations and Chaos **3**: 981–997.

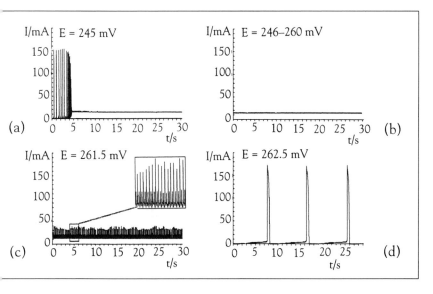

Figure 11.27 Current time series for increasing parameter.
Moving to the right along the solid a-axis in the subcritical Hopf bifurcation of
Figure 11.25. (a) Potential is 245 mV. Transient yields to steady state. (b) 246–260
mV show steady state. (c) 261.6 mV, oscillations of small amplitude and period.
(d) 262.6 mV, nine-second oscillations.

interpreted as being the result of a subcritical Hopf bifurcation, as seen earlier in
this chapter.

Referring to Figure 11.25, for large values of the parameter, the periodic
orbit is the only attractor, and so it is globally attracting. As the parameter
moves to the left, the trajectory continues its periodic behavior even as the Hopf
bifurcation point is passed. Only after the saddle-node bifurcation point is passed,
on the far left of Figure 11.25, does the system relax to the steady state. Thus
hysteresis is seen. As the parameter increases along the negative x-axis, we see
one system behavior, while when the parameter is decreased through the same
parameter range, we see a different behavior.

Figure 11.28 demonstrates this type of hysteresis in the electrochemical
experiment. The voltage parameter is decreased from part (a) to part (d), crossing
the same parameter range that was crossed in an increasing manner in Figure
11.27. This time, as expected for a subcritical Hopf bifurcation with hysteresis,
periodic behavior is recorded until the voltage is decreased sufficiently, to pass the
saddle-node bifurcation point. After this value, orbits are attracted to the steady
state as shown in Figure 11.28(d).

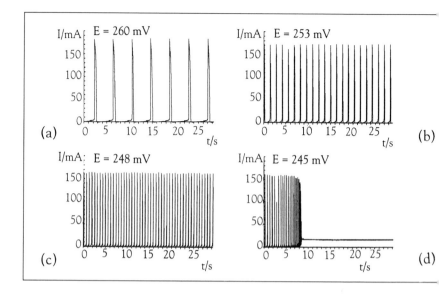

Figure 11.28   Current time series for decreasing parameter.
Moving to the left along the upper (periodic) branch of Figure 11.25. (a) Potential difference is 260 mV, periodic orbit. (b) 253 mV, periodic. (c) 248 mV, periodic. (d) 245 mV, periodic orbit disappears, convergence to steady state after transient.

The small aperiodic oscillations that are apparent in Figure 11.27(c) are unexplained by this theory and may be due to the presence of "noise" near the Hopf bifurcation point. The theory of the Hopf bifurcation is developed for finite-dimensional differential equations, and applies to the extent that this experiment is well-modeled by them. Model inaccuracies may matter at some times more than others. In particular, they may be magnified at parameter values where stability is in flux, such as in Figure 11.27(c), thereby limiting the explanatory power of the model.

CHAPTER TWELVE

Cascades

CASCADES OF period-doubling bifurcations have been seen in the great majority of low-dimensional systems that exhibit chaotic behavior. A "cascade" appears as an infinite sequence of period-doubling bifurcations. A stable periodic orbit is seen to become unstable as a parameter is increased or decreased and is replaced by a stable periodic orbit of twice its period. This orbit in turn becomes unstable and is replaced by a new stable orbit with its period again doubled, and the process continues through an infinity of such period-doubling bifurcations.

The goal of this chapter is to study properties of cascades, and to understand what general conditions imply their existence. We continue the study of bifurcations from Chapter 11 and develop a picture of families of orbits as connected sets, where the primary bifurcations are period doublings and saddle nodes. Our main focus will be families of one-dimensional maps, such as the quadratic family, and families of area-contracting planar maps, such as the Hénon family.

Cascades consist of stable periodic orbits and so are experimentally observable. An experimenter will only be able to see a few of these stages. Some state-of-the-art examples of cascades in experiments are shown in Lab Visit 12.

Renormalization theory has been extremely successful in showing that it is possible for the full infinite sequence to exist. There are a number of regularities of cascades, yielding universal numbers that are frequently observed in both physical and numerical experiments. The chapter begins with the Feigenbaum constant, one of these universal numbers. Challenge 12 is a more in-depth exploration of this remarkable universal behavior.

12.1 CASCADES AND 4.669201609...

Figure 12.1 hints at the complexity present within a single bifurcation diagram. The one-dimensional map used to generate the diagram is $f_a(x) = a - x^2$. Stable periodic orbits of periods one, two, four, and eight are clearly visible in the computer simulation. Factors of 2 beyond 8 are visible when the diagram is magnified.

In addition to the period-doublings that we can see in Figure 12.1, there are infinitely many that are not visible. Evidence of some of them can be seen in the magnifications of Figure 12.2. At a period-doubling bifurcation from a period-k orbit, two branches of period-$2k$ points emanate from a path of period-k points. When the branches split off, the period-k points change stability (going from attractor to repeller, or vice versa). This change is detected in the derivative of f^k which, when calculated along the path of fixed points, crosses -1.

Figure 12.2(a) shows a magnified view of the box drawn in Figure 12.1, and together with the further magnification in Figure 12.2(b), shows period-doubling bifurcations up to period 64. A period-doubling cascade can also occur from a periodic orbit of higher period. Magnifications of the period three window of Figure 12.1 are shown in Figure 12.2(c) A cascade beginning with sinks of periods 3, 6, 12, . . . can be seen. In Figure 12.2, the period five cascade exhibits sinks of periods 5, 10, 20, ...

The series of bifurcation diagrams shown in Figure 12.1, 12.2(a) and (b) suggests a scaling behavior in the cascade of the quadratic map. In 1978, M. Feigenbaum noted that the ratios of parameter distance between two successive period-

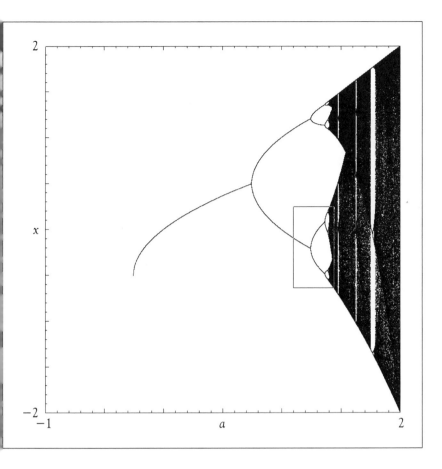

Figure 12.1 Bifurcation diagram for the one-dimensional quadratic family
$f_a(x) = a - x^2$.

The variable x is on the vertical axis, and the bifurcation parameter a is on the horizontal axis. When a is fixed at a value less than $-1/4$, the orbits of all initial conditions diverge to $-\infty$. There is a saddle-node bifurcation at $a = -1/4$, at which a period-one sink is created. It persists for $-1/4 < a < 3/4$. At $a = 3/4$, the period-one sink loses stability and is replaced with a period-two sink in a period-doubling bifurcation. Higher values of a show further bifurcations that create more complex attractors. For $a > 2$ there are no attractors.

doublings approach a constant as the periods increase to infinity. Moreover, this constant is universal in the sense that it applies to a variety of dynamical systems. Specifically, if the nth period-doubling occurs at $a = a_n$, then

$$\lim_{n\to\infty} \frac{a_n - a_{n-1}}{a_{n-1} - a_{n-2}} = 4.669201609\ldots, \tag{12.1}$$

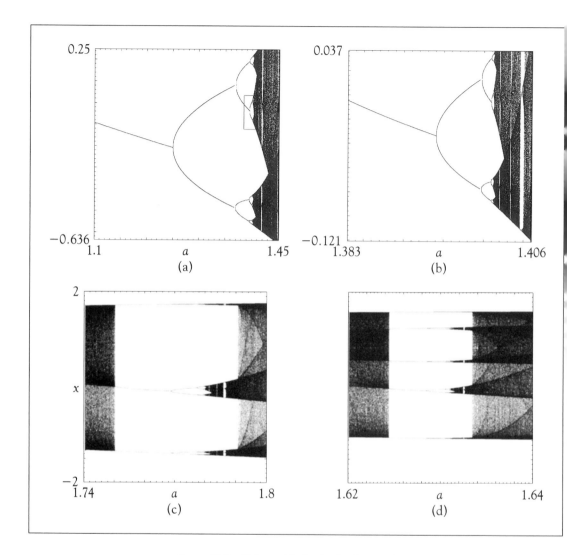

Figure 12.2 Universal behavior in the quadratic family.
(a) The box drawn in Figure 12.1 is enlarged to show many similarities to the full-size diagram. Cascades occur on increasingly fine scales. (b) Enlargement of the box in (a). (c) The period-three window in the quadratic family shows period-doubling cascades. (d) The period-five window.

a number now known as Feigenbaum's constant. Surprisingly, the limit is the same for any one-parameter family of unimodal maps with negative Schwarzian derivative. The first proof that this limit exists was given in (O. Lanford, 1982).

Table 12.1 shows a list of the parameters at which period-doublings occur in the quadratic map $f(x) = a - x^2$. The fixed point bifurcates into a fixed saddle

Period	Parameter a	Ratio
2	0.75	
4	1.25	
8	1.3680989	4.2337
16	1.3940462	4.5515
32	1.3996312	4.6458
64	1.4008287	4.6639
128	1.4010853	4.6682
256	1.4011402	4.6689

Period	Parameter a	Ratio
3	1.75	
6	1.7685292	
12	1.7772216	
24	1.7792521	4.2810
48	1.7796964	4.5698
96	1.7797923	4.6363
192	1.7798129	4.6524

Table 12.1   Feigenbaum's constant in the quadratic map.
(a) A list of parameters a_n at which the nth period-doubling bifurcation occurs in the period-one cascade, along with the ratio $(a_n - a_{n-1})/(a_{n-1} - a_{n-2})$. (b) Same for a period-three cascade of the quadratic map.

and a period-two attractor at $a = 0.75$, followed by a cascade of period-doublings. Also shown are the bifurcation values for the period-three cascade of the quadratic map.

A computer program to find the bifurcation parameter values uses a binary search method. To begin the determination of a period-doubling bifurcation value, say from period four to period eight, two values of the parameter a are chosen that bracket the bifurcation point. The period of the orbit at the midpoint of the bracketing interval is determined using the following simple method. First, a long trajectory is created, in order to approach the current attracting orbit as closely as possible. The period of the orbit is tested by comparing the current point to later iterates of the point, within a small tolerance. When the period is determined, either four or eight, the midpoint becomes a new endpoint of the bracketing interval, replacing the endpoint that has the same period. The length of the bracketing interval has been cut in two. By repeating this process, accurate estimates of the bifurcation values of a cascade can be determined.

Table 12.2 contains a list of bifurcation values for two other cascades. Note that the ratio (12.1) is repeated for the one-dimensional logistic family as well as for the two-dimensional Hénon family. This version of the Hénon map, given by $f(x, y) = (a - x^2 - 0.3y, x)$, is orientation-preserving (has positive Jacobian determinant).

Period	Parameter a	Ratio		Period	Parameter a	Ratio
2	3.0000000			2	1.2675000	
4	3.4494896			4	1.8125000	
8	3.5440903	4.7514		8	1.9216456	4.9933
16	3.5644073	4.6562		16	1.9452006	4.6337
32	3.5687594	4.6683		32	1.9502644	4.6516
64	3.5696916	4.6686		64	1.9513504	4.6630
128	3.5698913	4.6692		128	1.9515830	4.6678
256	3.5699340	4.6694		256	1.9516329	4.6688

Table 12.2 Feigenbaum's constant in the logistic map and Hénon map.
(a) Parameter values of period-doubling bifurcations for $f(x) = ax(1 - x)$. (b) The ratio also approaches Feigenbaum's constant for the Hénon map.

⇨ **COMPUTER EXPERIMENT 12.1**

Locate the period-doubling cascade for the orientation-reversing Hénon map $f(x, y) = (a - x^2 + 0.3y, x)$. Start with $a = 0$ and the fixed point $(x, y) = (0, 0)$. Calculate the ratio of successive period-doubling intervals as in Table 12.2.

12.2 SCHEMATIC BIFURCATION DIAGRAMS

Figure 12.3 shows cascades in the development of four distinct chaotic attractors occurring in the Poincaré map of the forced, damped pendulum. Notice that the attractors are simultaneously present for a certain range of the parameter. Studying the dynamics here is complicated by the fact that we have no explicit formula for the underlying Poincaré map. Our aim in this section is to develop a methodology for efficient analysis of bifurcation diagrams—to give a road map of essential features, even when the underlying equations are unknown.

It is convenient to think of a schematic "tinker toy" model, consisting of "sticks" and "sockets", for the periodic orbits in our bifurcation diagrams. The "sticks" are paths of hyperbolic orbits, all of one type—stable or unstable, while the "sockets" are bifurcation orbits. For the remainder of this chapter, we characterize hyperbolic orbits as stable or unstable, allowing the reader to translate these words within the particular system of interest. For one-parameter families of one-dimensional maps, the stable hyperbolic orbits are periodic sinks, while

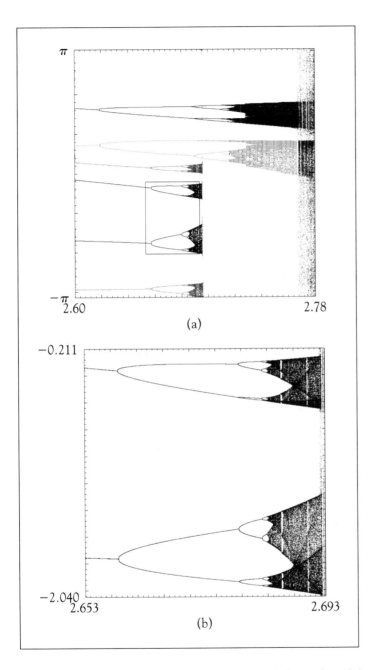

Figure 12.3   Period-doubling cascades in the forced, damped pendulum.
(a) The development of four chaotic attractors is shown, in four varying shades
of gray/black. Each of two period-two sinks at the bottom of the diagram evolve
through period-doublings to a two-piece chaotic attractor. Two fixed-point sinks at
the top evolve to one-piece chaotic attractors. Several crises can be observed. (b) A
magnification of the one of the cascades beginning from a period-two sink is shown.

the unstable hyperbolic orbits are periodic repellers. For one-parameter families of area-contracting planar maps, stable orbits are periodic sinks, while unstable orbits are saddles. These two examples are the primary settings in which we observe cascades in this chapter.

We will find our bookkeeping simplified by considering the several points of a periodic orbit as a single entity. This point of view sacrifices little, since our primary focus is on the stability of the orbit, which is a collective property. Recall that in order to determine the stability of a periodic orbit, all points in the orbit— together with their derivatives (or Jacobian matrices)—must be calculated. For example, assume that $\mathbf{x_1}$ is a period-two point of $\mathbf{f}$: $\mathbf{f(x_1)} = \mathbf{x_2}$ and $\mathbf{f(x_2)} = \mathbf{x_1}$. In order to find out the stability of the orbit, we must compute the eigenvalues of $\mathbf{Df^2(x_1)} = \mathbf{Df(x_2)Df(x_1)}$. Lemma 2 in Appendix A implies that the eigenvalues of $\mathbf{Df(x_2)Df(x_1)}$ are identical to the eigenvalues of $\mathbf{Df(x_1)Df(x_2)} = \mathbf{Df^2(x_2)}$. In the one-dimensional case (see, for example, the Stability Test for periodic orbits in Section 1.4), the derivative $(f^k)'$ is the same when evaluated at any of the k points in a period-k orbit.

Therefore, we will think of a periodic orbit as a single object, with all points in the orbit being represented by one point in our schematic model. The phrase "derivative of a periodic orbit" will refer to the derivative of f^k with respect to x evaluated at any point in the orbit. (For phase space dimensions greater than one, this phrase will have to be interpreted appropriately as an eigenvalue of $\mathbf{D_x f^k}$.) We always take k to be the *minimum* period of the orbit.

Figure 12.4 is a guide to the various types of periodic orbits we will encounter in cascades. An unstable periodic point p of period k is called a **regular repeller** if $(f^k)'(p) > 1$; it is called a **flip repeller** if $(f^k)'(p) < -1$. In other words, we have partitioned the set of hyperbolic periodic orbits into three subsets: stable, regular unstable, and flip unstable. We call these sets $\mathbf{S}$ (for stable), $\mathbf{U_+}$ (for regular unstable), and $\mathbf{U_-}$ (for flip unstable). Definition 12.1 makes precise what the "sticks" are in our tinker-toy model.

Definition 12.1   A maximal path of hyperbolic fixed points or periodic orbits is called a **schematic branch** (or just **branch**). In the case of a schematic branch of periodic orbits, one point on the branch represents all points in one orbit.

Figure 12.4(b) shows the bifurcation diagram of a family of maps with a period-three saddle node and a period-doubling bifurcation from the path of stable period-three orbits. A schematic bifurcation diagram representing the same family appears in (c), with the bifurcation orbits drawn as circles. The saddle node bifurcation is indicated schematically by a circle with a plus sign $(+)$ inside it,

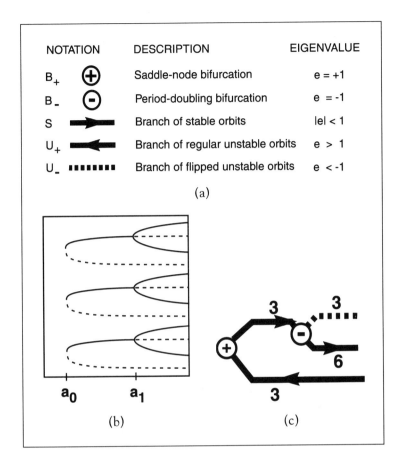

Figure 12.4 Elements of schematic diagrams.
(a) A saddle-node bifurcation is indicated schematically by a circle with a plus sign (+) inside it. The saddle node orbit has derivative +1 (or one eigenvalue equal to +1, in two or more dimensions). A period-doubling bifurcation is drawn with a minus sign (−). The map has derivative −1 (or one eigenvalue equal to −1) at this orbit. Solid segments represent branches of stable orbits or branches of regular unstable orbits, with the branches of stable orbits directed with an arrow pointed to the right and the branches of regular unstable orbits directed to the left. All points in an orbit are represented by one point in the schematic diagram. The dashed schematic branch represents the three paths of flip unstable orbits. Regular unstable orbits have exactly one eigenvalue larger than +1, while flip unstable orbits have exactly one eigenvalue smaller than −1. The absolute values of all eigenvalues of a stable orbit are smaller than 1. (b) The bifurcation diagram of a family of one-dimensional maps with a period-three saddle node at $a = a_0$ and a period-doubling bifurcation at $a = a_1$ from the stable period-three orbit is shown. A stable period-six orbit appears at a_1. (c) A schematic bifurcation diagram representing the family in (b). The number next to a schematic branch represents the period of orbits on the branch.

representing the $+1$ derivative of that orbit, while the period-doubling bifurcation is drawn with a minus sign $(-)$. A saddle node of period three occurs in the logistic family for a slightly less than 3.83. The bifurcation diagram for this family is given in Figure 1.8(b).

Notice that emanating from the saddle node in Figure 12.4(c) there are two solid schematic branches: one branch, with an arrow directing it to the right, represents all three paths of stable period-three orbits in (b); the other branch, directed to the left represents the three paths of unstable U_+ orbits in (b). Emanating from the period-doubling bifurcation orbit in (c) is a solid schematic branch directed to the right, representing the six paths of stable period-six orbits in (b). (The rationale for putting arrows on the branches is not apparent now, but will become clear in Section 12.3.) The dashed schematic branch represents the three paths of U_- orbits in (b).

✎ **EXERCISE T12.1**

Why do the three period-doubling bifurcation points in Figure 12.4(b) occur at the same parameter value?

In schematic diagrams, such as Figure 12.4, points on the branches represent fixed points or periodic orbits that are isolated. As we saw in Chapter 11, if p is a hyperbolic fixed point of f, then there is a neighborhood N in phase space that excludes other fixed points. What about higher-period orbits? Since p is also a fixed point of f^k, and $(f^k)'(p) \neq +1$, there is a neighborhood N_k in phase space (perhaps smaller than N) such that p is the only fixed point of f^k in N_k; that is, there are no period-k points in N_k.

✎ **EXERCISE T12.2**

Assume that p is a hyperbolic fixed point. Let $\{p_n\}$ be a sequence of periodic points such that $p_n \to p$, and, for each n, let t_n be the (minimum) period of p_n. Explain why $\lim_{n\to\infty} t_n = \infty$.

How do schematic branches end? Suppose we try to follow the branch containing a particular hyperbolic fixed point at $a = a_0$. Let G be the portion of this branch in $[a_0, \infty) \times \mathbb{R}$. Figure 12.5 shows portions of four such schematic branches. Now assume that there exists a parameter value a_1, with $a_1 > a_0$, such that G does not extend to a_1. What can happen to G between a_0 and a_1? One possibility is that G may become unbounded in x-space. In other words, the

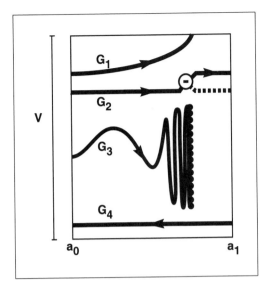

Figure 12.5 How schematic branches can end.
Schematic branch G_1 leaves the compact region $V \times [a_0, a_1]$ through the boundary of V; branch G_2 ends in a period-doubling bifurcation; branch G_3 ends in a line segment of fixed points; and G_4 extends through the parameter interval $[a_0, a_1]$.

absolute values of points on G go to infinity as or before G reaches a_1. Otherwise, there exists a bounded set V in x-space such that G remains in V; $G \subset [a_0, a_1] \times V$.

Since G is a path, it is defined as the graph of a function $\phi : J \to V$, where J is an interval of real numbers. That is, $G = \{(a, \phi(a)) : a \in J\}$. It follows that there exists a parameter value a_e such that $[a_0, a_e) \subset J \subset [a_0, a_e]$. We argue that J is open-ended on the right side. Suppose, on the contrary, that J contains the endpoint a_e. Then G has at least one fixed point p at $a = a_e$, according to Exercise T12.3. Since p is hyperbolic, G must extend beyond a_e (by Definition 11.5 and Theorem 11.6, Chapter 11), a contradiction.

The "half-branch" G must have at least one limit point at a_e. By continuity of the map, this point (or points) is fixed. By the above argument, any limit points of G that are not in G must be nonhyperbolic. Depending on the map, these nonhyperbolic fixed points may or may not be bifurcation orbits.

✎ **EXERCISE T12.3**

(a) Show that if G is a path of fixed points defined on $[a_0, a_e)$, and if G is bounded, then G has at least one limit point at $a = a_e$. (b) Show that any limit points of G are fixed points.

Figure 12.5 shows the fate of several schematic branches in a closed, bounded region $[a_0, a_1] \times V$: (1) G_1 leaves $[a_0, a_1] \times V$ through the boundary of V; (2) G_2 ends in a bifurcation orbit from which other branches emanate; (3) G_3 ends in an interval of fixed points; and (4) G_4 extends through the parameter interval $[a_0, a_1]$. We will concentrate below on families of maps where the nonhyperbolic fixed points and periodic orbits are all saddle-node or period-doubling bifurcations, thereby ruling out the fate of branch G_3.

12.3 GENERIC BIFURCATIONS

The word "generic" refers to behavior that is typical—behavior that we normally expect to see. In this section we explain the fact that for families of smooth maps on the real line, it is a generic property that the map has only saddle-node and period-doubling bifurcations. The point is that these two are the most commonly-seen bifurcations, and that other "nongeneric" bifurcations such as the pitchfork bifurcation occur rarely.

We will say that a property of smooth functions is **generic** if the set of functions that have the property is dense. This means that for each smooth f, whether it has the property or not, arbitrarily small perturbations of f have the property. We caution the reader that the term "generic" is often used in a more specialized form in the mathematical literature; consult a topology book for more details. In any case, the specialized usage implies the usage in this book.

An example of a generic property of smooth functions $f : [0, 1] \to \mathbb{R}$ is that the function has at most a finite number of zeros. (A **zero** of f is a number x for which $f(x) = 0$.) Not every smooth function has this property: The identically zero function has uncountably many. But functions arbitrarily close to the identically zero function, namely the constant functions $f(x) \equiv \epsilon$ for small ϵ, have no zeros. To show that this property is generic we need to find that these neighbors exist for *any* smooth function.

Before we check this fact, we mention another generic property. Let x_0 be a number in $[0, 1]$, and consider functions f with the following property:

$$(G_0) \quad f : [0, 1] \to \mathbb{R} \text{ and } f'(x_0) \neq 0 \text{ whenever } f(x_0) = 0.$$

Note that if $f(x_0) = 0$ and $f'(x_0) \neq 0$, then there is some neighborhood of x_0 in which it is the only zero. In fact, the Mean Value Theorem implies that

$$f(x) = f(x) - f(x_0) = f'(c)(x - x_0) \neq 0$$

for x close to but not equal to x_0, since c is trapped between x and x_0, and $f'(x_0) \neq 0$.

Notice that (G_0) implies that f has at most a finite number of zeros in a bounded set. If there were infinitely many zeros, we could choose a sequence x_i of zeros that converges to some point $\bar{x}$ in $[0, 1]$, which would also have to be a zero by the continuity of f. Because f satisfies (G_0), $f(\bar{x}) = 0$ and $f'(\bar{x}) \neq 0$, so that there must be some neighborhood of $\bar{x}$ in which it is the only zero, a contradiction.

It remains to explain why (G_0) is a generic property. This follows from Theorem 12.2, which can be found in (Hirsch, 1976). For a smooth function f, a critical point c satisfies $f'(c) = 0$.

Theorem 12.2 (Sard's Theorem) *Let $f : [0, 1] \to \mathbb{R}$ be a smooth function, and let C denote the set of critical points of f. Then $f(C)$ has measure zero.*

In this terminology, a function has the property (G_0) if and only if 0 does not belong to $f(C)$. The function $g_\epsilon(x) = f(x) + \epsilon$ has the same critical points as $f(x)$. Since $f(C)$ has measure zero, g_ϵ has property (G_0) for almost every real number ϵ, including values of ϵ arbitrarily near 0. Therefore functions g_ϵ arbitrarily near f are in (G_0), even if f isn't. This proves that property (G_0) is a generic property of smooth functions.

✎ **EXERCISE T12.4**

Let $g : \mathbb{R} \to \mathbb{R}$ be a smooth map, and assume that for each x, $g(x) = 0$ implies $g''(x) \neq 0$. Show that there are at most finitely many zeroes (points x for which $g(x) = 0$) in any bounded interval.

Another generic property for smooth functions h is

(G_1) $h : [0, 1] \to \mathbb{R}$ and $h''(x_0) \neq 0$ whenever $h'(x_0) = 0$.

Such a function h always has the property that there are only a finite number of x_0 for which $h'(x_0)$ is 0.

Next we look at a slightly different class:

(G_2) $k : [0, 1] \to \mathbb{R}$ and $k''(x_0) \neq 0$ whenever $k'(x_0) = 1$.

We can use this property to study certain bifurcation diagrams. Notice that if $k(x)$ satisfies (G_2), then $h(x) = x - k(x)$ satisfies (G_1). Also (G_2) guarantees that there are at most finitely many points x_0 at which $k'(x_0) = 1$.

We can also talk about generic properties of functions $k_a(x)$ that depend on a parameter. Define

$$k_a(x) = x + a - h(x),$$

where h satisfies (G_1). This is a rather special map depending on a parameter because a enters in a very simple way. Nonetheless, it is instructive because we can determine what the bifurcation diagram for fixed points of k must look like. In the bifurcation diagram, we plot the set $\{(a, x)\}$ of points at which $k_a(x) = x$.

This set is just $\{(h(x), x)\}$, the graph of $h(x)$ reflected about the 45° line: If we set $a = h(x)$ for a particular x, then (a, x) is on the graph of h and also $k_a(x) = x$. (Here we are ignoring all periodic orbits of period greater than 1.) At the saddle-node bifurcation points (a, x), $k_a(x) = x$ and the partial derivative $\partial k_a / \partial x(x) = 1$. Figure 12.6 shows the bifurcation diagram of fixed points for the function k, when $h(x) = x^3 - 9x^2 + 24x - 12$. Notice that the saddle nodes of k are the relative maximum and/or minimum points of h on the graph of $a = h(x)$. Since h satisfies (G_1), there are only a finite number of these points in a bounded region of (a, x)-space.

We have described one generic property of bifurcation orbits—namely, that there cannot be infinitely many (fixed-point) saddle nodes in a bounded region—and that argument is given for only a special set of maps. More generally, let $f : \mathbb{R} \times \mathbb{R} \to \mathbb{R}$ be a smooth family of scalar maps, and let $(\bar{a}, \bar{x})$ be a point for which $f(\bar{a}, \bar{x}) = \bar{x}$ and $\frac{\partial f}{\partial x}(\bar{a}, \bar{x}) = 1$; that is, $(\bar{a}, \bar{x})$ satisfies the conditions on f and $\frac{\partial f}{\partial x}$ necessary for a saddle node. If we specify further that both $\frac{\partial^2 f}{\partial x^2}(\bar{a}, \bar{x})$ and $\frac{\partial f}{\partial a}(\bar{a}, \bar{x})$ are nonzero, then there is a neighborhood of $(\bar{a}, \bar{x})$ in which it is the only (fixed-point) saddle node. (This property is verified in Theorem 11.9.) There are also conditions on the partial derivatives of f that guarantee that a given period-doubling bifurcation point will be similarly isolated from other such bifurcations of the same or lower period. (See Exercise 11.9 for these conditions.)

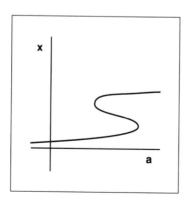

Figure 12.6   The bifurcation diagram of fixed points of $k_a(x) = x + a - h(x)$, where $h(x) = x^3 - 9x^2 + 24x - 12$.
The saddle node fixed points of k are the local maximum and minimum points of $a = h(x)$.

Thus we can conclude that having only a finite number of fixed point saddle-node and period-doubling bifurcations in a bounded region of (a, x)-space is a generic property. In particular, any smooth one-parameter family either has this property or can be closely approximated by a family that has it. By "closely approximated", we mean that a family g can be chosen so that $\sup_{(a,x) \in \mathbb{R} \times \mathbb{R}} | f(a, x) - g(a, x) |$ is as small as we like. The proof that the set of one-parameter families described below is generic (within the class of maps whose orbits have at most one unstable direction) is due to (Brunovsky, 1969).

Recall that a branch of orbits is a maximal path (in a) of hyperbolic fixed points or periodic orbits and that branches end only in nonhyperbolic orbits. For one-dimensional maps, nonhyperbolic fixed points have derivatives equal to $+1$ or -1. We denote these sets of orbits by $\mathbf{B}_+$ and $\mathbf{B}_-$, respectively. For higher-dimensional maps, the set $\mathbf{B}_+$ contains fixed points with exactly one eigenvalue equal to $+1$ and no other eigenvalues on or outside the unit circle, whereas the set $\mathbf{B}_-$ contains fixed points with exactly one eigenvalue equal to -1 and no other eigenvalues on or outside the unit circle. (By restricting $\mathbf{B}_+$ and $\mathbf{B}_-$ in maps of dimensions greater than one in this way, we are saying that nonhyperbolic fixed points can have at most one eigenvalue of absolute value greater than or equal to one.) So far we have ignored the cases in which an orbit is in $\mathbf{B}_+$ or $\mathbf{B}_-$ and a bifurcation does not occur, or the cases where nonhyperbolic fixed points are not isolated.

By assuming the following hypotheses, we require that all nonhyperbolic fixed points be either saddle-node or period-doubling bifurcations.

Definition 12.3 (Generic Bifurcation Hypotheses.)

(1) When p is in $\mathbf{B}_+$, then p is a saddle-node bifurcation orbit. Specifically, at an orbit in $\mathbf{B}_+$ two branches of orbits emanate: one branch of orbits in $\mathbf{S}$ (stable orbits) and one branch of orbits in $\mathbf{U}_+$ (regular unstable orbits). Orbits on both branches have the same period as the bifurcation orbit. In a bifurcation diagram (representing a compact subset of the domain), for each k, there are only finitely many orbits in $\mathbf{B}_+$. Schematic diagrams of all the generic saddle-node bifurcations are shown in Figure 12.7.

(2) When p is in $\mathbf{B}_-$, then p is a period-doubling bifurcation orbit. At an orbit in $\mathbf{B}_-$, three branches of orbits emanate. See Figure 12.4. Orbits on one of the branches have twice the period of the bifurcation orbit. Orbits on the other two branches have the same period: One of these two branches has orbits in $\mathbf{S}$ (stable orbits) and the other has orbits in $\mathbf{U}_-$ (flip unstable orbits). Again, in the bifurcation diagram, for each k, there are only finitely many orbits in $\mathbf{B}_-$. Schematic diagrams of all the generic period-doubling bifurcations are shown in Figure 12.10.

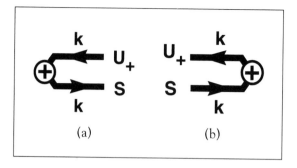

Figure 12.7   The generic saddle-node bifurcations.
At a period-k saddle node, two branches of orbits emanate: one branch contains stable (**S**) orbits and the other contains regular unstable (**U**$_+$) orbits. All orbits on both branches are period-k orbits. (a) The two branches appear to the right of the bifurcation parameter $a = a_+$; orbits on the branches exist for a greater than (and near) a_+. (b) The two branches appear to the left of a_+.

We will call a one-parameter family of maps **generic** if it satisfies these restrictions on orbits in **B**$_+$ and **B**$_-$. The saddle-node and period-doubling bifurcations are called **generic bifurcations**.

Generic maps form a large set in that they are dense in the set of one-parameter families of maps. Although other situations are possible (such as pitchfork bifurcations or orbits in **B**$_+$ or **B**$_-$ not being bifurcation orbits), they are not likely to occur. And if they do, there will be closely approximated by a family whose only bifurcations are saddle-node and period-doubling bifurcations.

For one-dimensional families, saddle-node bifurcation orbits connect schematic branches of stable orbits and regular unstable orbits. What types of schematic branches emanate from period-doubling bifurcation orbits? Following a path of fixed points through a period-doubling bifurcation, the derivative (or an eigenvalue) goes through -1 (either smaller to larger or vice versa) at the bifurcation orbit. Therefore, one branch of fixed points is stable (derivative has absolute value smaller than one) and the other is flip (derivative smaller than -1). To determine the third branch, the doubled-period branch, we appeal to a description of the graph of the function f_a^2 near the bifurcation point, as shown in Figure 12.8. As a branch of stable fixed points nears a period-doubling bifurcation (Figure 12.8(a)), the derivative of f_a^2 evaluated at orbits along the branch approaches $+1$. After the derivative has crossed $+1$ (Figure 12.8(b)), there are three fixed points. From the direction that the graph crosses the line $y = x$, we see that there are two new paths of stable fixed points. Suppose, on the other hand, that a single branch of flip fixed points approaches a period-doubling bifurcation

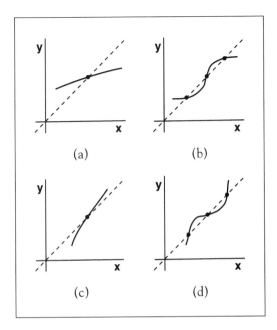

Figure 12.8 Graphs of f_a^2 as the function undergoes period-doubling bifurcations.

Since the map f_a has slope -1 at a period-doubling bifurcation, the slope of the second iterate f^2 is $+1$. If the parameter value a is chosen on one side of the bifurcation, as shown in (a), then the graph of f^2 looks like (b) for parameter values on the other side of the bifurcation. Similarly, the graph of f^2 may change from (c) to (d) at the bifurcation point.

(Figure 12.8(c)). Looking again at the graph of f_a^2 and at its direction of crossing at each intersection with the line $y = x$, we conclude that the continuation of the branch of flip fixed points after the bifurcation (Figure 12.8(d)) is a stable branch, while the two new paths of fixed points are regular unstable orbits.

We summarize these facts in Lemma 12.4.

Lemma 12.4 (Exchange of Stability Principle.) At a period-doubling bifurcation orbit of period k occurring at $a = a_b$, one period-k branch emanating from the bifurcation orbit contains flip unstable orbits and the other period-k branch consists of stable orbits; for one branch, $a < a_b$, and for the other, $a > a_b$. If the period-$2k$ branch lies on the same side as the period-k stable branch, then it has regular unstable orbits, while if it lies on the opposite side, it has stable orbits.

Although we do not prove this lemma, justification for one-dimensional families follows from diagrams of the possible generic period-doublings shown in Figure 12.9. For two-dimensional families, index theory is needed.

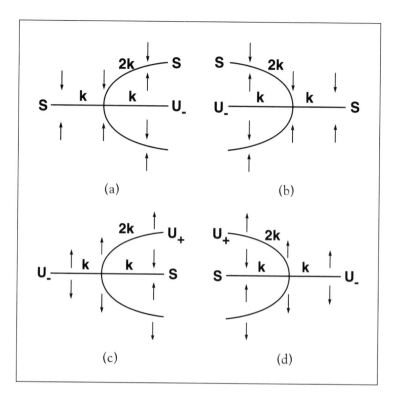

Figure 12.9 The generic period-doubling bifurcations for families of one-dimensional maps.
Bifurcation diagrams of all possible generic period doublings from period-k to period-$2k$ are shown. A single point of each period-k orbit appears, together with two points for each period-$2k$ orbit. The arrows indicate whether nearby points move toward or away from the periodic points.

Figure 12.10 shows the possible period-doubling bifurcations under the Generic Bifurcation Hypotheses of Definition 12.3. Again, branches of orbits in S are represented by solid line segments with arrows directed to the right, while branches of orbits in U_+ are represented by solid line segments with arrows directed to the left. Branches of orbits in U_- are represented by dashed line segments. In Figure 12.10(a), for example, one branch (a period-k branch) is directed toward the bifurcation orbit, one branch (a period-$2k$ branch) is directed away, and one branch is not directed.

Corollary 12.5 At each generic bifurcation orbit there is one branch that is directed toward the bifurcation orbit and one branch that is directed away.

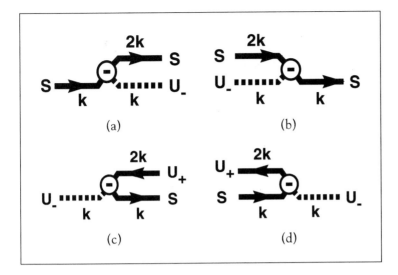

Figure 12.10 Schematic diagrams of the four generic period-doubling bifurcations.
Schematic diagrams of the bifurcations in Figure 12.9 are shown. These bifurcations are the only generic period-doubling bifurcations for families of one-dimensional maps or for families of two-dimensional area-contracting maps.

Proof: Lemma 12.4 admits only four possibilities of generic period-doubling bifurcations, and these are shown in Figure 12.10. Examining these cases gives the result for period-doubling bifurcations. For saddle-node bifurcations, Exercise 11.8 of Chapter 11 gives generic conditions under which there are two branches, both on the same side, with oppositely directed arrows, as shown in Figure 12.7.

✎ **EXERCISE T12.5**

Use the exchange of stability principle to verify that all the (generic) period-doubling bifurcations for a family of one-dimensional maps are shown in Figure 12.10.

In summary, our tinker-toy model for generic one-parameter maps contains "sockets", which are saddle-node and period-doubling bifurcations and "sticks", which are S branches, U_+ branches, or U_- branches. Each schematic branch that begins or ends within the given parameter range does so at a generic bifurcation orbit. Figure 12.11 shows a schematic bifurcation diagram illustrating these ideas.

In the next section, we use the theory of schematic diagrams to explain how chaos develops in one-parameter families of maps.

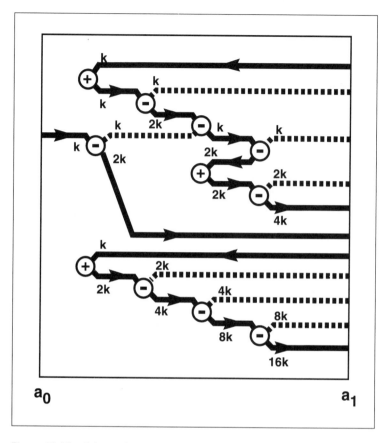

Figure 12.11 Schematic model of a bifurcation diagram.
The interconnections of branches of orbits are potentially quite complicated, even in generic cases. Here we show one such diagram, using only saddle-nodes and period-doubling bifurcations. Although we have included only a finite number of bifurcation points, in principle, we could have had period doublings (and even saddle nodes) of all the periods $k, 2k, 4k, 8k, \ldots$.

12.4 THE CASCADE THEOREM

Poincaré was the first to give a complete mathematical description of a period-doubling bifurcation, and (Myrberg, 1962) first described a cascade. Cascades became well known with the applications of (May, 1976). Figures 12.1 and 12.2 show cascades for the one-dimensional quadratic map, and Figure 12.3 shows cascades for the two-dimensional Poincaré map of the forced damped pendulum. Many distinct cascades are seen or are hinted at in this picture. Numerical studies reveal that cascades occur not in isolation but in great profusion, sprinkled throughout chaotic regimes.

The purpose of this section is to give an explanation of why cascades occur, and why they must occur in great profusion. The explanation is adequate only for chaotic systems that have at most one unstable direction at each point (at most one positive Lyapunov exponent). Indeed there seems to be no reason to expect to find many cascades in highly chaotic systems that are unstable in more than one direction. As an example of a class of systems with at most one positive Lyapunov exponent, we described two-dimensional dissipative systems in Chapter 11. A dissipative map satisfies $|\det \mathbf{Df}| \leq b < 1$, so that the two Lyapunov numbers of each trajectory satisfy $L_1 L_2 \leq b$. The Hénon map (for fixed $|b| < 1$) and the Poincaré map of the forced, damped pendulum are two examples of dissipative systems with constant Jacobian determinants. In these cases, $L_1(\mathbf{x})L_2(\mathbf{x}) = |\det \mathbf{Df}|$, for each initial point $\mathbf{x}$. Although other settings may satisfy the restriction that there be at most one unstable direction at each point, the primary applications of the Cascade Theorem will be to families of one-dimensional maps and families of two-dimensional dissipative systems.

The description is given in terms of a process of orbit creation followed by stability shedding and builds on two well-known phenomena: the coexistence of huge numbers of unstable periodic orbits in a chaotic attractor (for a fixed parameter value), and the general bifurcation phenomenon known as "the exchange of stabilities" (Lemma 12.4). As the parameter is increased, the collection of periodic orbits changes. For example, it might change from a single stable orbit at parameter value a_0 to the infinite hoard of unstable orbits one expects in a chaotic attractor at a_c. We will argue here that the only way in which these unstable orbits can arise (given the absence of stable orbits at a_c) is through the process of cascade formation. Close to half the orbits in a typical chaotic attractor are regular unstable ($\mathbf{U_+}$) orbits. (See Exercise 12.4.) We will show that for *each* of these $\mathbf{U_+}$ orbits there must be a *distinct cascade* occuring in some parameter range. Therefore, in changing from a single stable periodic orbit to a chaotic attractor, the system must have had not just one cascade but, in fact, infinitely many cascades.

We will assume throughout the remainder of this section that f is generic. While genericity is not a necessary assumption for the result we prove, it does significantly simplify the arguments. Any map that is not generic can be approximated by a generic map f. The approximation techniques needed to extend our arguments to the case without genericity assumptions can be found in (Yorke and Alligood, 1985).

Recall that all orbits on a branch (as shown in Figure 12.4) are the same type: either all stable $\mathbf{S}$, all regular unstable $\mathbf{U_+}$, or all flip unstable $\mathbf{U_-}$. These branches end at bifurcation orbits, and the bifurcation orbits serve to connect the different types of branches. We may therefore refer to branches as stable (or **S**) branches, regular unstable (or **U**$_+$) branches, etc.

Definition 12.6 A **cascade from period** k is a connected set of branches and bifurcation orbits containing stable branches of period k, $2k$, $4k$, . . . , all lying in a bounded region of parameter and phase space. The connected set will also contain unstable orbits.

Since we are interested in how chaotic sets develop, we assume Hypotheses 12.7 hold for some closed, bounded region V of phase space. For one-dimensional maps, V is an interval $[x_0, x_1]$, while for two-dimensional maps, V could be chosen to be a disk. We discuss only orbits all of whose points lie in V.

Hypotheses 12.7 **Hypotheses for the Cascade Theorem.**

1. There is a parameter value $a = a_0$ for which f has only stable fixed points or stable periodic points. In particular, it may have no periodic points.
2. There is a parameter value $a = a_c > a_0$ (c for "chaotic") for which f has unstable periodic points and for which the only periodic orbits are in $\mathbf{U}_+$ or $\mathbf{U}_-$. (Of course, in some cases the order of a_0 and a_c will be reversed, but we choose one case to make the discussion simpler.)
3. For a between a_0 and a_c, the periodic orbits of f all lie in a bounded region V of phase space. A weaker statement may also be used: there are no periodic orbits of f on the boundary of the set V.
4. The map f satisfies the Generic Bifurcation Hypotheses of Definition 12.3. In particular, all bifurcations look like those shown in Figures 12.7 and 12.10, and, for each k, there are only finitely many bifurcation orbits in $[a_0, a_c] \times V$.

In Theorem 12.8, we refer to a connected set of orbits. Thinking of a period-k orbit as a collection of k periodic points, we mean the connected set of branches together with endpoints (bifurcation orbits) to which a particular orbit belongs. For a period-k orbit, there will be k distinct connected sets of periodic *points* underlying this one connected set of *orbits*.

Theorem 12.8 (**The Cascade Theorem.**) *Assume Hypotheses 12.7. Let* p *be a regular unstable periodic orbit of f of period k at $a = a_c$. Then p is contained in a connected set of orbits that contains a cascade from period k. Distinct regular unstable orbits at $a = a_c$ yield distinct cascades: If p_1 and p_2 are two regular unstable orbits at $a = a_c$, then the two cascades associated with p_1 and p_2 have no branches of stable orbits in common.*

Before going into the proof of Theorem 12.8, we focus again on the type of situation to which the theorem applies: namely, a one-parameter family of maps for which at an initial value of the parameter there are no fixed points or periodic points or only sinks, and at a later parameter value there are only unstable fixed points or periodic points, such as in a chaotic set. We were introduced to cascades in the bifurcation diagram of the logistic map (see, for example, Figure 1.3 of Chapter 6). The one-dimensional quadratic family $f_a(x) = a - x^2$ satisfies Hypotheses 12.7 for (a, x) in $[-2, 2] \times [-2, 2]$. (See Exercise 12.1.) Our primary two-dimensional example is the formation of a horseshoe in a family of area-contracting maps. For example, (Devaney and Nitecki,1979) proved that for fixed $b, -1 < b < 0$, the Hénon family $\mathbf{H}(x, y) = (a - x^2 + by, x)$ develops a horseshoe as the parameter a is varied. In particular, for $a < \frac{-(1-b)}{4}$, $\mathbf{H}$ has no fixed or periodic points, and for $a > \frac{1}{4}(5 + 2\sqrt{5})(1 - b)^2$, a hyperbolic horseshoe exists. All periodic points and fixed points in this family are contained within a bounded region of the plane.

The proof of Theorem 12.8 follows from a few key ideas that we split off as lemmas for you to verify. First we isolate a path of orbits in (a, x)-space within the possibly vast and interconnected network of orbits that can occur even in the generic case. The path will enable us to follow cascades even when there are numerous saddle nodes, period doublings, and period halvings along the path.

Definition 12.9 For any point (a, x) on a periodic orbit that is not a flip unstable orbit, define the **snake** through (a, x) to be the collection of **S** branches and $\mathbf{U}_+$ branches and their endpoints (bifurcation orbits) that can be reached by a connected path of these branches and their endpoints from (a, x). See Figure 12.12.

⬧ EXERCISE T12.6

Verify that a snake passes through an orbit in **S**, $\mathbf{U}_+$, or a bifurcation orbit as a one-dimensional path of orbits.

Notice that a snake can be a closed loop of orbits. If it is not, however, then the snake will never go through any orbit twice. The following lemma says that under Hypotheses 12.7 regular unstable orbits cannot period double.

Lemma 12.10 When a period-k branch in a snake ends in a period-doubling bifurcation orbit of period k, then it is an **S** branch.

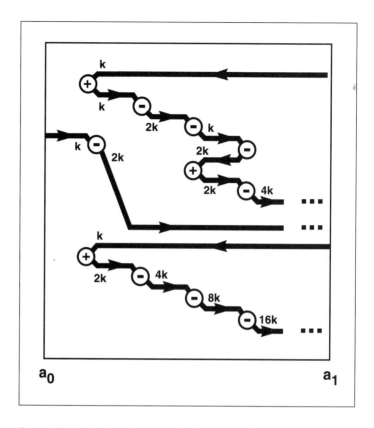

Figure 12.12 Snakes in a generic bifurcation diagram.
A snake is a connected path of orbits, none of which are flip unstable orbits. Thre
snakes are shown in this schematic diagram, which is constructed by eliminatin
the flip unstable branches in Figure 12.11.

Lemma 12.10 is proved by observing that it holds for the four cases i
Figure 12.10. Specifically, since the only branches on a snake are **S** branches an
$\mathbf{U}_+$ branches, the period-k branch must be a stable branch. (The fact that ther
are no generic period-doublings with $\mathbf{U}_+$ branches of period-k follows originall
from the assumption of area-contraction: If one eigenvalue crosses -1 at th
bifurcation point, the other cannot be greater than $+1$. For one-dimensiona
maps, the derivative must move through -1.)

Since snakes are composed only of **S** and $\mathbf{U}_+$ branches, the direction o
travel along a snake (whether a is increasing or decreasing) will change precisel
at a bifurcation orbit, when moving from an **S** branch to a $\mathbf{U}_+$ branch or vic
versa. See Figure 12.12, where **S** branches are indicated with arrows pointed t
the right, and $\mathbf{U}_+$ branches, with arrows pointed to the left. We may say tha
a snake enters (or leaves) a region if it has one orbit on the boundary and th

directed path enters (or leaves, according to the direction of arrows) the region at that orbit.

Lemma 12.11 implies that once a snake enters $[a_0, a_c] \times V$, entering either from the right at $a = a_c$ as a U_+ branch, or from the left at $a = a_0$ as an S branch, the snake is trapped inside $[a_0, a_c] \times V$.

Lemma 12.11   No snakes leave $[a_0, a_c] \times V$.

EXERCISE T12.7

Prove Lemma 12.11.

Lemmas 12.12, 12.13, and 12.14 together establish that a snake must contain infinitely many bifurcation orbits, only finitely many of which can be of a given period N or less. Therefore, the snake must contain bifurcation orbits of arbitrarily large periods. Since, for a generic family, the only way periods can increase along a snake is for it to contain period doublings, we obtain the desired cascade.

Lemma 12.12   Any snake that is not a closed loop of orbits must contain infinitely many branches and infinitely many bifurcation orbits.

EXERCISE T12.8

Prove Lemma 12.12.

Lemma 12.13 follows from the generic bifurcation hypotheses.

Lemma 12.13   Let $N > 0$ be given. A snake contains only finitely many bifurcation orbits of period N or less for parameter values between a_0 and a_c.

EXERCISE T12.9

Prove Lemma 12.13.

Lemma 12.14   The snake must pass through bifurcation orbits of periods $k, 2k, 4k, 8k, \ldots$, and stable branches of periods $k, 2k, 4k, 8k, \ldots$.

✎ **EXERCISE T12.10**

Prove Lemma 12.14.

To summarize: Assuming that there are only stable orbits at a_0 and no stable orbits at a_c, we have shown that each regular unstable orbit at a_c must lie on a connected set of orbits containing a cascade. Generically, an orbit in $\mathbf{U}_+$ is born in a saddle-node bifurcation, paired with an orbit in $\mathbf{S}$ of the same period. The stable orbit then sheds its stability through a period-doubling cascade.

Now we can translate the theorem within an appropriate system. For one-dimensional families, the stable orbits are attractors, and the unstable orbits are repellers, either regular or flip. Higher-dimensional systems where orbits have at most one unstable direction are also appropriate. For families of area-contracting maps of the plane, the stable orbits are attractors, while the unstable orbits are saddles, regular or flip.

As a final result, we restate the Theorem 12.8 within this setting.

Theorem 12.15 *Let $\mathbf{f}$ be a family of area-contracting maps of the plane that satisfy Hypotheses 12.7, and let $\mathbf{p}$ be a regular saddle of $\mathbf{f}$ of period k at $a = a_c$. Then $\mathbf{p}$ is contained in a connected set of orbits that contains a cascade from period k. Distinct saddles at $a = a_c$ yield distinct cascades: If $\mathbf{p}_1$ and $\mathbf{p}_2$ are two saddles at $a = a_c$, then the two cascades associated with these saddles have no branches of attractors in common.*

✎ **EXERCISE T12.11**

Explain which of the arguments in the proof of the Cascade Theorem depend on the assumption that orbits have at most one unstable direction. (This assumption underlies the structure of the generic bifurcations.)

We conclude with a brief discussion of moving from the generic to the general case. In particular, Theorem 12.8 holds for any smooth one-parameter family $\mathbf{f}$ of area-contracting maps of $\mathbb{R}^2$ for which hypotheses (1)–(3) of (12.7) hold (see (Yorke and Alligood, 1985) for the limit arguments). Families of area-preserving maps of the plane can also be approximated by maps in our generic set, although in the area-preserving case, the stable orbits are not attractors but elliptic orbits. Thus, for families of area-preserving maps of the plane that satisfy hypotheses (1)–(3) of (12.7), period-doubling cascades of elliptic orbits will occur.

Universality in Bifurcation Diagrams

THE STABILITY ARGUMENTS of this chapter have shown us why period-doubling cascades must occur in certain systems that develop chaos. Not only are distinct cascades qualitatively similar, as seen in Figure 12.2, but they also have metric properties in common, as we saw with Feigenbaum's discovery for one-parameter families of one-dimensional maps.

Self-similarity within a cascade follows from the basic idea that higher iterates of the map can be rescaled in a neighborhood of the critical point to resemble the original map. For example, Figure 12.13 shows the graphs of iterates of two maps in the quadratic family $g_a(x) = x^2 - a$. In Figure 12.13(a), g^2 is graphed for $g(x) = x^2 - 1$. At this parameter ($a = 1$), the origin is a period-two point of g, (a fixed point of g^2). In Figure 12.13(b), g^4 is graphed for $g(x) = x^2 - 1.3107$. The origin is a period-four point of g. In each case, the graph in a

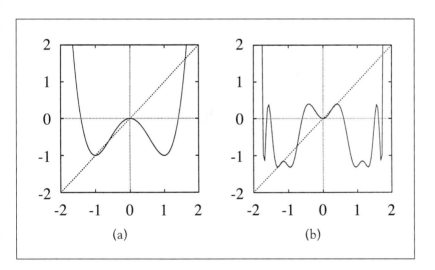

Figure 12.13 Iterates of maps within the quadratic family $g_a(x) = x^2 - a$.
(a) The origin is a period-two point of $g(x) = x^2 - 1$. The second iterate g^2 is graphed. (b) The origin is a period-four point of $g(x) = x^2 - 1.3107$. The fourth iterate g^4 is graphed. In each case, a neighborhood of the origin in phase space and of the appropriate a-value in parameter space can be rescaled so that the dynamic behavior of the iterate in the small scale mimics that of the whole quadratic family in the large.

neighborhood of the critical point can be rescaled to look like the graph of $x^2 -$ at $a = 0$. For example, the map g^2 in Figure 12.13(a) would be flipped about the horizontal axis and rescaled. As a is increased, each parabolic region mimics the behavior of $x^2 - a$. Thus sequences of bifurcations, such as cascades, could be expected to be the same for g_a^2 in a neighborhood of $x = 0$ and $a = 1$ and for g in a (still smaller) neighborhood of $x = 0$ and $a = 1.3107$ as for $g_a(x) = x^2 - a$.

In Challenge 12 we examine universal metric properties among different windows of periodic behavior within chaotic regions of parameter space. The arguments are given for one-parameter families of scalar unimodal maps, although the phenomena seems to occur within families of area-contracting planar maps as well. We show that, typically, the distances between bifurcations (both local and global) within a periodic window are well-approximated by the corresponding distances for the canonical map $g_a(x) = x^2 - a$ under a suitable linear change of coordinates. The procedure of rescaling to obtain universal properties is known as **renormalization**. Figure 12.14 shows first the complete bifurcation diagram for the map $g_a(x)$ followed by a period-nine window for g^9. The relative distances between corresponding bifurcations are strikingly similar.

A **period-n window** begins with a period-n saddle node at $a = a_s$. The stable period-n orbit then loses stability through a period-doubling cascade, eventually forming an n-piece chaotic attractor. The window ends at the crisis parameter

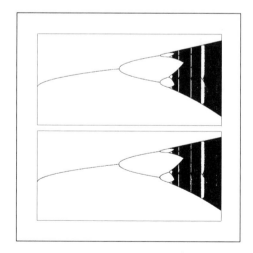

Figure 12.14   Universality in periodic windows.
The complete bifurcation diagram for the map $g_a(x) = x^2 - a$ is shown above while the bifurcations for g_a^9 within a period-nine window are shown below. In the top diagram, x goes from -2.0 to 2.0 and a is between -0.25 and 2.0, while in the bottom diagram x goes from -0.02 to 0.02 and a is between 1.5552567 and 1.5554906.

value $a = a_c$ at which the n-piece attractor disappears. As with a planar map, this crisis occurs when the attractor collides with a basin boundary orbit—in this case, with a period-n source. (See Chapter 10 for a discussion of the planar case.) The basin of the chaotic attractor includes the union of n disjoint open intervals, each of which contains a piece of the attractor and one of which contains the critical point of the map. Since the critical point must map to an endpoint of one of these intervals, the crisis occurs precisely when the critical point maps onto one of the period-n boundary sources.

Within a window, we denote the first period-doubling bifurcation parameter by $a = a_d$. Between the saddle-node at a_s and the period-doubling at a_d, the path of stable period-n orbits must pass through a parameter where the orbit has derivative 0. At this parameter, denoted $a = a_0$, the orbit is called **superstable** and contains a critical point x_0 of the map.

Step 1 Show that the ratio $R_g = (a_c - a_s)/(a_d - a_s)$ is $\frac{9}{4}$ for the period-one window of the canonical family $g_a(x) = x^2 - a$. [Hint: Find an interval that maps onto itself to determine a_c.]

Let k_a be a family of scalar maps that is linearly conjugate to the canonical family of maps g_a. Specifically, define $G : \mathbb{R}^2 \to \mathbb{R}^2$ to be $(a, x) \mapsto (a, g_a(x))$ and $K : \mathbb{R}^2 \to \mathbb{R}^2$ to be $(a, x) \mapsto (a, k_a(x))$. Then G and K are said to be **linearly conjugate** if there is a linear map $H : \mathbb{R}^2 \to \mathbb{R}^2$ such that $H \circ K = G \circ H$. The form of the linear map we use here is $H(a, x) = (C_1(a - a_0), C_2(x - x_0))$, for nonzero constants C_1, C_2.

Step 2 Show that k has a period-n saddle node or period-doubling bifurcation at (a, x) or a crisis at a_c if and only if g has a period-n saddle node or period-doubling bifurcation at $(C_1(a - a_0), C_2(x - x_0))$ or a crisis at $C_1(a_c - a_0)$, respectively.

Conclude for the map k that $R_k = (a_c - a_s)/(a_d - a_s)$ is $\frac{9}{4}$. Furthermore, any ratio of distances defined by conditions on periodic points and their derivatives is preserved under a linear conjugacy.

Now we focus on an arbitrary family f of unimodal maps that has a superstable period-n orbit, $\{x_0, \ldots, x_{n-1}\}$, at $a = a_0$. Again, let $F(a, x) = (a, f(a, x))$. We assume that x_0 is the critical point and show that, under certain assumptions, there is a neighborhood of (a_0, x_0) and a linear change of coordinates H for which HF^nH^{-1} is approximated by the canonical family G. The larger n, the period of the orbit, the closer the approximation, with the distance between them (and their partial derivatives up to a given order) going to 0 as n goes to infinity. The ratio R_{f^n} in periodic windows for which the hypotheses are well-satisfied is remarkably close to $R_g = 9/4$. We illustrate this correspondence at the end of this challenge with one of several published numerical studies of the phenomenon.

We say a function $k(n)$ is of **order** g, denoted $\mathcal{O}(g)$, if $\lim_{n \to \infty} \frac{k(n)}{g(n)}$ is bounded. For example, n^3 is $\mathcal{O}(2^n)$, but 2^n is not $\mathcal{O}(n^3)$.

When the orbit of x_0 is superstable, the map f will typically be quadratic near x_0, the critical point, and a neighborhood of x_0 will be folded over by f and then stretched by the next $n - 1$ iterates of f before returning close to x_0. See Figure 12.15. In order to get an intuitive idea of what is happening here, we describe an idealized f^n : there is a cycle of n disjoint intervals $J_0, \ldots, J_{n-1}$ that the map f permutes so that f is quadratic on J_0 and linear on the remaining J_i's. Specifically, $f^n(a, x) = L_{n-1} \cdots L_1 Q(a, x)$, for constants $L_1, \ldots, L_{n-1}$ which, in the ideal case, do not depend on a, for a quadratic map Q, which depends on both x and a, and for $x \in J_0$ and $a \in [a_c - a_s]$. Notice that under this simplifying assumption $f_a^n : J_0 \to J_0$ is quadratic. Ideally, the parameter extent of the window, $a_c - a_s$, is sufficiently small that the $n - 1$ points in the orbit remain far away from the critical point and that the constants $L_1, \ldots, L_{n-1}$ remain good approximations to the slopes of f_a near these points throughout the parameter window.

Let $S = \partial f^{n-1}/\partial x(a_0, x_1)$. (For our idealized map, S is the product of slopes $L_1 \cdots L_{n-1}$.) We make the following specific assumptions incorporating the ideas in the previous paragraph:

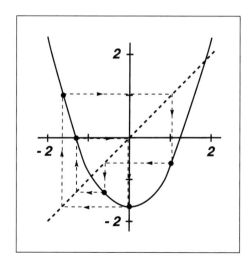

Figure 12.15   A superstable period-5 orbit.
The map is $f(x) = x^2 - 1.6254$.

(1) S grows exponentially with n; that is, there is a constant $r, r > 1$, such that S is $\mathcal{O}(r^n)$. (This assumption holds if the remaining points in the orbit, $\{x_1, \ldots, x_{n-1}\}$, are "far" from the critical point. How far, will determine in part how well the renormalization procedure succeeds in approximating the canonical map.)

(2) The second partial $\partial^2 f^n / \partial x^2$ is nonzero at (a_0, x_0).

(3) The partial $\partial f^n / \partial a(a_0, x_0)$ is nonzero.

By rescaling both x and a near (a_0, x_0), we can make the lowest order terms in the Taylor series for f^n match those of the canonical map. The hard part then is to show that the higher-order terms are made small by the rescaling. We give a heuristic argument here, based on estimating the order of the error term in a Taylor expansion of f^n. More technical estimates including bounds on higher order partial derivatives can be found in (Hunt, Gallas, Grebogi, and Yorke, 1995).

Step 3 Typically, all the kth order partial derivatives of f^{n-1} taken with respect to a and x near (a_0, x_1) are of order S^k. We illustrate this technical point with an example:

Show that $\partial^2 f^{n-1} / \partial x^2 (a_0, x_1)$ is of order S^2. [Hint:

$$\frac{\partial f^{n-1}}{\partial x}(a_0, x_1) = \frac{\partial f}{\partial x}(a_0, x_{n-1}) \frac{\partial f}{\partial x}(a_0, x_{n-3}) \ldots \frac{\partial f}{\partial x}(a_0, x_1).$$

Use the product rule (together with the chain rule) to obtain the second partial derivative.]

Step 4 Using Taylor's Theorem, verify:

(i) There is a nonzero constant p_1 such that

$$f^{n-1}(a, x) - x_0 = S(x - x_1) + p_1 S(a - a_0)$$
$$+ \mathcal{O}(S^2(|x - x_1|^2 + |a - a_0|^2)),$$

provided $|x - x_1|$ and $|a - a_0|$ are small compared with S^{-1}.

(ii) There are nonzero constants q and p_2 such that

$$f(a, x) - x_1 = q(x - x_0)^2 + p_2(a - a_0)$$
$$+ \mathcal{O}(|x - x_0|^3 + |x - x_0||a - a_0| + |a - a_0|^2),$$

for (a, x) near (a_0, x_0). Recall that x_1 is $f(a_0, x_0)$.

Step 5 Compose (i) and (ii) of Step 4 to obtain

(iii) $f^n(a, x) - x_0 = qS(x - x_0)^2 + pS(a - a_0)$
$$+ \mathcal{O}(S^2(|x - x_0|^4 + |a - a_0|^2))$$
$$+ \mathcal{O}(S(|x - x_0|^3 + |x - x_0||a - a_0|)), \qquad (12.2)$$

Window size	Period	a_0	R_f	$\|R_f - \frac{9}{4}\|/(\frac{9}{4})$
0.04506988	7	1.226617	1.635	0.27
0.00956120	7	1.299116	2.032	0.097
0.00197090	8	1.121835	2.229	0.0093
0.00084633	9	1.172384	2.245	0.0022
0.00049277	8	1.323307	2.248	0.00097
0.00033056	10	1.176765	2.244	0.0027
0.00032251	15	1.421811	1.637	0.27
0.00030118	9	1.402762	2.249	0.00036
0.00024599	13	1.353915	2.211	0.017
0.00022962	10	1.142882	2.246	0.0018
0.00022952	9	1.293955	2.246	0.0019

Table 12.3 **Values of the ratio R_f for the Hénon map $f(x_1, x_2) = (1 - ax_1^2 + x_2, 0.3x_1)$.**
Data is drawn from windows in the range $1.12 < a < 1.43$ with window widths greater than 2×10^{-4}.

for (a, x) near (a_0, x_0), where $p = p_1 + p_2$. To be precise, $q = \frac{1}{2}\partial^2 f/\partial x^2(a_0, x_0)$ and $p = S^{-1}\partial f^n/\partial a(a_0, x_0)$.

The rescaling is given by the linear change of coordinates

$$y = qS(x - x_0), \quad u = -pqS^2(a - a_0).$$

In the terminology of the discussion above, the conjugacy H is defined as

$$H(a, x) = (u, y) = (-pqS^2(a - a_0), qS(x - x_0)).$$

Step 6 Let $g(u, y)$ be the family of maps conjugate to $f^n(a, x)$ under this change of coordinates. From the expansion for f^n in Step 5, show that

$$g(u, y) = y^2 - u + \mathcal{O}(S^{-1}(|y|^3 + |y||u| + |u|^2)).$$

Since all cascades of the canonical map take place in a bounded region of (u, y)-space, say $(u, y) \in [-2.5, 2.5] \times [-2.5, 2.5]$, conclude that g converges to the canonical map as $n \to \infty$.

The universality of the ratio R_g appears to apply through a wide class of chaotic systems, including multidimensional systems that contain cascades. Table 12.3 shows a numerical study of periodic windows for the Hénon map $f(x_1, x_2) = (1 - ax_1^2 + x_2, 0.3x_1)$. The relative error between the calculated R_f and $\frac{9}{4}$ is seen to be typically quite small.

EXERCISES

12.1. Show that there are parameters a_0 and a_c such that the quadratic family $f_a(x) = a - x^2$ satisfies Hypotheses 12.7, the hypotheses of the Cascade Theorem.

12.2. Show that neither the quadratic family $f_a(x) = a - x^2$ nor the logistic family $g_a(x) = ax(1 - x)$ has a period-two saddle node. Show, on the other hand, that each must have a period-four saddle node.

12.3. Let $g_4(x) = 4x(1 - x)$, and let k be an odd number. Show that half the periodic orbits of period k are regular repellers and half are flip repellers. Similarly, for a map that forms a hyperbolic horseshoe, half the periodic orbits of period k are regular saddles and half are flip saddles. Do the same results hold for any even k?

12.4. Let g denote either the logistic map g_4 or the horseshoe map. Use the following step to show that for $n \geq 24$, at least 49% of the period n orbits of g are regular unstable orbits and at least 49% are flip unstable.

(a) Let N be the number of fixed points of g_4^n that are not period-n points. Show that the proportion of one type of orbits (flip or regular) is at least $\frac{2^{n-1}-N}{2^n} = \frac{1}{2} - \frac{N}{2^n}$. In the remaining two parts we show that $\frac{N}{2^n} < .01$, for $n \geq 24$.

(b) Show that the number N of fixed points of g_4^n of period less than n is at most

$$2^{n/2} + 2^{n/3} + 2^{n/4} + \cdots + 2 \leq n2^{n/2}.$$

(c) Show that the function $x2^{-x/2}$ is decreasing for $x > \frac{2}{\ln 2}$, and that it is less than .01 for $n \geq 24$.

12.5. Assume that a family of maps $\mathbf{f}_a$ satisfies Hypotheses 12.7 of the Cascade Theorem. Let $\mathbf{p}$ be a periodic orbit of period k at $a = a_c$. Show that each such orbit $\mathbf{p}$ is contained in a connected set of orbits that contains a cascade from period k.

   The result follows from the Cascade Theorem if $\mathbf{p}$ is a regular unstable orbit. We assume, therefore, that $\mathbf{p}$ is a flip-unstable orbit.

(a) Let G be the following set of orbits: $\mathbf{U}_-$ orbits of period k, and $\mathbf{U}_+$ and S orbits of period j, for all $j \geq 2k$. Show that there is a path of orbits of G through each saddle node of period $2k$ or higher and through each period-doubling bifurcation of period k or higher.

(b) Orient the branches of orbits in G to form a new type of "snake", as follows: $\mathbf{S} \rightarrow$, $\mathbf{U}_+ \leftarrow$, and $\mathbf{U}_- \leftarrow$. Show that Lemmas 12.10, 12.11, 12.12, 12.13, and 12.14 hold for these snakes to obtain the result. (Don't forget to show that period-k stable orbits are contained in the connected set of orbits.)

☞ LAB VISIT 12

Experimental Cascades

THE PERIOD-DOUBLING cascade is perhaps the most easily identifiable route to chaos for a dynamical system. For this reason, experimental researchers attempting to identify and study chaos in real systems are often drawn to look for cascades. Here we will survey the findings of laboratories in Lille, France and Berkeley, California. On the surface, the experiments have little in common—one involves nonlinear optics, and the other an electrical circuit, but the cascades that signal a transition to chaos are seen in both.

Scientists at the *Laboratoire de Spectroscopie Hertzienne* in Lille found period-doubling cascades in two different laser systems. The first, reported in 1991, was a CO_2 laser with modulated losses. The laser contains an electro-optic modulator in the laser cavity, that modulates the amplitude of the laser output intensity. The alternating current $C(t) = a + b \sin \omega t$ applied to the modulator can be controlled by the experimenter, and various behaviors are found as a, the dc bias, and b, the modulation amplitude, are varied. The modulation frequency ω is fixed at 640 kHz, the resonance frequency of the device.

Fixing b and using a as a bifurcation parameter yields Figure 12.16 (also shown as Color Plate 23). The modulation amplitude was set at $b = 3V$, and the dc bias was varied from 60V at the left side of the picture to 460V at the right side. To make this picture on the oscilloscope, the output intensity of the laser was sampled 640,000 times per second, the modulation frequency. The intensity makes a periodic orbit during each sampling period. A single value branch in Figure 12.16 corresponds to a periodic intensity oscillation in step with the modulation. The double branch that emanates from the single branch means that the oscillation takes two periods of the modulation to repeat, and so on. The orbit with doubled period is often called a subharmonic of the original orbit.

Lepers, C., Legrand, J., and Glorieux, P. 1991. Experimental investigation of the collision of Feigenbaum cascades in lasers. Physical Review A **43**:2573–5.

Bielawski, S., Bouazaoui, M., Derozier, D., and Glorieux, P. 1993. Stabilization and characterization of unstable steady states in a laser. Physical Review A **47**:3276–9.

Kocarev, L., Halle, K. S., Eckert, K., and Chua, L. O. 1993. Experimental observation of antimonotonicity in Chua's circuit. Int. J. of Bifurcations and Chaos 3:1051–5.

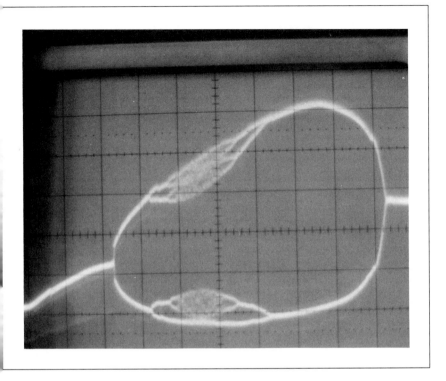

Figure 12.16 Period-doubling cascade from a CO$_2$ laser.
The dc bias of the modulator in the laser cavity is used as a bifurcation parameter, and is increased from the left side of the diagram to the right. The resulting bifurcation diagram of laser intensity shows a period-doubling cascade followed by a period-halving cascade.

Cascades originate from both the right and the left, colliding in the chaotic region in the middle. The orbits that are created in the initial cascade are systematically destroyed as the dc bias is further increased, resulting in a reverse cascade.

The same laboratory also produced Figure 12.17, which is a bifurcation diagram of the power output of an optical-fiber laser. The active medium is a 5-meter long silica fiber doped with 300 parts per million of trivalent neodymium, a rare-earth metallic element. The medium is pumped by a laser diode. The input power provided by the laser diode is the bifurcation parameter. Increasing the power from 5.5 milliwatts (mW) to 7 mW results in the oscilloscope picture shown here, which exhibits a period-doubling cascade ending in chaotic behavior for large values of the power. By adjusting the tilting of the mirrors in this experiment, various other nonlinear phenomena can be found, including Hopf bifurcations.

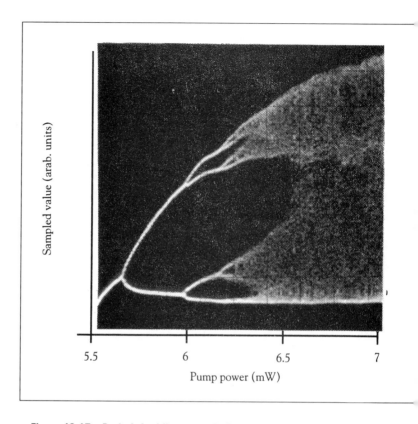

Figure 12.17 Period-doubling cascade from an optical-fiber laser.
The pump power is increased from the left of the diagram to the right, resulting in a bifurcation diagram showing a period-doubling cascade.

L. Chua's laboratory in Berkeley was the setting for studies of bifurcation phenomena in Chua's circuit, which we introduced in Chapter 9. The bifurcation parameter in this circuit was taken to be the capacitance C_1 across one of the capacitors. (This parameter is proportional to $1/c_1$ in the notation used for the differential equations (9.6).) Figure 12.18(a) shows the readout of an oscilloscope screen, exhibiting an experimental cascade as the parameter C_1 ranges from 3.92 nanoFarads (nF) to 4.72 nF. Each vertical division on the screen represents 200 millivolts.

Several different types of periodic and chaotic behavior can be seen in this picture, including cascades, periodic windows (parameter ranges with only periodic behavior), bubbles (collisions of cascades from the left and right) and reverse bubbles (periodic windows with cascades emanating to both left and right). A magnification of a part of the bifurcation diagram is shown in Figure 12.18(b). At least three odd-period windows are visible in this picture.

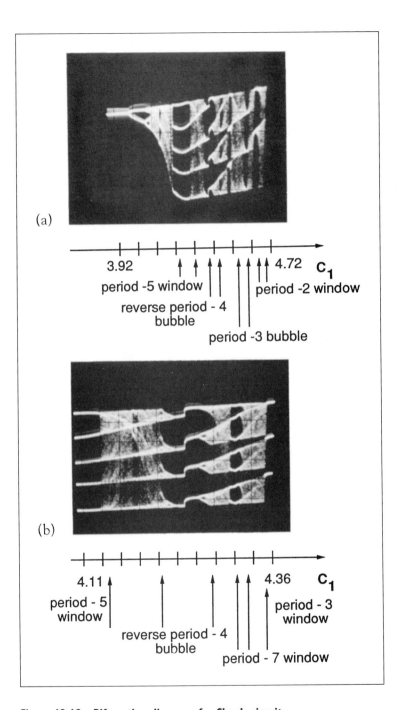

Figure 12.18 Bifurcation diagrams for Chua's circuit.
Voltage is graphed vertically as a function of the capacitance parameter C_1. (a) Period-doubling cascade followed by windows of period 5, 4, 3 and 2. (b) Magnification of part of (a).

Experimental bifurcation diagrams from both laboratories show **antimono tonicity**, the destruction of periodic orbits created in a cascade by a reverse cascade. Although this does not occur for the one-dimensional logistic map, an timonotonicity is typical of higher-dimensional systems which contain cascades

There are many other contributions to the scientific literature demonstrating period-doubling bifurcations and cascades in laboratory experiments. Many occurred shortly after Feigenbaum's published work in 1978.

In 1981, P. Linsay of MIT produced a cascade up to period 16 by varying the driving voltage in a periodically-driven RLC circuit (Linsay, 1981). His estimate of the Feigenbaum constant was 4.5 ± 0.6. At about the same time, a group of Italian researchers (Giglio, Musazzi, and Perini 1981), found a cascade up to period 16 from a Rayleigh-Bénard experiment in a heated chamber filled with water. Using the temperature gradient as a bifurcation parameter, they estimated the Feigenbaum constant at approximately 4.3. Cascades were first found in laser experiments by (Arecchi et al., 1982). Another source with many interesting examples of cascades and snakes in electronics is (Van Buskirk and Jeffries, 1985), who studied networks of p-n junctions.

CHAPTER THIRTEEN

State Reconstruction from Data

THE STATE of a system is a primitive concept that unifies the approach to many sciences. In this chapter we will describe the way states can be inferred from experimental measurements. In so doing we will revisit the Belousov-Zhabotinskii chemistry experiment from Lab Visit 3, the Couette-Taylor physics experiment from Lab Visit 4, and an example from insect physiology.

13.1 DELAY PLOTS FROM TIME SERIES

Our definition in Chapter 2 was that the state is all information necessary to tell what the system will do next. For example, the state of a falling projectile consists of two numbers. Knowledge of its vertical height and velocity are enough to predict where the projectile will be one second later.

The states of a periodic motion form a simple closed curve. Figure 13.1(a) shows a series of velocity measurements at a single point in the Couette-Taylor experiment, as described in Lab Visit 4. A plot of a system measurement versus time, as shown in Figure 13.1(a), is called a **time series**. For these experimental settings, the system settles into a periodic motion, so the time series of velocity measurements is periodic.

A useful graphical device in a situation like this is to plot the time series using delay coordinates. In Figure 13.1(b) we plot each value of the time series of velocities $\{S_t\}$ versus a time-delayed version, by plotting (S_t, S_{t-T}) for the fixed delay $T = 0.03$. This is called a delay-coordinate reconstruction, or **delay plot**.

In Figure 13.1, part (b) is made from part (a) as follows. At each time t, at point is plotted in part (b) using the two velocities $S(t)$ and $S(t - 0.03)$. At time $t = 2$, for example, the velocity is decreasing with $S(2) = 800$ and $S(1.97) = 1290$. The point $(800, 1290)$ is added to the delay plot. This is repeated for all t for which recording of the velocity were made, in this case, every 0.01 time unit. As these delay-coordinate points are plotted, they fit together in a loop that makes one revolution for each oscillation in the time series.

The interesting feature of a delay plot of periodic dynamics is that it can **reproduce the periodic orbit of the true system state space**. If we imagine our experiment to be ruled by k coupled autonomous differential equations, then the state space is $\mathbb{R}^k$. Each vector $\mathbf{v}$ represents a possible state of the experiment, and

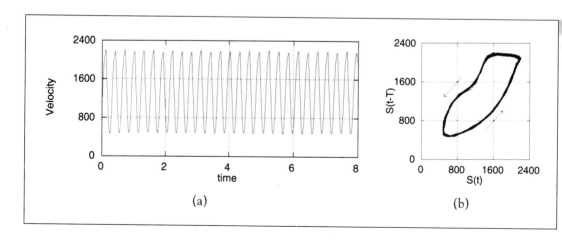

(a)   (b)

Figure 13.1   Periodic Couette-Taylor experimental data.
(a) Time series of velocities from the experiment. (b) Delay plot of data from (a), with time delay $T = 0.03$.

by solving the differential equations, the state $\mathbf{v}(t)$ moves continuously through the state space as time progresses.

We can imagine a set of coupled differential equations that governs the behavior of the Couette-Taylor experiment, but it is not easy to write them down. The state space dimension k would be very large; perhaps thousands of differential equations modeling the movement of fluid in small regions might need to be tracked. This is called "modeling from first principles". In contrast to the complicated differential equations of motion, the behavior we see in Figure 13.1 is fairly simple. Periodic motion means that trajectories trace out a one-dimensional curve of states through $\mathbb{R}^k$.

Of course, the idea of a real experiment being "governed" by a set of equations is a fiction. The Couette-Taylor experiment is composed of metal, glass, fluid, an electric motor, and many other things, but not equations. Yet science has been built largely on the success of mathematical models for real-world processes. A set of differential equations, or a map, may model the process closely enough to achieve useful goals.

Fundamental understanding of a scientific process can be achieved by building a dynamical model from first principles. In the case of Couette-Taylor, the basic principles include the equations of fluid flow between concentric cylinders, which are far from completely understood on a fundamental level. The best differential equation approximation for a general fluid flow is afforded by the Navier-Stokes equations, whose solutions are known only approximately. In fact, at present it has not been proved that solutions exist for all time for Navier-Stokes.

Can we answer questions about the dynamics of the system without understanding all details of the first principles equations? Suppose we would like to do time series prediction, for example. The problem is the following: Given information about the time series of velocity at the present time $t = 2$, predict the velocity at some time in the future, say 0.5 time units ahead. What information about the present system configuration do we need? Formally speaking, we need to know the state—by definition, that is the information needed to tell the system what to do next. But we don't even know the dimension of the state space, let alone what the differential equations are and how to solve them. To do time series prediction, we will use the **method of analogues**. That means we will try to identify what state the system is in, look to the past for similar states, and see what ensued at those times.

How do we identify the present state at $t = 2$? Knowing that $S(2) = 800$ is not quite enough information. According to the time series Figure 13.1(a), the velocity is 800 at two separate times; once when the velocity is decreasing, and once when it is increasing. If we look 0.5 time units into the future from the times that $S = 800$, we will get two quite different answers. We need a way of distinguishing between the upswings through 800 and downswings through 800, since they represent two different states of the system. The secret is to make use of delay coordinates. Instead of trying to represent the state by $S(t)$ alone, use the pair $(S(t), S(t - 0.03))$. If we use the current velocity $S(2) = 800$ and the time-delayed velocity $S(1.97) = 1290$ to identify the state at $t = 2$, we can identify the point as in a downswing. On the other hand, for $t = 1.8$, the pair $(S(1.8), S(1.77)) = (825, 560)$, which we identify as an upswing through 800. The two different states corresponding to $S(t) = 800$ are separately identified by the delay plot in Figure 13.1(b).

Figure 13.2 shows how to predict the future. Starting at $t = 2$, we collect analogues from the past, by which we mean states of form $(S(t), S(t - 0.03)) = (800, 1290)$, and find out what happened to the system 0.5 time units later. For example, the time series reveals that $(S(1.1), S(1.07)) = (808, 1294)$, and that 0.5 time units later, the velocity is $S(1.6) = 2125$. Also, $(S(1.4), S(1.37)) = (826, 1309)$ and the future velocity is $S(1.9) = 2123$. On the basis of this data we might predict that $S(2.5) \approx 2124$, the average of the two. The two analogues and the resulting prediction at $t = 2.5$ are shown in Figure 13.2. The prediction does a fairly good job of matching the actual velocity (shown in the dashed curve) at $t = 2.5$.

If the problem is to predict the velocity at some future time, the delay plot contains all information we need. If we know S_t and S_{t-T}, we can locate our current position on the curve. Using past observations, we can accurately predict

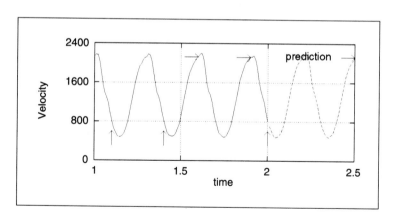

Figure 13.2 Prediction of the future by the method of analogues.
The vertical arrows at $t = 1.1$ and 1.4 show two past analogues for the Couette-Taylor system state at $t = 2$. In order to predict the velocity at $t = 2.5$, one can use the future evolution of the system encountered by the analogues (the horizontal arrows at $t = 1.6$ and 1.9) to predict that the velocity at $t = 2.5$ should be about 2124.

where on the curve the system will be in one additional second. We locate the points on the time series history (on the left of Figure 13.1) which correspond to points near (S_t, S_{t-T}) in the delay plot (on the right), and mark off the appropriate amount of time along the time series. If the system is deterministic, meaning that similar states result in similar outcomes, we will end up with several nearly equal predictions. In practice, we can average the predictions to improve the statistical quality of our final prediction, if that were our goal.

✎ **EXERCISE T13.1**

Let $x(t)$ be the height at time t of a falling projectile (it obeys the law $\ddot{x} = -g$). Decide whether it is always possible to predict $x(2)$ from the knowledge of $x(0)$ and $x(1)$ alone. Does the pair $[x(0), x(1)]$ uniquely determine the state of the projectile?

13.2 DELAY COORDINATES

Periodic orbits move through a one-dimensional set of states. The concept of dimension is important for quantifying the complexity of a process. The periodic Couette-Taylor process enters into infinitely many different states, but all states lie on a one-dimensional curve in the delay plot. In this motion, the system is restricted to a one-dimensional subset of its (possibly high-dimensional) state space.

Using the delay coordinates $[S(t), S(t - 0.03)]$, we were able to remove the self-intersections of the periodic Couette-Taylor attractor. However, the concept of delay coordinates is not limited to two-dimensional plots. In general, we can make the m-dimensional delay plot by graphing the vector of **delay coordinates**

$$[S(t), S(t - T), S(t - 2T), \ldots, S(t - (m - 1)T)]$$

for each time t in the time series. Attractors that are substantially more complicated than simple closed curves will require more dimensions to be untangled.

Figure 13.3 shows a second example of a periodic orbit. The time series is the measurement of the electrical impedance across the dorsal vessel, or what passes for a heart, of a living insect. The heart of *Periplaneta americana*, the American cockroach, is a slender pulsatile organ that extends as a long tube from the brain down the entire body. Electrodes are placed on either side of the tube, about 2 mm apart, and changes in impedance to radio frequencies are recorded which

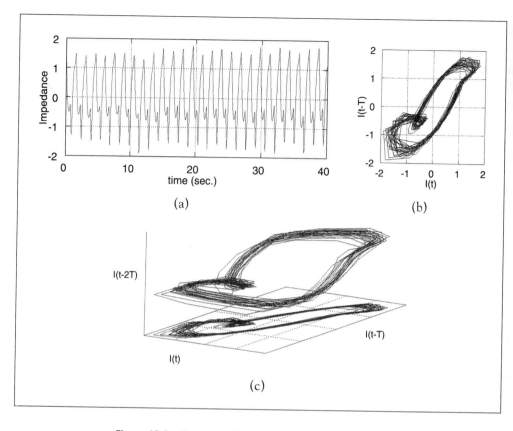

Figure 13.3 Cockroach heartbeat data.
(a) Time series of impedances tracks 26 beats. (b) Delay plot of series in (a) cannot be untangled in two dimensions. Time delay is $T = 1/12$ sec. (c) Three-dimensional delay plot shows successful embedding of the attractor and projection to the horizontal plane, which is the same as (b).

mirror the pumping of hemolymph from the abdomen into the head of the insect. This data comes from the laboratory of G. Birchard.

Although periodic motion is again evident for this system, there are two major differences between this time series and the series from the periodic Couette-Taylor experiment. First, the heartbeat is much noisier. By noisier, we don't mean in the auditory sense, but in the sense of experimental uncertainty. This is reflected both in the non-uniformity of peak heights and troughs in Figure 13.3(a), and in the extra width of the reconstructed cycle of Figure 13.3(b). The noise is characteristic of all real world processes. The cockroach heart, while simple as organs go, is a complicated multicellular system, and is not in a precisely periodic orbit. Furthermore, the electrodes measuring the impedance may be wobbling ever so ightly, measuring the impedance at slightly different points as time progresses. xperimental uncertainties like these also affect the Couette-Taylor experiment, ut are much easier to minimize using careful experimental technique than in a ving organism. All things considered, periodicity is still a good description for ne cockroach heartbeat.

EXERCISE T13.2

Assume that $S(t)$ is a periodic time series that has two local maxima per period, as shown for example in Figure 13.3(a). Show that for some values of T smaller than the period of S, the delay plot $(S(t), S(t - T))$ has a self-intersection.

The second difference from periodic Couette-Taylor is the crossing point ı Figure 13.3(b). The point at approximately $P = (-2/3, -1/3)$ is a point f self-intersection of the reconstruction plot. Our method of prediction via nalogues fails at this point, since the knowledge that the current pair of delay oordinates $(I(t), I(t - T))$ is P could mean either of two different states of the riginal system. Another measurement is needed to fully untangle the attractor. The three-dimensional plot of $(I(t), I(t - T), I(t - 2T))$ in Figure 13.3(c) shows hat it is possible to represent the curve in three dimensions without self-crossings.

A one-to-one continuous function from a compact set to $\mathbb{R}^m$ is called an mbedding of the set, or sometimes a **topological embedding**, to distinguish it rom other types. The number m is called the **embedding dimension** of the set. n Figure 13.1, The periodic orbit of states in the unseen state space is embedded n $\mathbb{R}^2$; the embedding dimension is 2. In Figure 13.3, the periodic orbit fails to mbed in $\mathbb{R}^2$ (because two points are mapped to the same point), but is embedded n $\mathbb{R}^3$ using three delay coordinates.

The achievement of a delay-coordinate embedding is the holy grail of lynamic data analysis, for the one-to-one property means that every state in he state space of the system can be uniquely represented by the measured data. A remarkable theorem says that a finite-dimensional attractor can always be embedded in some $\mathbb{R}^m$; in fact, the necessary embedding dimension m need be only a little larger than twice the dimension of the attractor.

To develop some intuition about this fact, we return to our example of a periodic orbit. Can we expect one-dimensional curves to be embedded in $\mathbb{R}^3$? The answer is yes, if by "expect" we mean what *usually* occurs. Figure 13.4 shows the basic intuition. A closed curve can have a self-intersection, as in (a), in either $\mathbb{R}^2$ or $\mathbb{R}^3$. The difference is that a small bump will dislodge the intersection in $\mathbb{R}^3$,

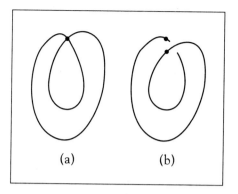

Figure 13.4 Self-intersection of a curve.
(a) The image of a one-dimensional curve by a function to the plane may have a self-intersection that cannot be removed by small perturbations of the function. (b) For a function to three dimensional space, small perturbations remove the self-intersection.

while in $\mathbb{R}^2$ it will only move the intersection somewhere else. This is a key point. If the curve represents the series of states of a dynamical system, reconstructed from measurements, then a slight perturbation in the system dynamics or the measurement would "typically", or "generically", cause the self-intersection in $\mathbb{R}^3$ to disappear. Although it is certainly possible for curves to have self-intersections in $\mathbb{R}^k$ for $k \geq 3$, we should view them as exceptional cases and expect them to occur with essentially zero probability.

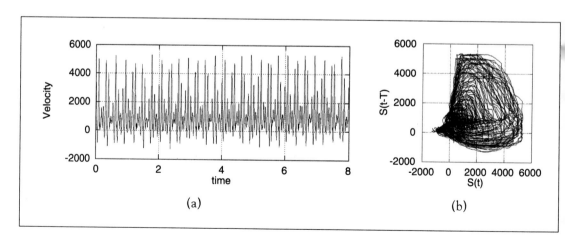

Figure 13.5 Chaotic Couette-Taylor experiment.
(a) Time series of velocities from the experiment. (b) Delay plot of data from (a).

The set of mathematical facts that underlie this point is investigated in Challenge 13. There we find that two sets of box-counting dimensions d_1 and d_2 sitting within $\mathbb{R}^m$ typically fail to intersect if $d_1 + d_2 < m$. If we imagine for a moment that the two strands of the curve in Figure 13.4 are separate manifolds, then they should generically fail to intersect if $1 + 1 < m$. If $m = 3$, there should be no intersection.

If all motion were periodic, and all attractors were curves, the story would be finished. Attractors could be generically embedded in three dimensional delay plots. However, we already know that attractors can have higher dimension. Moreover, they can be fractals. Figure 13.5 shows a time series of measurements from the Couette-Taylor experiment when it is in chaotic motion. Can we embed, examine, and exploit this chaotic attractor as we did the periodic attractor?

13.3 EMBEDOLOGY

It is time to be clearer about the connections between state space, the measurements that comprise a time series, and the reconstructed state space. First suppose that $\mathbb{R}^k$ is the state space of a dynamical system, and trajectories are attracted to a d-dimensional manifold A. Assume that we have a way of making m simultaneous independent measurements of the system at any given time—not just one, as in a time series. For each state, then, our measurements yield a vector in $\mathbb{R}^m$. We make the measurements at several different instants, thereby collecting several points in $\mathbb{R}^m$, each one representing m simultaneous measurements. We think of the measuring process as a function $\mathbf{F}$ from $\mathbb{R}^k$ to $\mathbb{R}^m$. At any time, the state is a point of A in $\mathbb{R}^k$, and we can evaluate $\mathbf{F}$ at that point by doing the m measurements and making a vector out of them. The next theorem says that we should expect $\mathbf{F}(A)$ to uniquely represent all states that were in the original manifold A. The proofs of this theorem and the others that we present in this Chapter are too difficult to be presented here. However, Challenge 13 explores the main concepts, and gives directions for those who want to pursue the proofs.

Theorem 13.1 *Assume that A is a d-dimensional manifold in $\mathbb{R}^k$. If $m > 2d$ and $\mathbf{F} : \mathbb{R}^k \to \mathbb{R}^m$ is generic, then $\mathbf{F}$ is one-to-one on A.*

This means that if $\mathbf{x} \neq \mathbf{y}$ are points on A, then $\mathbf{F}(\mathbf{x}) \neq \mathbf{F}(\mathbf{y})$ in $\mathbb{R}^m$. Two different states in A remain different when mapped into $\mathbb{R}^m$, or in other words, $\mathbf{F}(A)$ has no self-intersections. Note that Theorem 13.1 does not rule out an embedding dimension of less than $2d + 1$; it simply guarantees that $2d + 1$ is sufficient in generic cases. Figure 13.4(b) shows the case $d = 1, m = 3$.

The meaning of generic in Theorem 13.1 was introduced in Chapter 1
Think of it this way: Even if the image $F(A)$ does have self-intersections, other
functions which are extremely small perturbations of F have no self-intersections
To be more precise, we can say the following. For any $\mathbf{F} : \mathbb{R}^k \to \mathbb{R}^m$, define
$\mathbf{F}_M : \mathbb{R}^k \to \mathbb{R}^m$ by $\mathbf{F}_M = \mathbf{F} + M$, where M is the linear map defined by the $m \times$
matrix with real entries between -1 and 1. For all but a measure zero set of
choices from the unit cube in $\mathbb{R}^{mk}$, the function $\mathbf{F}_M$ will be one-to-one on A.

Theorem 13.1 is one of the conclusions of the Whitney Embedding Theo-
rem (Whitney, 1936). The statement requires that the coordinates of $\mathbf{F}$ are independent. Later, it was shown (Takens, 1981) that it is sufficient to choose $\mathbf{F}$ from
the special class of functions formed strictly from delay coordinate reconstructions, using the time series of a single measurement. If we call the measurement
function $h : \mathbb{R}^k \to \mathbb{R}$, then the delay coordinate function is

$$\mathbf{F}(\mathbf{x}) \equiv [h(\mathbf{x}), h(\mathbf{g}_{-T}(\mathbf{x})), \ldots, h(\mathbf{g}_{-(m-1)T}(\mathbf{x}))].$$

Here $\mathbf{g}$ denotes the dynamical system for which A is the attractor. It can be
either an invertible map, in which $\mathbf{g}_{-T}$ denotes T steps of the inverse map, or a
differential equation, in which case it denotes the state T time units ago. Figure
13.6 is a schematic view of the dynamics $\mathbf{g}$, the scalar measurement function h
and the delay coordinate function F.

Figure 13.7 shows the result of a delay coordinate reconstruction for the
Lorenz attractor. Figure 13.7(a) is the original state space $\mathbb{R}^k$ ($k = 3$) of xyz-

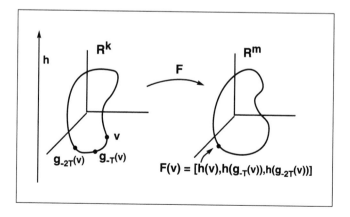

Figure 13.6 Delay coordinate reconstruction function.
With each state $\mathbf{v}$ in the state space $\mathbb{R}^k$ is associated a vector $\mathbf{F}(\mathbf{v})$ in reconstruction
space $\mathbb{R}^m$. The function h is the measurement function, which is a scalar function
of the state space.

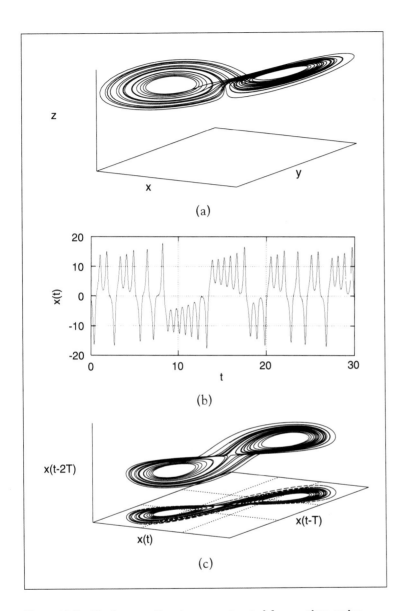

Figure 13.7   The Lorenz attractor reconstructed from a time series.
(a) A trajectory from the Lorenz system, 30 time units long. (b) The x-coordinate of
the trajectory in (a) is plotted as a function of t. (c) A delay-coordinate embedding
of the Lorenz attractor, using three delayed values of the time series in (b). The
projection to the first two coordinates is also shown. The delay is $T = 0.1$.

variables. The x-coordinate alone is measured and graphed as a time series in (b).
Even though the y and z variables have been thrown away, the three dimensional
graph of $[x(t), x(t - T), x(t - 2T)]$ in Figure 13.7(c) gives a fairly faithful visual
reproduction of the original attractor.

Theorem 13.2   *Assume that A is a d-dimensional submanifold of $\mathbb{R}^k$ which
is invariant under the dynamical system* **g**. *If $m > 2d$ and* **F** $: \mathbb{R}^k \to \mathbb{R}^m$ *is a delay
coordinate reconstruction function with a generic measurement function h and generic
time delay T, then F is one-to-one on A.*

Takens' Theorem 13.2 says that if the attractor dimension is the integer d, then
for generic delay plots, the embedding dimension is at most $2d + 1$.

✎   **EXERCISE T13.3**

The genericity requirement on T in Theorem 13.2 is necessary. (a) Show
that if A is a periodic orbit whose period equals the time delay T, then no
delay coordinate reconstruction function can be one-to-one on A. (b) Show
that the same is true if the period is $2T$. [Note: There is no such restriction
for a period $3T$ orbit.]

Takens' Theorem 13.2 triggered an avalanche of research, as scientists tried
to interpret time series collected from experiments by drawing delay plots in
enough dimensions to untangle the attractor. In fact, delay coordinate plots were
advocated independently in the physics literature by (Packard et al., 1980). This
technique was one of few available techniques for analyzing potentially chaotic
data. A one-to-one reconstruction means that the method of analogues can be
used to predict short-term future behavior of a system, even in case it is chaotic.

As we already saw in Lab Visit 4, the Couette-Taylor experiment can
exhibit nonperiodic dynamics. Figure 13.5 shows a chaotic Couette-Taylor time
series and reconstructed trajectory. This plot raises the question whether there
is a fact similar to Theorem 13.2 for attractors A which are fractal rather than
manifolds. It turns out that Theorems 13.1 and 13.2 are true for non-manifolds,
as long as the dimension d is interpreted as box-counting dimension.

Theorem 13.3   *Assume that A is a subset of $\mathbb{R}^k$ with box-counting dimension
d. If $F : \mathbb{R}^k \to \mathbb{R}^m$ is generic and $m > 2d$, then F is one-to-one on A.*

Theorem 13.4   *Assume that A is a subset of $\mathbb{R}^k$ with box-counting dimension
d, which is invariant under the dynamical system* **g**. *If $m > 2d$ and* **F** $: \mathbb{R}^k \to \mathbb{R}^m$ *is*

delay coordinate reconstruction with a generic measurement function h and a generic delay T, then F is one-to-one on A.

In Chapter 4 we found the box-counting dimension of the Hénon attractor of Example 4.11 to be approximately 1.27. Since $2d \approx 2.54$, we expect the attractor to be embedded in three dimensions, but not necessarily in two. We test this conclusion in Figure 13.8. Define a "measurement" function on the xy state space of the Hénon map by $H(x, y) = y - 3\sin(x + y)$. Figure 13.8 should be read in the order (b) → (a) → (c). After each (x, y) point is produced on the attractor in (b), the value of H is plotted in the times series (a). Then a delay map in (c) is made from (a), using a delay of $T = 1$ iteration. There are self-crossings of the image attractor in (c), corresponding to the fact that the assumption $m > 2d$ is not satisfied. In (d), a three-dimensional reconstruction, the image does not cross itself. For this example, the attractor dimension is 1.27, and the embedding dimension is 3.

✎ EXERCISE T13.4

A set with a small box-counting dimension need not lie in a Euclidean space of low dimension. For each $m > 1$, find a subset A of $\mathbb{R}^m$ such that boxdim(A) < 1, but A does not lie in a smooth surface of dimension less than m. [Hint: Find a subset B of $\mathbb{R}^1$ such that boxdim(B) < 1/m. Then let $A = \{(x_1, \ldots, x_m) : x_i \text{ in } B, i = 1, \ldots, m\}$.]

The Belousov-Zhabotinskii reaction, the subject of Lab Visit 3, was one of the first real experiments to be subjected to delay coordinate reconstruction as an analysis tool. Figure 13.9(a) shows a time series of the measured bromide ion concentration from a fixed spot in the reaction apparatus. The time series shows a certain amount of structure, but it is clearly aperiodic.

Two- and three-dimensional reconstructions of the BZ attractor from the time series are shown in Figure 13.9(b) and (c). The embedding dimension appears to be three, using delay $T = 0.05$. In panels (d), (e), and (f) of Figure 13.9 the effect of changing the time delay T is explored. As T is varied, one gets geometrically different but topologically equivalent fractal sets.

For relatively large $T = 0.4$, as in panel (f), the reconstructed set begins to resemble spaghetti. For experimentally measured data from chaotic systems such as the Belousov-Zhabotinskii reaction, there is a practical upper limit on the time delay T. Since nearby states are diverging exponentially in time, an experimental error eventually grows to be comparable to the size of the attractor. If this can

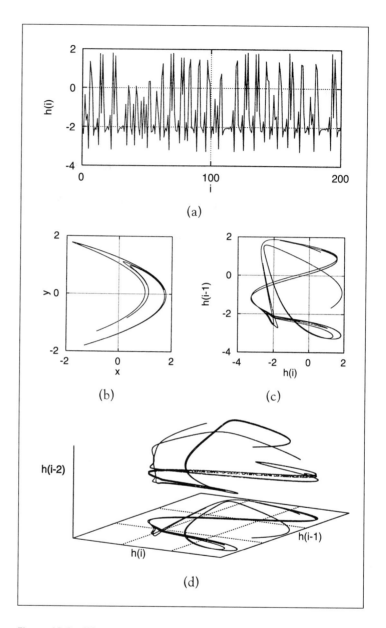

Figure 13.8 Hénon attractor reconstructed from a time series.
(a) A time series of the function $h(i) = y_i - 3\sin(x_i + y_i)$, where x_i and y_i are taken from the Hénon attractor in (b). (c) A two-dimensional delay reconstruction of the attractor. (d) A three-dimensional reconstruction. The projection of this set onto the horizontal plane is seen in (c).

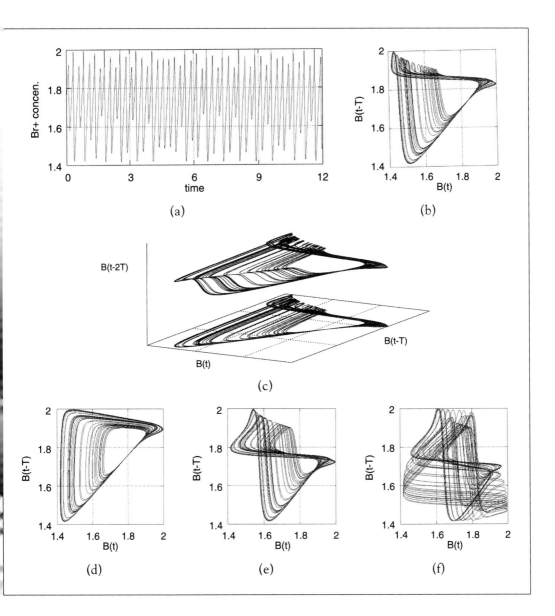

Figure 13.9 Belousov-Zhabotinskii reaction data.
(a) Time series of bromide ion concentration. (b) Two-dimensional delay reconstruction of the BZ attractor, with time delay $T = 0.05$. (c) Three-dimensional
reconstruction whose projection is (b). The remaining three panels show delay plots
with four different choices of delay time T. (d) $T = 0.02$ (e) $T = 0.1$ (f) $T = 0.4$

happen during the time spanned by a single delay coordinate vector, the state that the vector specifies will have limited value for applications like prediction.

Entire books have been written about state reconstruction from data and the scientific applications that follow. Ott, Sauer, and Yorke, 1994 is a starting point for further exploration.

CHALLENGE 13

Box-Counting Dimension and Intersection

How large does a space have to be so that two sets can move around and generally avoid one another? The x-axis and y-axis of the plane intersect at the origin. If we move either or both of them slightly, they still intersect, at a point somewhere near the origin. But if we add another dimension, they hardly ever intersect. The x and y axes in $\mathbb{R}^3$ intersect at the origin, but a small push of either one in almost any direction (except a push precisely along the xy-plane) eliminates the intersection.

In Challenge 13 we want to explore and eventually prove the fact that two sets of dimension d_1 and d_2 sitting in $\mathbb{R}^m$ usually don't intersect if $d_1 + d_2 < m$. By dimension, we mean box-counting dimension. There are interesting applications of this fact for fractal sets. By "usually", we mean that intersections are possible but that they are unusual cases in some sense—a little push, or spatial translation, causes the intersections to disappear. The x and y axes have dimensions $d_1 = 1$ and $d_2 = 1$, so $d_1 + d_2 = 2$. Two lines in a Euclidean space of dimension three or more should intersect only if they are placed just so.

Certainly we can make any two nonempty sets intersect by moving them so that they have a point in common. If the intersection is robust, the sets should still intersect if they are moved a small amount. This is the situation for two lines in the plane, but not for two lines in $\mathbb{R}^3$.

What do we mean by moving the sets? We could allow many kinds of rigid motions, that change the location of the set without changing the relative positions of the points in the set. To keep things simple, let's restrict our movement to spatial translations, which are the result of vector addition of a single point to each of the sets. Adding the vector $(1, 1)$ to a set in the plane moves the set one unit up and one unit to the right. We might as well let one of the two sets stay fixed and move the other, to simplify matters.

Let S be a subset of $\mathbb{R}^m$, and let $\mathbf{v}$ be a point in $\mathbb{R}^m$. We will call the set $S + \mathbf{v} = \{\mathbf{s} + \mathbf{v} : \mathbf{s} \text{ in } S\}$ the **translation** of the set S by $\mathbf{v}$. The translated set is found by adding $\mathbf{v}$ to each point in S, in other words, moving S rigidly to a new location specified by $\mathbf{v}$. For two subsets S_1 and S_2 of $\mathbb{R}^m$, we will say that the **translates of S_1 almost never intersect** S_2 if the set $\{\mathbf{v} \text{ in } \mathbb{R}^m : (S_1 + \mathbf{v}) \cap S_2 \neq \varnothing\}$ has m-dimensional volume zero. This definition is symmetric in S_1 and S_2; it is equivalent that the translates of S_2 almost never intersect S_1. We can now restate our goal in these more precise terms:

Theorem 13.5 *Let S_1 and S_2 be subsets of $\mathbb{R}^m$ of box-counting dimensions d_1 and d_2, respectively. If $d_1 + d_2 < m$, then the translates of S_1 almost never intersect S_2.*

In this sense, the x and y axes always intersect in $\mathbb{R}^2$, but almost never intersect in $\mathbb{R}^3$. To see the latter, notice that of all possible translations in $\mathbb{R}^3$ only those of form $\mathbf{v} = (x, y, 0)$ preserve the intersection. This exceptional set of translations corresponds to the xy-plane, which is a set of three-dimensional volume 0.

Challenge 13 has three parts. The first two concern self-intersection inside the real line, which refers to the case $S_1 = S_2$ and $m = 1$. The Cantor middle-third set has dimension greater than $1/2$, so the Theorem does not apply. In fact, you will show that *every* translation of the Cantor set $S_1 + v$ intersects the (untranslated) Cantor set S_2, provided $-1 < v < 1$. Second, we consider the middle three-fifths Cantor set, which has dimension less than $1/2$. A pair of these sets almost never intersect under translation, which is consistent with Theorem 13.5. Finally, you will put together a proof of the theorem.

We learned earlier that the box-counting dimension of the middle-third Cantor set S is $\ln 2 / \ln 3 \approx 0.63$. Because $2 \times 0.63 > 1$, we do not expect two translated middle-third Cantor sets to be able to avoid one another within the real line. In fact, Step 3 shows that for any $0 < v < 1$, the translated Cantor set $S + v$ has a point in common with the Cantor set S. Recall that in base 3 arithmetic, the middle-third Cantor set consists of all expansions that can be made using only the digits 0 and 2.

Step 1 Explain the fact that if x is in the middle-third Cantor set, then so is $1 - x$.

Step 2 Show that if x is in $[0, 1]$, then x can be written as the sum of two base 3 expansions each of which use only the digits 0 and 1.

Step 3 Let v be any number in $[0, 1]$. Then $(v + 1)/2$ is also in $[0, 1]$. Use Step 2 to show that $v + 1$ is the sum of two numbers belonging to the middle-third Cantor set S. Use Step 1 to conclude that there is a number in S which when added to v gives another number in S. Therefore for each number v in $[0, 1]$, the set $S + v$ intersects with S.

The next three steps refer to the **middle-3/5 Cantor set** $K(5)$. This set is analogous to the middle-third Cantor set, except that we remove the middle $3/5$

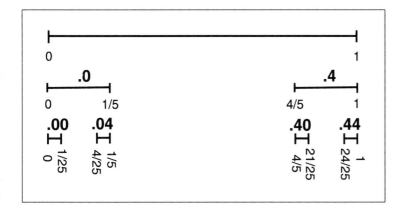

Figure 13.10 The middle three-fifths Cantor set construction.
At each step, the middle three-fifths of each remaining interval is removed. This
Cantor set contains all numbers whose base 5 expansion can be made with only
the digits 0 and 4. The set denoted .04 consists of all numbers in the interval [0, 1]
whose expansion begins with .04.

 each remaining interval instead of the middle one-third. See Figure 13.10 for
 schematic view of the construction.

Step 4 Verify that the middle-3/5 Cantor set consists of all numbers in
 , 1] possessing a base 5 expansion containing only the digits 0 and 4. Find the
 ox-counting dimension of $K(5)$. Show that if S_1 and S_2 are both the middle-3/5
 antor set, then the dimensions satisfy $d_1 + d_2 < 1$.

Step 5 Let v be a number in [0, 1] whose base 5 expansion contains a 22
 wo adjacent base 5 digits are 2). Show that $S_1 + v$ does not intersect S_2.

Step 6 Show that the set of numbers whose base 5 expansion does not
 ontain a 22 is a length 0 set. Conclude that translates of the middle-3/5 Cantor
 t almost never intersects itself. [Hint: Sketch the subinterval .22, which consists
  1/25 of the interval [0, 1]. Each of the remaining subintervals $.s_1 s_2$ of length
 /25 have a subinterval $.s_1 s_2 22$, and so on. Sum up the lengths.]

Next you will prove Theorem 13.5. Assume that S_1 and S_2 are bounded
 ıbsets of $\mathbb{R}^m$ of box-counting dimensions d_1 and d_2, respectively. Put down an
 -grid over $\mathbb{R}^m$ for some $\epsilon > 0$. Consider two ϵ-cubes, one containing points of
  and the other containing points of S_2. To begin, Step 7 asks you to find out
 ow likely it is for the two cubes to intersect under translation.

Step 7 Let C_1 and C_2 be m-dimensional cubes of side ϵ chosen from th grid in $\mathbb{R}^m$. Show that the subset of points $\mathbf{v}$ such that the translation $C_1 +$ intersects C_2 is less than $C\epsilon^m$, for some constant C. [Hint: Denote by P_1 and the centers of the cubes. The cubes can't intersect unless the distance betwee $P_1 + \mathbf{v}$ and P_2 is less than $\sqrt{m}\epsilon$. Show that the cubes don't intersect as long as avoids the ball of radius $\sqrt{m}\epsilon$ centered at $P_2 - P_1$.]

Step 8 Let $N_1(\epsilon)$ be the number of ϵ-cubes covering S_1, and $N_2(\epsilon)$ be th number covering S_2. The number of pairs of ϵ-cubes, one from S_1 and one fro S_2, is $N_1(\epsilon)N_2(\epsilon)$. Denote by $B_m(r)$ the set of points in $\mathbb{R}^m$ within a distance of the origin. Use Step 7 to conclude that the subset of $B_m(r)$ corresponding vectors $\mathbf{v}$ that cause any of these pairs to collide under translation by $\mathbf{v}$ has volum at most $CN_1(\epsilon)N_2(\epsilon)\epsilon^m$, where C is a constant.

Step 9 Use the definition of box-counting dimension and the assumptic that $d_1 + d_2 < m$ to prove that the volume in Step 8 goes to zero as $\epsilon \to 0$. The the volume of translating vectors within $B_m(r)$ which cause any intersections the translated S_1 with S_2 is zero. This is true no matter how large r is, so th translates of S_1 almost never intersect S_2.

Postscript. Theorem 13.5 and its proof in Step 9 are closely related to Theorem 13.3 and 13.4. To achieve the embedding in Theorem 13.3 it is necessary for the imag of distant sections of the attractor in state space to almost always miss each other reconstruction space $\mathbb{R}^m$. It can be shown that small linear perturbations of the functic $\mathbf{F} : \mathbb{R}^k \to \mathbb{R}^m$ in Theorem 13.3 have a similar effect as rigid translations in Theorem 13. Since the distant sections each have box-counting dimension d, the necessary conditic of Theorem 13.5 amounts to $2d < m$ in the notation of Theorem 13.3. Theorem 13 follows from showing that small perturbations of the scalar measurement function h alor are sufficient to restrict self-intersections to occur with probability zero. See (Sauer, York and Casdagli, 1991) for full details.

Matrix Algebra

IN THIS APPENDIX we review important concepts from matrix algebra.

A.1 EIGENVALUES AND EIGENVECTORS

Definition A.1 Let A be an $m \times m$ matrix and x a nonzero m-dimensional real or complex vector. If $Ax = \lambda x$ for some real or complex number λ, then λ is called an **eigenvalue** of A and x the corresponding **eigenvector**.

For example, the matrix $A = \begin{pmatrix} 1 & 3 \\ 2 & 2 \end{pmatrix}$ has an eigenvector $\begin{pmatrix} 1 \\ 1 \end{pmatrix}$, and corresponding eigenvalue 4.

Eigenvalues are found as roots λ of the **characteristic polynomial** $\det(A - \lambda I)$. If λ is an eigenvalue of A, then any nonzero vector in the nullspace of $A - \lambda I$ is an eigenvector corresponding to λ. For this example,

$$\det(A - \lambda I) = \det \begin{pmatrix} 1 - \lambda & 3 \\ 2 & 2 - \lambda \end{pmatrix} = (\lambda - 1)(\lambda - 2) - 6$$

$$= (\lambda - 4)(\lambda + 1), \tag{A.}$$

so the eigenvalues are $\lambda = 4, -1$. The eigenvectors corresponding to $\lambda = 4$ a found in the nullspace of

$$A - 4I = \begin{pmatrix} -3 & 3 \\ 2 & -2 \end{pmatrix} \tag{A.}$$

and so consist of all nonzero multiples of $\begin{pmatrix} 1 \\ 1 \end{pmatrix}$. Similarly, the eigenvectors corr sponding to $\lambda = -1$ are all nonzero multiples of $\begin{pmatrix} 3 \\ -2 \end{pmatrix}$.

If A and B are $m \times m$ matrices, then the set of eigenvalues of the produ matrix AB is the same as the set of eigenvalues of the product BA. In fact, let be an eigenvalue of AB, so that $AB\mathbf{x} = \lambda\mathbf{x}$ for some $\mathbf{x} \neq \mathbf{0}$. If $\lambda = 0$, then

$$0 = \det(AB - \lambda I) = \det(AB) = (\det A)(\det B) = \det(BA - \lambda I),$$

so $\lambda = 0$ is also an eigenvalue of BA. If $\lambda \neq 0$, then $AB\mathbf{x} = \lambda\mathbf{x}$ implies $BA(B\mathbf{x})$ $\lambda B\mathbf{x}$. Note that $B\mathbf{x} \neq \mathbf{0}$ because $\lambda \neq 0$ and $\mathbf{x} \neq \mathbf{0}$. Therefore $B\mathbf{x}$ is an eigenvect of the matrix BA with eigenvalue λ.

A generalization of this fact is Lemma A.2.

Lemma A.2 If $A_1, \ldots, A_n$ are $m \times m$ matrices, then $A_1 \cdots A_n$ and th cyclic permutation $A_{r+1} \cdots A_n A_1 \cdots A_r$ have the same set of eigenvalues, whe $1 \leq r \leq n$.

This follows from the previous paragraph by setting $A = A_1 \cdots A_r$ and B $A_{r+1} \cdots A_n$.

Definition A.3 The $m \times m$ matrices A_1 and A_2 are **similar**, denote $A_1 \sim A_2$, if there exists an invertible $m \times m$ matrix S such that $A_1 = SA_2 S^{-1}$.

Similar matrices have identical eigenvalues, because their characterist polynomials are identical:

$$A_1 - \lambda I = SA_2 S^{-1} - \lambda I = S(A_2 - \lambda I)S^{-1} \tag{A.}$$

implies that

$$\det(A_1 - \lambda I) = (\det S)\det(A_2 - \lambda I)\det S^{-1} = \det(A_2 - \lambda I). \qquad (A.4)$$

If a matrix A has eigenvectors that form a basis for R^m, then A is similar to diagonal matrix, and A is called **diagonalizable**. In fact, assume $A\mathbf{x}_i = \lambda_i \mathbf{x}_i$ for $= 1, \ldots, m$, and define the matrix

$$S = \begin{pmatrix} \mathbf{x}_1 & \cdots & \mathbf{x}_m \end{pmatrix}.$$

Then one can check that the matrix equation

$$AS = S \begin{pmatrix} \lambda_1 & & \\ & \ddots & \\ & & \lambda_m \end{pmatrix} \qquad (A.5)$$

holds. The matrix S is invertible because its columns span R^m. Therefore A is similar to the diagonal matrix containing its eigenvalues.

Not all matrices are diagonalizable, even in the 2×2 case. In fact, all 2×2 matrices are similar to one of the following three types:

1. $A_1 = \begin{pmatrix} a & 0 \\ 0 & b \end{pmatrix}$

2. $A_2 = \begin{pmatrix} a & 1 \\ 0 & a \end{pmatrix}$

3. $A_3 = \begin{pmatrix} a & -b \\ b & a \end{pmatrix}$

Remember that eigenvalues are identical for similar matrices. A matrix is similar to case 1 if there are two eigenvectors that span $\mathbb{R}^2$; a matrix is similar to case 2 if there is a repeated eigenvalue with only one dimensional space of eigenvectors; and to case 3 if it has a complex pair of eigenvalues.

The proof of the fact that the three types suffice for 2×2 matrices follows from the Cayley-Hamilton Theorem, which states that a matrix satisfies its characteristic equation.

Theorem A.4 (Cayley-Hamilton) *If $P(\lambda)$ is the characteristic polynomial of the matrix A, then $P(A) = 0$ (as a matrix equation).*

There are three parts to the above classification. Let A be a 2×2 re[al] matrix.

Fact 1. If A has real distinct eigenvalues a and b, or if $A = aI$, then $A \sim \begin{pmatrix} a & 0 \\ 0 & b \end{pmatrix}$

Proof: If $A = aI$ we are done and $a = b$. If the eigenvalues are distinc[t] then A is diagonalizable. To see this, choose eigenvectors $\mathbf{v}$ and $\mathbf{x}$ satisfyin[g] $A\mathbf{v} = a\mathbf{v}$ and $A\mathbf{x} = b\mathbf{x}$. Note that $\mathbf{v}$ and $\mathbf{x}$ are not multiples of one another sinc[e] $a \neq b$, so that the matrix whose columns are $\mathbf{v}$ and $\mathbf{x}$ is invertible. Then

$$A \begin{pmatrix} \mathbf{v} & \mathbf{x} \end{pmatrix} = \begin{pmatrix} a\mathbf{v} & b\mathbf{x} \end{pmatrix} = \begin{pmatrix} \mathbf{v} & \mathbf{x} \end{pmatrix} \begin{pmatrix} a & 0 \\ 0 & b \end{pmatrix}.$$

Now A is of form SDS^{-1}, where $D = \text{diag}\{a, b\}$.

Fact 2. If A has a repeated eigenvalue $\lambda = a$ and $A \neq aI$, then $A \sim \begin{pmatrix} a & 1 \\ 0 & a \end{pmatrix}$.

Proof: $(A - aI)^2 = 0$. Since $A \neq aI$, there exists a vector $\mathbf{x}$ such tha[t] $\mathbf{v} = (A - aI)\mathbf{x} \neq 0$. Then $(A - aI)\mathbf{v} = (A - aI)^2\mathbf{x} = 0$, so $\mathbf{v}$ is an eigenvecto[r] of A. Note that $\mathbf{v}$ and $\mathbf{x}$ are not linearly dependent, since $\mathbf{v}$ is an eigenvector of A and $\mathbf{x}$ is not. The facts $A\mathbf{x} = a\mathbf{x} + \mathbf{v}$ and $A\mathbf{v} = a\mathbf{v}$ can be written

$$A \begin{pmatrix} \mathbf{v} & \mathbf{x} \end{pmatrix} = \begin{pmatrix} \mathbf{v} & \mathbf{x} \end{pmatrix} \begin{pmatrix} a & 1 \\ 0 & a \end{pmatrix}.$$

Fact 3. If A has eigenvalues $a \pm bi$, with $b \neq 0$, then $A \sim \begin{pmatrix} a & -b \\ b & a \end{pmatrix}$.

Proof: $(A - (a + bi)I)(A - (a - bi)I) = 0$ can be rewritten as $(A - aI)^2 =$ $-b^2 I$. Let $\mathbf{x}$ be a (real) nonzero vector and define $\mathbf{v} = \frac{1}{b}(A - aI)\mathbf{x}$, so that $(A - aI)\mathbf{v} = -b\mathbf{x}$. Since $b \neq 0$, $\mathbf{v}$ and $\mathbf{x}$ are not linearly dependent because $\mathbf{x}$ is no[t] an eigenvector of A. The equations $A\mathbf{x} = b\mathbf{v} + a\mathbf{x}$ and $A\mathbf{v} = a\mathbf{v} - b\mathbf{x}$ can b[e] rewritten

$$A \begin{pmatrix} \mathbf{x} & \mathbf{v} \end{pmatrix} = \begin{pmatrix} \mathbf{x} & \mathbf{v} \end{pmatrix} \begin{pmatrix} a & -b \\ b & a \end{pmatrix}.$$

The similarity equivalence classes for $m > 2$ become a little more com[-]plicated. Look up *Jordan canonical form* in a linear algebra book to investigate further.

A.2 COORDINATE CHANGES

We work in two dimensions, although almost everything we say extends to higher dimensions with minor changes. A vector in $\mathbb{R}^2$ can be represented in many different ways, depending on the coordinate system chosen. Choosing a coordinate system is equivalent to choosing a basis of $\mathbb{R}^2$; then the coordinates of a vector are simply the coefficients that express the vector in that basis.

Consider the standard basis

$$\mathbf{B}_1 = \left\{ \begin{pmatrix} 1 \\ 0 \end{pmatrix}, \begin{pmatrix} 0 \\ 1 \end{pmatrix} \right\}$$

and another basis

$$\mathbf{B}_2 = \left\{ \begin{pmatrix} 1 \\ 0 \end{pmatrix}, \begin{pmatrix} 1 \\ 1 \end{pmatrix} \right\}.$$

The coefficients of a general vector $\begin{pmatrix} x_1 \\ x_2 \end{pmatrix}$ are x_1 and x_2 in the basis $\mathbf{B}_1$, and are $x_1 - x_2$ and x_2 in the basis $\mathbf{B}_2$. This is because

$$x_1 \begin{pmatrix} 1 \\ 0 \end{pmatrix} + x_2 \begin{pmatrix} 0 \\ 1 \end{pmatrix} = \begin{pmatrix} x_1 \\ x_2 \end{pmatrix} = (x_1 - x_2) \begin{pmatrix} 1 \\ 0 \end{pmatrix} + x_2 \begin{pmatrix} 1 \\ 1 \end{pmatrix}, \qquad \text{(A.6)}$$

or in matrix terms,

$$\begin{pmatrix} x_1 \\ x_2 \end{pmatrix} = \begin{pmatrix} 1 & 1 \\ 0 & 1 \end{pmatrix} \begin{pmatrix} x_1 - x_2 \\ x_2 \end{pmatrix} = S \begin{pmatrix} x_1 - x_2 \\ x_2 \end{pmatrix}. \qquad \text{(A.7)}$$

This gives us a convenient rule of thumb. To get coordinates of a vector in the second coordinate system, multiply the original coordinates by the matrix S^{-1}, where S is a matrix whose columns are the coordinates of the second basis vectors written in terms of the first basis. Therefore in retrospect, we could have computed

$$\begin{pmatrix} x_1 - x_2 \\ x_2 \end{pmatrix} = S^{-1} \begin{pmatrix} x_1 \\ x_2 \end{pmatrix} = \begin{pmatrix} 1 & -1 \\ 0 & 1 \end{pmatrix} \begin{pmatrix} x_1 \\ x_2 \end{pmatrix} \qquad \text{(A.8)}$$

as the coordinates, in the second coordinate system $\mathbf{B}_2$, of the vector $\begin{pmatrix} x_1 \\ x_2 \end{pmatrix}$ in the original coordinate system $\mathbf{B}_1$. For example, the new coordinates of $\begin{pmatrix} 1 \\ 1 \end{pmatrix}$ are

$S^{-1} \begin{pmatrix} 1 \\ 1 \end{pmatrix} = \begin{pmatrix} 0 \\ 1 \end{pmatrix}$, which is equivalent to the statement that

$$1 \begin{pmatrix} 1 \\ 0 \end{pmatrix} + 1 \begin{pmatrix} 0 \\ 1 \end{pmatrix} = 0 \begin{pmatrix} 1 \\ 0 \end{pmatrix} + 1 \begin{pmatrix} 1 \\ 1 \end{pmatrix}. \tag{A.9}$$

Now let F be a linear map on $\mathbb{R}^2$. For a fixed basis (coordinate system), we find a matrix representation for F by building a matrix whose columns are the images of the basis vectors under F, expressed in that coordinate system.

For example, let F be the linear map that reflects vectors through the diagonal line $y = x$. In the coordinate system $\mathbf{B}_1$ the map F is represented by

$$A_1 = \begin{pmatrix} 0 & 1 \\ 1 & 0 \end{pmatrix} \tag{A.10}$$

since $F \begin{pmatrix} 1 \\ 0 \end{pmatrix} = \begin{pmatrix} 0 \\ 1 \end{pmatrix}$ and $F \begin{pmatrix} 0 \\ 1 \end{pmatrix} = \begin{pmatrix} 1 \\ 0 \end{pmatrix}$. In the second coordinate system $\mathbf{B}_2$ the vector $F \begin{pmatrix} 1 \\ 0 \end{pmatrix} = \begin{pmatrix} 0 \\ 1 \end{pmatrix}$ has coordinates

$$S^{-1} \begin{pmatrix} 0 \\ 1 \end{pmatrix} = \begin{pmatrix} 1 & -1 \\ 0 & 1 \end{pmatrix} \begin{pmatrix} 0 \\ 1 \end{pmatrix} = \begin{pmatrix} -1 \\ 1 \end{pmatrix},$$

and F fixes the other basis vector $\begin{pmatrix} 1 \\ 1 \end{pmatrix}$, so the map F is represented by

$$A_2 = \begin{pmatrix} -1 & 0 \\ 1 & 1 \end{pmatrix}. \tag{A.11}$$

The matrix representation of the linear map F depends on the coordinate system being used. What is the relation between the two representations? If $\mathbf{x}$ is a vector in the first coordinate system, then $S^{-1}\mathbf{x}$ gives the coordinates in the second system. Then $A_2 S^{-1}\mathbf{x}$ applies the map F, and $SA_2 S^{-1}\mathbf{x}$ returns to the original coordinate system. Since we could more directly accomplish this by multiplying $\mathbf{x}$ by A_1, we have discovered that

$$A_1 = SA_2 S^{-1}, \tag{A.12}$$

or in other words, that A_1 and A_2 are similar matrices. Thus similar matrices are those that represent the same map in different coordinate systems.

A.3 MATRIX TIMES CIRCLE EQUALS ELLIPSE

In this section we show that the image of the unit sphere in $\mathbb{R}^m$ under a linear map is an ellipse, and we show how to find that ellipse.

Definition A.5 Let A be an $m \times n$ matrix. The **transpose** of A, denoted A^T, is the $n \times m$ matrix formed by changing the rows of A into columns. A square matrix is **symmetric** if $A^T = A$.

In terms of matrix entries, $A_{ij}^T = A_{ji}$. In particular, if the matrices are column vectors $\mathbf{x}$ and $\mathbf{y}$, then $\mathbf{x}^T \mathbf{y}$, using standard matrix multiplication, is the dot product, or scalar product, of $\mathbf{x}$ and $\mathbf{y}$. It can be checked that $(AB)^T = B^T A^T$.

It is a standard fact found in elementary matrix algebra books that for any symmetric $m \times m$ matrix A with real entries, there is an orthonormal eigenbasis, meaning that there exist m real-valued eigenvectors $\mathbf{w}_1, \ldots, \mathbf{w}_m$ of A satisfying $\mathbf{w}_i^T \mathbf{w}_j = 0$ if $i \neq j$, and $\mathbf{w}_i^T \mathbf{w}_i = 1$, for $1 \leq i, j \leq m$.

Now assume that A is an $m \times m$ matrix that is not necessarily symmetric. The product $A^T A$ is symmetric (since $(A^T A)^T = A^T (A^T)^T = A^T A$), so it has an orthogonal basis of eigenvectors. It turns out that the corresponding eigenvalues must be nonnegative.

Lemma A.6 Let A be an $m \times m$ matrix. The eigenvalues of $A^T A$ are nonnegative.

Proof: Let $\mathbf{v}$ be a unit eigenvector of $A^T A$, and $A^T A \mathbf{v} = \lambda \mathbf{v}$. Then

$$0 \leq |A\mathbf{v}|^2 = \mathbf{v}^T A^T A \mathbf{v} = \lambda \mathbf{v}^T \mathbf{v} = \lambda.$$

These ideas lead to an interesting way to describe the result of multiplying a vector by the matrix A. This approach involves the eigenvectors of $A^T A$. Let $\mathbf{v}_1, \ldots, \mathbf{v}_m$ denote the m unit eigenvectors of $A^T A$, and denote the (nonnegative) eigenvalues by $s_1^2 \geq \cdots \geq s_m^2$. For $1 \leq i \leq m$, define $\mathbf{u}_i$ by the equation $s_i \mathbf{u}_i = A\mathbf{v}_i$ if $s_i \neq 0$; if $s_i = 0$, choose $\mathbf{u}_i$ as an arbitrary unit vector subject to being orthogonal to $\mathbf{u}_1, \ldots, \mathbf{u}_{i-1}$. The reader should check that this choice implies that $\mathbf{u}_1, \ldots, \mathbf{u}_n$ are pairwise orthogonal unit vectors, and therefore another orthonormal basis of $\mathbb{R}^m$. In fact, $\mathbf{u}_1, \ldots, \mathbf{u}_n$ forms an orthonormal eigenbasis of AA^T.

Figure A.1 shows a succinct view of the action of a matrix on the unit circle in the $m = 2$ case. There are a pair of orthonormal coordinate systems, with bases $\{\mathbf{v}_1, \mathbf{v}_2\}$ and $\{\mathbf{u}_1, \mathbf{u}_2\}$, so that the matrix acts very simply: $\mathbf{v}_1 \mapsto s_1 \mathbf{u}_1$ and

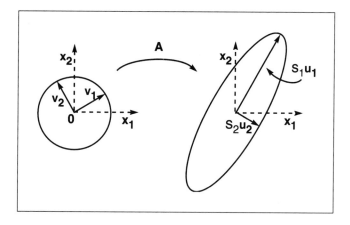

Figure A.1 The ellipse associated to a matrix.
Every 2×2 matrix A can be viewed in the following simple way. There is a coordinate system $\{v_1, v_2\}$ for which A sends $v_1 \to s_1 u_1$ and $v_2 \to s_2 u_2$, where $\{u_1, u_2\}$ is another coordinate system and s_1, s_2 are nonnegative numbers. Geometrically speaking, a matrix A can be broken down into a dilation (stretching or shrinking in various directions) followed by a rotation. This picture extends to $\mathbb{R}^m$ for an $m \times m$ matrix.

$v_2 \mapsto s_2 u_2$. This means that the unit circle of vectors is mapped into an ellipse with axes $\{s_1 u_1, s_2 u_2\}$. In order to find where Ax goes for a vector x, we can write $x = a_1 v_1 + a_2 v_2$ (where $a_1 v_1$ (resp. $a_2 v_2$) is the projection of x onto the direction v_1 (resp. v_2)), and then $Ax = a_1 s_1 u_1 + a_2 s_2 u_2$. Summarizing, we have proved Theorem A.7.

Theorem A.7 *Let A be an $m \times m$ matrix. Then there exist two orthonormal bases of $\mathbb{R}^m$, $\{v_1, \ldots, v_m\}$ and $\{u_1, \ldots, u_m\}$, and real numbers $s_1 \geq \cdots \geq s_m \geq 0$ such that $Av_i = s_i u_i$ for $1 \leq i \leq m$.*

We conclude from this theorem that the image of the unit sphere of vectors is an ellipsoid of vectors, centered at the origin, with semi-major axes $s_i u_i$.

The foregoing discussion is sometimes summarized in a single matrix equation. Define S to be a diagonal matrix whose entries are $s_1 \geq \cdots \geq s_m \geq 0$. Define U to be the matrix whose columns are $u_1, \ldots, u_m$, and V to be the matrix whose columns are $v_1, \ldots, v_m$. Notice that $USV^T v_i = s_i u_i$ for $i = 1, \ldots, m$. Since the matrices A and USV^T agree on the basis $v_1, \ldots, v_m$, they are identical $m \times m$ matrices. The equation

$$A = USV^T \tag{A.13}$$

called the **singular value decomposition** of A, and the s_i are called the **singular values**. The matrices U and V are called **orthogonal** matrices because they satisfy $U^T U = I$ and $V^T V = I$. Orthogonal matrices correspond geometrically to rigid body transformations like rotations and reflections.

Computer Solution of ODEs

SOME ORDINARY DIFFERENTIAL EQUATIONS can be solved explicitly. The initial value problem

$$\dot{x} = ax$$

$$x(0) = x_0 \qquad\qquad (B.1)$$

has solution

$$x(t) = x_0 e^{at}. \qquad\qquad (B.2)$$

The solution can be found using the separation of variables technique introduced in Chapter 7. The fact that the problem yields an explicit solution makes it atypical among differential equations. In the large majority of cases, there does not exist an expression for the solution of the initial value problem in terms of elementary functions. In these cases, we have no choice but to use the computer to approximate a solution.

B.1 ODE SOLVERS

A computational method for producing an approximate solution to an initial value problem (IVP) is called an **ODE solver**. Figure B.1(a) shows how a typical ODE solver works. Starting with the initial value x_0, the method calculates approximate values of the solution $x(t)$ on a grid $\{t_0, t_1, \ldots, t_n\}$.

The simplest ODE solver is the **Euler method**. For the initial value problem

$$\dot{x} = f(t, x)$$

$$x(t_0) = x_0, \tag{B.3}$$

the Euler method produces approximations by an iterative formula. Choosing a step size $h > 0$ determines a grid $\{t_0, t_0 + h, t_0 + 2h, \ldots\}$. Denote the correct values of the solution $x(t)$ by $x_n = x(t_n)$. The Euler method approximations at the grid points $t_n = t_0 + nh$ are given by

$$w_0 = x_0$$

$$w_{n+1} = w_n + hf(t_n, w_n) \quad \text{for } n \geq 1. \tag{B.4}$$

Each w_n is the Euler method approximation for the value $x_n = x(t_n)$ of the solution.

It is easy to write a simple program implementing the Euler method. The code fragment

```
w[0]  = w0;
t[0]  = t0;
for (n=0;n<=N;n++) {
   t[n+1]  = t[n]+h;
   w[n+1]  = w[n]+h*f(t[n],w[n]);
}
```

together with a defined function f from the differential equation will generate the grid of points $\{t_n\}$ along with the approximate solutions $\{w_n\}$ at those points. Remember to set the step size h. The smaller the value for h, the smaller the error, which is the subject of the next section.

A far more powerful method that requires just a few more lines is the **Runge-Kutta method** of order 4. A code fragment implementing this method is

```
[0] = x0;
[0] = t0;
or(n=0;n<=N;n++){
  t[n+1] = t[n]+h;
  k1 = h*f(t[n],w[n]);
  k2 = h*f(t[n]+h/2,w[n]+k1/2);
  k3 = h*f(t[n]+h/2,w[n]+k2/2);
  k4 = h*f(t[n+1],w[n]+k3);
  w[n+1] = w[n]+(k1 + 2*k2 + 2*k3 + k4)/6;
```

To run a simulation of a three-dimensional system such as the Lorenz attractor, it is necessary to implement Runge-Kutta for a system of three differential
equations. This involves using the above formulas for each of the three variables
in parallel. A possible implementation:

```
x[0] = x0;
y[0] = y0;
z[0] = z0;
[0] = t0;
or(n=0;n<=N;n++){
  t[n+1] = t[n]+h;
  kx1 = h*fx(t[n],wx[n],wy[n],wz[n]);
  ky1 = h*fy(t[n],wx[n],wy[n],wz[n]);
  kz1 = h*fz(t[n],wx[n],wy[n],wz[n]);
  kx2 = h*fx(t[n]+h/2,wx[n]+kx1/2,wy[n]+ky1/2,wz[n]+kz1/2);
  ky2 = h*fy(t[n]+h/2,wx[n]+kx1/2,wy[n]+ky1/2,wz[n]+kz1/2);
  kz2 = h*fz(t[n]+h/2,wx[n]+kx1/2,wy[n]+ky1/2,wz[n]+kz1/2);
  kx3 = h*fx(t[n]+h/2,wx[n]+kx2/2,wy[n]+ky2/2,wz[n]+kz2/2);
  ky3 = h*fy(t[n]+h/2,wx[n]+kx2/2,wy[n]+ky2/2,wz[n]+kz2/2);
  kz3 = h*fz(t[n]+h/2,wx[n]+kx2/2,wy[n]+ky2/2,wz[n]+kz2/2);
  kx4 = h*fx(t[n+1],wx[n]+kx3,wy[n]+ky3,wz[n]+kz3);
  ky4 = h*fy(t[n+1],wx[n]+kx3,wy[n]+ky3,wz[n]+kz3);
  kz4 = h*fz(t[n+1],wx[n]+kx3,wy[n]+ky3,wz[n]+kz3);
  wx[n+1] = wx[n]+(kx1 + 2*kx2 + 2*kx3 + kx4)/6;
  wy[n+1] = wy[n]+(ky1 + 2*ky2 + 2*ky3 + ky4)/6;
  wz[n+1] = wz[n]+(kz1 + 2*kz2 + 2*kz3 + kz4)/6;
```

Here it is necessary to define one function for each of the differential equations:

```
fx(t,x,y,z) = 10.0*(y-x);
fy(t,x,y,z) = -x*z + 28*x - y;
fz(t,x,y,z) = x*y - (8/3)*z;
```

This code can be economized significantly using object-oriented programming ideas, which we will not pursue here. Any other three-dimensional system can be approximated using this code, by changing the function calls `fx`, `fy` and `fz`. Higher-dimensional systems require only a simple extension to more variables.

So far we have given no advice on the choice of step size h. The smaller is, the better the computer approximation will be. For the Lorenz equations, the choice $h \approx 10^{-2}$ is small enough to produce representative trajectories.

Coding time can be reduced to a minimum by using a software package with ODE solvers available. Matlab is a general-purpose mathematics software package whose program `ode45` is an implementation of the Runge-Kutta method in which step size h is set automatically and monitored to keep error within reasonable bounds. To use this method one needs only to create a file named `f.m` containing the differential equations; `ode45` does the rest. Type `help ode45` in Matlab for full details. Other generally available mathematical software routines such as Maple and Mathematica have similar capabilities.

Finally, the software package *Dynamics* has a built-in graphical user interface that runs Runge-Kutta simulations of the Lorenz equations and several other systems with no programming required. See (Nusse and Yorke, 1995).

B.2 ERROR IN NUMERICAL INTEGRATION

As suggested in Figure B.1(a), there is usually a difference between the true solution $x(t)$ to the IVP and the approximation. Denote the **total error** at t_n by $E_n = x_n - w_n$. Error results from the fact that an approximation is being made by discretizing the differential equation. For Euler's method, this occurs by replacing $\dot{x}$ by $(x_{n+1} - x_n)/h$ in moving from (B.3) to (B.4).

The initial value problem (B.1) has an explicit solution, so we can work out the exact error to illustrate what is going on. Set $t_0 = 0$, so that $t_n = nh$. The Euler method formula from (B.4) is

$$w_0 = x_0$$

$$w_{n+1} = w_n + ahw_n \quad \text{for } n \geq 0. \tag{B.5}$$

For example, assume $a = 1$, $x_0 = 1$, and the stepsize is set at $h = 0.1$. Then $w_1 = 1 + 0.1 = 1.1$, which is not too far from the correct answer $x_1 = e^{0.1} \approx 1.105$. The error made in one step of the Euler method is ≈ 0.005. A formula for the one-step error is $e_1 = w_0(e^{ah} - (1 + ah))$.

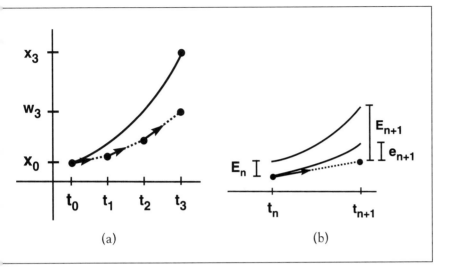

Figure B.1   Output from an ODE solver.
(a) The Euler method follows a line segment with the slope of the vector field at
the current point to the next point. The upper curve represents the true solution to
the differential equation. (b) The total error E_{n+1} is the total of the one-step error
e_{n+1} and accumulated error from previous steps.

After one step, the total error E_1 is the same as the one-step error e_1. The
total error E_2 after two steps has two contributions. First, because of the error E_1,
the right side f of the differential equation is going to be sampled at an incorrect
point w_1 instead of x_1. Second, a new one-step error e_2 will be made because of
the discretization of the derivative. As Figure B.1(b) shows, the new total error
E_2 will have a contribution from the accumulated error from the previous step as
well as a new one-step error.

The output of the Euler method is $w_n = x_0(1 + ah)^n$ for $n \geq 0$. The true
solution is $x_n = x_0 e^{ahn}$, so that the difference is

$$E_n = x_n - w_n$$

$$= x_0 e^{ahn} - x_0(1 + ah)^n$$

$$= x_0 e^{ahn} - x_0(1 + ah)^{n-1}e^{ah} + x_0(1 + ah)^{n-1}e^{ah} - x_0(1 + ah)^n$$

$$= e^{ah} \cdot x_0(e^{ah(n-1)} - (1 + ah)^{n-1}) + w_{n-1}(e^{ah} - (1 + ah)). \qquad \text{(B.6)}$$

We recognize E_{n-1}, the total error at step $n - 1$, and e_n, the one-step error at
step n, in the above equation. If we set $A = e^{ah}$, we can write the result in the
following general form.

TOTAL ERROR FOR AN ODE SOLVER

$$E_n = AE_{n-1} + e_n$$

In this expression, the term A is the **amplification factor**. It depends on the method being used and the particular differential equation. If $A \approx 1$, one expects the total error to be approximately the sum of the one-step errors. For IVPs that are sensitive to initial conditions, however, the amplification factor is commonly greater than one. For system (B.1), this case holds when $a > 0$.

The order of an ODE solver is a measure of its relative accuracy. Let t_0 be the initial time of the IVP, and let $t_e > t_0$ be a later time. By definition, the **order** of an ODE solver is k if the total error evaluated at t_e is

$$E \approx Ch^k \tag{B.7}$$

for small h. More precisely, one requires

$$\lim_{h \to 0} \frac{E}{h^k} < \infty.$$

Order measures the dependence of the total error on the stepsize h, and is helpful in a relative way. It gives no absolute estimate of the error, but it tells us that the error of a second-order method, for example, would decrease by a factor of 4 if we cut the step size in half (replace h by $h/2$ in (B.7)). This concept of order is used to rank methods by accuracy.

The order of an ODE solver can be informally determined by expressing the one-step error in terms of h. Using the differential equation (B.3), the Taylor expansion of the solution at t_0 can be written

$$x(t_0 + h) = x_0 + h\dot{x}(t_0) + \frac{h^2}{2}\ddot{x}(t_0) + \cdots$$

$$= x_0 + hf(t_0, x_0) + \frac{h^2}{2}\left[\frac{\partial f}{\partial x}(t_0, x_0) + \frac{\partial f}{\partial t}\right] + \cdots. \tag{B.8}$$

Comparing with the Euler's method approximation (B.4), we find the one-step error e_1 to be proportional to h^2, where terms of higher degree in h are neglected for small h. **To find the order of the method, subtract one from the power of h in the one-step error.** The reasoning is as follows. Assuming that the amplification factor is approximately one, the simplest case, the total error made by the ODE solver between t_0 and t_e will be approximately the sum of the one-step errors.

he number of steps needed to reach the desired time t_e grows as h decreases: It proportional to h^{-1}. The total error for the Euler method, as a function of h, nould therefore be proportional to $h^2 \cdot h^{-1} = h^1$. On this basis, the Euler method of order one. Cutting the step size in half results in cutting the total error in alf.

Higher order methods can be derived by more elaborate versions of the easoning used in the Euler method. The **modified-Euler method** is

$$z_0 = x_0$$

$$z_{n+1} = z_n + h \frac{s_1 + s_2}{2} \quad \text{for } n \geq 1, \tag{B.9}$$

here $s_1 = f(t_n, z_n)$ and $s_2 = f(t_n + h, z_n + hs_1)$. It can be checked that the one-ep error for this method is $\approx h^3$ (see, for example, (Burden and Faires, 1993)). igure B.2(a) shows one step of the method along with the roles of s_1, which is ne slope of f as in the Euler method, and s_2, which is the slope at t_{n+1}, where the uler method is used as a guess for x_{n+1}. Instead of taking the Euler method step, ne two slopes s_1 and s_2 are averaged, and a step is taken with the averaged slope. his is a simple **predictor-corrector** method, in which the Euler method is used) predict the new solution value, followed by a more accurate correction. Note nat the correction cannot be made without knowing the prediction.

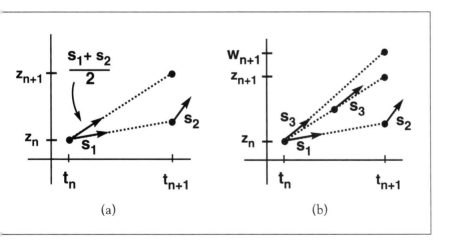

(a) (b)

Figure B.2 The geometry of a higher-order one-step method.
(a) The Modified Euler method takes an Euler step whose slope is the average of the two slopes shown. (b) A third-order method first uses modified Euler to find the vector field in the middle of the time interval, and translates the slope s_3 determined there back to t_n. Then an Euler step is taken with slope s_3 to produce (t_{n+1}, w_{n+1}).

B.3 ADAPTIVE STEP-SIZE METHODS

Thus far we have treated the step size h as a constant throughout the calculation. We have given no advice on how to choose h, except to say that the smaller the h, the smaller the error. Adaptive methods find the best step size automatically and constantly recalculate the optimal size as the calculation proceeds.

Matlab's ode23 program is an adaptive method using second-order modified Euler along with the third-order method

$$w_0 = x_0$$

$$w_{n+1} = w_n + h\frac{s_1 + 4s_3 + s_2}{6} \quad \text{for } n \geq 1, \tag{B.10}$$

where s_1 and s_2 are as defined for the modified Euler method and $s_3 = f(t_n + h/2, w_n + (h/2)(s_1 + s_2)/2)$. As sketched in Figure B.2(b), the average slope used to calculate z_{n+1} in the Euler-type step in the modified Euler method is used instead to sample the differential equation for a new slope at $t_n + h/2$. This slope is used for the Euler-type step from t_n to t_{n+1}.

To get the program to adjust the step size h automatically, the user must set a tolerance TOL for the one-step error. For the new step, both the second-order approximation z_{n+1} from (B.9) and the third-order approximation w_{n+1} from (B.10) are calculated. Since the third-order approximation is so much better, the difference between the two is a good approximation for the one-step error of z_{n+1}, that is, $e(h) \approx |z_{n+1} - w_{n+1}|$. Denote by c the factor by which we want to change the step size. Then the new step size should be the ch that satisfies TOL $= e(ch)$ which can be approximated as follows:

$$\text{TOL} = e(ch) \approx c^3 e(h) \approx c^3 |z_{n+1} - w_{n+1}|. \tag{B.11}$$

Here we have used the fact that the one-step error of the (second-order) modified Euler method is proportional to h^3. Now (B.11) can be solved for c:

$$c = \left(\frac{\text{TOL}}{|z_{n+1} - w_{n+1}|}\right)^{1/3}. \tag{B.12}$$

Two further points make the automatic choice of step size conservative. First, the new h is set to be $0.9ch$ instead of ch. Second, although the step size is being set to keep one-step error within the preset tolerance for the second-order method, the third-order approximation w_{n+1} is accepted as the new value of the solution.

Matlab also provides a higher-order adaptive method called ode45. This is an implementation of the **Runge-Kutta-Fehlberg method**. It uses a variation

f the fourth-order Runge-Kutta method described above together with a fifth-
rder method that can be accomplished while reusing some of the fourth-order
alculations, in the same spirit as `ode23`. This method, often denoted RK45 in
ther sources, is very popular for applications in which running time is not a
ritical factor.

Depending on your intended use, we have either completely solved the
roblem of computational solution of differential equations, or have barely
cratched the surface. One-step methods are fine for the simulations we have
utlined in this book, but are too slow for industrial-strength applications. A
ough way to compare the computational effort required for ODE methods is to
ount the number of times per step that f, the right-hand side of the equation,
eeds to be evaluated.

As we saw in the above code fragment, fourth-order Runge-Kutta requires
our function evaluations per step, for a one-dimensional system. It is not hard
o derive multistep methods that are fourth order but require only one function
valuation per step. A **multi-step ODE method** uses not only the previous esti-
ate w_n, but several previous estimates $w_n, w_{n-1}, \ldots$ to produce w_{n+1}. In fact,
t is fairly wasteful of one-step methods to ignore this available information. If
ost of the computational effort lies in the evaluation of f, we would expect a
ourth-order multistep method to run four times as fast as RK4.

Multistep methods have some liabilities. The most obvious is that starting
he method is nontrivial. Usually, a one-step method is used to intialize the pre-
ious w_i estimates that are needed. Second, when adaptive step sizing is used, the
act that the previous estimates $w_n, w_{n-1}, \ldots$ are not equally spaced complicates
he formulas significantly. A new generation of methods called **multivalue ODE
methods** has recently been introduced to alleviate some of these problems. We
efer you to a current numerical analysis textbook, such as (Burden and Faires,
993) to read more about these issues.

Answers and Hints to Selected Exercises

T1.2   Hint: Mimic the proof of Part 1.

T1.3   The points in the open intervals $(-1, 0)$ and $(0, 1)$ satisfy $|f(x)| > |x|$, and converge to sinks -1 and 1, respectively. The points $|x| > \sqrt{5}$ satisfy $|f(x)| > |x|$, and because they increase the distance from the origin on each iteration they cannot converge to either of the sinks.

T1.4   One condition is that f lies strictly between the lines $y = p$ and $y = x$ in $(p - \epsilon, p + \epsilon)$.

T1.5 $\{1 + \sqrt{2}, 1 - \sqrt{2}\}$.

T1.7 $\{(5 + \sqrt{5})/8, (5 - \sqrt{5})/8\}$.

T1.8 f^k has 2^k fixed points. Some are also fixed points of period less than k, but since $2^k - (2^{k-1} + 2^{k-2} + \ldots + 2^1) = 2$, not all can have period less than k.

T1.10 If x is eventually periodic, then $3^m x = 3^n x \pmod 1$. Since $3^{m-n}x$ is an integer p, $x = p/3^{m-n}$ is a rational number. Conversely, assume $x = p/q$, where p and are integers. The integers $3^m p$ cannot all be distinct modulo q, since there a only q possibilities. For some integers m and n, $3^m p = 3^n p \pmod q$, which impli $3^m p/q = 3^n p/q \pmod 1$.

T1.14 (a) Any number between $(2 - \sqrt{2 + \sqrt{2}})/4$ and $(2 - \sqrt{2})/4$ will do.

T1.16 (a) Greater than $1/2$ (b) Less than $1/2$.

1.1 l has an attracting fixed point if and only if $-1 < a < 1$, and has a repellin fixed point if and only if $a < -1$ or $a > 1$. There is a fixed point that is neith attracting nor repelling if $a = -1$, and many such if $a = 1$ and $b = 0$. If $a =$ and $b \neq 0$, the map l has no fixed point.

1.3 $x = 0$ is a source.

1.5 Source.

1.9 (a) $\sqrt{2}$. (b) $-\sqrt{2}$ and all preimages of -1. Hints: Draw the graph and show th except for the undefined orbit with $x_0 = -1$, all orbits that ever reach outsi $[-1.5, -1.4]$ converge to $\sqrt{2}$. Note that the endpoints of this interval both ma to -1, so the remainder of the orbit (and limit) is undefined. Next show th any point inside this interval is repelled from $-\sqrt{2}$ at the rate of more than factor of 4 per iterate, so that no orbit other than the fixed orbit $\{-\sqrt{2}\}$ ca avoid eventually leaving the interval. Most leave the interval and converge t $\sqrt{2}$; a few unlucky ones land on -1.5 or -1.4 on the way out and end up wit undefined orbits.

1.11 (b) Hint: Show that $(f^2)'(x) \geq 0$ for all x.

1.15 The only fixed point is $x = 0$. Orbits with initial conditions in $[-1, 0]$ a attracted to 0; all other orbits diverge.

1.16 Hint: Check for $n = 0$. To prove the formula for $n = 1$, you may need th double-angle formula: $\cos 2x = 1 - 2\sin^2 x$.

T2.4 Hint: Write the vector in polar coordinates and use the sin and cos addition law

T2.7 (b) -1. (d) -1.

T2.8 The inverse map is $g(x, y) = (y, (x - a + y^2)/b)$. If $b = 0$, then the points $(1, 0$ and $(1, 1)$ map to the same point, so the map is not one-to-one.

T2.9 (b) Part (a) confirms that the orbit of any point on S stays on S, a parabola throug 0. Since the x-coordinate of the orbit is halved each iteration, it converges to 0 and the y coordinate must follow.

T2.10 The axes of AN are of length $4/3$ and $1/3$ along the directions $(2, 1)$ and $(1, -2$ respectively.

.1   (a) Source (b) Saddle (c) Sink.

.2 $(0, 0)$.

.3 $(0, 0)$ and $(3, 9)$ are saddle fixed points.

.4   (a) $(-0.7, -0.7)$ is a fixed sink, $(-0.8, -0.8)$ is a fixed saddle; there are no period-two orbits. (b) $(-0.7, -0.7)$ and $(0.3, 0.3)$ are saddles and $\{(0.5, -0.1),$ $(-0.1, 0.5)\}$ is a period-two sink orbit.

.7   (a) $-0.1225 < a < 0.3675$ (b) $0.3675 < a < 0.9125$

.8   (a) Image ellipse has one axis of length $2\sqrt{2}$ in the direction $(1, 1)$, and another of length $1/\sqrt{2}$ in the $(-1, 1)$ direction. Area is 2. (b) Ellipse has axes of length 3 and 2, and area of 6.

T3.2   Hint: Find the Lyapunov exponents of all fixed points, and then show that all orbits either are unbounded or converge to one of the fixed points.

T3.3   If f has a period-n orbit, then $f^n(x_0) = x_0 + nq = x_0 \pmod 1$, so that nq is an integer. That is not possible since q is irrational. The Lyapunov exponent of the orbit of any x_0 is $\lim_{n\to\infty} \frac{1}{n}[\ln 1 + \ldots + \ln 1] = 0$.

T3.4   The set of points which share the same length-k itinerary is a single subinterval of length 2^{-k}. An infinite itinerary corresponds to the nested intersection of subintervals of length $2^{-k}, k \geq 1$, which is a single point.

T3.7   First check that $T^n(x) = x$ if and only if $G^n C(x) = C(x)$, for any positive integer n. If x is a period-k point for T, then $T^k(x) = x$ implies $G^k C(x) = C(x)$. The equality $G^n C(x) = C(x)$ cannot hold for any $n < k$ because it would imply $T^n(x) = x$, which is not true—x is a period-k point. Therefore $C(x)$ is a period-k point for G.

T3.8   The first statement of Theorem 3.11 is Exercise T3.7. Secondly, apply the chain rule to the equation $g^k(C(x)) = C(f^k(x))$ to get $(g^k)'(C(x))C'(x) = C'(f^k(x))(f^k)'(x)$. Since $f^k(x) = x$ and $C'(x) \neq 0$, cancelling yields $(g^k)'(C(x)) = (f^k)'(x)$.

T3.10   (a) The itinerary $\mathbf{LRR}\cdots\mathbf{R}$, consisting of one $\mathbf{L}$ and $k - 1\,\mathbf{R}$'s, is not periodic for any period less than k. (b) According to Corollary 3.18, the interval $\mathbf{LRR}\cdots\mathbf{RL}$ contains a fixed point of f^k. By part (a), that point must be part of a period-k orbit.

T3.12   The ten distinct periodic orbits of period ≤ 5 are: $\overline{\mathbf{L}}$, $\overline{\mathbf{KL}}$, $\overline{\mathbf{JKL}}$, $\overline{\mathbf{KLL}}$, $\overline{\mathbf{JKLL}}$, $\overline{\mathbf{KLLL}}$, $\overline{\mathbf{IJKL}}$, $\overline{\mathbf{JKLLL}}$, $\overline{\mathbf{KLLLL}}$, and $\overline{\mathbf{IJKLL}}$.

3.1   (a) $a_1 = -1/4$. (b) $-\infty$. (c) $a_2 = 3/4$. (d) $a_3 = 5/4$. (e) First find an interval which maps onto itself. Then find a partition as was done for the logistic map G.

3.3 (a) G maps $[0, 1]$ onto itself and g maps another interval onto itself. C will be one-to-one continuous map between $[0, 1]$ and the other domain interval.

3.4 Hint: See Exercise T3.2.

3.5 (b) $[1, 2]$ and $[2, 3]$. (c) The transition graph is the fully connected graph on two symbols. There exists periodic points of all periods.

3.8 (a) All positive integers except 3, 5, and 7. (b) All positive integers.

3.9 (c) All positive integers.

3.10 All positive integers except for 3.

3.11 (b) Hint: Explain why either f or f^2 must have points $0 \le x_1 < x_2 \le 1$ such that x_1 maps to 0 and x_2 maps to 1.

3.12 Hint: Find a positive lower bound for f' on $[0, 1]$.

T4.2 The expansions end in $0\overline{2}$.

T4.7 Complete the square by replacing $w = x - 1/2$.

T4.8 Hint: Count the number of boxes needed to cover a square contained in (respectively, containing) the disk.

T4.9 $1 + \ln 2 / \ln 3$.

T4.10 (b) 1.

T4.11 (a) 0. (b) 1.

4.1 (a) Hint: Define $f(x) = 3x$. Show that f is a one-to-one function from one set onto the other.

4.2 (a) 1. (b) In base 4 arithmetic, all numbers whose expansions use only 0 and 3. (c) $1/5 = .\overline{03}$, and $17/21$ is also in $K(4)$. (d) $a = 4$.

4.4 Hint: Consider a family $\{Q_n\}$ of intervals such that the nth rational is contained in Q_n and the length of Q_n is $\epsilon/(2^n)$, for each n, $n \ge 1$.

4.8 $\ln 2 / \ln 3$.

4.11 (a) True. Prove for a union of two sets, then generalize. (b) False.

4.12 (a) Hint: Find out the distance between points $1/(n + 1)$ and $1/n$. Show that $N(\epsilon) = 2n$ for $\epsilon = 1/n(n + 1)$.

4.13 (a) $1/(1+p)$. (b) 0.

T5.1 $(2/3)^n$.

T5.4 Cat map: 2; Ikéda map: $1 + .51/.72 \approx 1.7$.

T5.6 $(3/4, 1/3)$ and $(1/4, 2/3)$; there are no others.

5.1   Since $J_n = \mathbf{Df}(\mathbf{v}_{n-1}) \cdots \mathbf{Df}(\mathbf{v}_0)$, the matrix $J_n J_n^T$ has determinant D^{2n}. (The determinant is unchanged under the transpose operation.) Since the product of the m eigenvalues of $J_n J_n^T$ is D^{2n}, the product of the semimajor axes of $J_n U$ is $r_1^n \cdots r_m^n = |D|^n$. Therefore $(r_1^n)^{1/n} \cdots (r_m^n)^{1/n} = |D|$ for $n \geq 1$, and $L_1 \cdots L_m = [\lim_{n \to \infty} (r_1^n)^{1/n}] \cdots [\lim_{n \to \infty} (r_m^n)^{1/n}] = |D|$, assuming the Lyapunov numbers exist.

5.2   The Lyapunov numbers will be the absolute values of the eigenvalues of the symmetric defining matrix, which are $(1 \pm \sqrt{5})/2$. The Lyapunov exponents are $\ln(1 + \sqrt{5})/2$ and $\ln(\sqrt{5} - 1)/2$. These are smaller than the cat map Lyapunov exponents by a factor of 2 because the cat map matrix is the square of this one.

5.6   Hint: Design a two-sided symbol sequence that contains every finite subsequence of L and R on both sides. There is a positive Lyapunov exponent as explained in Section 5.6.

5.7   (b) The unit circle $r = 1$. (c) Orbits on the unit circle have Lyapunov exponents $\ln 2$ and $\ln 2$. Orbits inside the unit circle have Lyapunov numbers 0 and 0, so the Lyapunov exponents do not exist.

T6.2   Yes; consider $G(x) = 4x(1 - x)$.

T6.3   (b) Hint: You will need to find the symbol sequence of a dense orbit. Any neighborhood of a point in the invariant set of the horseshoe contains a block determined by specifying a finite number of symbols to the left and to the right of the decimal point. Proceed, as in the construction of a dense orbit for the logistic map g_4, by listing every finite block of symbols, ordered by length.

T6.6   Hint: Points in the attractor A can be assigned two–sided symbol sequences just as points in the chaotic set of the horseshoe map. Notice that the orbit of a point in the solenoid depends entirely on its t coordinate. Specifically, code points symbolically by the following rule: the ith symbol is 0 if the t coordinate of $\mathbf{f}^i(t, z)$ is in $[0, \frac{1}{2})$; otherwise, the ith symbol is 1. Show that any neighborhood of a point in A can be represented by a symbol block with a finite number of symbols specified to the right and to the left of the decimal point. (The symbols to the right determine a neighborhood of the t coordinate, while those to the left determine on which local branch the point is located.) Then follow the construction for a dense orbit in the horseshoe.

T6.7   Hint: Construct an orbit similar to the dense orbit in the tent map T_2 or the logistic map g_4, which includes all allowable symbol sequences, ordered by length.

6.1   Hint: See Exercise T6.2.

6.2   Figure 6.10(a): All positive integers. Figure 6.13(a): All positive integers. Figure 6.14(a): All even positive integers.

6.3   The fixed points $(-1, 0), (0, 0), (1, 0)$ are forward limit sets; $(-1, 0)$ and $(1, 0)$ are attracting fixed points. The left half-plane is the basin of $(-1, 0)$ and the right half-plane is the basin of $(1, 0)$. The basin boundary is the y-axis. These points tend to the saddle $(0, 0)$ on iteration.

6.4 $F(x_0, \{0\}) = \begin{cases} 0 & \text{if } x_0 \neq 0 \\ 1 & \text{if } x_0 = 0 \end{cases}$. $F(x_0, N(r, \{0\})) = 1$ for all $x_0, r > 0$. $F(x_0, [0, \infty)) = $

$\begin{cases} 1/2 & \text{if } x_0 \neq 0 \\ 1 & \text{if } x_0 = 0 \end{cases}$. The natural measure μ_f is the atomic measure located at 0.

6.7   Hint: This map is conjugate to the tent map. See Exercise 3.3 and Challenge 6.

6.8   (e) $p(x) = \begin{cases} 2/3 & \text{if } 0 \leq x \leq 1/4 \\ 4/3 & \text{if } 1/4 \leq x \leq 3/4 \\ 2/3 & \text{if } 3/4 \leq x \leq 1 \end{cases}$

T7.2   (a) The characteristic equation of A is $f(\lambda) = (\lambda - 3)^2$. Solving $(A - 3I)\mathbf{v} = 0$ for $\mathbf{v}$ yields only $\mathbf{v} = (1, 0)$. (b) The first equation is $\dot{x} - 3x = y = c_1 e^{3t}$. Multiplying through by e^{-3t} yields $(xe^{-3t})' = c_1$, or $xe^{-3t} = c_1 t + c_2$. Therefore $(x(t), y(t)) = (c_1 t e^{3t} + c_2 e^{3t}, c_1 e^{3t})$.

T7.4   The ellipses are $y^2 - xy + \frac{5}{2}x^2 = y_0^2$, where y_0 is the y-intercept. Solutions are $x(t) = -\frac{2}{3}y_0 \sin 3t$, $y(t) = y_0 \cos 3t - \frac{y_0}{3} \sin 3t$.

T7.7 $L = \sqrt{a^2 + b^2 + c^2 + d^2}$ works. This constant is not optimal. The smallest Lipschitz constant for A is the square root of the largest eigenvalue of AA^T

T7.9   The existence and uniqueness Theorem 7.14 applies to (7.32). Solutions with positive $x(0)$ are of form $x(t) = (2t + 1/x_0^2)^{-1/2}$; those with negative $x(0)$ are $x(t) = -(2t + 1/x_0^2)^{-1/2}$. All approach the equilibrium $x = 0$ monotonically as $t \to 0$, so $x = 0$ is asymptotically stable.

T7.11   The orbit starting at $(x, \dot{x}) = (0, 0)$; the orbit that begins along one wall and has just enough initial energy to reach $(0, 0)$; etc.

T7.14   Hint: Try the function $E = x^2 + y^2$.

T7.16   (c) First, determine which points on the x-axis belong to the largest possible W. Then extend W to the plane so that it will be a forward invariant set.

T7.17   The constant K is set so that $E(\frac{c}{d}, \frac{a}{b}) = 0$. Show that $E(\frac{c}{d}, \frac{a}{b})$ is a relative minimum of E. (A test for finding relative maximum/minimum values that involves the second-order partial derivatives can be found in most standard calculus texts.)

7.3   The repeller $(0, 0)$ and the saddle $(-8, -16)$ are both unstable.

7.4   (b) $x = (b - cKe^{a(b-c)t})/(1 + Ke^{a(b-c)t})$, where $K = (b - x_0)/(x_0 - c)$.

    (c) $T = \dfrac{\ln \frac{|x_0 - b|}{|x_0 - c|}}{a(c - b)}$.

.7   (a) Hint: Setting $y = \dot{x}$, try $E(x, y) = y^2/2 + x^2/2 + x^4/4$.

.8   Hint: Use the LaSalle Corollary.

.9   (a) $(x_1, 0)$ and $(x_2, 0)$ are stable; there is an unstable equilibrium $(x_3, 0)$ in between. (b) $(x_1, 0)$ stable, $(x_2, 0)$ unstable.

.14 $x \equiv 0$ and $x(t) = \begin{cases} 0 & \text{if } x < 0 \\ 2t^{3/2}/3 & \text{if } x \geq 0 \end{cases}$

⁻8.4 Let z be in the orbit of $\mathbf{u}$. Then there is a number t_1 such that $z = F(t_1, \mathbf{y})$. Since $\mathbf{u}$ is in $\omega(\mathbf{v_0})$, there is a sequence $\{F(t_n, \mathbf{v_0})\}$, $\{t_n\} \to \infty$ as $n \to \infty$, converging to $\mathbf{u}$. By Theorem 7.16, given $\epsilon > 0$, there is an $N > 0$, such that $F(t_1, F(t_n, \mathbf{v_0}))$ is in $N_\epsilon(z)$, for $n > N$. Now use the composition property to get these points on the orbit of $\mathbf{v_0}$.

⁻8.6 Assume that there are distinct points $\mathbf{v_1}$ and $\mathbf{v_2}$ in $\omega(\mathbf{v_0})$ such that $V(\mathbf{v_1}) = a$ and $V(\mathbf{v_2}) = b$, where $a < b$. Reach a contradiction by showing that the value of V along $F(t, \mathbf{v_0})$ must increase (with t) from values near a to values near b.

⌈8.7 If $\mathbf{v}(t_1) = \mathbf{v}(t_2)$, $t_1 < t_2$, and if $\mathbf{v}(t)$ is not an equilibrium, then it can be shown (by the composition property of autonomous equations) that $\mathbf{v}(t + (t_2 - t_1)) = \mathbf{v}(t)$, for any t. In this case, $t_2 - t_1$ must be an integer multiple of the period of the orbit.

⒊2   The ω-limit sets of $(0, 0)$ and $(1/2, 0)$ are both the origin. The ω-limit set of $(1, 0)$ is the circle $r = 1$ and of $(2, 0)$ is the circle $r = 3$.

⒊3   (a) Refer to the similarity types given in Appendix A.

⒊6   D.

⒊7   A, C, D, P, Y.

⒊9   Hint: Set $E(x, y) = y^2/2 + \int g(x)\, dx$ where $y = \dot{x}$. Show that $\dot{E}(x, y) \leq 0$, with equality possible only when $y = 0$.

8.10 Yes.

8.12 (a) Hint: Use the Monotonicity Lemma 8.12.

8.16 If $\theta(t)$ is the direction of travel of $x(t)$, measured in radians, show that $|\dot{\theta}| \leq L$.

T9.3 Hints: Find some r_0 and $\epsilon > 0$ depending only on the constants a_i, b_i, and c such that (9.2) implies $\dot{V}(\mathbf{u}) < -\epsilon$ whenever $|\mathbf{u}| \geq r_0$. Define m to be the maximum of the function $V(\mathbf{u})$ on the ball of radius r_0. The set $S = \{\mathbf{u} : V(\mathbf{u}) \leq m\}$ is forced to be a bounded set by the hypotheses. Trajectories can get into S from the outside but no trajectory can leave S, because all points of the boundary are outside radius r_0. Choose B large enough so that a ball of radius B contains S.

9.1   Hint: Set $x = y = 0$ in (9.1) and solve.

9.3   Area contraction rate is $e^{-2\pi c}$ per iterate; the system is dissipative if $c > 0$.

9.6   Both are 0.

9.7   Note that all orbits are identical except for a rescaling of time. The Lyapunov exponents of $\dot{\mathbf{v}} = c\mathbf{f}(\mathbf{v})$ are obtained by adding $\ln c$ to each of the Lyapunov exponents of $\dot{\mathbf{v}} = \mathbf{f}(\mathbf{v})$.

9.9   (a) $\dot{x} = y, \dot{y} = -x$. (b) $\dot{x} = y, \dot{y} = x$. (c) $J_1(x_0, y_0) = [(\cos 1)x_0 + (\sin 1)y_0$ $(-\sin 1)x_0 + (\cos 1)y_0]$, Lyapunov exponents are $\{0, 0\}$. (d) $J_1(x_0, y_0) = [\frac{1}{2}(e + e^{-1})x_0 + \frac{1}{2}(e - e^{-1})y_0, \frac{1}{2}(e - e^{-1})x_0 + \frac{1}{2}(e + e^{-1})y_0]$, Lyapunov exponents are $\{1, -1\}$. (e) The stable equilibria $(2n\pi, 0)$ for integer n have Lyapunov exponents $\{0, 0\}$. The saddle equilbria $\{(2n + 1)\pi, 0\}$ for integer n have Lyapunov exponents $\{1, -1\}$, as do the connecting arcs converging to them. All periodic orbits have Lyapunov exponents $\{0, 0\}$.

T10.2   Any point on a stable (resp., unstable) manifold maps to another point on the stable (resp., unstable) manifold under both $\mathbf{f}$ and $\mathbf{f}^{-1}$. Therefore a homoclinic point, which lies on both, remains homoclinic under mapping by either $\mathbf{f}$ or $\mathbf{f}^{-1}$.

T10.3   When $c_2 \neq 0$, the inverse of the Ikéda map $f(z) = c_1 + c_2 z \exp[i(c_3 - \frac{a}{1 + |z|^2})]$ is $f^{-1}(w) = \frac{w - c_1}{c_2} \exp[i(a/(1 + |\frac{w - c_1}{c_2}|^2) - c_3)]$.

T10.4   Let Q_a and $\mathbf{f}(Q_a)$ denote the two pieces of the attractor for $a < a_c$. Each piece attracts a branch of $\mathcal{U}(\mathbf{p})$. As a crosses a_c, Q_a crosses one branch of the stable manifold of $\mathbf{p}$, while $\mathbf{f}(Q_a)$ simultaneously crosses the other. For each $a > a_c$, each point in $\mathcal{U}(\mathbf{p})$ is a limit point of forward iterates of Q_a. Notice that Q_a is invariant under $\mathbf{f}^2$. There is a crisis at $a = a_c$ for the map $\mathbf{f}^2$ in which the attractor Q_a suddenly jumps in size to contain $\mathbf{f}(Q_a)$.

10.3   (a) Hint: Show that curves g_1 and g_2 intersect at a nonzero angle at $\mathbf{x}$ if and only if $\mathbf{f}(g_1)$ and $\mathbf{f}(g_2)$ intersect at a nonzero angle at $\mathbf{f}(\mathbf{x})$.

10.6   Hint: Sufficiently large forward iterates of an endpoint or a crossing point must be on the local stable manifold.

10.7   The unstable manifold of the cat map is the line $y = [(1 + \sqrt{5})/2]x$, treated modulo one. In the unit square, it consists of infinitely many parallel line segments dense in the square; on the torus, it is a curve which wraps around infinitely many times without intersecting itself. The stable manifold $y = [(1 - \sqrt{5})/2]x$ does the same, crossing the unstable manifold at right angles because the vectors $(1, (1 + \sqrt{5})/2)$ and $(1, (1 - \sqrt{5})/2)$ are perpendicular.

10.9   The part of the stable manifold lying within the unit square consists of the horizontal length-one segments whose y-coordinates are middle-third Cantor set endpoints—in other words, $0 \leq x \leq 1, y = p/3^k$ for integers $k \geq 0$ and integers $0 \leq p \leq 3^k$. These segments are connected via the part of the stable manifold lying outside the unit square in the order $y = 0, y = 1, y = 2/3, y = 1/3, y = 2/9, y = 7/9, y = 8/9, y = 1/9, \ldots$. The part of the unstable manifold lying within the unit square consists of the vertical length-one segments whose x-coordinates are middle-third Cantor set endpoints. The second fact follows from the first by noticing that the unstable manifold of $\mathbf{f}$ is the stable manifold of the inverse map

$$\mathbf{f}^{-1}(x, y) = \begin{cases} (3x, y/3) & \text{if } 0 \leq x \leq 1/3 \\ (3 - 3x, 1 - y/3) & \text{if } 2/3 \leq x \leq 1, \end{cases}$$

which is $\mathbf{f}$ with the roles of x and y reversed. (In fact, $\mathbf{f}^{-1}$ is conjugate to $\mathbf{f}$ via the conjugacy $C(x, y) = (y, x)$).

11.1   Yes. The chain rule shows that $f^{2\prime} = f'(f(x))f'(x)$, and $f^{2\prime\prime} = f''(f(x))f'(x)^2 + f'(f(x))f''(x) = f''(x)[f'(x)^2 + f'(x)] = f''(x)[1 - 1] = 0$.

11.3   (a) See, for fixed $d \neq 0$, how the family of graphs $y = f_{a,d}(x)$ varies with a.

11.5   Hint: See the proof of Theorem 11.9.

11.8   (a) Show that $e_1 + e_2 = m_{11} + m_{22}$ and $e_1 e_2 = \det A = m_{11}m_{22} - m_{21}m_{12}$.

11.14 (b) $0 < a < 2$.

11.15 (a) First show that for $r < 1$, the sequence is increasing and bounded and, therefore, must have a limit. Conclude that the limit is the fixed point $r = 1$. A similar argument works for $r > 1$. (b) $e^{-2\pi b}$. (c) $r \geq 0$.

11.16 Linked. The outside edge of a Möbius strip, regarded as a closed curve, is linked with the equator.

11.17 Hint: Let g be a periodic orbit and let L be a line segment which crosses g transversally. A period-doubling bifurcation from g would produce a periodic orbit which crosses L on both sides of g, violating uniqueness of solutions.

11.2   There is a period-doubling bifurcation of the origin for $a = 1$. The period-two orbit created is a sink for $1 < a < \sqrt{6} - 1$, when it loses stability in a period-doubling bifurcation to a period-four sink.

11.3   Period-doubling bifurcation of the origin for $a = -1$.

11.5   (a) Saddle-node at $a = 1$, no period-doubling. (b) No saddle-node, period-doubling at $a = 1$. (c) Saddle-node at $a = e$, period-doubling at $a = -1/e$. (d) Period-doubling at $a = 2$.

11.6   The origin is a fixed point for all a. The other path of fixed points $(-(1 - a)^3, -(1 - a)^2)$ crosses the origin at $a = 1$, at which point the origin changes from attracting to repelling.

11.10   (b) Hint: 0 is attracting for g if and only if 0 is attracting for $g \circ g$. Find the power series expansion for $g \circ g$, and apply part(a) of this problem.

11.12   The minus sign can be moved to the denominator of the left side of the equation. Reverse the direction of time.

11.14   (c) In Chapter 7, $H(x, y) = -c \ln x + dx - a \ln y + by$ was shown to be Lyapunov function for the equilibrium $(\frac{c}{d}, \frac{a}{b})$ when $r = 0$. Calculate $\frac{dH}{dt} = \frac{\partial H}{\partial x}\frac{dx}{dt} + \frac{\partial H}{\partial y}\frac{dy}{dt}$ for $r \neq 0$.

T12.1   Hint: If $\{x_1, x_2, x_3\}$ is a periodic orbit of f_a, then $f_a^{3\,'}(x_1) = f_a^{3\,'}(x_2) = f_a^{3\,'}(x_3)$ by the Chain Rule.

T12.5   Assume that there are infinitely many zeros of g in the interval. First show that g' can be zero at only a finite subset of these zeros. Is this fact compatible with the hypotheses?

T12.8   Hint: Check the direction of the snake that returns to $a = a_c$ to obtain a contradiction.

T12.9   The snake cannot have an endpoint.

T12.11   First argue that the snake must pass through period–doubling bifurcation orbits of these periods.

12.1   Use the fact that f_2 is conjugate to the logistic map $g(x) = 4x(1 - x)$, and hence to the tent map T_2, to show that all periodic orbits of f_2 are (hyperbolic) repellers.

12.3   Hint: Use symbol sequences to classify regular and flip unstable orbits.

Bibliography

Abraham, R. H. and Shaw, C. D. 1988. Dynamics: The Geometry of Behavior. Aerial Press, Santa Cruz, CA.

Abraham, R. H. and Marsden, J. E. 1978. Foundations of Mechanics. Benjamin/Cummings, Reading, MA.

Alexander, J. and Yorke, J. A. 1978. Global bifurcation of periodic orbits, American J. Math. 100:263–292.

Alligood, K. T. and Yorke, J. A. 1992. Accessible saddles on fractal basin boundaries. Ergodic Theory and Dynamical Systems 12:377–400.

Arecchi, F. T., Meucci, R., Puccioni, G., and Tredicce, J. 1982. Experimental evidence of subharmonic bifurcations, multistability, and turbulence in a Q-switched gas laser. Phys. Rev. Lett. 49, 1217.

Arnold, V. I. 1973. Ordinary Differential Equations. M.I.T. Press, Cambridge, MA.

Arnold, V. I. 1978. Mathematical Models of Classical Mechanics. Springer-Verlag, New York, Heidelberg, Berlin.

Arnold, V. I. and Avez, A. 1968. Ergodic Problems of Classical Mechanics. W. A. Benjamin, New York, NY.

Barnsley, M. F. 1988. Fractals Everywhere. Academic Press, Orlando, FL.

Benedicks, M. and Carleson, L. 1991. The dynamics of the Hénon map, Annals of Math. **133**:73–169.

Benedicks, M. and Young, L. S. 1996. SBR measures for certain Hénon maps.

Bergé, P., Pomeau, Y., and Vidal, C. 1984. Order Within Chaos: Towards a Deterministic Approach to Turbulence. Wiley, New York, NY.

Bielawski, S., Bouazaoui, M., Derozier, D., and Glorieux, P. 1993. Stabilization and characterization of unstable steady states in a laser. Physical Review A **47**:3276–9.

Blanchard, P., Devaney, R., and Hall, G. 1996. Differential Equations. Prindle, Weber, Schmidt.

Block, L., Guckenheimer, J., Misiurewicz, M., and Young, L.-S. 1979. Periodic points of one-dimensional maps. Springer Lecture Notes in Math. **819**:18–34.

Braun, M. 1978. Differential Equations and their Applications. Springer-Verlag, New York, Heidelberg, Berlin.

Brunovsky, P. 1969. One parameter families of diffeomorphisms. Symposium on differential equations and dynamical systems, University of Warwick, Springer Lecture Notes **206**. Springer-Verlag, Berlin, Heidelberg, New York.

Burden, R. L. and Faires, J. D. 1993. Numerical Analysis. PWS Publishing, Boston, MA.

Busenberg, S., Fisher, D., and Martelli, M. 1989. Minimal periods of discrete and smooth orbits. American Math. Monthly **96**:5–17.

Cartwright, M. L. and Littlewood, J. E. 1951. Some fixed point theorems. Ann. of Math. (2) **54**:1–37.

Chinn, W. G. and Steenrod, N. E. 1966. First Concepts of Topology. Mathematical Association of America, Washington, DC.

Chow, S. N. and Hale, J. K. 1982. Methods of Bifurcation Theory. Springer-Verlag, New York, Heidelberg, Berlin.

Coddington, E. A. and Levinson, N. 1955. Theory of Ordinary Differential Equations. McGraw-Hill, New York, NY.

Coffman, K. G., McCormick, W. D., Noszticzius, Z., Simoyi, and R. H., Swinney, H. L. 1987. Universality, multiplicity, and the effect of iron impurities in the Belousov-Zhabotinskii reaction. J. Chemical Physics **86**:119–129.

Collet, P., and Eckmann, J.-P. 1980. Iterated Maps on the Interval as Dynamical Systems. Birkhäuser, Boston.

Conley, C. 1978. Isolated Invariant Sets and the Morse Index. CBMS Regional Conferences in Mathematics, Vol. 38. AMS Publications, Providence, RI.

Costantino, R. F., Cushing, J. M., Dennis, B., and Desharnais, R. A. 1995. Experimentally induced transitions in the dynamic behavior of insect populations. Nature **375**:227–230.

Cvitanovic, P., ed. 1984. Universality in Chaos. Adam Hilger Ltd., Bristol.

evaney, R. 1986. Introduction to Chaotic Dynamical Systems. Benjamin Cummings, Menlo Park, CA.

evaney, R. 1992. A First Course in Chaotic Dynamical Systems: Theory and Experiment. Addison-Wesley, Reading, MA.

evaney, R., Keen, L., eds. 1989. Chaos and Fractals: The mathematics behind the computer graphics. American Mathematical Society, Providence, RI.

evaney, R. and Nitecki, Z. 1979. Shift automorphisms in the Hénon mapping. Commun. Math. Phys., **67**:137–148.

itto, W., Rauseo, S. N., and Spano, M. L. 1990. Experimental control of chaos. Phys. Rev. Lett. 65:3211–4.

ressler, U., Ritz, T., Schenck zu Schweinsberg, A., Doerner, R., Hübinger, B., and Martienssen, W. 1995. Tracking unstable periodic orbits in a bronze ribbon experiment. Physical Review E **51**:1845–8.

dgar, G. A., ed. 1993. Classics on Fractals. Addison-Wesley, Reading, MA.

rmentrout, B. 1990. Phaseplane, Version 3.0. Brooks-Cole.

alconer, K. 1990. Fractal Geometry. John Wiley and Sons, Chichester.

eder, J. 1988. Fractals. Plenum Press, NY

eigenbaum, M. J. 1978. Quantitative universality for a class of nonlinear transformations. J. Statistical Physics **19**:25–52.

eigenbaum, M. J. 1979. The universal metric properties of nonlinear transformations. J. Statistical Physics **21**:669–706.

itzpatrick, P. M. 1996. Advanced Calculus. PWS Publishing Co, Boston.

arfinkel, A., Spano, M., Ditto, W., and Weiss, J. 1992. Controlling cardiac chaos. Science 257:1230.

iglio, M., Musazzi, S., and Perini, V., 1981. Transition to chaotic behavior via a reproducible sequence of period-doubling bifurcations. Phys. Rev. Lett. 47:243–6.

ills, Z., Iwata, C., Roy, R., Schwartz, I., and Triandaf, I. 1992. Tracking unstable steady states: Extending the stability regime of a multimode laser system. Phys. Rev. Lett. 69:3169–73.

lass, L. and Mackey, M. C. 1988. From Clocks to Chaos: The Rhythms of Life. Princeton University Press, Princeton, NJ.

leick, J. 1987. Chaos: The Making of a New Science. Viking, New York, NY.

rassberger, P. and Procaccia, I. 1983. Characterization of strange attractors. Physics Review Letters **50**:346.

rebogi, C., Hammel, S., Yorke, J. A., and Sauer, T. 1990. Shadowing of physical trajectories in chaotic dynamics. Physical Review Letters **65**:1527–30.

uckenheimer, J., Moser J., and Newhouse, S. 1980. Dynamical Systems. Birkhäuser, Boston.

uckenheimer, J. and Holmes, P. 1983. Nonlinear Oscillations, Dynamical Systems, and Bifurcations of Vector Fields. Springer-Verlag, New York, Heidelberg, Berlin.

uillemin, V. and Pollack, A. 1974. Differential Topology. Prentice-Hall, Engelwood Cliffs, N.J.

Gulick, D. 1992. Encounters with Chaos. McGraw-Hill, New York, NY.

Guttman, R., Lewis, S., and Rinzel, J., 1980. Control of repetitive firing in squid axo membrane as a model for a neuroneoscillator. J. Physiology 305:377–395.

Haken, H., 1983. Synergetics, 3rd edition. Springer-Verlag, Berlin.

Hale, J. and Kocak, H. 1991. Dynamics and bifurcations. Springer-Verlag, New York.

Hao, B.-L. 1984. Chaos. World Scientific, Singapore.

Hartman, P. 1964. Ordinary Differential Equations. J. Wiley and Sons, New York.

Heagy, J., Carroll, T., and Pecora, L. 1994. Experimental and numerical evidence fo riddled basins in coupled chaotic systems. Physical Review Letters 73:3528–31.

Hirsch, M. 1976. Differential Topology. Springer-Verlag, New York, Heidelberg, Berlin.

Hirsch, M. and Smale, S. 1974. Differential Equations, Dynamical Systems, and Linea Algebra. Academic Press, New York.

Hocking, J. and Young, G. 1961. Topology. Addison-Wesley, Reading, MA.

Holden, A. V. 1986. Chaos. Manchester University Press, Manchester.

Hubbard, J. H. and West, B. H. 1991. Differential Equations: A Dynamical System Approach, Part I. Springer-Verlag, New York, Heidelberg, Berlin.

Hubbard, J. H. and West, B. H. 1992. MacMath: A Dynamical Systems Software Packag for the Macintosh. Springer-Verlag, New York, Heidelberg, Berlin.

Hübinger, B., Doerner, R., Martienssen, W., Herdering, M., Pitka, R., and Dressler, U. 1994 Controlling chaos experimentally in systems exhibiting large effective Lyapuno exponents. Physical Review E 50:932–948.

Hunt, B. R., Gallas, J. A. C., Grebogi, C., Kocak, H., and Yorke, J. A. 1996. Bifurcation rigidity.

Jackson, E. A. 1989. Perspectives of Nonlinear Dynamics, I and II. Cambridge University Press, Cambridge.

Jalife, J. and Antzelevitch, C. 1979. Phase resetting and annihiliation of pacemaker activity in cardiac tissue. Science 206:695–7.

Kantz, H., Schreiber, T., Hoffman, I., Buzug, T., Pfister, G., Flepp, L. G., Simonet, J., Badii, R., and Brun, E., 1993. Nonlinear noise reduction: A case study on experimental data. Physical Review E 48:1529–1538.

Kaplan, D. and Glass, L. 1995. Understanding Nonlinear Dynamics. Springer-Verlag, New York, Heidelberg, Berlin.

Kaplan, J. L. and Yorke, J. A. 1979. Preturbulence: A regime observed in a fluid flow model of Lorenz. Communications in Math. Physics 67:93–108.

Kennedy, J. A. and Yorke, J. A. 1991. Basins of Wada. Physica D 51:213–225.

Kocak, H. 1989. Differential and Difference Equations Through Computer Experiments, 2nd ed. Springer-Verlag, New York, Heidelberg, Berlin.

Kocarev, L., Halle, K. S., Eckert, K., and Chua, L. O. 1993. Experimental observation of antimonotonicity in Chua's circuit. Int. J. of Bifurcations and Chaos 3:1051–5.

Kostelich, E. and Armbruster, D. 1996. Introductory Differential Equations. Addison-Wesley, Reading, MA.

nford, O. 1982. A computer-assisted proof of the Feigenbaum conjecture. Bulletin Amer. Math. Soc. **6**:427–434.

askar, J. 1989. A numerical experiment on the chaotic behaviour of the solar system. Nature **338**:237–8.

askar, J. and Robutel, P. 1993. The chaotic obliquity of the planets. Nature **361**:608–612.

asota, A. and Yorke, J. A. 1973. On the existence of invariant measures for piecewise monotonic transformations, Transactions Amer. Math. Soc. **186**:481–488.

epers, C., Legrand, J., and Glorieux, P. 1991. Experimental investigation of the collision of Feigenbaum cascades in lasers. Physical Review A **43**:2573–5.

evi, M. 1981. Qualitative Analysis of the Periodically Forced Relaxation Oscillations. Mem. Amer. Math. Soc. No. 214.

i, T. Y. and Yorke, J. A., 1975. Period three implies chaos. American Mathematical Monthly **82**:985–992.

i, T.Y. and Yorke, J. A. 1978. Ergodic transformations from an interval into itself. Transactions Amer. Math. Soc. **235**:183–192.

ichtenberg, A. J. and Lieberman, M. A. 1992. Regular and Chaotic Dynamics, 2nd ed. Springer-Verlag, New York, Heidelberg, and Berlin.

insay, P., 1981. Period doubling and chaotic behavior in a driven anharmonic oscillator. Phys. Rev. Lett. 47:1349–52.

ivdahl, T. P. and Willey, M. S. 1991. Propects for an invasion: Competition between *Aedes albopictus* and native *Aedes triseriatus*. Science **253**:189–91.

orenz, E. 1963. Deterministic nonperiodic flow. J. Atmospheric Science **20**:130–141.

orenz, E. 1993. The Essence of Chaos. The University of Washington Press, Seattle.

Mackay, R. S. and Meiss, J. D., eds. 1987. Hamiltonian Dynamical Systems, a reprint volume. Adam Hilger Ltd., Bristol.

Malthus, T.R. 1798. An essay on the principle of population. Reprint. Penguin, New York, 1970.

Mandelbrot, B. 1977. Fractals: Form, Chance, and Dimension. Freeman, San Francisco, CA.

Mandelbrot, B. 1982. The Fractal Geometry of Nature. Freeman, San Francisco, CA.

Marsden, J. E. and McCracken, M. 1976. The Hopf Bifurcation and Its Applications. Springer-Verlag, New York, Heidelberg, Berlin.

Martien, P., Pope, S. C., Scott, P. L., and Shaw, R. S. 1985. The chaotic behavior of the leaky faucet. Physics Letters A **110**:399–404.

Matsumoto, T., Chua, L.O., and Komuro, M. 1985. IEEE Trans. Circuits Syst. **23**:798.

May, R. M. 1976. Simple mathematical models with very complicated dynamics. Nature **261**:459–467.

Mayer-Kress, G., ed. 1986. Dimensions and Entropies in Chaotic Systems: Quantification of complex behavior. Springer-Verlag, Heidelberg.

Meyer, K. R. and Hall, G. R. 1992. Introduction to Hamiltonian Dynamical Systems and the N-Body Problem. Springer-Verlag, New York, Heidelberg, Berlin.

Milnor, J. 1985. On the concept of attractor, Commun. Math. Phys. **99**:177

Moon, F. 1992. Chaotic and Fractal Dynamics: An Introduction for Scientists and Engineers. Wiley, New York, NY.

Moore, A. W. 1995. A brief history of infinity. Scientific American **272**:112–116.

Murray, J. D. 1989. Mathematical Biology. Springer-Verlag, New York, Heidelberg, Berlin.

Myrberg, P. J. 1962. Sur l'iteration des polynomes reels quadratiques. J. Math. Pures Appl **41**:339–351.

Nusse, H. E. and Yorke, J. A. 1994. Dynamics: Numerical Explorations. Springer-Verlag, New York, Heidelberg, Berlin.

Nusse, H.E. and Yorke, J. A. 1996. The structure of basins of attraction and their trapping regions. Ergodic Theory and Dynamical Systems.

Ott, E. 1993. Chaos in Dynamical Systems. Cambridge University Press, Cambridge.

Ott, E., Grebogi, C., and Yorke, J. A. 1990. Controlling chaos. Phys. Rev. Lett. 64:1196–9

Ott, E., Sauer, T., and Yorke, J. A. 1994. Coping with Chaos: Analysis of chaotic data and the exploitation of chaotic systems. J. Wiley, New York.

Packard, N., Crutchfield, J., Farmer, D., and Shaw, R. 1980. Geometry from a time series Physical Review Letters **45**:712.

Palis, J. and de Melo, F. 1982. Geometric Theory of Dynamical Systems: An Introduction Springer-Verlag, New York, Heidelberg, Berlin.

Palis, J. and Takens, F. 1993. Hyperbolicity and sensitive chaotic dynamics at homoclinic bifurcations. Cambridge University Press, Cambridge.

Pecora, L. M. and Carroll, T. 1990. Synchronization in chaotic systems. Physics Review Letters **64**:821–4.

Peitgen, H.-O. and Richter, P. H. 1986. The Beauty of Fractals. Springer-Verlag, New York Heidelberg, Berlin.

Perko, L. 1991. Differential Equations and Dynamical Systems. Springer-Verlag, New York Heidelberg, Berlin.

Quinn, G. D. and Tremaine, S. 1992. On the reliability of gravitational N-body integrations. Monthly Notices of the Royal Astronomical Society, **259**:505–518.

Peak, D. and Frame, M. 1994. Chaos under Control: The Art and Science of Complexity. W. H. Freeman, New York.

Petrov, V., Gáspár, V., Masere, J., and Showalter, K. 1993. Controlling chaos in the Belousov-Zhabotinsky reaction. Nature **361**:240–3.

Poincaré, H. 1897. Les Méthodes Nouvelles de la Méchanique Céleste. Vol. I and II. Gauthier-Villars, Paris.

Pomeau, Y. and Manneville, P. 1980. Intermittent transition to turbulence in dissipative dynamical systems, Commun. Math. Phys. **74**:189.

Rasband, S. N. 1989. Chaotic Dynamics of Nonlinear Systems. John Wiley and Sons, New York.

Robinson, C. 1995. Dynamical Systems: Stability, Symbolic Dynamics, and Chaos. CRC Press, Boca Raton.

Rössler, O. 1976. An equation for continuous chaos. Phys. Lett. A **57**:397.

ux, J.-C., Simoyi, R. H., and Swinney, H. L. 1983 Observation of a strange attractor. Physica D **8**:257–266.

y, R., and Thornburg, K.S. 1994. Experimental synchronization of chaotic lasers. Physical Review Letters **72**:2009–2012.

elle, D. 1989. Chaotic Evolution and Strange Attractors. Cambridge University Press, Cambridge.

elle, D. and Takens, F. 1971. On the nature of turbulence, Commun. Math. Phys. **20**:167.

lzmann, B. 1962. Finite amplitude free convection as an initial value problem. J. Atmos. Sci. **19**:239–341.

rtorelli, J. C., Goncalves, W. M., and Pinto, R. D. 1994. Crisis and intermittence in a leaky-faucet experiment. Physical Review E **49**: 3963–3975.

uer, T., Yorke, J. A., and Casdagli, M. 1991. Embedology. J. Statistical Physics **65**:579.

zou, D., Karantonis, A., and Pagitsas, M. 1993. Generalized Hopf, saddle-node infinite period bifurcations and excitability during the electrodissolution and passivation of iron in a sulfuric acid solution. Int. J. Bifurcations and Chaos **3**:981–997.

hiff., S., Jerger, K., Duong, D., Chang, T., Spano, M., and Ditto, W. 1994. Controlling chaos in the brain. Nature 370:615–620.

aw, R. 1984. The Dripping Faucet as a Model Chaotic System. Aerial Press.

il'nikov, L. 1994. Chua's circuit: Rigorous results and future problems. Int. Journal of Bifurcation and Chaos **4**:489–519.

male, S. 1967. Differentiable dynamical systems, Bulletin Am. Math. Soc **73**:747.

ommerer, J. 1994. Fractal tracer distributions in complicated surface flows: an application of random maps to fluid dynamics. Physica D **76**:85–98.

parrow, C. 1982. The Lorenz Equations: Bifurcations, Chaos, and Strange Attractors. Springer-Verlag, New York, Heidelberg, Berlin.

tewart, I. 1989. Does God Play Dice? The Mathematics of Chaos. Basil Blackwell.

trang, G. 1988. Linear Algebra and its Applications. Harcourt Brace Jovanovich, San Diego (1988).

trogatz, S. H. 1994. Nonlinear Dynamics and Chaos. Addison-Wesley, Reading, MA.

trogatz, S. H. 1985. Yeast oscillations, Belousov-Zhabotinsky waves, and the nonretraction theorem. Math. Intelligencer **7 (2)**:9.

ussman, G. J. and Wisdom, J. 1988. Numerical evidence that the motion of Pluto is chaotic. Science **241**:433–7.

ussman, G. J. and Wisdom, J. 1992. Chaotic evolution of the solar system. Science **257**:56–62.

abor, M. 1989. Chaos and Integrability in Nonlinear Dynamics: An Introduction. Wiley-Interscience, New York, NY.

akens, F. 1981. Detecting strange attractors in turbulence. Lecture Notes in Mathematics **898**, Springer-Verlag.

hompson, J.M.T. and Stewart, H. B. 1986. Nonlinear Dynamics and Chaos. John Wiley and Sons, Chichester.

Touma, J. and Wisdom, J. 1993. The chaotic obliquity of Mars. Science **259**:1294–129

Tufillaro, N. B., Abbott, T., and Reilly, J. 1992. An Experimental Approach to Nonline Dynamics and Chaos. Addison-Wesley, Redwood City.

Van Buskirk, R. and Jeffries, C. 1985. Observation of chaotic dynamics of coupled nonline oscillators. Phys. Rev. A 31:3332–57.

Vincent, T. L., Mees, A., and Jennings, L., ed. 1990. Dynamics of Complex Interconnect Biological Systems. Birkhauser, Boston, MA.

Whitelaw, W. A., Derenne, J.-Ph., and Cabane, J. 1995. Hiccups as a dynamical diseas Chaos **5**:14.

Whitney, H. 1936. Differentiable manifolds. Annals of Math. **37**:645.

Wiggins, S. 1990. Introduction to Applied Nonlinear Dynamical Systems and Chao Springer-Verlag, New York, Heidelberg, Berlin.

Winfree, A. T. 1980. The Geometry of Biological Time. Springer-Verlag, New Yor Heidelberg, Berlin.

Yorke, J. A., Alligood, K. T. 1985. Period doubling cascades of attractors: a prerequisite fo horseshoes. Communications in Mathematical Physics **101**:305–321.

Yorke, J. A., Yorke, E. D., and Mallet-Paret, J. 1987. Lorenz-like chaos in a partial diffe ential equation for a heated fluid loop. Physica D **24**:279–291.

You, Z.-P., Kostelich, E. J., and Yorke, J. A. 1991. Calculating Stable and Unstable Man folds. International Journal of Bifurcation and Chaos, **1**:605–623.

Young, L.-S. 1982. Dimension, entropy and Lyapunov exponents. Ergodic Th. and Dyr Sys. **2**:109.

ndex

TEXTBOOKS IN MATHEMATICAL SCIENCES

CHAOS

AN INTRODUCTION TO DYNAMICAL SYSTEMS

KATHLEEN T. ALLIGOOD
TIM D. SAUER
JAMES A. YORKE

CHAOS: AN INTRODUCTION TO DYNAMICAL SYSTEMS was developed and class-tested by a distinguished team of authors at two universities through their teaching of courses based on the material. Intended for courses in nonlinear dynamics offered in either Mathematics or Physics, the text requires only calculus, differential equations, and linear algebra as prerequisites.

Spanning the wide reach of nonlinear dynamics throughout mathematics, natural and physical science, CHAOS: AN INTRODUCTION TO DYNAMICAL SYSTEMS develops and explains the most intriguing and fundamental elements of the topic, and examines their broad implications.

Among the major topics included are: discrete dynamical systems, chaos, fractals, nonlinear differential equations, and bifurcations. The text also features Lab Visits, short reports that illustrate relevant concepts from the physical, chemical, and biological sciences, drawn from the scientific literature. There are Computer Experiments throughout the text that present opportunities to explore dynamics through computer simulation, designed to b used with any standard software package. An each chapter ends with a Challenge, whic provides students a tour through an advance topic in the form of an extended exercise.

CONTENTS

1. **One-Dimensional Maps**
2. **Two-Dimensional Maps**
3. **Chaos**
4. **Fractals**
5. **Chaos in Two-Dimensional Maps**
6. **Chaotic Attractors**
7. **Differential Equations**
8. **Periodic Orbits and Limit Sets**
9. **Chaos in Differential Equations**
10. **Stable Manifolds and Crises**
11. **Bifurcations**
12. **Cascades**
13. **State Reconstruction of Data**

APPENDICES A & B
ANSWERS & HINTS TO SELECTED EXERCISES
BIBLIOGRAPHY
INDEX

Springer

ISBN 0-387-94677-2

ISBN 0-387-94677-2

EAN

9 780387 946771 >